AF344393

ADVANCES IN OPTICAL FIBER TECHNOLOGY: FUNDAMENTAL OPTICAL PHENOMENA AND APPLICATIONS

Edited by **Moh Yasin, Hamzah Arof and Sulaiman Wadi Harun**

Advances in Optical Fiber Technology: Fundamental Optical Phenomena and Applications
http://dx.doi.org/10.5772/58517
Edited by Moh Yasin, Hamzah Arof and Sulaiman Wadi Harun

Contributors

Ashok K Sood, Oleg Morozov, Gennady Morozov, Ilnur Nureev, Anvar Talipov, Hadi Guna, Krzysztof Perlicki, Liang Dong, Weiwen Zou, Xin Long, Jianping Chen, Yuri Barmenkov, Alexander Kir'Yanov, Renxian Li, Lixin Guo, Bing Wei, Chunying Ding, Zhensen Wu, Silvio Abrate, Stefano Straullu, Roberto Gaudino, Yoshito Shuto, David Sánchez Montero, Plinio Jesús Pinzón Castillo, Alberto Tapetado Moraleda, Pedro Contreras Lallana, Carmen Vázquez García, Isabel Pérez Garcilópez, Michal Lucki, Tomas Zeman, Joao M. P. Coelho, Catarina Silva, Marta Nespereira, Manuel Abreu, José Rebordão

CBS, Edition **2017**

Published by InTech

Janeza Trdine 9, 51000 Rijeka, Croatia

Notice

Publishing Process Manager Danijela Duric

Technical Editor InTech DTP team

Cover InTech Design team

Additional hard copies can be obtained from orders@intechopen.com

Advances in Optical Fiber Technology: Fundamental Optical Phenomena and Applications, Edited by Moh Yasin, Hamzah Arof and Sulaiman Wadi Harun

p. cm.

ISBN 978-953-51-1742-1

Contents

Preface

This book is a compilation of works presenting recent developments and practical applications in optical fiber technology. It provides an overview of recent advances in optical fiber related researches of fundamental topics and applications. It is divided into three sections of seven, four and three chapters respectively.

The first section presents recent studies on various optical phenomena such as scattering, dispersion, polarization, interference, fuse phenomena and optical manipulation. Chapter 1 introduces the basic principle and recent advances of Brillouin scattering in optical fibers. The working mechanism, different interrogation techniques, challenges and breakthroughs of Brillouin based distributed optical fiber sensors are also reported in this chapter. Chapter 2 reviews the principle and applications of the poly-harmonic (two-frequency or four-frequency) CW laser systems for characterization of Mandelstam-Brillouin gain contour, Raman scattering contours and FBG reflection spectra. Chapter 3 discusses various design approaches for dispersion compensating fiber and offers a guideline to flexible design optical fibers used in telecommunication systems. This chapter also describes how to control the location and shape of the chromatic dispersion curves in photonic crystal fibers (PCFs). Chapter 4 explains the description of polarized light, polarization phenomena in optical fiber links, modeling of polarization phenomena and polarizing component. The understanding on polarization effects is important to comprehend the signal propagation in modern long haul light-wave communication networks. Chapter 5 presents a simulation study on fuse phenomena in an optical fiber. In this study, the threshold power of fiber fuse propagation in hole-assisted fibers (HAFs) is evaluated using the finite-difference method. Chapter 6 reports a theoretical study on fiber-based cylindrical vector beams and its applications to optical manipulation.

The second section covers optical fiber applications in fiber laser and sensor. The emergence of PCFs in the late nineties gave new impetus to the mode area scaling of single-mode optical fibers. Chapter 7 gives a brief introduction to the key approaches to effective mode area scaling in PCF that show great potentials in future high power fiber lasers. Chapter 8 reviews some of the well-known nonlinear-optical effects that affect the oscillation regimes of Erbium-doped fiber lasers (EDFLs). The development of optical fiber gratings (OFGs) have brought about astounding advances in research and development of optical communications and sensors. Among OFGs, long period fiber gratings (LPFGs) are one of the most important fiber-based sensors. Chapter 9 addresses the application of CO2 laser radiation in writing LPFGs and the physical principles involved in the process. Chapter 10 covers recent advances in SiGe based detector technology, including device operation, fabrication processes, and various optoelectronic applications. Optical sensing technology is critical for defense and commercial applications including telecommunications, which requires near-infrared (NIR) detection in the 1300-1550 nm wavelength range.

The third section comprises three chapters related to PON technology. Fiber-optic access networks are necessary for a real broadband delivery, allowing the fiber to arrive closer to the final customer, eventually up to the premises equipment. Such infrastructures, depending on the depth of reach of the fiber, are usually referred to as FTTX (Fiber To The X), where X stands for H (Home), B (Building), C (Curb) or Cab (Cabinet). In this section, a new development of passive optical network (PON) and plastic optical fibers technologies are presented. Chapter 11 provides a general description of the self-coherent reflective PON architecture as a possible technological approach to the NG-PON2 (Next Generation PON 2) requirements in terms of performance, cost, wavelength control and data transmission capability, among others. Chapter 12 reports the state-of-the art, description and experimental validation of different POF-based key devices that provide an easy-reconfigurable performance for wavelength division multiplexing (WDM) applications. The last chapter demonstrates a POF based solutions in WDM network and some effects due to the placement of color filters as a demultiplexer for the In-Car Entertainment System.

Moh Yasin
Faculty of Science and Technology, Airlangga Univesity,
Indonesia

Sulaiman Wadi Harun
Department of Electrical Engineering, University of Malaya,
Malaysia

Hamzah Arof
Department of Electrical Engineering, University of Malaya,
Malaysia

Scattering, Dispersion, Polarisation, Interference, Fuse Phenomena and Optical Manipulation

Brillouin Scattering in Optical Fibers and Its Application to Distributed Sensors

Weiwen Zou, Xin Long and Jianping Chen

Additional information is available at the end of the chapter

http://dx.doi.org/10.5772/59145

1. Introduction

Brillouin based distributed optical fiber sensors have been studied for more than two decades because they have incomparable abilities over the pointed or multiplexed fiber-optic sensors based on fiber Bragg grating and/or inline Fabry-Perot resonator. They originated from the intrinsic fiber-optic nonlinearity in optical fibers, i.e. Brillouin scattering, and have many distinguished advantages, such as high accuracy due to the frequency revolved interrogation, multiple sensitivities of measurands (strain, temperature etc.), no dead zones of sensing location due to the distributed sensing ability, and immunity to the electro-magnetic interference. Nowadays, they have been thought as great potentials in industrial applications to smart materials and smart structures.

This chapter introduces the basic principle and recent advances of Brillouin scattering in optical fibers. The working mechanism, different interrogation techniques, difficulty or challenge of the sensing ability, and recent breakthroughs of Brillouin based distributed optical fiber sensors are demonstrated, respectively.

2. Brillouin scattering in optical fibers

2.1. Principle

Light scattering phenomena in optical fibers occur regardless of how intense the incident optical power is. They can be basically categorized into two groups, i.e. spontaneous scattering and stimulated scattering[1]. Spontaneous scattering refers to the process under conditions such that the material properties are unaffected by the presence of the incident optical fields.

For input optical fields of sufficient intensities spontaneous scattering becomes quite intense and stimulated scattering starts. The nature of the stimulated scattering process grossly modifies the optical properties of the material system and vice versa. Spontaneous and stimulated scattering in optical fibers are composed of Rayleigh, Raman, and Brillouin scattering processes. Each scattering process is always present in optical fibers since no fiber is free from microscopic defects or thermal fluctuations which originate the three processes. For a monochromatic incident lightwave of frequency f_0 at $\lambda_0 \approx$ 1550 nm (telecom wavelength), three processes are schematically described by the spectrum of the scattered light as shown in Fig. 1. The components, whose frequency is beyond f_0, correspond to anti-Stokes while those below f_0 correspond to Stokes.

Brillouin scattering is a "photon-phonon" interaction as annihilation of a pump photon creates a Stokes photon and a phonon simultaneously. The created phonon is the vibrational modes of atoms, also called a propagation density wave or an acoustic phonon/wave. In a silica-based optical fiber, Brillouin Stokes wave propagates dominantly backward [2] although very partially forward[3]. The frequency (~9-11 GHz) of Stokes photon at ~1550-nm wavelength is in quantity dramatically different from or smaller by three orders of magnitude than Raman scattering (see Fig. 1) and is dominantly down-shifted due to Doppler shift associated with the forward movement of created acoustic phonons. In a polymer optical fiber, the frequency is ~2-3 GHz due to the different phonon property[4].

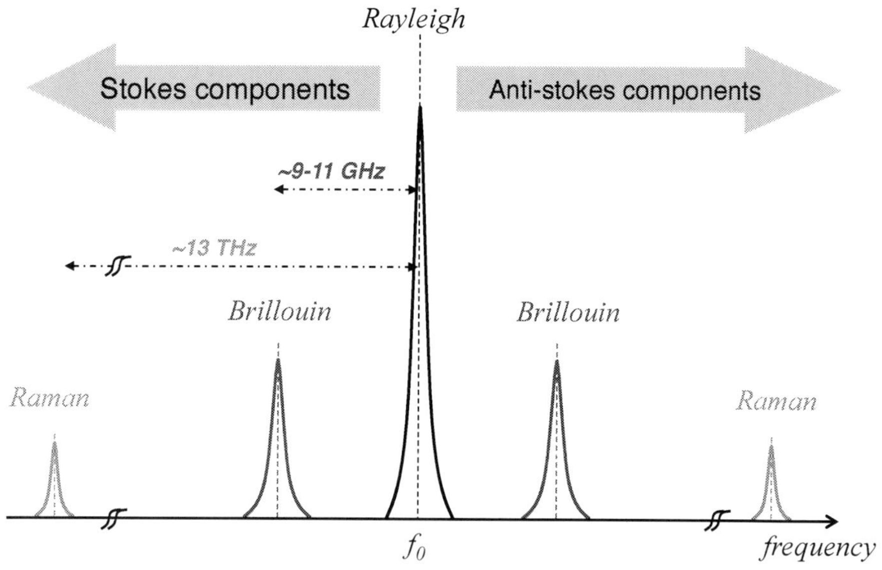

Figure 1. Schematic spectrum of scattered light resulting from three scattering processes in optical fibers.

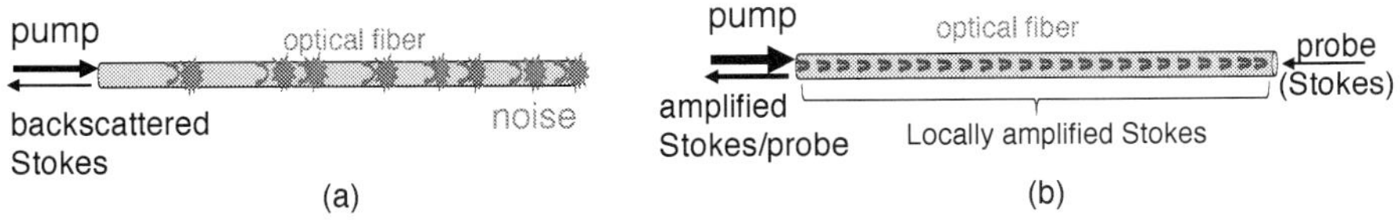

Figure 2. Comparison of (a) spontaneous Brillouin scattering (SpBS) and (b) stimulated Brillouin scattering (SBS) in optical fibers.

Figure 2 illustrates the difference between spontaneous Brillouin scattering (SpBS) and stimulated Brillouin scattering (SBS) in optical fibers. In principle, the SpBS (see Fig. 2(a)) is started from a noise fluctuation and influences the pump wave (E_p); the SBS (see Fig. 2(b)) occurs when the pump power for SpBS is beyond the so-called Brillouin threshold value (P_{th}) or when two coherent waves with a frequency difference equivalent to the phonon's frequency are counter-propagated. Brillouin scattering dynamics in optical fibers are generally governed by the following coupling equations [5, 6]:

$$\left(\frac{1}{v_g}\frac{\partial}{\partial t}+\frac{\partial}{\partial z}\right)E_p = -\frac{\alpha}{2}E_p + i\kappa_1\rho E_s, \tag{1}$$

$$\left(\frac{1}{v_g}\frac{\partial}{\partial t}-\frac{\partial}{\partial z}\right)E_s = -\frac{\alpha}{2}E_s + i\kappa_1\rho^* E_s, \tag{2}$$

$$\left(+\frac{\partial}{\partial t}+\frac{\Gamma_B}{2}+2\pi i v_B\right)\rho = i\kappa_2 E_p E_s^* + N, \tag{3}$$

where E_p and E_s stand for the normalized slowly-varying fields of pump and Stokes (or probe) waves, respectively; and $P_p=|E_p|^2$ and $P_s=|E_s|^2$ correspond to their optical powers; ρ denotes the acoustic (or phonon) field in terms of the material density distribution; N represents the random fluctuation or white noise in position and time [5]; v_g is the group light velocity in the fiber; α is the fiber's propagation loss; Γ_B is the damping rate of the acoustic wave, which equals to the reciprocal of the phonon's lifetime ($1/\Gamma_B=\tau_p=\sim10$ ns) and is related to the acoustic linewidth ($\Delta v_B=\Gamma_B/\pi$) [7]; κ_1 and κ_2 are the coupling coefficients among E_p, E_s, and ρ. If SpBS is considered, it is reasonable to assume E_s is sufficiently small so that the second term of N dominates in the right side of Eq. (3). In contrast, the first term of $i\kappa_2 E_p E_s^*$ dominates for SBS.

Taken into account the SBS power transfer between P_p and P_s under the assistance of the acoustic wave and the so-called acousto-optic effect, Eqs. (1-3) can be rewritten as

$$\left(\frac{1}{v_g}\frac{\partial}{\partial t}+\frac{\partial}{\partial z}+\alpha\right)P_p = -g(v)P_pP_s, \tag{4}$$

$$\left(\frac{1}{v_g}\frac{\partial}{\partial t}-\frac{\partial}{\partial z}+\alpha\right)P_s = +g(v)P_pP_s, \tag{5}$$

where the sign difference between the right hands of Eq. (4) and Eq. (5) means that the pump power is reduced or depleted but the probe (Stokes) power is increased or amplified.

$g(v)$ is called Brillouin gain spectrum (BGS), the key phraseology to represent Brillouin scattering in optical fibers. It denotes the spectral details of the light amplification from strong pump wave to weak counter-propagating probe/Stokes wave in SBS or those of the noise-initialized scattered phonons in SpBS. The BGS is generally expressed by [8-10]

$$g(v) = \sum_l g^{(l)}(v), \tag{6}$$

$$g^{(l)}(v) = \left\{ g_{B0} \cdot \frac{v_B \cdot \Delta v_B}{v_{ac}^{(l)} \cdot \Delta v_{ac}^{(l)}} \right\} \cdot \left\{ \frac{(\Delta v_{ac}^{(l)}/2)^2}{[v-(f_0-v_{ac}^{(l)})]^2 + (\Delta v_{ac}^{(l)}/2)^2} \right\} \cdot \left\{ \frac{1}{A_{ao}^{(l)}} \right\}, \tag{7}$$

where Eq. (6) means that the BGS is the summation of all the longitudinal acoustic modes' gain spectra and Eq. (7) corresponds to the lth-order one assigned the subscript of "l". In Eq. (7), $\Delta v_{ac}^{(l)}$ is the lth-order linewidth or full width at half magnitude (FWHM) which can be assumed to be approximately the same for all the acoustic modes; $v_{ac}^{(l)}$ is the effective acoustic velocity; $A_{ao}^{(l)}$ (in μm^2) is the so-called acousto-optic effective area of the lth-order one. $v_{ac}^{(l)}$ and $A_{ao}^{(l)}$ are qualitatively different among all acoustic modes.

There are two basic methods to theoretically and numerically analyze the BGS in optical fibers [8, 11]. One method [11] is based on Bessel or modified Bessel functions for optical fibers with regular geometric and dopant distribution, such as step-index optical fibers. The other one [8] is called two-dimensional finite-element-method (2D-FEM) modal analysis of BGS for optical fibers with complicated or arbitrary distribution. The 2D-FEM modal analysis has been used to study a Panda-type polarization-maintaining optical fiber (PMF) [8], a SMF with arbitrary residual stress [12], a w-shaped triple-layer fiber [13], or optical fibers with non-uniform optical/acoustic profiles such as solid or microsctrucuted photonic crystal fibers (PCF) [14-21].

The contribution of the fundamental acoustic mode to the entire BGS is basically dominant, which has a Lorentzian feature as schematically depicted in Fig. 3(a). Besides, it modulates the refractive index of optical fiber and changes the group velocity of optical fields in a profile shown in the inset of Fig. 3(a), which has been adopted for Brillouin slow or fast light[22, 23].

There are three basic parameters of Brillouin frequency shift (BFS, v_B), Brillouin gain peak (g_{B0}), and Brillouin linwidth (Δv_B) in the main-peak BGS (i.e. the fundamental acoustic mode or $l=1$ in Eq. (7)). v_B is defined as

$$v_B \equiv v_{ac}^{(1)} = \frac{2}{\lambda_0} \cdot n_{eff} \cdot V_a,$$
(8)

where λ_0 is the light wavelength ($\lambda_0 = c/f_0$ with c the light speed in vacuum), n_{eff} is the effective refractive index of the fiber, and V_a is the effective acoustic velocity of the fundamental acoustic mode. n_{eff} and V_a in Eq. (8) as well as $A_{ao}^{(l)}$ in Eq. (7) are all determined by the respective waveguide structures of the optical modes (n_0 and n_1) and those of the longitudinal acoustic modes (V_{11}, V_{10}), relative to the silica dopant materials and distributions in the cross section [24]. Figure 3(b) illustrates a simple example of step-index single-mode optical fiber (SMF), even for which the BGS comprises of several (typically, four) longitudinal acoustic modes due to the different contrast of optical and acoustic waveguides [8, 11]. The measured BGS of the SMF and a high-delta nonlinear optical fiber at 1550nm are depicted in Fig. 3(c)[13]. It is worth noting that v_B in optical fibers suffers strong influence from the residual elastic and inelastic strains induced by different draw tensions during fiber fabrication[12].

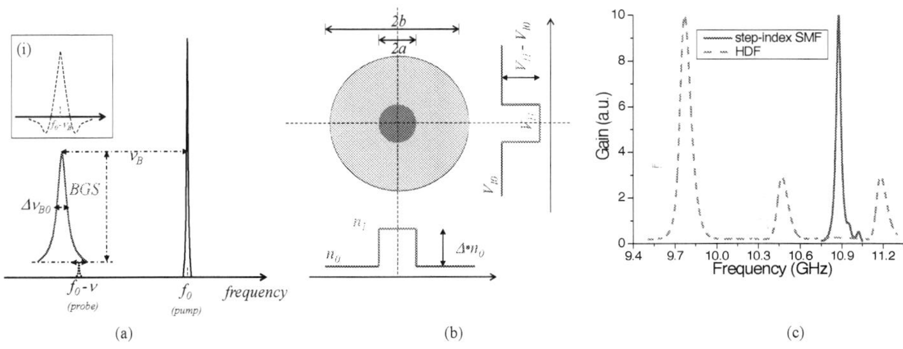

Figure 3. (a) Brillouin gain spectrum (BGS) in optical fibers. The inset "i" denotes the change of group velocity of optical fields. (b) Cross section of a step-index SMF. $\Delta=(n_1-n_0)/n_0$ is the relative index difference between the core (n_1) and the cladding (n_0). (c) Measured BGS of a step-index SMF (solid curve) and of a 17.0-mol% high-delta fiber (dashed curve). (Fig. 3(c) after Ref. [13]; © 2008 OSA.)

g_{B0} in Eq. (7) is determined by

$$g_{B0} = \frac{4\pi n_{eff}^8 p_{12}^2}{\lambda_0^3 \rho_0 c v_B \Delta v_{B0}},$$
(9)

where ρ_0 is the density of silica glass (~2202 kg/m^3) and p_{12} the photo-elastic constant (~0.271). In most silica-based fibers, the peak gain value of g_{B0} lies in the range of 1.5~3× 10^{-11} m/W [25].

Δv_{B0} in silica optical fibers with a typical value of 30~40 MHz is characteristic of SpBS. However, in the SBS process, it was theoretically proved that Δv_B strongly depends on the pump power, which is expressed as follows [5, 26]:

$$\Delta v_B = \Delta v_{B0} \sqrt{\frac{\ln 2}{G_s}},$$ (10)

where G_s is the single pass gain experienced by the weak probe wave from the strong pump wave, defined by

$$G_s = \frac{g_{B0} P_{pump} L_{eff}}{K \cdot A_{eff}^{ao}},$$ (11)

where K (=1~2) is a polarization factor (=2 for a complete polarization scrambling process), L_{eff}=[1-exp(-αL)]/α is the effective length of the fiber with α the optical loss (m^{-1}) and L the fiber length, P_{pump} the pump power, and A_{eff}^{ao}=$A_{ao}^{(1)}$ is the acoustic-optic effective area of the fundamental acoustic mode. It is noted that the exponential (G_e) or logarithmic (G_{dB}, in dB) gain of the weak probe power (P_{probe}) are presented by

$$G_e = \frac{\Delta P_{probe}}{P_{probe}} = \exp(G_s) - 1,$$ (12)

$$G_{dB} = 10 \log_{10}(G_e) \approx 4.342 G_s.$$ (13)

Eq. (13) is valid when G_e>>1.

From Eq. (10), one could estimate that the linewidth goes gradually to zero for very high gain, which can be obtained by either increasing the pump power or the interaction length of the fiber (see Eq. (11)). A zero linewidth corresponds to acoustic oscillation with an infinite time. However, pump depletion always occurs when the single pass gain (G_s) increases, which results in a limited effective single pass gain and in turn leads to a finite linewidth instead of zero one.

The experimental characterization of the phenomenon of the Brillouin linwidth's narrowing in three SMFs are depicted in Fig. 4 [27]. When the pump power is intensified to a high value of above ~24 dBm (~250 mW), the Brillouin main-peak linewidth of 180-m-long SMF becomes increasing. This is because the pump power depletes much faster than its contribution to the single-pass gain G_s since the probe wave experiences an amplification of more than G_{dB}=20 dB

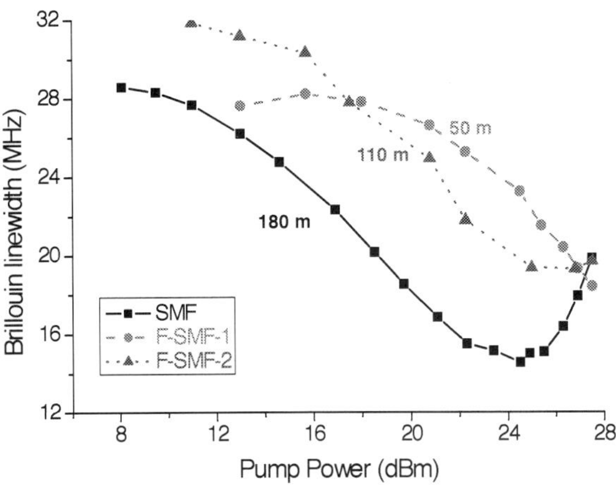

Figure 4. Measured Brillouin linewidth in three SMF varying with increase of Brillouin pump power. F-SMF denotes a SMF with pure-silica core and F-doped-silica cladding. (After Ref. [27]; © 2008 OSA.)

for a greater pump power than ~24 dBm. From the other point of view, during the pump-probe-based BGS measurement, which will be described later, the Brillouin probe wave with a down-shifted frequency of just v_B feels more significantly the depletion of the pump power than the one with a downshifted frequency of a finite offset from v_B. More details of the effect of pump depletion in SBS will be demonstrated in **Section 2.3**.

From Eq. (11), one can derivate the so-called pump threshold value of SBS originating from SpBS (also called Brillouin generator). It is given by

$$P_{th} = \kappa \frac{A_{\text{eff}}^{ao}}{L_{\text{eff}}} \frac{K}{g_{B0}}, \tag{14}$$

where κ is a numerical factor (=~21) [28] that may change in terms of the fiber length [29]. If a long-enough SMF is considered, α=0.2 dB/km (or 0.046 /km) meaning L_{eff}=21.7 km. $A_{\text{eff}}^{ao} \approx A_{\text{eff}}$=100 µm² and g_{B0}=2× 10⁻¹¹ m/W. For a perfectly-linearized pump wave, $P_{th} \approx$ 3.2 mW; for a completely polarization-scrambled pump wave, $P_{th} \approx$ 4.6 mW.

2.2. Experimental characterization

The experimental characterization of BGS in optical fibers can be implemented by two individual ways that depend on which principle of SpBS or SBS is based on. The SpBS based configuration is illustrated in Fig. 5(a). A pump wave is amplified by an erbium doped fiber amplifier (EDFA) and its polarization is optimized by a polarization controller (PC), scrambled by a polarization scrambler (PS) or switched by a polarization switcher (PSW). It is launched through an optical coupler or circulator into the fiber under test (FUT). A weak Stokes wave with downshifted frequency around ν_B is backscattered towards the coupler/circulator and can be detected by three different schemes, which are depicted in Fig. 5(b). First, an optical filter (Etalon, Fabry-Perot filter or FBG) is inserted before a photo-detector (PD) so as to eliminate the influence of Rayleigh scattering on Stokes wave and spectrally analyzed by an electrical spectrum analyzer (ESA) [2, 30]. Second, Stokes wave is optically mixed with a part of the pump wave (serving as an optical oscillator) and then detected or heterodyne-detected by a high-speed PD or a high-speed balanced PD, which is directed to the ESA [31]. Third, heterodyne detection can be carried out at an intermediate frequency (IF) range by tuning the frequency of the optical oscillator or further using of a local microwave (RF) oscillator before ESA [32].

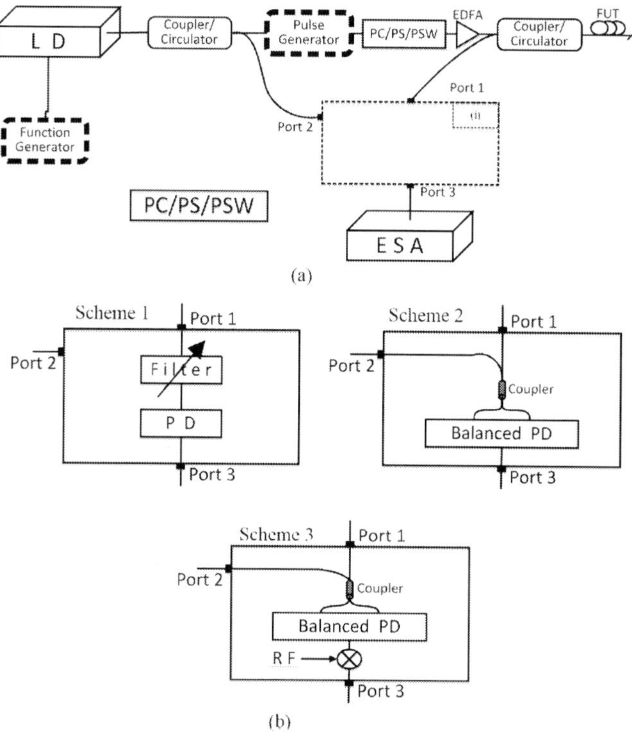

Figure 5. Experimental setup of BGS measurement based on spontaneous Brillouin scattering (SpBS). (a) Basic configuration. LD: laser diode; PC: polarization controller; PS: polarization scrambler; PSW: polarization switcher; EDFA: erbium-doped fiber amplifier; ESA: electrical spectrum analyzer. The dashed boxes of Function Generator or Pulse generator are used for distributed SpBS measurement. (b) Three different methods to detect the weak Stokes wave, corresponding to the dotted box "i" in (a) with two optical ports (port 1 and port 2) and one electrical port (port 3).

The pump-probe-based experimental configuration, depicted in Fig. 6(a), is more attractive to investigate the BGS in optical fibers (especially with very short length) since SBS process occurs and high Brillouin gain can be utilized. Two light waves from a laser unit are the optical sources: the one with larger optical frequency (f_0) serves as SBS pump and the other with lower optical frequency (f_0-ν) works as SBS probe wave. They are launched into the opposite ends of FUT so as to ensure their counter-propagation and generate intense SBS interaction. The pump wave transfers intense energy to the probe wave, called Brillouin gain; in contrast, the probe wave absorbs energy from the pump wave, called Brillouin loss. The magnitude of both Brillouin gain and loss depends on the frequency offset (ν) between the pump and probe waves, determined by the Lorentzian feature of BGS (see Eq. (7)). Subsequently, the BGS can be characterized by monitoring the power of the probe wave at PD1 (i.e. Brillouin gain) [8] or that of the pump wave at PD2 (i.e. Brillouin loss) [33] as a function of ν (provided by the laser unit), respectively. Recently, a scheme based on a combination of Brillouin gain and loss was newly proposed to enhance the signal to noise ratio (SNR) of the BGS measurement [34]. It can be realized by periodic switching of the pump and probe wave and detected at either PD1 or PD2. Besides, the simultaneous detection of PD1 and PD2 followed by a subtraction may work equally.

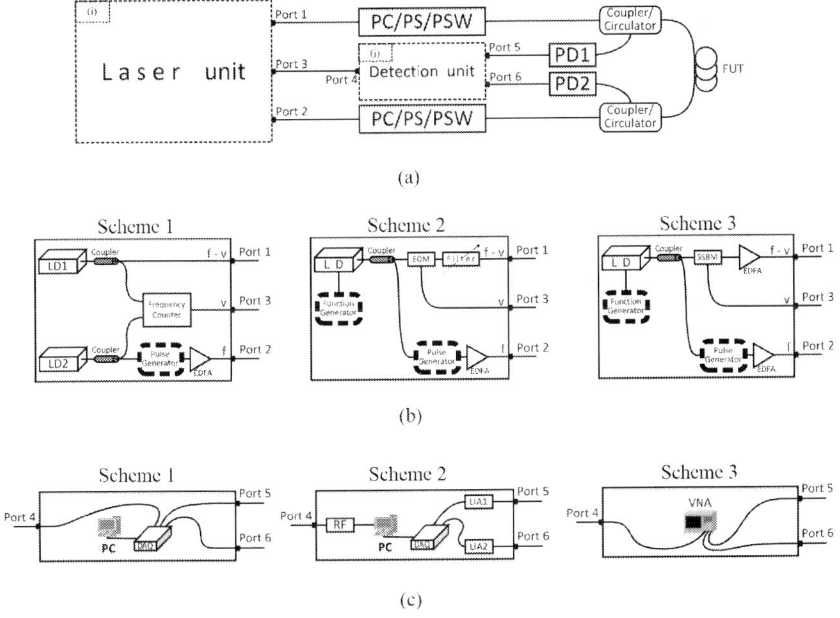

Figure 6. Experimental setup of BGS measurement based on stimulated Brillouin scattering (SBS). (a) Basic configuration. (b) Three different methods of laser unit (dotted box "i" in (a)) with two optical ports and one electrical port. The dashed boxes of Function Generator or Pulse generator are used for distributed SBS measurement. (c) Three schemes of detection unit shown in dotted box "j" in (a).

The laser unit shown in Fig. 6(a) comprises three ports (two corresponding to optical fields and the other to the revealed value of ν or a microwave/RF input), which has three different schemes as illustrated in Fig. 6(b). First, one can utilize two individual lasers under frequency/phase locking and frequency countering [33]. Second, one laser is divided into two parts. One part is amplified by an EDFA working as pump wave; the second part serving as probe wave is modulated by an electro-optic intensity modulator (EOM) to generate two sidebands working as probe wave[25]. An optical filter is inserted before launching into the FUT or laid after PD1 so as to cut off the influence of the frequency-upshifted sideband. Third, the second part can be also modulated by a single-sideband modulator [35] to get well-suppressed frequency-downshifted sideband directly serving as probe wave. There are also three schemes, depicted in Fig. 6(c), to realize the detection unit shown in Fig. 6(a). The first and simple scheme is related to the first scheme of the laser unit. A personal computer with a multi-channel data acquisition card (DAQ) can catch the value of ν and record the data of PD1 and/or PD2 so as to pick up the Brillouin signal (gain and/or loss) as a function of ν. The second and third schemes in Fig. 6(c) can be used for either the second and/or third laser unit in Fig. 6(b), respectively. For instance, a high-cost vector network analyzer provides a frequency-tuned RF signal to modulators and simultaneously detect the Brillouin signal [36]. Alternatively, the RF signal can be achieved from a microwave synthesizer and the data of PD1 and/or PD2 can be picked up by a DAQ with or without a lock-in amplifier (LIA). It is notable that the use of LIA for detection unit requires an intensity-chopping of the pump wave by an additional EOM [10] or periodic switching of upshifted or downshifted sideband at the SSBM [34], which is advantageous for characterization of very weak BGS or a short-length FUT due to its high SNR and accuracy[12, 34, 37].

Figure 7(a) depicts a high-accuracy experimental setup of pump-probe SBS-based BGS characterization by use of SSBM and LIA for a short-length FUT [37]. An EDFA is inserted after the SSBM to increase the probe power, which is aimed to reduce the impact of Rayleigh scattering or splicing/crack induced reflection of the pump wave in the FUT. The optical lights after a circulator include the following components:

$$P_{tot} = P_{probe}^{0} + \delta P_{probe} + P_{pump}^{R},$$

(15)

where P_{probe}^{0} denotes the probe power experiencing no Brillouin amplification, ΔP_{probe} the amplified probe power, and P_{pump}^{R} the reflected pump power. Thanks to the lock-in detection, the component of P_{probe}^{0} is effectively cut off by the LIA since it has no relationship with the chopped pump power given by

$$P_{pump}(t) = P_{pump}^{0} \cdot \cos(2\pi f_{ch} t),$$

(16)

where f_{ch} is the chopping frequency. Simply assuming that there is no depletion for pump power and no optical propagation loss for either probe or pump light, ΔP_{probe} can be expressed by

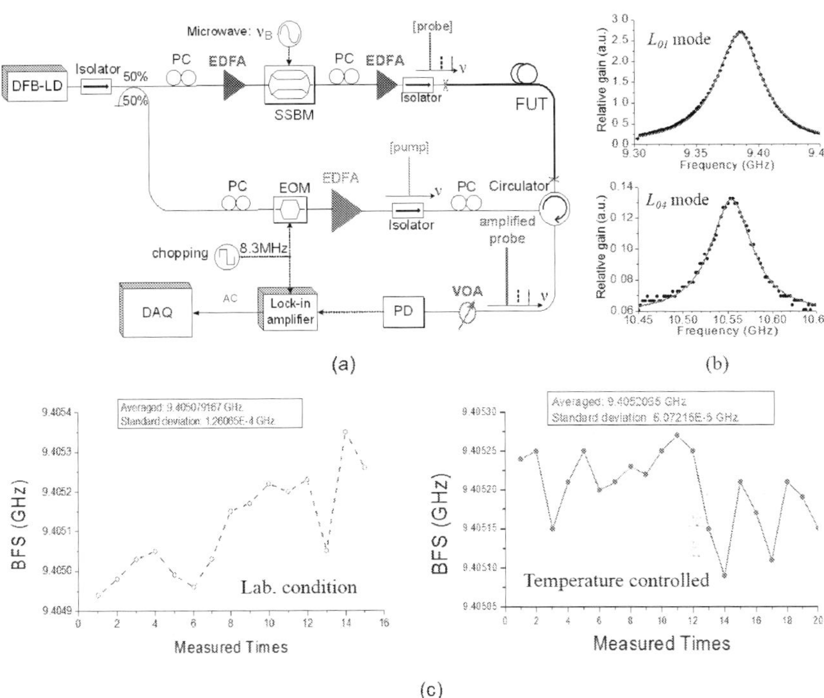

Figure 7. High accuracy pump-probe-based BGS characterization. (a) Experimental setup. (b) Characterized BGS. (c) Measurement accuracy. ((a) and (b) after Ref. [37]; © 2007 OSA.)

$$\delta P_{probe} = g(\nu) \cdot P_{probe}^0 (f_0 - \nu) \cdot P_{pump}, \tag{17}$$

If R denotes the reflectivity of pump power arising from both Rayleigh scattering and some reflection points, the reflected pump power can be expressed:

$$P_{pump}^R = [P_{pump}^0 - P_{loss}] \cdot R, \tag{18}$$

where P_{loss} is the so-called Brillouin loss of pump power during Brillouin interaction which is approximately equal to ΔP_{probe} as

$$P_{loss} = \delta P_{probe} = g(\nu) \cdot P_{probe}^0 \cdot P_{pump}^0 \cdot \cos(2\pi f_{ch} t), \tag{19}$$

The demodulated electric amplitude via a LIA at f_{ch} is given by

$$P_{de} = \left[P^0_{probe} \cdot g(\nu) \cdot (1-R) + R \right] \cdot P^0_{pump}, \tag{20}$$

where the part of $\left[P^0_{probe} \cdot g(\nu) \cdot (1-R) \right] \cdot P^0_{pump}$ is the signal to be detected by the LIA and the rest part of $R \cdot P^0_{pump}$ is the noise level. From it, one can deduce the signal-to-noise ratio (SNR):

$$SNR = \frac{P^0_{probe} \cdot g(\nu) \cdot (1-R)}{R}, \tag{21}$$

which is independent on the pump level, but just determined by the probe power of $P_{probe}{}^0$ and the reflection rate of R as well as the BGS of $g(\nu)$. In other words, an increase of probe power can drastically enhance SNR and also improve the system accuracy. As an example, Figure 7(b) shows a characterized fundamental-order or higher-order resonance BGS in a w-shaped high-delta fiber with fluorine inner cladding (F-HDF) [37]. The measurement system has a high accuracy of 0.13-MHz standard deviation at laboratory condition or 0.05 MHz for well-temperature-controlled condition, as illustrated in Fig. 7(c).

2.3. Pump depletion effect

As mentioned above (see Fig. 4), pump depletion effect influences the linewidth of pump-probe-based BGS. Early in 2000 [38], it was first observed that the spectrum broadening and hole burning occurs in a SBS generator (i.e. noise-started spontaneous Brillouin scattering). The reason was thought as the waveguide interaction among different angular components of the pump and backscattered Stokes signals. Besides, during the application of SBS-based amplifier, two coherent optical waves with precise frequency difference equal to Brillouin frequency shift are launched into optical fibers; then a frequency-scanned weak signal could suffer non-uniform amplification if the two waves' powers are too high [39, 40]. In SBS-based distributed fiber optical sensor, which will be introduced in **Section 3**, two coherent waves (pulse and/or continuous wave (CW)) are injected into the two opposite ends of the sensing fiber. Recently, it was found that pump depletion of pump-probe-based system configuration could induce a significant measurement error of the local Brillouin frequency shift in the far end of the probe (Stokes) wave [41].

Assuming that CW probe wave, $P_s(0)$, is injected at the near end of the fiber ($z=0$) while CW pump wave, $P_p(l)$, is launched at the far end of the fiber ($z=l$ with l the fiber length). Considering the steady-state condition and neglecting the transmission loss of the fiber, the coupling equations of Eq. (4) and Eq. (5), describing the SBS interaction, can be modified to the dimensionless equations [42]:

$$\frac{dQ_p}{dx} = kQ_p Q_s, \tag{22}$$

$$\frac{dQ_s}{dx} = kQ_pQ_s,\tag{23}$$

where $Q_p=P_p/P_s(0)$ and $Q_s=P_s/P_s(0)$ represent the normalized pump and probe waves with respect to the injected probe wave of $P_s(0)$, $x=z/l$ is the normalized position, and $k=G\bullet l\bullet P_s(0)$ is the normalized Brillouin gain with $G=g(v)/A_{eff}$. Figure 8(a) illustrates the normalized BGS at different positions of an arbitrary-length SMF, which are numerically calculated according to Eqs. (22) and (23). It is found that BGS gradually gets broadened, saturated, and hole-burned when the position moves from the far end ($z=l$) towards the near end of the fiber ($z=0$). We define the BGS saturation as the critical condition of spectral hole burning phenomenon.

Further introduce the injected power ratio between pump and probe waves, defined as $\gamma=Q_p(x=1)=P_p(l)/P_s(0)$. Since $dQ_p/dx=dQ_s/dx$, the difference between Q_p and Q_s maintains a constant (A), i.e. $A=Q_p-Q_s$, which is determined by k and γ. Consequently, the analytical solutions to Eqs. (22) and (23) are derived as

$$Q_p = \frac{A(A+1)e^{-kAx}}{(A+1)e^{-kAx}-1},\tag{24}$$

$$Q_s = \frac{A}{(A+1)e^{-kAx}-1}.\tag{25}$$

The critical condition of the spectral hole burning phenomenon can be theoretically expressed by:

$$\frac{dQ_s}{dv} = 0.\tag{26}$$

By numerically solving Eqs. (24)-(26), one can interpret the critical condition by two different ways: (1) the critical position x_c for the fixed pump and probe power; (2) the critical powers for a specific position of the fiber. Figure 8(b) depicts the calculated relation of x_c to k and γ, which indicates that x_c moves towards the fiber far end when k and γ reach higher values (i.e. the fiber gets longer or the injected powers are stronger). It means that the pump depletion gets worse since much longer segments in the fiber suffer spectral hole burning. It is notable to address that the physical nature of the critical powers is essentially the same as that of the critical position because they can be also deduced by the contour (i.e. k-γ curve) at a fixed position x in Fig. 8(b):

$$P_s(0) = k / GL,\tag{27}$$

$$P_p(L) = \gamma k / GL. \tag{28}$$

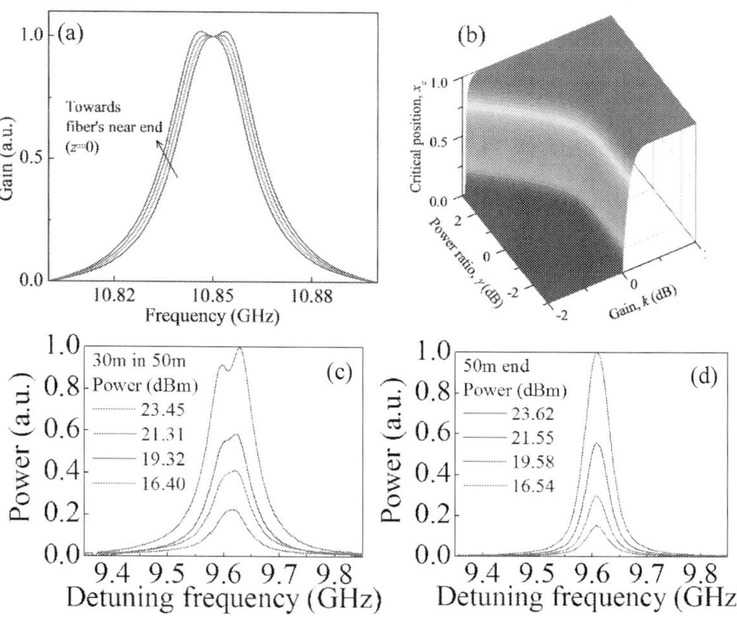

Figure 8. (a) Simulated BGS at different positions in the fiber. (b) Critical position x_c, determined by the normalized gain k and injected power ratio γ. The critical positions divide the fiber into two parts with and without the spectral hole burning phenomenon. Measured BGS in the middle (c) or at the end (d) of a 50-m-long fiber. (after Ref. [42]; © 2014 JJAP.)

Figure 8(c) and 8(d) illustrates the measured BGS under different pump power for two different positions (in the middle and at the far end, respectively) of a 50-m-long dispersion compensated fiber (DCF). The probe power is fixed at 9.8 dBm. At the far end, the BGS [see Fig. 8(d)] rises with the power increased but always preserves the Lorentz shape. While in the middle, the experimental result [see Fig. 8(c)] is in a qualitative accordance with the numerical analysis [see Fig. 8(a)]. The Brillouin gain keeps rising with the increase of optical power, while the peak at the local Brillouin frequency shift seems to be saturated gradually and a hollow starts appearing when it reaches ~20 dBm, which is just the spectral hole burning phenomenon. The hollow in the BGS may introduce great errors to pump-probe-based Brillouin distributed sensors since it disables the peak-searching of the Brillouin frequency shift.

The pump power leading to the BGS saturation is approximately characterized as the critical pump power (for instance, 21.3 dBm at z=30 m). The measured critical powers for two positions (z=20 m or 30 m) of 50-m-long DCF are depicted in Fig. 9, where the simulated critical powers are compared. It illustrates that the critical pump powers approximately measured for several probe powers (open symbols) have very similar trend as the theoretical analysis (curves). It is clear that the position of bigger z requires greater critical powers. This is because the pump depletion is weaker and the distortion of the BGS is less serious if the position is much closer to the fiber far end.

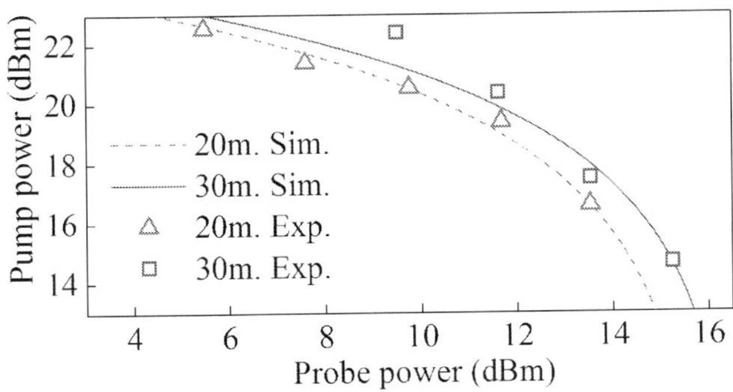

Figure 9. Critical powers for two different positions in a 50-m-long DCF. Solid and dashed curves, simulation; open symbols: experiment. (after Ref. [42]; © 2014 JJAP.)

3. Brillouin–based distributed sensors

3.1. Sensing of measurands

The first report of Brillouin based distributed optical fiber sensors [43] was based on the same principle as that of optical time domain reflectometry (OTDR) or Raman based OTDR (ROTDR) technique as a non-destructive attenuation measurement technique for optical fibers. In that proposal [33], SBS process was performed by injecting an optical pulse source and a continuous-wave (CW) light into two ends of FUT. When the frequency difference of the pulse pump and CW probe is tuned offset around v_B of the FUT, the CW probe power experiences Brillouin gain from the pulse light through SBS process. Similarly like the case of OTDR, the SBS distributed measurement could measure attenuation distribution along the fiber having no break from an interrogated optical power as a function of time, but it has much higher signal-to-noise ratio (more than ~10 dB) than OTDR due to SBS high gain. Later, Horiguchi and co-researchers found that this non-destructive can be extended into **a frequency-resolved technique** because v_B of optical fibers has linear dependence on measurands of strain and temperature as follows [44, 45]:

$$v_B - v_{B0} = A \cdot \delta\varepsilon + B \cdot \delta T, \tag{29}$$

where v_{B0} is measured at room temperature (25°C) and in the "loose state" as a reference point, $\Delta\varepsilon$ the applied strain and ΔT the temperature change. The "loose state" means that the FUT is laid freely in order to avoid any artificial disturbances. A (or $Cv\varepsilon$) is the strain coefficient in a unit of MHz/$\mu\varepsilon$ and B (or $C_T\varepsilon$) is the temperature coefficient in a unit of MHz/°C. Figure 10 illustrates the characterized strain or temperature dependence in a standard SMF under the experimental setup of Fig. 7(a), where the BGS always moves towards higher v_B and its gain reduces or increases when $\Delta\varepsilon$ or ΔT is increased, respectively. At 1550 nm, A=0.04~0.05 MHz/$\mu\varepsilon$ and B=1.0~1.2 MHz/°C, which depends on the fiber's structure and jackets. **Note that Eq. (29) is the basic sensing mechanism of Brillouin-based distributed sensors.**

The nowadays telecom optical fibers (ITU-T G.651, G.652, G.653, and G.655) mostly have GeO_2-doped fiber cores [46] and pure-silica (or other-doped-silica) cladding. Naturally, the GeO_2 doping induces the reduction of the longitudinal acoustic velocity in GeO_2-doped core V_{l1} with respect to that in pure-silica cladding V_{l2} (i.e. $V_{l1} < V_{l2}$) [24, 47]. It provides a waveguide of longitudinal acoustic modes in the core region as schematically depicted in Fig. 3(b). A recent study further proves that the acoustic modes sense better confinement than the optical modes in a GeO_2-doped optical fiber [13]. The enhanced confinement results in the existence of multiple L_{0l} acoustic modes in a single-mode optical fiber (SMF) [8], and also leads to that the first-order L_{01} acoustic mode among all L_{0l} modes is best confined in the core and even better confined than the fundamental LP_{01} optical mode [13]. Furthermore, the enhanced confinement shows that the effective acoustic velocity of L_{01} mode (V_a) is close to V_{l1} (i.e. $V_a \approx V_{l1}$), the longitudinal acoustic velocity in the core. Therefore, the change of the L_{01} mode's effective acoustic velocity V_a is dominantly due to the change of the core's acoustic velocity V_{l1} but negligibly (less than 1%) due to that of the cladding's acoustic velocity V_{l2}, even though the core's acoustic velocity V_{l1} and the cladding's acoustic velocity V_{l2} vary equally [12].

The longitudinal acoustic velocity V_{l1} in the GeO_2-doped core (approximately, the L_{01} mode's effective acoustic velocity V_a) is determined by the Young's modulus (E_1) and the density (ρ_1) [48]:

$$V_a \approx V_{l1} = \sqrt{E_1 / \rho_1}, \tag{30}$$

For convenience, we introduce a normalized strain coefficient ($A'=A/v_{B0}$, in a unit of $10^{-6}/\mu\varepsilon$) and a normalized temperature coefficient ($B'=B/v_{B0}$, in a unit of $10^{-6}/$°C). The normalized strain coefficients include three respective factors[49]:

$$A' \equiv \frac{A}{v_{B0}} = A'_{neff} + A'_{\rho} + A'_{E}, \tag{31}$$

$$B' \equiv \frac{B}{v_{B0}} = B'_{neff} + B'_{\rho} + B'_{E}. \tag{32}$$

Each three parts in the right sides of Eq. (31) and Eq. (32) are determined by relative change rates in n_{eff}, E_1, and ϱ_1 due to the applied strain $\Delta\varepsilon$ or the temperature change ΔT. A'_{neff} and B'_{neff} are determined by the elasto-optic and thermo-optic effects; A'_ρ and B'_ρ are subject to the strain-induced distortion and the thermal expansion; A'_E and B'_E are decided by the strain-induced second-order nonlinearity of Young's modulus and the thermal-induced second-order nonlinearity of Young's modulus.

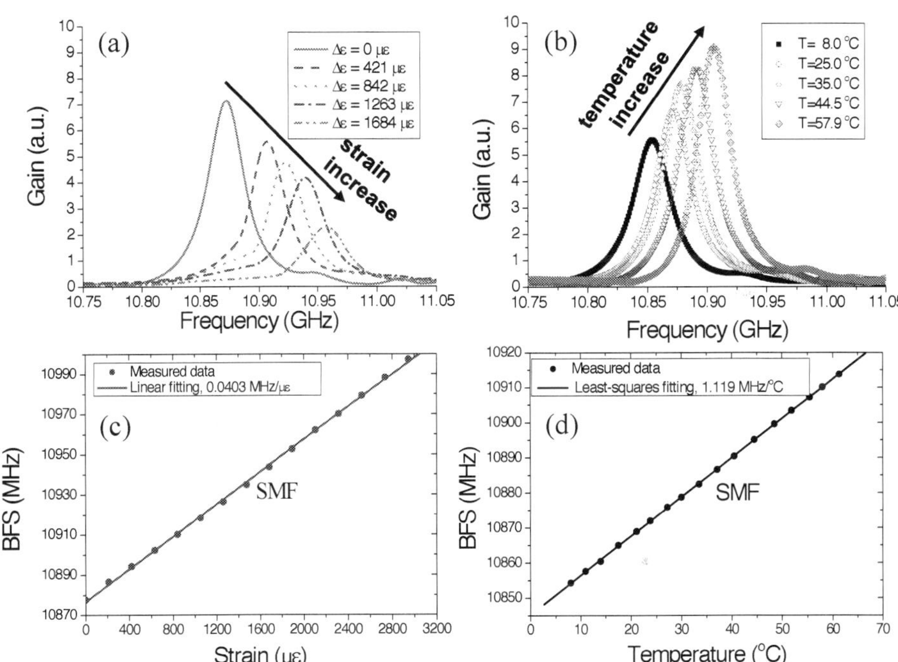

Figure 10. (a) Stain and (b) temperature dependences of BGS in SMF; (c) Strain and (d) temperature dependences of Brillouin frequency shift ν_B in SMF.

Strict experimental characterization on a series of optical fibers with different GeO_2 concentration is depicted in Fig. 11 [49]. The BFS has linear dependence on the GeO_2 concentration in the fiber's core (i.e.-87.3 MHz/mol%), which corresponds to ν_{B0} change of-87.3 MHz regarding 1-mol% increase of GeO_2 concentration in the core (i.e. an incremental Δ of 0.1 %). It specifies the previously reported values [25, 50, 51]. Besides, the frequency spacing between neighbouring acoustic modes increases by orders when the GeO_2 concentration is enhanced, for example, ~50-60 MHz for Fiber-A (SMF, 3.65 mol%) versus ~700-720 MHz for Fiber-C (HNF, 17.0 mol%).

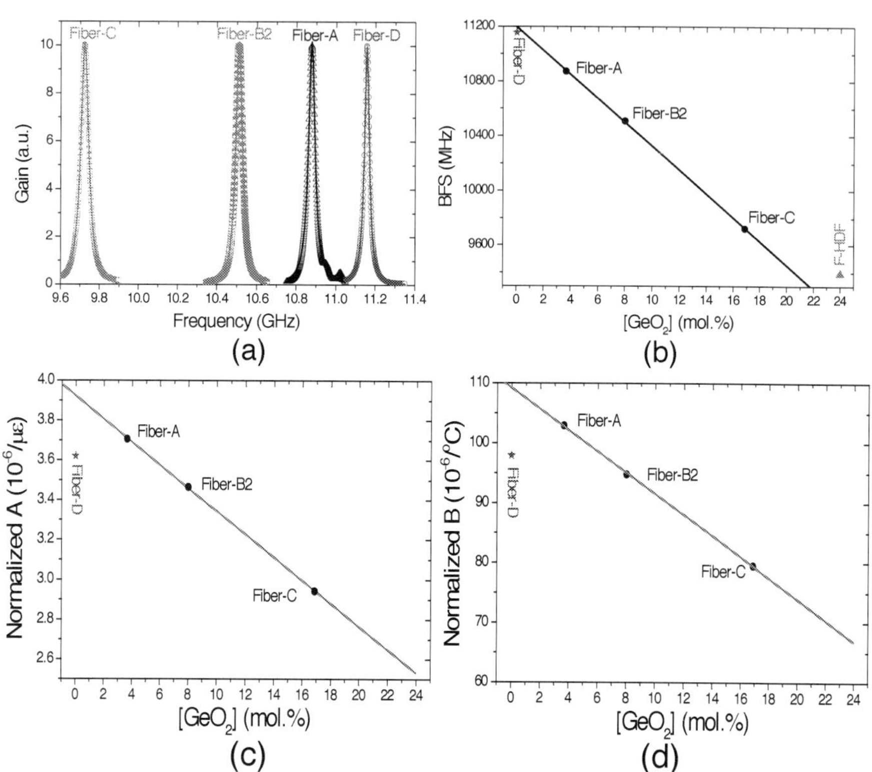

Figure 11. (a) BGS, (b) BFS, (c) normalized strain coefficient, and (d) normalized temperature coefficients in silica optical fibers with different GeO$_2$ concentration. (After Ref. [49]. © 2008 OSA/IEEE.)

The normalized strain and temperature coefficients defined in Eq. (31) and Eq. (32) were characterized by repeating the BGS measurement under different applied strain and temperature change. It shows a linear dependence of A' or B' on GeO$_2$ concentration with slope of-1.48 %/mol% or-1.61 %/mol%, which denote that A' and B' are relatively decreased by-1.48 % and-1.61 % for an incremental Δ of 0.1 %. The theoretical study further indicates that both the strain and temperature dependences in Eq. (29) are dominantly (~92%) responsible from the strain-induced and thermal-induced second-order nonlinearities of Young's modulus, that is, Eq. (31) and Eq. (32) [49].

3.2. Sensing of location

Besides the sensing of measurands (see Eq. (29)), the mapping of spontaneous or stimulated Brillouin scattering process (not just non-destructive attenuation measurement [43]) is another key issue to realize distributed optical fiber sensing [52-54]. Two different mapping ways, as schematically illustrated in Fig. 12, were proposed. One is to repeat the localized BGS in scanned positions along the FUT; the other is to repeat the Brillouin interaction under different frequency offset.

There are three different mapping or position-interrogation techniques, including time domain [33, 52-56], frequency domain [57, 58], and correlation domain [59-61]. Regarding the injection ways of optical fields, there are two opposite groups, i.e. analysis versus reflectometry. The analysis is two-end injection based on SBS; while the reflectometry is one-end injection based on SpBS. Comparably, the analysis has much higher SNR than the reflectometry. Note that there is an additional method between analysis and reflectometry, called one-end analysis [55, 62, 63]. Its only difference from the traditional (two-end) analysis is the one-end injection and its SBS process occurs between the forward pump and the backward probe wave that is reflected at the far end of FUT.

The basic principle of time-domain sensing technique is the "time-of-flight" phenomenon in FUT. For two-end or one-end analysis, named Brillouin optical time domain analysis (BOTDA) [33, 52, 54], one of pump and probe waves is pulsed in time and the other is continuous wave (CW). Subsequently, they are successively interacted along the FUT during the time-of-flight of the pulsed wave. In contrast, for one-end reflecometry, called Brillouin optical time domain reflectometry (BOTDR) [53, 56], the pump wave is pulsed in time and the SpBS Stokes wave is reflected along the FUT during the pump's time-of-flight. The basic experimental configuration of BOTDR or BOTDA can be simply carried out in Fig. 5 or Fig. 6, respectively. The required modification is to insert an optical pulse generator (for example, an electro-optic intensity modulator driven by an electric pulse generator). The spatial resolution (ΔZ_{TD}) of time-domain distributed sensing is physically determined by the pulse width (τ)[43]:

$$\Delta Z_{TD} = \frac{\tau \cdot c}{2n}, \tag{33}$$

where c is the light speed in vacuum and n the group velocity of the pulse. The BGS mapping is realized by repeating the above measurement when the spectrum of the reflected Stokes in BOTDR is processed or the optical frequency offset between the pump and probe in BOTDA is tuned around the BFS ν_B.

There are two kinds of correlation-domain sensing techniques, nominated Brillouin optical correlation domain analysis (BOCDA) [59, 60] and Brillouion optical correlation domain reflectometry (BOCDR) [61, 64]. Both of them originate from the so-called synthesis of optical coherence function (SOCF) [65, 66]. Nevertheless, the SOCF in BOCDA or BOCDR is generated between the pump and probe waves or between the pump-scattered Stokes wave and the optical oscillator, respectively. In experiment, the BOCDR and BOCDA can be executed by substituting a distributed feedback laser diode (DFB-LD) driven by a function generator (such as in a sinusoidal function) for the light source in Fig. 5 and Fig. 6, respectively. Thanks to the current-frequency transferring effect of DFB-LD [67], the optical frequencies of the light

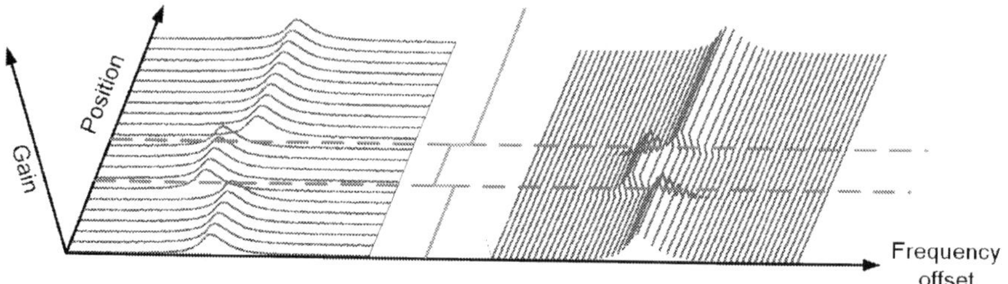

Figure 12. Schematic of sensing of location or mapping of BGS.

sources are simultaneously modulated also in a sinusoidal function. Subsequently, the optical frequency offset between pump and probe or between scattered Stokes and optical oscillator changes with time as well as position, deviating from the preset constant frequency offset around the BFS v_B. Only at some particular locations (called correlation peaks), the frequency offset is maintained as the constant frequency offset because of the in-phase condition so that the local SBS interaction or the beating of the local Stokes and oscillator is **constructive**. At other locations rather than correlation peaks, the frequency offset is always vibrating with time, which leads to a broadened and **destructive** SBS or SpBS. The spatial resolution of BOCDA and BOCDR are both determined by[59]

$$\Delta Z_{CD} = \frac{c}{2nf_m} \cdot \frac{\Delta v_B}{\pi \Delta f},\tag{34}$$

where f_m is the modulation frequency of the sinusoidal function, Δf the modulation depth, and Δv_B the Brillouin linewidth defined in Eq. (10). Since the SOCF is naturally realized by an integral or summation signal processing in photonics or electronics, all SBS or SpBS along the entire FUT should be accumulated together (as an example shown in Fig. 6(a), accumulated by a LIA). Consequently, the maximum measurement length (or sensing range, L_{CD}) is decided by the distance between two neighboring correlation peaks [59]:

$$L_{CD} = \frac{c}{2nf_m}.\tag{35}$$

Because of the difference of the physical pictures between time domain and correlation domain, their sensing performance is different. For example, the spatial resolution of BOTDA/BOTDR was typically limited to be ~1 m by the lifetime of acoustic phonons (10 ns) and the nature of intrinsic Brillouin linewidth. However, BOCDA/BOCDR is of CW nature free from this limitation, and their spatial resolution can be ~cm-order [60, 68] or even ~mm-order [69]. Since BOTDA/BOTDR carries out the whole mapping of BGS along the FUT during the time-of-the-flight while BOCDA/BOCDR realizes the distributed sensing by sweeping the modulation frequency (or correlation peak), the sensing speed is different. The entire sensing speeds for both BOTDA/BOTDR and BOCDA/BOCDR are time-consuming due to the tuning of pump-probe frequency offset, averaging of mapping, and signal processing of data fitting. However, the sensing position of BOCDA/BOCDR can be random accessed [59, 61], and the dynamic sensing with high speed at the random accessed position is possible [70, 71]. The detailed difference of other performances will be described in **Section 4** and **Section 5**.

4. Challenges in Brillouin based distributed optical fiber sensors

4.1. Simultaneous measurement of strain and temperature

As explained in **Section 3**, all Brillouin based distributed optical fiber sensors interrogate the Brillouin frequency shift so as to deduce strain and temperature information based on Eq. (29). It naturally gives a physical challenge, i.e. how to distinguish the response of strain from the response of temperature based on the single parameter of BFS interrogation in a single piece of sensing fiber. In current industrial practices, two individual fibers or two fibers in a fiber cable are used to discriminate the strain and the temperature: the first one is embedded or bonded at the target material/structure to feel the total responses of strain and temperature, while the second fiber is placed beside the first one and kept in loose condition so that it feels the response of temperature only. Another way is to use two distributed sensing systems with two individual fibers [72-75]: one Raman-based or Rayleigh-based sensor is to monitor the temperature; the other Brillouin sensor to monitor the temperature and strain. After distributed sensing measurements, the strain and temperature responses can be calculated by mathematics. However, the above practices make the entire sensing system complicated and the calculated responses of strain and temperate change with service time.

Practical applications of Brillouin based distributed optical fiber sensors require a method to effectively discriminate them by use of two intrinsic parameters (denoted by y_1 and y_2) in one sensing fiber. Their changes (Δy_1 and Δy_2) depend on simultaneously the applied strain ($\Delta \varepsilon$) and temperature change (ΔT), which are governed by the following matrix:

$$\begin{pmatrix} \Delta y_1 \\ \Delta y_2 \end{pmatrix} = \begin{pmatrix} A_1 & B_1 \\ A_2 & B_2 \end{pmatrix} \begin{pmatrix} \Delta \varepsilon \\ \Delta T \end{pmatrix}, \tag{36}$$

where A_1 (A_2) and B_1 (B_2) are the strain and temperature coefficients of y_1 (y_2), respectively. Both $\Delta \varepsilon$ and ΔT can be deduced from Eq. (36), given by

$$\begin{pmatrix} \Delta \varepsilon \\ \Delta T \end{pmatrix} = \frac{1}{A_1 B_2 - B_1 A_2} \begin{pmatrix} B_2 & -B_1 \\ -A_2 & A_1 \end{pmatrix} \begin{pmatrix} \Delta y_1 \\ \Delta y_2 \end{pmatrix}, \tag{37}$$

It is obvious to know that the condition that the strain and temperature can be successfully distinguished is determined by

$$A_1 B_2 \neq B_1 A_2. \tag{38}$$

In fact, the Brillouin-based distributed sensing system always suffer a measurement uncertainty (Δy_1 and Δy_2), which is in a linear proportional relation with the discrimination errors in strain ($\Delta\varepsilon$) and temperature (ΔT), also given by Eq. (37).

A possible solution using one fiber is to monitor two acoustic resonance peaks at different orders of Brillouin gain spectrum (BGS) in a specially-designed optical fiber [13, 37, 76, 77]. So far, this method cannot ensure accurate discrimination because all the acoustic resonance frequencies exhibit similar behaviors in their dependences on strain and temperature (see Fig. 11) [49]. There is another kind of method reported for discrimination that relies on the possibility that the peak amplitude and BFS of the BGS could have quantitatively different dependences on strain and temperature [78-81]. Its accuracies is not sufficient (e.g., several degrees Celsius and hundreds of micro-strains), which is mainly due to the low signal-to-noise ratio in the BGS peak-amplitude measurement particularly for distributed sensing where troublesome noise from non-sensing locations is accumulated.

4.2. System limitation of time-domain or correlation-domain technique

There are several system limitations in time-domain BOTDA/BOTDR and correlation-domain BOCDA/BOCDR, which comes from their individual sensing techniques. For example, BOTDA/BOTDR suffers a typical limitation of spatial resolution (~1 m) mainly determined by the linewidth of BGS or the lifetime (~10 ns) of acoustic phonons. Narrower pulse width corresponding to higher spatial resolution according to Eq. (33) weakens the acoustic phonons due to the lifetime of the acoustic phonons and leads to broader BGS as well as lower frequency accuracy due to the convolution between the intrinsic BGS and broader spectrum of the pulse [82, 83]. Moreover, although the time-of-the-flight feature of BOTDA/BOTDR is suitable for long distance sensing, the nature of pump depletion and fiber transmission loss confines the maximum of measurement range within several tens of kilometers [84].

On the other hand, BOCDA/BOCDR can provide extremely high spatial resolution of cm order or mm order with a cost of system complexity. However the correlation-domain sensing nature means that there intrinsically exist periodic correlation peaks in the fiber. Besides, the nominal definitions of spatial resolution and measurement range (see Eq. (34) and Eq. (35)) show that they both depend on the modulation frequency and thus they are in a tradeoff relation with each other [59]. The accumulation of the entire BGS along the FUT corresponding to the measured BGS at the sensing location should include a high-magnitude background of the BGS at the uncorrelated positions, which makes it difficult achieve large range of strain or temperature since higher strain or temperature change shifts the measured BGS closer to the background. As introduced in **Section 3.2**, the access ability of BOCDA/BOCDR is random and the sensing speed in one location is high. However, the sensing speed along the entire FUT is still low and just comparable to BOTDA/BOTDR because the scanning of the sensing location is realized by changing the modulation frequency (see Eq. (35)), which needs quite long time to restart the communication among electronic devices (specially, function generator).

5. Advances in Brillouin based distributed optical fiber sensors

5.1. Concept of Brillouin dynamic grating

Dynamic grating can be generated by use of gain saturation effect in rare-earth-metal-doped optical fibers [85-87] or stimulated Brillouin scattering (SBS) process in optical fibers [88-93] and even in a photonic chip [94]. Dynamic grating is more advantageous for certain applications than fiber Bragg grating (FBG) [95] because it can be dynamically constructed using two coherent pump waves while FBG is static after fabrication. In comparison, the SBS-generated dynamic grating, also called Brillouin dynamic grating (BDG), is superior to the saturation gain grating due to its elasto-optic nature and lack of quantum noise [1]. In addition, the BDG is much easier to experimentally characterize [88, 90-93] while the saturation gain grating needs sophisticated double lock-in detection [86].

Up to date, there are various methods to generate BDG in optical fibers, which are schematically compared in Fig. 13 [93]. The basic principle of BDG in optical fibers is quite similar, which is shown in Fig. 13(a). Two coherent optical waves, i.e. the pump and probe (or Stokes) waves in the SBS process, are launched from the two opposite ends of optical fibers. When their optical frequency offset ($v_1 = f_1 - f_1'$) is equal to the BFSv_B as well as the resonance frequency of the fundamental acoustic mode ($v_{ac}^{(1)}$) defined by Eq. (8):

$$v_1 \equiv f_1 - f_1' = v_B = v_{ac}^{(1)}, \tag{39}$$

where f_1 and f_1' are the optical frequencies of the pump and probe waves, a strong acoustic wave of the fundamental acoustic mode (the so-called BDG) is optically generated. As long as the third optical wave (i.e. the readout wave) is injected from the same end as the pump wave, there is a diffracted/reflected optical wave originating from the BDG. The diffraction or reflection efficiency (also called BDG reflectivity) is determined by the phase-matching condition, under which the pump/probe and readout wave can efficiently couple their energy via the BDG.

The method of BDG generation and detection can be classified into two different cases, which depends on the used optical fibers. In the first case of polarization maintaining fiber (PMF) [88-90] or few-mode fiber (FMF) [91] (see Fig. 13(b)), the BDG generation is separated from the BDG detection by use of orthogonal polarization states or different optical modes, respectively. The phase-matching condition means that the BFS of the BDG generation and detection should be unique as Eq. (39), which results in a frequency difference determined by the PMF's

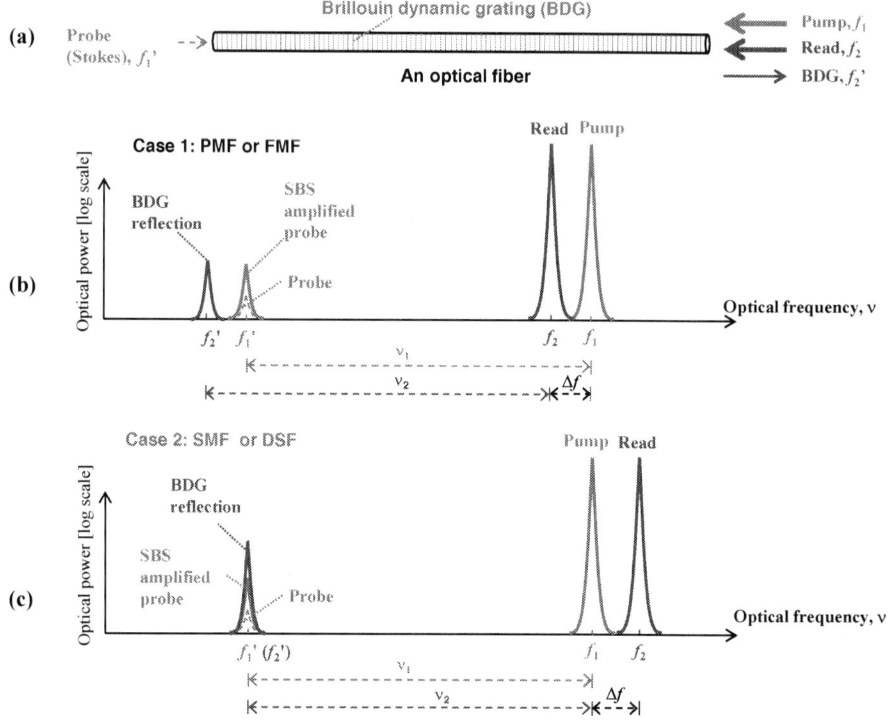

Figure 13. Principle of BDG in an optical fiber. (a): Orientation of optical injection. (b) and (c): Two different cases of the optical frequency relation among the pump, probe (Stokes), readout, and BDG reflection. (After Ref. [93]; © 2013 OSA.)

birefringence (B) or the FMF's modal refractive index difference (n_{eff}-n_{eff}' with n_{eff}' the higher-order modal refractive index):

$$\Delta f \equiv f_2 - f_1 = \frac{B}{n_{eff}} \cdot f_1,$$

(40)

$$\Delta f \equiv f_2 - f_1 = \frac{n_{eff} - n_{eff}'}{n_{eff}} \cdot f_1,$$

(41)

where f_2 is the optical frequency of the readout wave.

As shown in Fig. 13 (c), the BDG in a SMF [92] or dispersion shifted fiber (DSF) [93] can be generalized into the second case. If the readout wave with the optical frequency of f_2 is launched for BDG detection, multiple-peak Stokes wave is intrinsically backscattered via SpBS. The ith-peak Stokes wave is downshifted in frequency from the readout wave by

$$v_2^{(i)} = \frac{2n_{eff}}{c} \cdot V_a^{(i)} \cdot f_2, \tag{42}$$

where $V_a^{(i)}$ is the acoustic velocity of the ith-order acoustic wave. The phase-matching condition of the BDG generation and detection turns to be determined by the frequency difference between the pump and readout wave:

$$\Delta f \equiv f_2 - f_1 = \frac{v_1^{(i)} - v_1^{(1)}}{1 - 2n \cdot V_a^{(i)} / c} \approx v_1^{(i)} - v_1^{(1)}, \tag{43}$$

where $v_1^{(i)}$ is the ith-order resonance frequency of the BGS measured by the pump-probe SBS process, also given by Eq. (42) except that f_2 is replaced by f_1. The approximation in Eq. (43) is reasonable since the acoustic velocity ($V_a^{(i)}$=~5300-5900 m/s) in silica-based fibers is far smaller than the optical velocity (c=3.0× 10^8 m/s) [11]. Note that the BDG observed in a SMF [92] can be regarded as one special example of the generalized second case with$\Delta f = 0$ or $f_2 = f_1$ in Eq. (43) because the BDG generation and detection share the same fundamental acoustic mode.

In comparison, the method based on a PMF is more attractive because the BDG generation and readout are oriented and separated in two orthogonal polarization states [88-90]. The frequen‐ cy-deviation property provides an additional degree of freedom to precisely characterize the birefringence according to Eq. (28). Figure 14 depicts the optical spectra of the BDG reflection measured by an optical spectrum analyzer (OSA), including four components (leaked pump and probe waves, BDG reflected wave, and Rayleigh scattered wave from left to right). The BDG property is qualitatively confirmed by the great enhancement of the third component (BDG reflected wave), since it is transferred from weak SpBS to strong SBS process under the assistance of the BDG generated by pump and probe waves. The frequency deviation can be roughly estimated to be 44.0 GHz by a wavelength meter, which gives the birefringence value of 3.28*10^{-4}. However, the resolution is limited to about 1*10^{-6} due to 0.1 GHz-level resolution of the wavelength meter.

Most recently, a heterodyne detection was demonstrated to straightforwardly characterize the physical BDG property in a high-delta PMF [96]. Figure 15(a) summarizes a 3D distribution of the heterodyne-detected electronic spectra between the BDG reflection and the readout wave while scanning the pump–probe frequency offset (v_1 or f_1-f_1') around the BFS of 10.510 GHz. The peak frequency and power dependence of the on f_1-f_1' are shown in Fig. 15(b) and Fig. 15(c), respectively. The frequency dependence is a linear relation because the acoustic reso‐ nance frequency of the BDG is determined by f_1-f_1' so that the diffraction wave suffers the identical frequency downshift from the readout wave. The Brillouin gain determined by the

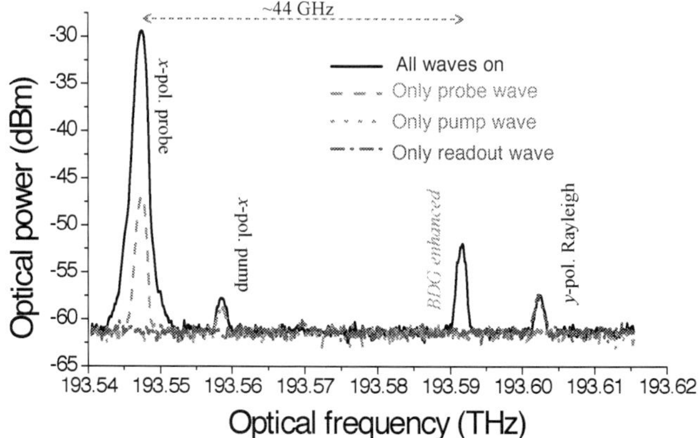

Figure 14. Qualitative characterization (optical spectra) of BDG reflection in a PMF with $B=3.3*10^{-4}$. Black-solid, all pump, probe and readout waves are launched; blue-dash-dotted, only readout wave; red-dashed, only probe wave; green-dotted, only pump wave.

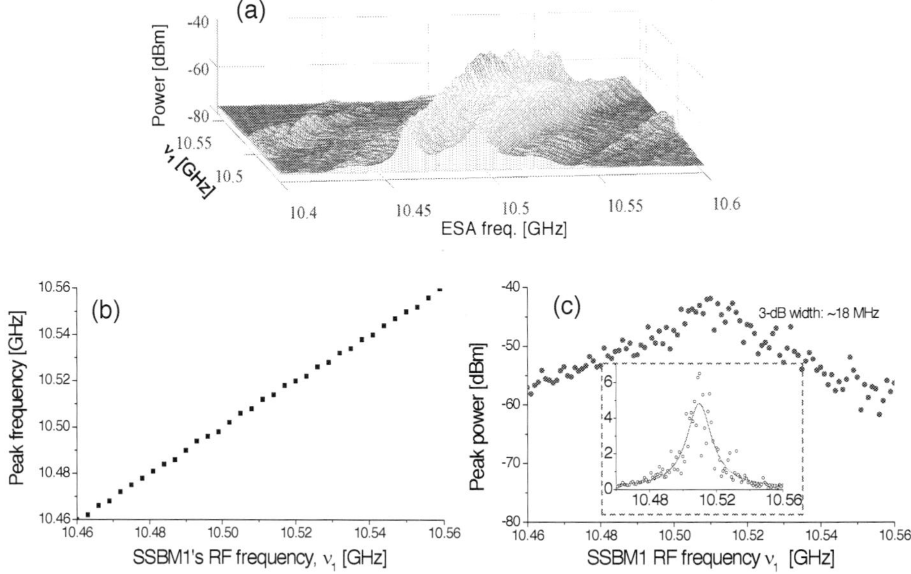

Figure 15. Characterization of the BDG reflection based on heterodyne detection. (a) 3D plot of electronic spectra of heterodyne detection when the pump-probe frequency offset $(f_1\text{-}f_1')$ is scanned. Dependence of peak frequency (b) and power (c) of each spectrum on $f_1\text{-}f_1'$. The dashed inset denotes Lorentz fitting (solid curve) to the linear vertical scaled symbols in (c). (After Ref. [96]; © 2013 JJAP.)

pump-probe-based SBS process is well known to be changed when f_1-f_1' is swept [12], which is herein reflected by the measured power dependence [see Fig. 15(c)] since the diffraction/reflection wave sees the same change of the induced Brillouin gain. As shown in the inset of Fig. 15(c), Lorentz fitting provides the central frequency of ~10.510 GHz (just equal to the BFS), and Brillouin intrinsic linewidth of ~18 MHz. All experimental observation matches well the theoretical analysis of the BDG [97].

5.2. Complete discrimination of strain and temperature

As mentioned in **Section 5.1**, the BDG in a PMF can be generated by two coherent pump and probe waves in one principal polarization while readout by another wave deviated in frequency and separated spatially in the other principal polarization. This feature enables the BDG in a PMF working as a dynamic reflector for any optical wavelength of the readout wave by simply tuning the wavelengths of the pump and probe waves. The location of the BDG generation can be also dynamically assigned by changing the fiber's longitudinal structure [98] or programing the interaction position of the pump and probe waves [90, 99-103]. Up to date, the BDG in a PMF has been used for many applications in microwave photonics, all-optical signal processing, and Brillouin-based distributed sensors. For example, the BDG programmed in position or spectrum is very useful in microwave photonics of tunable optical delays [98, 104-106] or programmable microwave photonic filter [107]. Besides, it can also find significant applications in all-optical signal processing such as storing and compressing light [108], ultrawideband communications [109], and all-optical digital signal processing [110].

The first, but most successful, application of the BDG in a Panda-type PMF was demonstrated for complete discrimination of strain and temperature responses for Brillouin based distributed optical fiber sensing applications [89]. Figure 16 shows the basic experimental configuration of the high-precision BDG characterization, which can be used to precisely measure the birefringence of a PMF and to completely discriminate strain and temperature. The BDG generation is based on the pump-probe scheme, which is also the high-accuracy BGS characterization shown in Fig. 7(a). The BDG measurement is realized by the lock-in detection of the BDG reflection since the BDG is periodically chopped due to the chopping of the pump wave. The birefringence-determined frequency deviation defined in Eq. (40) is characterized within a standard error of Δf_{yx}=4 MHz, corresponding to a high-accuracy birefringence of ΔB=3 × 10^{-8}.

The principle of the complete discrimination is based on the dependence of the BFS on strain and temperature as introduced in Eq. (29) and the orthogonal dependence of the birefringence (B) or its determined frequency deviation on strain and temperature. This is because the residual tensile stress (σ_{xy}) determining the Panda-type PMF's birefringence scales with the ambient temperature (T_i):

$$B \propto \sigma_{xy} = k \cdot (\alpha_3 - \alpha_2) \cdot (T_{fic} - T_i), \tag{44}$$

where T_{fic} denotes the fictive temperature (e.g., 850 °C) of silica glass, α_3 (α_2) the thermal coefficient of B$_2$O$_3$-doped-silica stress-applying parts (pure-silica cladding), and k a constant determined by the geometrical location of stress-applying parts in the fiber [111]. When temperature increases ($\Delta T=T_i-25 > 0$), the residual stress is released and thus the birefringence decreases as

$$\Delta B^T = -B_0 \cdot \frac{\Delta T}{T_{fic} - 25},$$

(45)

where B_0 is the intrinsic birefringence at room temperature (T_i=25 °C).

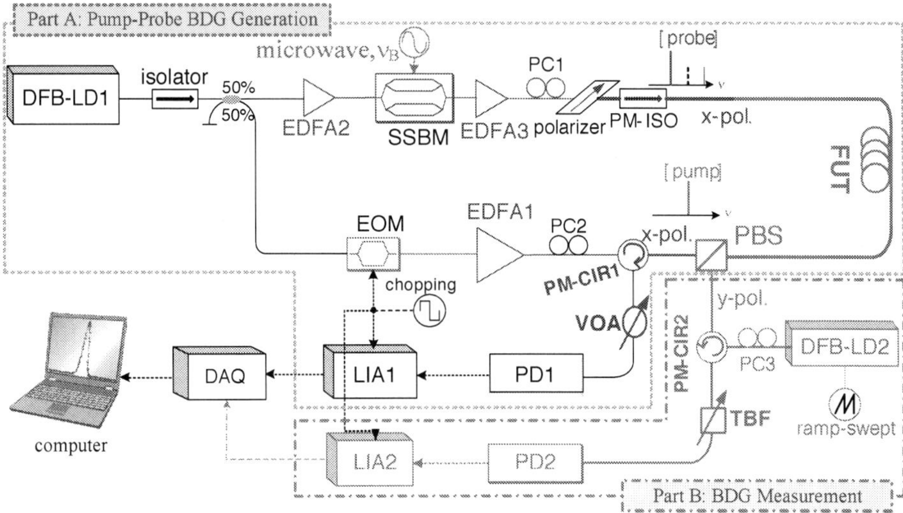

Figure 16. Configuration of the BDG characterization and strain-temperature discrimination. Part A, Pump-probe scheme to measure the BFS along *x*-axis and generate the BDG. Part B, Detection of the BDG diffraction spectrum to *y*-polarized readout wave. (After Ref. [89]; © 2009 OSA.)

In contrast, when an axial strain $\Delta\varepsilon$ is applied upon the fiber, additional stress is generated because the stress-applying parts and the cladding contract in the lateral direction differently due to their different Poisson's ratios ($\gamma_3 > \gamma_2$) [111], the birefringence is enlarged with applied strain as

$$\Delta B^\varepsilon = +B_0 \cdot \frac{(\gamma_3 - \gamma_2)}{(\alpha_3 - \alpha_2)(T_{fic} - 25)} \cdot \Delta\varepsilon.$$

(46)

Consequently, the birefringence-determined frequency deviation (Δf) varies linearly with respect to temperature increase and to applied strain. Suppose that $C_f\varepsilon$ and C_f^T are the strain

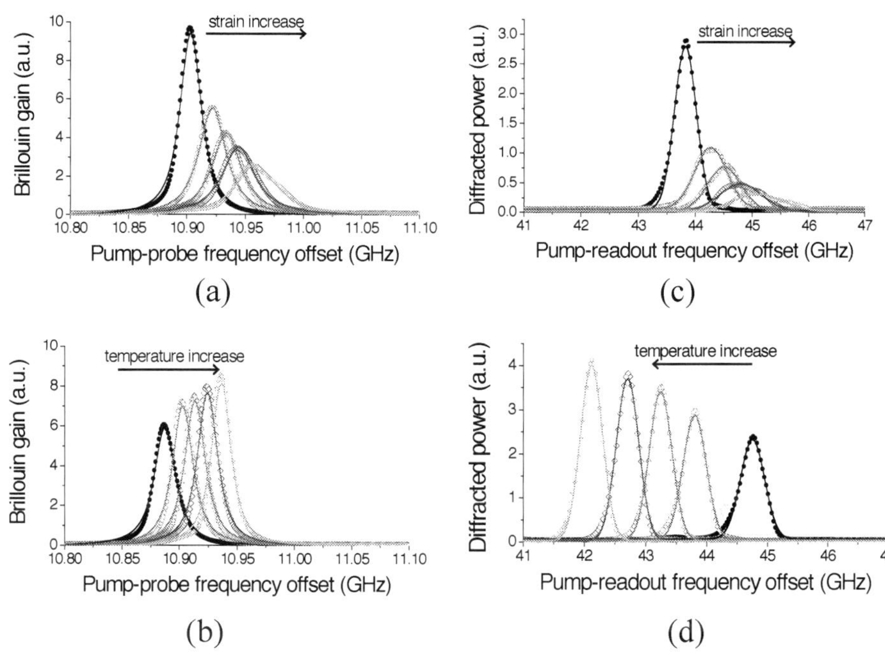

Figure 17. BGS (a, b) and BDG reflection (c, d) measured at various strain (a, c) and temperature (b, d). (After Ref. [89]; © 2009 OSA.)

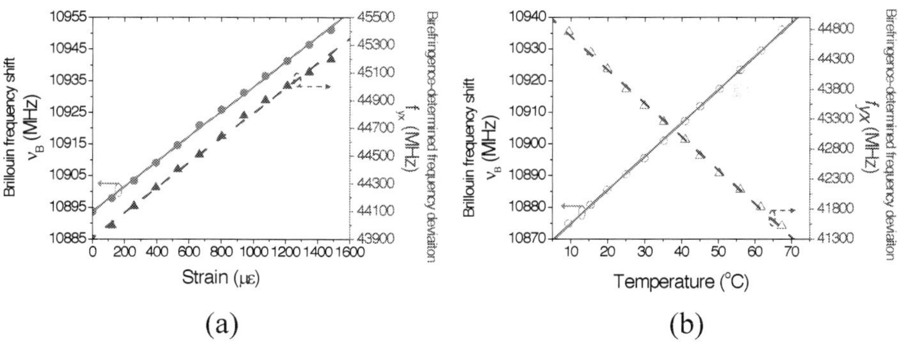

Figure 18. Brillouin frequency shift and birefringence-determined frequency deviation measured as functions of strain (a) and temperature (b). (After Ref. [89]; © 2009 OSA.)

coefficient and the temperature coefficient of the birefringence-determined frequency deviation, which can be deduced from Eqs. (40), (44), (45) and (46) as follows:

$$C_f^\varepsilon = +\Delta f_0 \cdot \frac{(\gamma_3 - \gamma_2)}{(\alpha_3 - \alpha_2)(T_{fic} - 25)}$$

$$C_f^T = -\Delta f_0 \cdot \frac{1}{(T_{fic} - 25)},$$

$$(47)$$

where Δf_0 is the frequency deviation at 25°C and in loose condition.

By jointly considering the BFS (ν_B) and the frequency deviation (Δf), one can deduce the strain ($\Delta\varepsilon$) and temperature (ΔT) referred to Eq. (37):

$$\begin{pmatrix} \Delta\varepsilon \\ \Delta T \end{pmatrix} = \frac{1}{C_\nu^\varepsilon \cdot C_f^T - C_\nu^T \cdot C_f^\varepsilon} \begin{pmatrix} C_f^T & -C_\nu^T \\ -C_f^\varepsilon & C_\nu^\varepsilon \end{pmatrix} \begin{pmatrix} \nu_B - \nu_{B0} \\ \Delta f - \Delta f_0 \end{pmatrix},$$

$$(48)$$

where Δf_0 and ν_{B0} are the frequency deviation and the BFS at room temperature and in loose state.

In physics, the two phenomena/quantities, i.e. the ν_B and Δf of the fiber, are inherently independent. In mathematics, C_f^T has a sign opposite to those of other three coefficients, so that the denominator ($C\nu\varepsilon\, C_f^T$-$C\nu^T\, C_f\varepsilon$) of Eq. (48) has a significant value. The experimental results are depicted in Fig. 17 and Fig. 18, which give the two groups of coefficients ($C\nu\varepsilon$=+0.03938 MHz/$\mu\varepsilon$ and $C\nu^T$=+1.0580 MHz/°C; and $C_f\varepsilon$=+0.8995 MHz/$\mu\varepsilon$ and C_f^T=-55.8134 MHz/ °C). Putting above strain/temperature coefficients into Eq. (48), and taking the standard errors of the measurement system ($\Delta\nu_B$=0.1 MHz and Δf_{yx}=4 MHz, respectively) into account, the accuracy of the discrimination is given as high as $\Delta\varepsilon$=± 3.1 $\mu\varepsilon$ and ΔT=± 0.078 °C. Therefore, a complete discrimination of strain and temperature based on simultaneous measurement of the two quantities is ensured.

For distributed discrimination of strain and temperature, the localized BDG generation and readout in the PMF should be firstly proved to be effective. A correlation-based continuous-wave technique based on the BOCDA system [99] is used for random access and a pulse-based time-domain technique based on the BOTDA system [100] is employed for continuous access. It was found that the generation and readout waves based on the BOCDA system should be synchronously frequency-modulated because of the dispersion properties of all four waves (see Fig. 19) [112], including pump and probe waves, readout wave and acoustic wave (BDG as well).

The preliminary success of distributed discrimination of strain and temperature was realized by use of several lasers based on the BOCDA system [113] or the BOTDA system [114]. In [113], all pump, probe, and readout waves are synchronously modulated in frequency by sinusoidal functions to the two laser diodes. The measurement range of the distributed BGS and BDG is

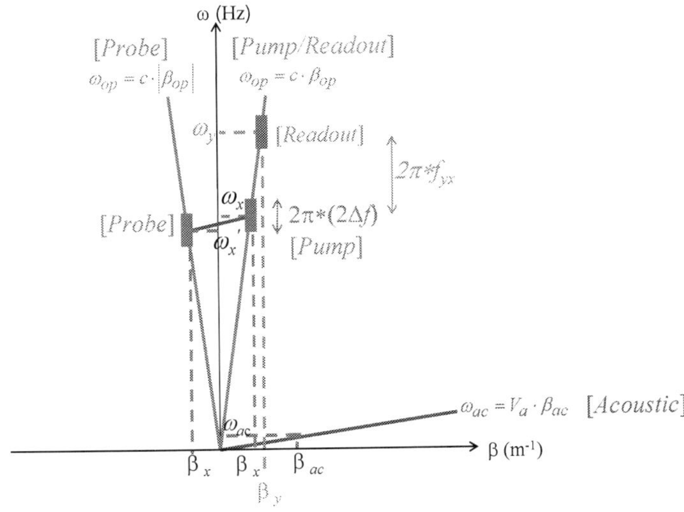

Figure 19. Dispersion properties of all optical (pump, probe and readout) and acoustic wave (Brillouin dynamic grating). (After Ref. [112]; © 2011 OSA.)

commonly given by the neighboring correlation peaks of the BOCDA system as defined in Eq. (35). Although the spatial resolution of the BGS measurement is still given by Eq. (34), that of the BDG reflection was thought to be determined by the BDG bandwidth (Δf_{yx}) as follows:

$$\Delta Z_{BDG} = \frac{c}{2nf_m} \cdot \frac{\Delta f_{yx}}{\pi \Delta f}. \tag{49}$$

The feasibility of distributed discrimination of strain and temperature was experimentally demonstrated with 10-cm spatial resolution. The f_m=12.429 MHz determines the nominal measurement range as d_m=8.35 m according to Eq. (35). For local BGS and BDG measurement, the Δf_B=1.5 GHz and Δf_D=10 GHz correspond to a nominal spatial resolution Δz_B=5 cm and Δz_D=8 cm [see Eq. (34) and Eq. (49)], respectively. As shown in Fig. 20(a), a ~8-m PMF sample is prepared, which consists of nine (A-I) cascaded fiber portions of 10-16 cm in length. The A, C, G and I portions were loosely laid at 25.1 °C for reference, while the B, D, F, and H portions were loosely inserted into a temperature-controlled water bath with 0.1-°C accuracy. The E portion was also inserted into the water bath and glued to a set of translation stages to load strain. The measured distribution of the changes of Δv_B and Δf_{yx} are summarized in Figs. 20(b) and 20(c), respectively. Referred to the characterized coefficients in Fig. 18 and the cross-sensitivity matrix in Eq. (48), the deduced distribution of temperature and strain along the fiber is depicted in Figs. 20(d) and 20(e), which clearly shows the feasibility of distributed discrimination of strain and temperature. In [114], all pump, probe, and readout waves are pulsed in time domain; in turn, the BDG generation and readout as well as the BGS and BDG reflection are continuously localized by control of their relative delay towards the FUT and thus the local v_B and Δf are detected for distributed discrimination of the strain and temperature responses.

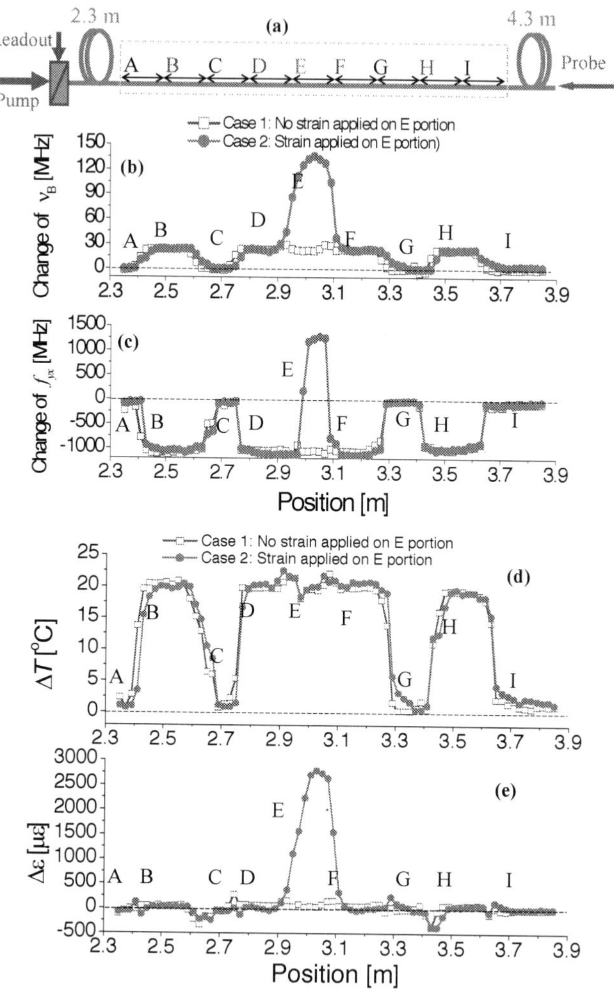

Figure 20. Preliminary experiment of distributed discrimination of strain and temperature based on two lasers modulated in frequency. (a) FUT configuration. Measured distribution of Brillouin frequency shift (b) and the birefringence-determined frequency deviation (c). Deduced distribution of temperature (d) and strain (e). (After Ref. [113]; © 2010 IEEE.)

One-laser-based Brillouin correlation-domain distributed discrimination system [112] by use of the sideband-generation technique was recently demonstrated to overcome the frequency fluctuation among the free-running lasers for pump, probe, and readout waves, and thus improve the accuracy of distributed discrimination of strain and temperature. Figure 21 represents the experimental setup. A 40-GHz intensity modulator (IM2) laid after the laser diode is driven by a radio frequency synthesizer (RF2 at ν_{RF2}) with a proper dc bias so as to generate double sidebands with suppressed carrier (DSB-SC). The optical filtering (FBG and tunable band-pass filter) is used to separate the two sidebands for the BDG generation and readout. By control of the RF1 (similar to Fig. 16), the BFS can be precisely measured and then fixed; by tuning of the RF2, the BDG reflection can be also precisely characterized; by simply change of the modulation frequency of the one laser diode, the location of the BGS and BDG can be swept for distributed measurement.

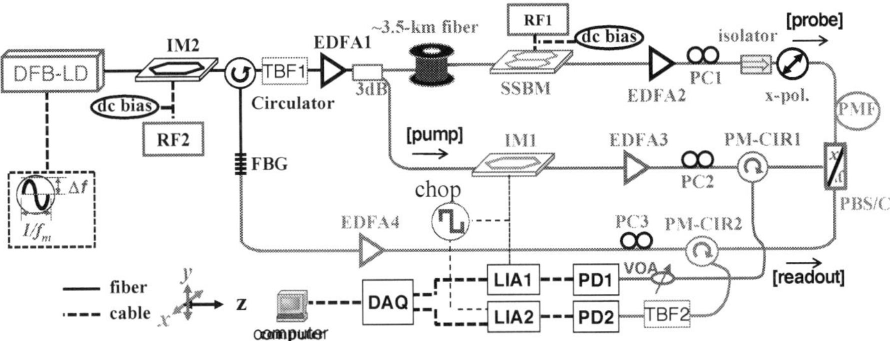

Figure 21. Experimental setup of the one-laser-based Brillouin correlation-domain distributed discrimination system. (After Ref. [112]; © 2011 OSA.)

Figure 22 shows the higher stability and accuracy (several MHz) of the one-laser scheme when compared to the two-laser scheme (several hundreds of MHz) both under no averaging process. Note that the one-laser scheme can also provide higher speed in the measurement of BGS and BDG and simpler measurement without sophisticated synchronization. Its distributed discrimination of strain and temperature was confirmed with the spatial resolution of ~10 cm and measurement range of ~5 m, which is depicted in Fig. 23 when the fiber was heated from 25 °C to 30 °C or/and the strain (ε=2000 $\mu\varepsilon$) was applied both at the location of 3.1 m. The measured results match well with the setting situation.

In order to overcome the tradeoff between the spatial resolution and measurement range always existing in the BOCDA system, a temporal gating [115] or a dual frequency modulation scheme [116] with a simple modification in Fig. 21 was used to elongate the measurement range of distributed discrimination of strain and temperature. For temporal gating scheme, the pulse modulation of RF2 makes the frequency-modulated pump, probe and readout waves optically pulsed in time and only one of the multiple correlation peaks are effectively generated in the

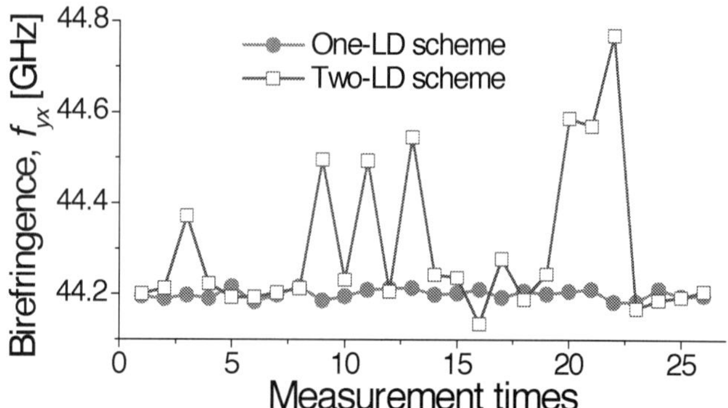

Figure 22. Comparison of the stability and accuracy of one-laser (solid dots) and two-laser (dashed squares) schemes of the BDG reflection. (After Ref. [112]; © 2011 OSA.)

FUT. For dual frequency modulation scheme, two sinusoidal functions are combined together to simultaneously modulate the optical frequencies of the pump, probe and readout waves. The greater modulation frequency ensures the higher spatial resolution while the lower modulation frequency realizes the longer measurement range. A 20 times [115] or 7 times [116] enlargement of the ratio between the measurement range and the spatial resolution was successfully demonstrated. It is expectable to achieve Brillouin optical correlation-domain distributed discrimination of strain and temperature having both a higher spatial resolution (better than 10 cm) and a longer measurement range (better than 1,000 m) by combining the dual frequency modulation scheme with the temporal gating scheme, which is now under study. Most recently, an apodization method under the assistance of intensity modulation was proposed to suppress the sidelobe of SOCF and enhance the spatial resolution of the strain-temperature discrimination by 4.5 times [117].

5.3. System improvement of sensing techniques

Many works have been involved in improving the system performance of BOTDA/BOTDR and BOCDA/BOCDR in terms of spatial resolution, measurement range, sensing speed and accuracy. In 1995, Bao *et al.* developed a Brillouin-loss-based BOTDA [118] by reversing the functions of the pulse laser and CW light. In other words, a strong CW light acts as a Brillouin pump wave and a pulse light with a scanned down-shifted frequency from that of the CW light works as a probe wave [119]. After monitoring the optical loss profile of the pump wave due to Brillouin interaction between the two light waves as a function of time or position along the fiber, a numerical signal processing of the poor SNR was used to achieve ~25 cm spatial resolution with a strain resolution of ~40 μɛ [120]. A pulse-pre-pump BOTDA (called PPP-BOTDA) was proposed to realize cm-order spatial resolution by using a wide pulse (larger than 10 ns) followed by a narrow pulse (smaller than 1 ns) [121], which is in principle similar to the BOTDA with a pulse generated by a finite extinction ratio [122]. A dark-pulse-based BOTDA was later presented to hopefully obtain 2-cm spatial resolution [123, 124], which

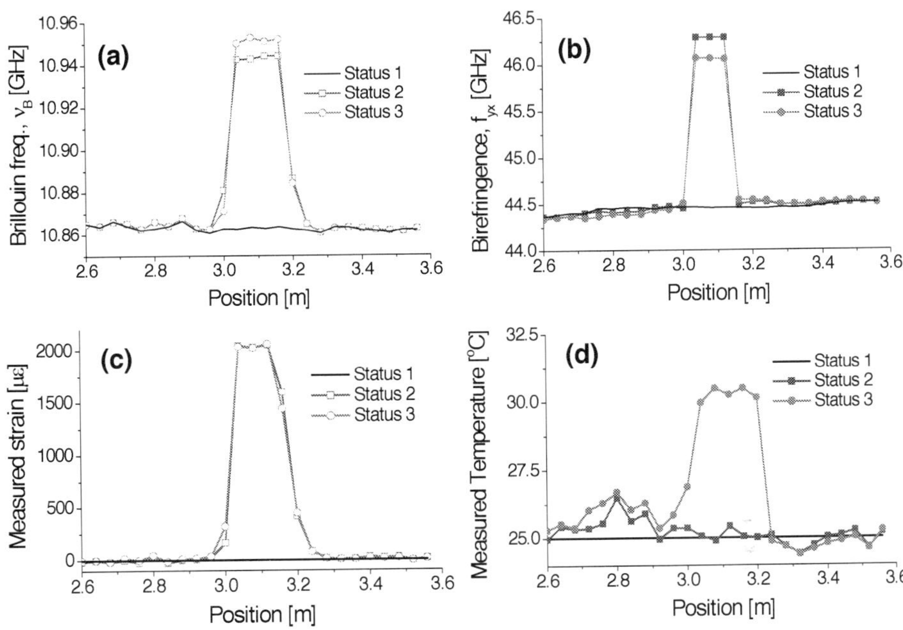

Figure 23. Experimental results of high-accuracy distributed discrimination of strain and temperature based on one-laser scheme. Distribution of Brillouin frequency shift (a), the birefringence-determined frequency deviation (b), strain (c), and temperature (d). (After Ref. [112]; © 2011 OSA.)

suffers a neglectful influence of the acoustic lifetime (~10 ns) but an experimental difficulty of high-qualify dark pulse and a scarified measurement range due to the pump depletion of the dc base. A new scheme by combination of the PPP-BOTDA and dark-pulse-based BOTDA [125] was demonstrated to achieve higher spatial resolution as well as better frequency resolution. Taking the similar principle of the PPP-BOTDA technique, a BDG-based BOTDA was proposed to obtain cm-order or sub-cm-order spatial resolution [126-128]. The only difference lies on the fact that the Brillouin interaction in the BDG-based BOTDA is generated by a long pulse along one principal polarization state (the BDG generation process) and detected by a short pulse along the orthogonal polarization state (the BDG readout process). Another type of the modified BOTDA with higher spatial resolution of less than 1 meter is based on a group of pulses, called differential pulse-width pair BOTDA (DPP-BOTDA) [129-132] or Brillouin echo BOTDA [133, 134]. In DPP-BOTDA, a pair of pulses with a small difference of the pulse widths are successively launched into the FUT; the Brillouin interaction is recorded for twice and a subtraction is performed to achieve the sensing trace with the high spatial resolution determined by the small pulse-width difference. The spatial resolution of BOTDR has been also more or less improved by an experimental optimization or signal processing process [135-137].

Although BOTDA and BOTDR are excellent for long sensing range (such as kilometers or tens of km), they still suffer the physical limitation of maximum range due to the nature of fiber loss and/or the Brillouin depletion effect. There are two typical methods, i.e. Raman-assisted BOTDR[138-140] or Raman-assisted BOTDA[141-145] and coded BOTDR[146] or coded BOTDA[147-151], to improve the poor SNR and achieve a very long sensing range. The best performance of the sensing range (longer than 120 km) [152, 153] with an acceptable spatial resolution (1 m or 2 m) has been renovated by combination of Raman assistance and coding although the system becomes extremely complicated. Most recently, specially-designed EDFA repeaters were used to extend the sensing measurements of BOTDA to more than 300 km [154]. Some efforts were also made to study the influence of Brillouin depletion on the maximum range of BOTDA [41] and to avoid it to some extent by use of Stokes together with anti-Stokes wave as Brillouin probe [155, 156].

BOCDA and BOCDR systems have natural advantages of high spatial resolution without any dependence on the acoustic lifetime and random programmable accessibility of the sensing location. Except for the great innovation of Brillouin optical correlation-domain distributed discrimination of strain and temperature introduced in **Section 5.2**, advances in BOCDA and BOCDR systems have also boosted in the past decade. The polarization disturbance along the FUT has been effectively solved by use of the polarization diversity scheme to the BOCDA system [157]. A complicated double-lock-in detection was proposed to improve the SNR of the BOCDA system [69, 158] although a modified lock-in detection based on variable chopping frequency [159] or a simplified but equivalent BOCDA system based on combination of Brillouin gain and loss [34] was later proposed. The existence of a big noise floor originated from the uncorrelated locations strongly limits the maximum strain or temperature change to be detected, which has been eliminated by use of intensity modulation for SOCF apodization [160, 161] or differential measurement scheme based on external phase modulation [162]. The measurement range of BOCDA [116, 163, 164] or BOCDR [165, 166] was extended by use of temporal gating or double frequency modulation scheme, respectively. Besides, combination of time-domain and correlation domain techniques [167, 168] has been proposed to enlarge the measurement range of the BOCDA [169] based on external phase modulation. The distributed sensing speed with cm-order spatial resolution [170-172] has been substantially increased to several Hertz along the entire FUT by optimizing the position sweeping and the BGS mapping although the local sensing speed of the BOCDA [173-175] or BOCDR [176] was well-known to be high at the random-accessed sensing location.

6. Conclusions

We have presented an essential overview of Brillouin scattering in optical fibers and Brillouin based distributed optical fiber sensors. Started from the basic principle of Brillouin scattering in optical fibers, the basic mechanism of Brillouin based distributed optical fiber sensors (linear dependence of Brillouin frequency shift on strain and temperature) and the two different groups of Brillouin based distributed optical fiber sensors (time domain: BOTDA/BOTDR; correlation domain: BOCDA/BOCDR) were described in detail. The difficulties and challenges of how to simultaneously sense strain and temperature were demonstrated and the physical limitation of the sensing abilities (spatial resolution, measurement range, accuracy etc.) were introduced, respectively. Finally, we summarized recent advances of this field towards the solutions to those difficulties and challenges.

It is valuable to address that Brillouin based distributed optical fiber sensors are nowadays in a high technical level, which have been attracting industrial companies to commercialize for structural health monitoring in civil structures, aerospace, energy (gas, oil) pipeline, and engineers (power supply).

Acknowledgements

This work was partially supported by National Natural Science Foundation of China (Grant Nos. 61007052 and 61127016), Shanghai Pujiang Program (Grant No. 12PJ1405600), and by the State Key Lab Project of Shanghai Jiao Tong University under Grant GKZD030033. Professor Kazuo Hotate at the University of Tokyo and Professor Zuyuan He at Shanghai Jiao Tong University are gratefully acknowledged for their contributions in many relevant works presented in this chapter.

Author details

Weiwen Zou*, Xin Long and Jianping Chen

*Address all correspondence to: wzou@sjtu.edu.cn

State Key Laboratory of Advanced Optical Communication Systems and Networks, Department of Electronic Engineering, Shanghai Jiao Tong University, Shanghai, China

References

[1] G. P. Agrawal, *Nonlinear Fiber Optics,* 5th ed. (Academic Press, 2012).

[2] E. Ippen and R. Stolen, "Stimulated Brillouin scattering in optical fibers," Appl. Phys. Lett. 21, 539-541 (1972).

[3] R. Shelby, M. Levenson, and P. Bayer, "Resolved forward Brillouin scattering in optical fibers," Phys. Rev. Lett. 54, 939-942 (1985).

[4] Y. Mizuno and K. Nakamura, "Brillouin Scattering in Polymer Optical Fibers: Fundamental Properties and Potential Use in Sensors," Polymers 3, 886-898 (2011).

[5] A. L. Gaeta and R. W. Boyd, "Stochastic dynamics of stimulated Brillouin scattering in an optical fiber," Phys. Rev. A 44, 3205-3209 (1991).

[6] R. B. Jenkins, R. M. Sova, and R. I. Joseph, "Steady-state noise analysis of spontaneous and stimulated Brillouin scattering in optical fibers," J. Lightwave Technol. 25, 763-770 (2007).

[7] A. S. Pine, "Brillouin Scattering Study of Acoustic Attenuation in Fused Quartz," Physical Review 185, 1187-1193 (1969).

[8] W. Zou, Z. He, and K. Hotate, "Two-dimensional finite-element modal analysis of Brillouin gain spectra in optical fibers," Photonics Technology Letters, IEEE 18, 2487-2489 (2006).

[9] A. Kobyakov, M. Sauer, and D. Chowdhury, "Stimulated Brillouin scattering in optical fibers," Advances in optics and photonics 2, 1-59 (2010).

[10] A. Kobyakov, S. Kumar, D. Q. Chowdhury, A. B. Ruffin, M. Sauer, S. R. Bickham, and R. Mishra, "Design concept for optical fibers with enhanced SBS threshold," Opt. Express 13, 5338-5346 (2005).

[11] Y. Koyamada, S. Sato, S. Nakamura, H. Sotobayashi, and W. Chujo, "Simulating and designing brillouin gain spectrum in single-mode fibers," J. Lightwave Technol. 22, 631-639 (2004).

[12] W. Zou, Z. He, A. D. Yablon, and K. Hotate, "Dependence of Brillouin frequency shift in optical fibers on draw-induced residual elastic and inelastic strains," Photonics Technology Letters, IEEE 19, 1389-1391 (2007).

[13] W. Zou, Z. Y. He, and K. Hotate, "Acoustic modal analysis and control in w-shaped triple-layer optical fibers with highly-germanium-doped core and F-doped inner cladding," Opt. Express 16, 10006-10017 (2008).

[14] L. Tartara, C. Codemard, J.-N. Maran, R. Cherif, and M. Zghal, "Full modal analysis of the Brillouin gain spectrum of an optical fiber," Opt. Commun. 282, 2431-2436 (2009).

[15] B. Ward and J. Spring, "Finite element analysis of Brillouin gain in SBS-suppressing optical fibers with non-uniform acoustic velocity profiles," Opt. Express 17, 15685-15699 (2009).

[16] Y. S. Mamdem, X. Phéron, F. Taillade, Y. Jaoüen, R. Gabet, V. Lanticq, G. Moreau, A. Boukenter, Y. Ouerdane, and S. Lesoille, "Two-dimensional FEM analysis of Brillouin Gain Spectra in acoustic guiding and antiguiding single mode optical fibers," in *COMSOL Conference*, 2010),

[17] S. Dasgupta, F. Poletti, S. Liu, P. Petropoulos, D. J. Richardson, L. Gruner-Nielsen, and S. Herstrøm, "Modeling Brillouin Gain Spectrum of solid and microstructured

optical fibers using a finite element method," Lightwave Technology, Journal of 29, 22-30 (2011).

[18] X. Qian, B. Han, and Q. Wang, "Numerical Research on Gain Spectrum of Stimulated Brillouin Scattering in Photonics Crystal Fiber," Instrumentation Science & Technology 41, 175-186 (2013).

[19] L. Dong, "Formulation of a complex mode solver for arbitrary circular acoustic waveguides," J. Lightwave Technol. 28, 3162-3175 (2010).

[20] P. D. Dragic and B. G. Ward, "Accurate modeling of the intrinsic Brillouin linewidth via finite-element analysis," Photonics Technology Letters, IEEE 22, 1698-1700 (2010).

[21] C. G. Carlson, R. B. Ross, J. M. Schafer, J. B. Spring, and B. G. Ward, "Full vectorial analysis of Brillouin gain in random acoustically microstructured photonic crystal fibers," Phys. Rev. B 83, 235110 (2011).

[22] Y. Okawachi, M. S. Bigelow, J. E. Sharping, Z. M. Zhu, A. Schweinsberg, D. J. Gauthier, R. W. Boyd, and A. L. Gaeta, "Tunable all-optical delays via Brillouin slow light in an optical fiber," Phys. Rev. Lett. 94(2005).

[23] K. Y. Song, M. G. Herraez, and L. Thevenaz, "Observation of pulse delaying and advancement in optical fibers using stimulated Brillouin scattering," Opt. Express 13, 82-88 (2005).

[24] N. Shibata, K. Okamoto, and Y. Azuma, "Longitudinal acoustic modes and Brillouin-gain spectra for GeO< sub> 2</sub>-doped-core single-mode fibers," JOSA B 6, 1167-1174 (1989).

[25] M. Nikles, L. Thevenaz, and P. A. Robert, "Brillouin gain spectrum characterization in single-mode optical fibers," Lightwave Technology, Journal of 15, 1842-1851 (1997).

[26] A. Yeniay, J.-M. Delavaux, and J. Toulouse, "Spontaneous and Stimulated Brillouin Scattering Gain Spectra in Optical Fibers," J. Lightwave Technol. 20, 1425 (2002).

[27] W. Zou, Z. He, and K. Hotate, "Experimental study of Brillouin scattering in fluorine-doped single-mode optical fibers," Opt. Express 16, 18804-18812 (2008).

[28] R. G. Smith, "Optical Power Handling Capacity of Low Loss Optical Fibers as Determined by Stimulated Raman and Brillouin Scattering," Appl. Optics 11, 2489-2494 (1972).

[29] V. I. Kovalev and R. G. Harrison, "Threshold for stimulated Brillouin scattering in optical fiber," Opt. Express 15, 17625-17630 (2007).

[30] N. Rowell, P. Thomas, H. Van Driel, and G. Stegeman, "Brillouin spectrum of single mode optical fibers," Appl. Phys. Lett. 34, 139-141 (1979).

[31] S. M. Maughan, H. H. Kee, and T. P. Newson, "Simultaneous distributed fibre temperature and strain sensor using microwave coherent detection of spontaneous Brillouin backscatter," Measurement Science and Technology 12, 834 (2001).

[32] Y. Koyamada, "Proposal and simulation of double-pulse brillouin optical time-domain analysis for measuring distributed strain and temperature with cm spatial resolution in km-long fiber," IEICE Trans. Commun. E90B, 1810-1815 (2007).

[33] X. Bao, D. J. Webb, and D. A. Jackson, "32-km distributed temperature sensor based on Brillouin loss in an optical fiber," Opt. Lett. 18, 1561-1563 (1993).

[34] W. Zou, C. J. Jin, and J. P. Chen, "Distributed Strain Sensing Based on Combination of Brillouin Gain and Loss Effects in Brillouin Optical Correlation Domain Analysis," Appl. Phys. Express 5(2012).

[35] K. Higuma, S. Oikawa, Y. Hashimoto, H. Nagata, and M. Izutsu, "X-cut lithium niobate optical single-sideband modulator," Electron. Lett. 37, 515-516 (2001).

[36] A. Loayssa, R. Hernández, D. Benito, and S. Galech, "Characterization of stimulated Brillouin scattering spectra by use of optical single-sideband modulation," Opt. Lett. 29, 638-640 (2004).

[37] W. Zou, Z. Y. He, M. Kishi, and K. Hotate, "Stimulated Brillouin scattering and its dependences on strain and temperature in a high-delta optical fiber with F-doped depressed inner cladding," Opt. Lett. 32, 600-602 (2007).

[38] V. I. Kovalev and R. G. Harrison, "Observation of inhomogeneous spectral broadening of stimulated Brillouin scattering in an optical fiber," Phys. Rev. Lett. 85, 1879 (2000).

[39] Y. Takushima and K. Kikuchi, "Spectral gain hole burning and modulation instability in a Brillouin fiber amplifier," Opt. Lett. 20, 34-36 (1995).

[40] L. Stépien, S. Randoux, and J. Zemmouri, "Origin of spectral hole burning in Brillouin fiber amplifiers and generators," Phys. Rev. A 65, 053812 (2002).

[41] L. Thévenaz, S. F. Mafang, and J. Lin, "Effect of pulse depletion in a Brillouin optical time-domain analysis system," Opt. Express 21, 14017-14035 (2013).

[42] X. Long, W. Zou, H. Li, and J. Chen, "Critical condition for spectrum distortion of pump–probe-based stimulated Brillouin scattering in an optical fiber," Appl. Phys. Express 7, 082501 (2014).

[43] T. Horiguchi and M. Tateda, "BOTDA-nondestructive measurement of single-mode optical fiber attenuation characteristics using Brillouin interaction: Theory," Lightwave Technology, Journal of 7, 1170-1176 (1989).

[44] T. Horiguchi, T. Kurashima, and M. Tateda, "Tensile strain dependence of Brillouin frequency shift in silica optical fibers," Photonics Technology Letters, IEEE 1, 107-108 (1989).

[45] T. Kurashima, T. Horiguchi, and M. Tateda, "Thermal effects of Brillouin gain spectra in single-mode fibers," Photonics Technology Letters, IEEE 2, 718-720 (1990).

[46] http://www.itu.int/rec/T-REC-G/en.

[47] C. K. Jen, C. Neron, A. Shang, K. Abe, L. Bonnell, and J. Kushibiki, "Acoustic characterization of silica glasses," Journal of the American Ceramic Society 76, 712-716 (1993).

[48] S. P. Timoshenko and J. Goodier, *Theory of elasticity* (2011).

[49] W. Zou, Z. He, and K. Hotate, "Investigation of Strain-and Temperature-Dependences of Brillouin Frequency Shifts in GeO< sub> 2</sub>-Doped Optical Fibers," J. Lightwave Technol. 26, 1854-1861 (2008).

[50] R. W. Tkach, A. R. Chraplyvy, and R. Derosier, "Spontaneous Brillouin scattering for single-mode optical-fibre characterisation," Electron. Lett. 22, 1011-1013 (1986).

[51] N. Shibata, R. G. Waarts, and R. P. Braun, "Brillouin-gain spectra for single-mode fibers having pure-silica, GeO 2-doped, and P 2O5-doped cores," Opt. Lett. 12, 269-271 (1987).

[52] T. Horiguchi, T. Kurashima, and M. Tateda, "A technique to measure distributed strain in optical fibers," Photonics Technology Letters, IEEE 2, 352-354 (1990).

[53] T. Kurashima, T. Horiguchi, H. Izumita, S.-i. Furukawa, and Y. Koyamada, "Brillouin optical-fiber time domain reflectometry," IEICE Trans. Commun. 76, 382-390 (1993).

[54] T. Horiguchi, T. Kurashima, and Y. Koyamada, "Measurement of temperature and strain distribution by Brillouin frequency shift in silica optical fibers," in *Fibers' 92*, (International Society for Optics and Photonics, 1993), 2-13.

[55] M. Niklès, L. Thévenaz, and P. A. Robert, "Simple distributed fiber sensor based on Brillouin gain spectrum analysis," Opt. Lett. 21, 758-760 (1996).

[56] K. Shimizu, T. Horiguchi, Y. Koyamada, and T. Kurashima, "Coherent self-heterodyne Brillouin OTDR for measurement of Brillouin frequency shift distribution in optical fibers," Lightwave Technology, Journal of 12, 730-736 (1994).

[57] D. Garcus, T. Gogolla, K. Krebber, and F. Schliep, "Brillouin optical-fiber frequency-domain analysis for distributed temperature and strain measurements," Lightwave Technology, Journal of 15, 654-662 (1997).

[58] D. Garus, K. Krebber, F. Schliep, and T. Gogolla, "Distributed sensing technique based on Brillouin optical-fiber frequency-domain analysis," Opt. Lett. 21, 1402-1404 (1996).

[59] K. Hotate and T. Hasegawa, "Measurement of Brillouin Gain Spectrum Distribution along an Optical Fiber Using a Correlation-Based Technique--Proposal, Experiment and Simulation," IEICE Trans. Electron. 83, 405-412 (2000).

[60] K. Hotate and M. Tanaka, "Distributed fiber Brillouin strain sensing with 1-cm spatial resolution by correlation-based continuous-wave technique," Photonics Technology Letters, IEEE 14, 179-181 (2002).

[61] Y. Mizuno, W. Zou, Z. He, and K. Hotate, "Proposal of Brillouin optical correlation-domain reflectometry (BOCDR)," Opt. Express 16, 12148-12153 (2008).

[62] K.-Y. Song and K. Hotate, "Brillouin optical correlation domain analysis in linear configuration," Photonics Technology Letters, IEEE 20, 2150-2152 (2008).

[63] W. Zou, Z. He, and K. Hotate, "Single-End-Access Correlation-Domain Distributed Fiber-Optic Sensor Based on Stimulated Brillouin Scattering," J. Lightwave Technol. 28, 2736-2742 (2010).

[64] Y. Mizuno, W. W. Zou, Z. Y. He, and K. Hotate, "Operation of Brillouin Optical Correlation-Domain Reflectometry: Theoretical Analysis and Experimental Validation," J. Lightwave Technol. 28, 3300-3306 (2010).

[65] K. Hotate and O. Kamatani, "Optical coherence domain reflectometry by synthesis of coherence function," Lightwave Technology, Journal of 11, 1701-1710 (1993).

[66] K. Hotate and T. Okugawa, "Optical information processing by synthesis of the coherence function," Lightwave Technology, Journal of 12, 1247-1255 (1994).

[67] R. Lang and K. Kobayashi, "External optical feedback effects on semiconductor injection laser properties," Quantum Electronics, IEEE Journal of 16, 347-355 (1980).

[68] Y. Mizuno, Z. He, and K. Hotate, "One-end-access high-speed distributed strain measurement with 13-mm spatial resolution based on Brillouin optical correlation-domain reflectometry," Photonics Technology Letters, IEEE 21, 474-476 (2009).

[69] K. Y. Song, Z. Y. He, and K. Hotate, "Distributed strain measurement with millimeter-order spatial resolution based on Brillouin optical correlation domain analysis," Opt. Lett. 31, 2526-2528 (2006).

[70] K. Hotate and S. S. Ong, "Distributed dynamic strain measurement using a correlation-based Brillouin sensing system," Photonics Technology Letters, IEEE 15, 272-274 (2003).

[71] K.-Y. Song and K. Hotate, "Distributed fiber strain sensor with 1 kHz sampling rate based on Brillouin optical correlation domain analysis," in *Optics East 2007*, (International Society for Optics and Photonics, 2007), 67700J-67700J-67708.

[72] M. Alahbabi, Y. Cho, and T. Newson, "Simultaneous temperature and strain measurement with combined spontaneous Raman and Brillouin scattering," Opt. Lett. 30, 1276-1278 (2005).

[73] G. Bolognini, M. A. Soto, and F. Di Pasquale, "Simultaneous distributed strain and temperature sensing based on combined Raman–Brillouin scattering using Fabry–Perot lasers," Measurement Science and Technology 21, 094025 (2010).

[74] M. Taki, A. Signorini, C. Oton, T. Nannipieri, and F. Di Pasquale, "Hybrid Raman/Brillouin-optical-time-domain-analysis-distributed optical fiber sensors based on cyclic pulse coding," Opt. Lett. 38, 4162-4165 (2013).

[75] D.-P. Zhou, W. Li, L. Chen, and X. Bao, "Distributed Temperature and Strain Discrimination with Stimulated Brillouin Scattering and Rayleigh Backscatter in an Optical Fiber," Sensors 13, 1836-1845 (2013).

[76] C. Lee, P. Chiang, and S. Chi, "Utilization of a dispersion-shifted fiber for simultaneous measurement of distributed strain and temperature through Brillouin frequency shift," Photonics Technology Letters, IEEE 13, 1094-1096 (2001).

[77] L. F. Zou, X. Y. Bao, V. S. Afshar, and L. Chen, "Dependence of the Brillouin frequency shift on strain and temperature in a photonic crystal fiber," Opt. Lett. 29, 1485-1487 (2004).

[78] T. Parker, M. Farhadiroushan, V. Handerek, and A. Rogers, "Temperature and strain dependence of the power level and frequency of spontaneous Brillouin scattering in optical fibers," Opt. Lett. 22, 787-789 (1997).

[79] M. Alahbabi, Y. T. Cho, and T. P. Newson, "Comparison of the methods for discriminating temperature and strain in spontaneous Brillouin-based distributed sensors," Opt. Lett. 29, 26-28 (2004).

[80] M. Belal and T. P. Newson, "Experimental Examination of the Variation of the Spontaneous Brillouin Power and Frequency Coefficients Under the Combined Influence of Temperature and Strain," J. Lightwave Technol. 30, 1250-1255 (2012).

[81] X. Y. Bao, Q. R. Yu, and L. Chen, "Simultaneous strain and temperature measurements with polarization-maintaining fibers and their error analysis by use of a distributed Brillouin loss system," Opt. Lett. 29, 1342-1344 (2004).

[82] H. Naruse and M. Tateda, "Trade-off between the spatial and the frequency resolutions in measuring the power spectrum of the Brillouin backscattered light in an optical fiber," Appl. Optics 38, 6516-6521 (1999).

[83] S.-B. Cho, Y.-G. Kim, J.-S. Heo, and J.-J. Lee, "Pulse width dependence of Brillouin frequency in single mode optical fibers," Opt. Express 13, 9472-9479 (2005).

[84] M. A. Soto and L. Thévenaz, "Modeling and evaluating the performance of Brillouin distributed optical fiber sensors," Opt. Express 21, 31347-31366 (2013).

[85] A. Minassian, G. J. Crofts, and M. J. Damzen, "Spectral filtering of gain gratings and spectral evolution of holographic laser oscillators," Quantum Electronics, IEEE Journal of 36, 802-809 (2000).

[86] X. Fan, Z. He, Y. Mizuno, and K. Hotate, "Bandwidth-adjustable dynamic grating in erbium-doped fiber by synthesis of optical coherence function," Opt. Express 13, 5756-5761 (2005).

[87] R. Elsner, R. Ullmann, A. Heuer, R. Menzel, and M. Ostermeyer, "Two-dimensional modeling of transient gain gratings in saturable gain media," Opt. Express 20, 6887-6896 (2012).

[88] K.-Y. Song and K. Hotate, "All-optical dynamic grating generation based on Brillouin scattering in polarization maintaining fiber," in *19th International Conference on Optical Fibre Sensors*, (International Society for Optics and Photonics, 2008), 70043T-70043T-70044.

[89] W. Zou, Z. He, and K. Hotate, "Complete discrimination of strain and temperature using Brillouin frequency shift and birefringence in a polarization-maintaining fiber," Opt. Express 17, 1248-1255 (2009).

[90] Y. Dong, L. Chen, and X. Bao, "Truly distributed birefringence measurement of polarization-maintaining fibers based on transient Brillouin grating," Opt. Lett. 35, 193-195 (2010).

[91] S. Li, M.-J. Li, and R. S. Vodhanel, "All-optical Brillouin dynamic grating generation in few-mode optical fiber," Opt. Lett. 37, 4660-4662 (2012).

[92] K. Y. Song, "Operation of Brillouin dynamic grating in single-mode optical fibers," Opt. Lett. 36, 4686-4688 (2011).

[93] W. Zou and J. P. Chen, "All-optical generation of Brillouin dynamic grating based on multiple acoustic modes in a single-mode dispersion-shifted fiber," Opt. Express 21, 14771-14779 (2013).

[94] R. Pant, E. Li, C. G. Poulton, D.-Y. Choi, S. Madden, B. Luther-Davies, and B. J. Eggleton, "Observation of Brillouin dynamic grating in a photonic chip," Opt. Lett. 38, 305-307 (2013).

[95] K. O. Hill and G. Meltz, "Fiber Bragg grating technology fundamentals and overview," Lightwave Technology, Journal of 15, 1263-1276 (1997).

[96] W. Zou and J. Chen, "Spectral Analysis of Brillouin Dynamic Grating Based on Heterodyne Detection," Appl. Phys. Express 6, 122503 (2013).

[97] D. P. Zhou, L. Chen, and X. Y. Bao, "Polarization-decoupled four-wave mixing based on stimulated Brillouin scattering in a polarization-maintaining fiber," J. Opt. Soc. Am. B-Opt. Phys. 30, 821-828 (2013).

[98] K. Y. Song, K. Lee, and S. B. Lee, "Tunable optical delays based on Brillouin dynamic grating in optical fibers," Opt. Express 17, 10344-10349 (2009).

[99] W. Zou, Z. He, K.-Y. Song, and K. Hotate, "Correlation-based distributed measurement of a dynamic grating spectrum generated in stimulated Brillouin scattering in a polarization-maintaining optical fiber," Opt. Lett. 34, 1126-1128 (2009).

[100] K. Y. Song, W. Zou, Z. He, and K. Hotate, "Optical time-domain measurement of Brillouin dynamic grating spectrum in a polarization-maintaining fiber," Opt. Lett. 34, 1381-1383 (2009).

[101] Y. Antman, N. Primerov, J. Sancho, L. Thévenaz, and A. Zadok, "Localized and stationary dynamic gratings via stimulated Brillouin scattering with phase modulated pumps," Opt. Express20 (7), 7807-7821 (2012).

[102] M. Santagiustina and L. Ursini, "Dynamic Brillouin gratings permanently sustained by chaotic lasers," Opt. Lett. 37, 893-895 (2012).

[103] Y. Antman, L. Yaron, T. Langer, M. Tur, N. Levanon, and A. Zadok, "Experimental demonstration of localized Brillouin gratings with low off-peak reflectivity established by perfect Golomb codes," Opt. Lett. 38, 4701-4704 (2013).

[104] Y. Antman, N. Primerov, J. Sancho, L. Thevenaz, and A. Zadok, "Long variable delay and distributed sensing using stationary and localized Brillouin dynamic gratings," in *Optical Fiber Communication Conference*, (Optical Society of America, 2012),

[105] Y. Antman, N. Primerov, J. Sancho, L. Thévenaz, and A. Zadok, "Variable delay using stationary and localized Brillouin dynamic gratings," in *SPIE OPTO*, (International Society for Optics and Photonics, 2012), 82730C-82730C-82738.

[106] S. Chin and L. Thevenaz, "Tunable photonic delay lines in optical fibers," Laser Photon. Rev. 6, 724-738 (2012).

[107] J. Sancho, N. Primerov, S. Chin, Y. Antman, A. Zadok, S. Sales, and L. Thévenaz, "Tunable and reconfigurable multi-tap microwave photonic filter based on dynamic Brillouin gratings in fibers," Opt. Express20 (6), 6157-6162 (2012).

[108] H. G. Winful, "Chirped Brillouin dynamic gratings for storing and compressing light," Opt. Express 21, 10039-10047 (2013).

[109] L. Ursini and M. Santagiustina, "Applications of the Dynamic Brillouin Gratings to Ultrawideband Communications," IEEE Photonics Technol. Lett. 25, 1347-1349 (2013).

[110] M. Santagiustina, S. Chin, N. Primerov, L. Ursini, and L. Thévenaz, "All-optical signal processing using dynamic Brillouin gratings," Scientific reports 3(2013).

[111] K. Chiang, D. Wong, and P. Chu, "Strain-induced birefringence in a highly birefringent optical fibre," Electron. Lett. 26, 1344-1346 (1990).

[112] W. Zou, Z. He, and K. Hotate, "One-laser-based generation/detection of Brillouin dynamic grating and its application to distributed discrimination of strain and temperature," Opt. Express 19, 2363-2370 (2011).

[113] W. Zou, Z. He, and K. Hotate, "Demonstration of Brillouin distributed discrimination of strain and temperature using a polarization-maintaining optical fiber," Photonics Technology Letters, IEEE 22, 526-528 (2010).

[114] Y. Dong, L. Chen, and X. Bao, "High-spatial-resolution time-domain simultaneous strain and temperature sensor using Brillouin scattering and birefringence in a polarization-maintaining fiber," Photonics Technology Letters, IEEE 22, 1364-1366 (2010).

[115] R. K. Yamashita, W. Zou, Z. He, and K. Hotate, "Measurement range elongation based on temporal gating in Brillouin optical correlation domain distributed simultaneous sensing of strain and temperature," Photonics Technology Letters, IEEE 24, 1006-1008 (2012).

[116] W. Zou, Z. He, and K. Hotate, "Range Elongation of Distributed Discrimination of Strain and Temperature in Brillouin Optical Correlation-Domain Analysis Based on Dual Frequency Modulations," Sensors Journal, IEEE 14, 244-248 (2014).

[117] R. K. Yamashita, H. Zuyuan, and K. Hotate, "Spatial Resolution Improvement in Correlation Domain Distributed Measurement of Brillouin Grating," Photonics Technology Letters, IEEE 26, 473-476 (2014).

[118] X. Bao, J. Dhliwayo, N. Heron, D. J. Webb, and D. A. Jackson, "Experimental and theoretical studies on a distributed temperature sensor based on Brillouin scattering," Lightwave Technology, Journal of 13, 1340-1348 (1995).

[119] J. Smith, A. Brown, M. DeMerchant, and X. Bao, "Pulse width dependance of the Brillouin loss spectrum," Opt. Commun. 168, 393-398 (1999).

[120] A. W. Brown, M. D. DeMerchant, X. Bao, and T. W. Bremner, "Spatial resolution enhancement of a Brillouin-distributed sensor using a novel signal processing method," J. Lightwave Technol. 17, 1179 (1999).

[121] K. Kishida, C. Li, S. Lin, and K. I. NISHIGUCHI, "Pulse pre-pump method to achieve cm-order spatial resolution in Brillouin distributed measuring technique," Technical report of IEICE, 15-20 (2004).

[122] S. Afshar V, G. A. Ferrier, X. Bao, and L. Chen, "Effect of the finite extinction ratio of an electro-optic modulator on the performance of distributed probe-pump Brillouin sensorsystems," Opt. Lett. 28, 1418-1420 (2003).

[123] A. W. Brown, B. G. Colpitts, and K. Brown, "Distributed sensor based on dark-pulse Brillouin scattering," Photonics Technology Letters, IEEE 17, 1501-1503 (2005).

[124] A. W. Brown, B. G. Colpitts, and K. Brown, "Dark-pulse Brillouin optical time-domain sensor with 20-mm spatial resolution," J. Lightwave Technol. 25, 381-386 (2007).

[125] F. Wang, X. Bao, L. Chen, Y. Li, J. Snoddy, and X. Zhang, "Using pulse with a dark base to achieve high spatial and frequency resolution for the distributed Brillouin sensor," Opt. Lett. 33, 2707-2709 (2008).

[126] K. Y. Song, S. Chin, N. Primerov, and L. Thévenaz, "Time-domain distributed fiber sensor with 1 cm spatial resolution based on Brillouin dynamic grating," J. Lightwave Technol. 28, 2062-2067 (2010).

[127] K. Y. Song and H. J. Yoon, "High-resolution Brillouin optical time domain analysis based on Brillouin dynamic grating," Opt. Lett. 35, 52-54 (2010).

[128] S. Chin, N. Primerov, and L. Thevenaz, "Sub-centimeter spatial resolution in distributed fiber sensing based on dynamic Brillouin grating in optical fibers," Sensors Journal, IEEE 12, 189-194 (2012).

[129] W. Li, X. Bao, Y. Li, and L. Chen, "Differential pulse-width pair BOTDA for high spatial resolution sensing," Opt. Express 16, 21616-21625 (2008).

[130] Y. Dong, X. Bao, and W. Li, "Differential Brillouin gain for improving the temperature accuracy and spatial resolution in a long-distance distributed fiber sensor," Appl. Optics 48, 4297-4301 (2009).

[131] Y. Dong, H. Zhang, L. Chen, and X. Bao, "2 cm spatial-resolution and 2 km range Brillouin optical fiber sensor using a transient differential pulse pair," Appl. Optics 51, 1229-1235 (2012).

[132] A. Minardo, R. Bernini, and L. Zeni, "Differential techniques for high-resolution BOTDA: an analytical approach," Photonics Technology Letters, IEEE 24, 1295-1297 (2012).

[133] S. Foaleng-Mafang, J.-C. Beugnot, and L. Thévenaz, "Optimized configuration for high resolution distributed sensing using Brillouin echoes," in *20th International Conference on Optical Fibre Sensors*, (International Society for Optics and Photonics, 2009), 75032C-75032C-75034.

[134] S. M. Foaleng, M. Tur, J.-C. Beugnot, and L. Thévenaz, "High spatial and spectral resolution long-range sensing using Brillouin echoes," J. Lightwave Technol. 28, 2993-3003 (2010).

[135] Y. Koyamada, Y. Sakairi, N. Takeuchi, and S. Adachi, "Novel technique to improve spatial resolution in Brillouin optical time-domain reflectometry," Photonics Technology Letters, IEEE 19, 1910-1912 (2007).

[136] Y. Yao, Y. Lu, X. Zhang, F. Wang, and R. Wang, "Reducing Trade-Off Between Spatial Resolution and Frequency Accuracy in BOTDR Using Cohen's Class Signal Processing Method," Photonics Technology Letters, IEEE 24, 1337-1339 (2012).

[137] F. Wang, W. Zhan, X. Zhang, and Y. Lu, "Improvement of spatial resolution for BOTDR by iterative subdivision method," J. Lightwave Technol. 31, 3663-3667 (2013).

[138] Y. T. Cho, M. N. Alahbabi, M. J. Gunning, and T. P. Newson, "Enhanced performance of long range Brillouin intensity based temperature sensors using remote Raman amplification," Meas. Sci. Technol. 15, 1548-1552 (2004).

[139] K. De Souza and T. P. Newson, "Signal to noise and range enhancement of a Brillouin intensity based temperature sensor," Opt. Express 12, 2656-2661 (2004).

[140] Y. T. Cho, M. N. Alahbabi, G. Brambilla, and T. P. Newson, "Distributed Raman amplification combined with a remotely pumped EDFA utilized to enhance the per-

formance of spontaneous Brillouin-based distributed temperature sensors," IEEE Photonics Technol. Lett. 17, 1256-1258 (2005).

[141] X. H. Jia, Y. J. Rao, L. A. Chang, C. Zhang, and Z. L. Ran, "Enhanced Sensing Performance in Long Distance Brillouin Optical Time-Domain Analyzer Based on Raman Amplification: Theoretical and Experimental Investigation," J. Lightwave Technol. 28, 1624-1630 (2010).

[142] S. Martin-Lopez, M. Alcon-Camas, F. Rodriguez, P. Corredera, J. D. Ania-Castanon, L. Thevenaz, and M. Gonzalez-Herraez, "Brillouin optical time-domain analysis assisted by second-order Raman amplification," Opt. Express 18, 18769-18778 (2010).

[143] M. A. Soto, G. Bolognini, and F. Di Pasquale, "Optimization of long-range BOTDA sensors with high resolution using first-order bi-directional Raman amplification," Opt. Express 19, 4444-4457 (2011).

[144] X. Angulo-Vinuesa, S. Martin-Lopez, P. Corredera, and M. Gonzalez-Herraez, "Raman-assisted Brillouin optical time-domain analysis with sub-meter resolution over 100 km," Opt. Express 20, 12147-12154 (2012).

[145] X. Angulo-Vinuesa, S. Martin-Lopez, J. Nuno, P. Corredera, J. D. Ania-Castanon, L. Thevenaz, and M. Gonzalez-Herraez, "Raman-Assisted Brillouin Distributed Temperature Sensor Over 100 km Featuring 2 m Resolution and 1.2 degrees C Uncertainty," J. Lightwave Technol. 30, 1060-1065 (2012).

[146] M. A. Soto, G. Bolognini, and F. Di Pasquale, "Analysis of optical pulse coding in spontaneous Brillouin-based distributed temperature sensors," Opt. Express 16, 19097-19111 (2008).

[147] M. A. Soto, G. Bolognini, F. Di Pasquale, and L. Thévenaz, "Long-range Brillouin optical time-domain analysis sensor employing pulse coding techniques," Measurement Science and Technology 21, 094024 (2010).

[148] M. A. Soto, G. Bolognini, F. Di Pasquale, and L. Thévenaz, "Simplex-coded BOTDA fiber sensor with 1 m spatial resolution over a 50 km range," Opt. Lett. 35, 259-261 (2010).

[149] M. A. Soto, P. K. Sahu, G. Bolognini, and F. Di Pasquale, "Brillouin-based distributed temperature sensor employing pulse coding," IEEE Sens. J. 8, 225-226 (2008).

[150] M. A. Soto, S. Le Floch, and L. Thevenaz, "Bipolar optical pulse coding for performance enhancement in BOTDA sensors," Opt. Express 21, 16390-16397 (2013).

[151] M. Taki, Y. Muanenda, C. J. Oton, T. Nannipieri, A. Signorini, and F. Di Pasquale, "Cyclic pulse coding for fast BOTDA fiber sensors," Opt. Lett. 38, 2877-2880 (2013).

[152] M. A. Soto, G. Bolognini, and F. D. Pasquale, "Long-range simplex-coded BOTDA sensor over 120km distance employing optical preamplification," Opt. Lett. 36, 232-234 (2011).

[153] M. A. Soto, M. Taki, G. Bolognini, and F. Di Pasquale, "Simplex-Coded BOTDA Sensor Over 120-km SMF With 1-m Spatial Resolution Assisted by Optimized Bidirectional Raman Amplification," IEEE Photonics Technol. Lett. 24, 1823-1826 (2012).

[154] F. Gyger, E. Rochat, S. Chin, M. Niklès, and L. Thévenaz, "Extending the sensing range of Brillouin optical time-domain analysis up to 325 km combining four optical repeaters," in *OFS2014 23rd International Conference on Optical Fiber Sensors*, (International Society for Optics and Photonics, 2014), 91576Q-91576Q-91574.

[155] R. Bernini, A. Minardo, and L. Zeni, "Long-range distributed Brillouin fiber sensors by use of an unbalanced double sideband probe," Opt. Express 19, 23845-23856 (2011).

[156] D. M. Nguyen, B. Stiller, M. W. Lee, J.-C. Beugnot, H. Maillotte, A. Mottet, J. Hauden, and T. Sylvestre, "Distributed Brillouin Fiber Sensor With Enhanced Sensitivity Based on Anti-Stokes Single-Sideband Suppressed-Carrier Modulation," IEEE Photonics Technol. Lett. 25, 94-96 (2013).

[157] K. Hotate, K. Abe, and K. Y. Song, "Suppression of signal fluctuation in Brillouin optical correlation domain analysis system using polarization diversity scheme," IEEE Photonics Technol. Lett. 18, 2653-2655 (2006).

[158] K. Y. Song and K. Hotate, "Enlargement of measurement range in a Brillouin optical correlation domain analysis system using double lock-in amplifiers and a single-sideband modulator," IEEE Photonics Technol. Lett. 18, 499-501 (2006).

[159] J. H. Jeong, K. Lee, K. Y. Song, J. M. Jeong, and S. B. Lee, "Variable-frequency lock-in detection for the suppression of beat noise in Brillouin optical correlation domain analysis," Opt. Express 19, 18721-18728 (2011).

[160] K. Y. Song, Z. He, and K. Hotate, "Optimization of Brillouin optical correlation domain analysis system based on intensity modulation scheme," Opt. Express 14, 4256-4263 (2006).

[161] K. Y. Song, Z. Y. He, and K. Hotate, "Effects of intensity modulation of light source on Brillouin optical correlation domain analysis," J. Lightwave Technol. 25, 1238-1246 (2007).

[162] J. H. Jeong, K. Lee, K. Y. Song, J. M. Jeong, and S. B. Lee, "Differential measurement scheme for Brillouin Optical Correlation Domain Analysis," Opt. Express 20, 27094-27101 (2012).

[163] M. Kannou, S. Adachi, and K. Hotate, "Temporal gating scheme for enlargement of measurement range of Brillouin optical correlation domain analysis for optical fiber distributed strain measurement," in *Proc. 16th Int. Conf. Optical Fiber Sensors*, 2003), 454-457.

[164] K. Hotate and H. Arai, "Enlargement of measurement range of simplified BOCDA fiber-optic distributed strain sensing system using a temporal gating scheme," in *Brug-*

es, Belgium-Deadline Past, (International Society for Optics and Photonics, 2005), 184-187.

[165] Y. Mizuno, Z. He, and K. Hotate, "Measurement range enlargement in Brillouin optical correlation-domain reflectometry based on temporal gating scheme," Opt. Express 17, 9040-9046 (2009).

[166] Y. Mizuno, Z. Y. He, and K. Hotate, "Measurement range enlargement in Brillouin optical correlation-domain reflectometry based on double-modulation scheme," Opt. Express 18, 5926-5933 (2010).

[167] D. Elooz, Y. Antman, N. Levanon, and A. Zadok, "High-resolution long-reach distributed Brillouin sensing based on combined time-domain and correlation-domain analysis," Opt. Express 22, 6453-6463 (2014).

[168] A. Denisov, M. A. Soto, and L. Thévenaz, "1'000'000 resolved points along a Brillouin distributed fibre sensor," in *OFS2014 23rd International Conference on Optical Fiber Sensors*, (International Society for Optics and Photonics, 2014), 9157D9152-9157D9152-9154.

[169] A. Zadok, Y. Antman, N. Primerov, A. Denisov, J. Sancho, and L. Thevenaz, "Random-access distributed fiber sensing," Laser Photon. Rev. 6, L1-L5 (2012).

[170] W. Zou, Z. He, and K. Hotate, "Distributed dynamic-strain sensing based on brillouin optical correlation domain analysis," in *Lasers & Electro Optics & The Pacific Rim Conference on Lasers and Electro-Optics, 2009. CLEO/PACIFIC RIM'09. Conference on*, (IEEE, 2009), 1-2.

[171] W. Zou, Z. He, and K. Hotate, "Realization of high-speed distributed sensing based on Brillouin optical correlation domain analysis," in *Conference on Lasers and Electro-Optics*, (Optical Society of America, 2009),

[172] K. Y. Song, M. Kishi, Z. Y. He, and K. Hotate, "High-repetition-rate distributed Brillouin sensor based on optical correlation-domain analysis with differential frequency modulation," Opt. Lett. 36, 2062-2064 (2011).

[173] K. Hotate and S. S. L. Ong, "Distributed dynamic strain measurement using a correlation-based brillouin sensing system," IEEE Photonics Technol. Lett. 15, 272-274 (2003).

[174] K. Y. Song and K. Hotate, "Distributed fiber strain sensor with 1-kHz sampling rate based on Brillouin optical correlation domain analysis," IEEE Photonics Technol. Lett. 19, 1928-1930 (2007).

[175] T. Yamauchi and K. Hotate, "Distributed and dynamic strain measurement by BOCDA with time-division pump-probe generation scheme," in *Conference on Lasers and Electro-Optics*, (Optical Society of America, 2004),

[176] Y. Mizuno, Z. Y. He, and K. Hotate, "One-End-Access High-Speed Distributed Strain Measurement with 13-mm. Spatial Resolution Based on Brillouin Optical Correlation-Domain Reflectometry," IEEE Photonics Technol. Lett. 21, 474-476 (2009).

Chapter 2

Poly-harmonic Analysis of Raman and Mandelstam-Brillouin Scatterings and Bragg Reflection Spectra

Oleg G. Morozov, Gennady A. Morozov,
Ilnur I. Nureev and Anvar A. Talipov

Additional information is available at the end of the chapter

http://dx.doi.org/10.5772/59144

1. Introduction

The chapter is devoted to applications and construction principles of poly-harmonic (two-frequency or four-frequency) cw laser systems for characterization of different nonlinear scattering effects in fibers and reflection of devices based on fiber Bragg gratings (FBG) in telecommunication lines and sensor nets. In particular, we'll speak about evaluation of Mandelstam-Brillouin gain contour, Raman scattering contours and FBG reflection spectra characterization. Investigation methods and approaches are based on the unity of resonant structures of generated fiber responses on exciting and probing radiation or external physical fields for all given effects. The main decision is based on poly-harmonic probing of formed resonance responses.

At a certain level of the laser power, exciting the optical fiber, resonant contours of Mandel-stam-Brillouin and Raman scattering are formed [1]. The first are based on periodic photon-phonon interactions, the second – the photon-photon ones. Similarly, spectral reflection characteristics of Bragg gratings, based on periodic variation of the refractive index in the fiber core, can be described by the resonant contour [2]. In telecommunication lines Mandelstam-Brillouin and Raman scatterings are undesirable effects, but in sensor nets there are the main sources of measuring information. FBG, as known, is the powerful instruments as for tele-communication lines and so on sensor nets design. Thus, characterization of Mandelstam-Brillouin gain and Raman scattering contours and FBG reflection spectra is the actual and important task for scientists and designers. The typical characteristics of given effects are discussed in the first part of the chapter.

To convert the information about the spectral contours characteristics are classically used broadband or tunable over a wide range optoelectronic means (optical spectrum analyzers, tunable lasers, optical reflectometers in time (OTDR) and frequency (OFDR) domains, scanning Fabry-Perot interferometers, diffraction gratings with CCD, and etc.) and complex "fit" algorithms for determining a desired value of accuracy of the central wavelength of the contours [3]. In recent years, significant progress in terms of accuracy and resolution of measurements, as well as practicality, was registered in the use of narrowband poly-harmonic technology [4] for characterization of contour spectrums that makes them competitive for the above mentioned methods in metrological characteristics, ease and cost of implementation. Their main advantage is that no measurements in the resonance region of spectral characteristics are necessary, that allows eliminating the influence of power instability of probing laser radiation and to detect information on the difference inter-harmonic frequency in the region with small noise level of photo-detector.

Second part of the chapter is devoted to poly-harmonic characterization of Mandelstam-Brillouin gain contours, Raman scattering contours and FBG reflection spectra for telecommunication applications. For example, Mandelstam-Brillouin frequency shift for silica fibers is about 10-20 GHz, and Mandelstam-Brillouin gain is observed in a bandwidth of 20-100 MHz [1]. The main parameters are the central frequency of a gain spectrum, its Q-factor and gain coefficient. The classical method for characterization of stimulated Mandelstam-Brillouin scattering (SMBS) gain spectrum is based on use of two lasers: one – for SMBS pumping, and another – for probing of generated gain spectrum [5]. The disadvantage of this method is need in strong control of a frequencies difference of two sources. Not long ago the measurement system which is free from this drawback was presented [6]. It is based on SMBS gain spectrum conversion from optical to the electrical field by single-sideband amplitude modulated radiation, in which the upper sideband is suppressed. Despite advantages, realization of this method is not always effective; because relevant low sensitivity of measurements is remained, similar to measurements by double-band amplitude modulated probing radiation in a wide bandwidth. A new method for SMBS gain spectrum characterization in single-mode optical fiber was presented in our papers [7-8]. It is based on use of advantages of single-sideband modulation and two-frequency probing radiation, which gives possibility of transfer the data signal's spectrum in the low noise region of a photo-detector. Also this radiation is characterized by effective procedure of received spectral information processing by the envelope's characteristics (phase and amplitude) of two spectral components beats. In the end of second part we also go back to the issues of poly-harmonic analysis of the FBG reflection spectra affected by us in [9-10].The modernization of previously used methods [4] will be presented, which based on the four-frequency probing radiation and only amplitude analysis of the envelopes [11]. Consideration of stimulated Raman scattering (SRS) in this part will be carried out only in general terms because of its negligible impact on the performance of telecommunication none-WDM lines [1].

Third part of the chapter is devoted to poly-harmonic characterization of Mandelstam-Brillouin gain contours, Raman scattering contours and FBG reflection spectra for sensor nets applications. An illustrative example of the relevance of considered issues is the use of all three given physical mechanisms in distributed and the quasi-distributed sensor nets for down-hole telemetry [12,13]. If Mandelstam-Brillouin and Raman scatterings are carrying distributed information of the measured parameters (temperature, pressure), the FBG gratings allows to receive its point localization and flow velocity data [14,15]. For more information OTDR with Rayleigh scattering can be used and characterized to the Mandelstam-Brillouin ones by Landau-Placzek relation [16]. So way it seems reasonable to use a single radiation source for getting fiber-forming response to external stimuli, forming its special signal shape or spectra, optimized for recording spectrally separated responses from various nonlinear effects and reflections, and monitoring their bound to the central wavelength of the radiation. Some pair wise effects of such an implementation are known in practice of Weatherford, Schlumberger, Halliburton and other companies. Comprehensive option to generate and use of responses from the three types of scattering and reflection from the Bragg gratings not yet been studied. Its implementation could put information redundancy in the process of measuring, the use of which would improve the metrological characteristics systems being developed.

In fourth part perspective systems and their elements are presented, describing and discussing the methods, tools and systems parameters of means to get poly-harmonic radiation on the base of dual-drive MZM, scanning and poly-harmonic (more than four harmonics) methods for SMBS characterization based on technology transfer from LIDAR systems, designing of notch filters for blocking of elastic Rayleigh scattering in SRS and SMBS measurements. Additionally Raman and Mandelstam-Brillouin scattering in sensor nets so as a FBG reflection carry vast amount information about fiber conditions but sometime have low energy level. That's why it's one more cause to detect these types of scattering with high SNR and determine their properties. Applying of photo mixing allows significantly increase the reflectometeric system sensitivity under the condition of weak signals and receives information from frequency pushing of backscattered signal spectrum [56]. We offer in [4] to use two-frequency heterodyne and the second nonlinear receiver in the structure of reflectometers and now discuss its advantages comparatively to other methods of SNR increasing in the end of the fourth part.

In conclusion we'll resume results of above mentioned researches, mark it practical implementation and show new tasks in Mandelstam-Brillouin gain contours, Raman scattering contours and FBG reflection spectra characterization.

2. Spectral characteristics of SMBS gain, SRS scattering and FBG reflection contours

2.1. Spectral characteristics of SMBS gain and SRS scattering contours

Two important nonlinear effects in optical fibers, known as SRS and SMBS, are related to vibration excitation modes of silica and fall in the category of stimulated inelastic scattering in which the optical field transfers part of its energy to the nonlinear medium [1]. Even though SRS and SMBS are very similar in their origin, the main difference between the two is that optical phonons participate in SRS while acoustic phonons participate in SMBS. In a simple quantum-mechanical picture applicable to both SRS and SMBS, a photon of the incident field (called the pump) is annihilated to create a photon at a lower frequency (belonging to the Stokes wave) and a phonon with the right energy and momentum, to conserve the energy and the momentum. Of course, a higher-energy photon at the so-called anti-Stokes frequency can also be created if a phonon of right energy and momentum is available. Different dispersion relations for acoustic and optical phonons lead to some basic differences between the two. The fundamental one is that SMBS in single-mode fibers occurs only in the backward direction whereas SRS can occur in both directions.

Both the Raman-gain spectrum $g_R(v)$ and the Mandelstam-Brillouin-gain spectrum $g_{MB}(v)$ have been measured experimentally for silica fibers. The Raman-gain spectrum is found to be very broad, extending up to 40 THz [17]. The peak gain $g_R \approx 6\times10^{-14}$ m/W at pump wavelength near 1.5 µm and occurs for a spectral shift of about 13.1 THz. In contrast, the Mandelstam-Brillouin-gain spectrum is extremely narrow and has a bandwidth of <100 MHz. The peak value of Mandelstam-Brillouin-gain occurs, for the Stokes shift of ~10 GHz, for pump wavelengths near 1.5 µm. The peak gain $g_{MB} \approx 6\times10^{-11}$ m/W for a narrow-bandwidth pump [18] and decreases by a factor of $\Delta v_P/\Delta v_{MB}$ for a broad-bandwidth pump, where Δv_P is the pump bandwidth and Δv_{MB} is the Mandelstam-Brillouin-gain bandwidth. As the Mandelstam-Brillouin-gain coefficient g_{MB} is larger by nearly three orders of magnitude compared with g_R, typical values of SMBS threshold are ~1 mW and for SRS threshold are ~1 W.

Although a complete description of SRS $g_R(v)$ in optical fibers is quite involved, the spectral characteristics for SMBS $g_{MB}(v)$ can be described by a simple relation. Little reminding – SMBS is a result of scattering of light on acoustic waves (acoustic phonons), that are excited by thermal fields and produce periodic modulation of the refractive index of fiber [1]. As a result of Bragg diffraction the induced grating of refractive index scatters the pumping radiation. As the scattered light undergoes a Doppler effect, the frequency shift v_{MB}, caused by SMBS, depends on acoustic velocity and is given by

$$v_{MB} = \frac{2nV_A}{\lambda_P},\tag{1}$$

where V_A acoustic velocity in the fiber, n refractive index, λ_P pump wavelength. The shape of the SMBS gain spectrum is determined by strong attenuation of sound waves in silica. Growth of Stokes wave's intensity is characterized by gain coefficient $g_{MB}(v)$, maximum at $v=v_{MB}$. Because of exponential decay of the acoustic waves, the gain spectrum $g_{MB}(v)$ will have the Lorentzian spectral profile

$$g_{MB}(v) = g_0 \frac{\left(\Delta v_{MB}/2\right)^2}{\left(v - v_{MB}\right)^2 + \left(\Delta v_{MB}/2\right)^2},\tag{2}$$

where Δv_{MB} spectrum full width at half maximum. The maximum gain is given by

$$g_{MB}(v_{MB}) = g_0 = \frac{2\pi n^7 p_{12}^2}{c\lambda_P^2 \rho_0 V_A \Delta v_{MB}},$$ (3)

where p_{12} longitudinal acousto-optic coefficient, ρ_0 density of material, c light speed in vacuum.

SMBS using allows to measure temperature (frequency shift of v_{MB} about 1 MHz /°C) and strain (frequency shift of v_{MB} about 493 MHz/%) of fiber so provide distributed technologies in sensor nets [1].

Let's return to the description of SRS $g_R(v)$ in optical fibers [1]. A weak Stokes signal launched into a fiber with a stronger pump will be amplified due to SRS as discussed in this chapter. The amplification of the signal is described through the following equation

$$g_R(v) = \sigma_0(v)\frac{\lambda_s^3}{c^2 \eta n^2(v)},$$ (4)

where η is Planck's constant, λ_s is the Stokes wavelength, $n(v)$ is the refractive index, which is frequency dependent. The spontaneous Raman cross section $\sigma_0(v)$ is defined as the ratio [1] of radiated power at the Stokes wavelengths to the pump power at temperature 0 °K, which can be obtained with the measured Raman cross-section $\sigma_T(v)$ at temperature T °K, the thermal population factor $N(v, T)$ and Boltzmann constant κ_B:

$$\sigma_0(v) = \sigma_T(v)/\left(N(\Delta v, T) + 1\right),$$ (5)

$$N(\Delta v, T) = 1 \left/ \exp\left(\frac{\eta c \Delta v}{\kappa_B T} - 1\right)\right..$$ (6)

The $g_R(v)$ in (4) is the spontaneous Raman gain coefficient in bulk glass and is uniform in all directions. For the first time, the Gaussian decomposition technique was proposed earlier in [19] for a spontaneous Raman spectrum. It is known that the Raman gain profile differs from the spontaneous Raman spectrum, especially in the low frequency region [21]. So for SRS the Raman gain coefficient at the Stokes frequency $v_S = v_P - v$ with some assumptions can be written as follows [20]

$$g_R(v) \approx \frac{v\tilde{A}}{\left(v_S^2 - v^2\right)^2 + v^2\tilde{A}^2}, \tag{7}$$

where v_S and v are the resonance and phonon angular frequencies, respectively, Γ is the phonon damping constant.

Model analysis of the Raman spectrum with Gaussian profiles is based on the following expression [21]:

$$g_R(v) = g_R \sum_{i=1}^{N_m} A_i \exp\left[-\left(v - v_{v,i}\right)^2 / \Gamma_i^2\right], \tag{8}$$

where N_m is the number of modes used for decomposition, v,i is the central frequency of i-th Gaussian profile, $\Gamma_i = \mathrm{FWHM}_i / \left(2\sqrt{\ln 2}\right) \approx 0.6\ \mathrm{FWHM}_v$, where FWHM_i is the full width at the half maximum of i-th Gaussian profile. Amplitudes Γ_i together with v,i and Γ_i are used as parameters in the nonlinear fitting procedure. There are two important aspects of the developed decomposition procedure. Firstly, this approach is seen as a possible method of subdividing the density of states according to specific contributions. Secondly, function $g_R(v)$ that gives the best fit of the experimental Raman gain profile is constructed [21]. We can see SMBS gain spectra and its parameters on the fig. 1,a with the results of Lorentzian fitting [22], so as SRS gain spectra on the fig. 1,b with the results of triangle (linear) and Gaussian (nonlinear) fitting [23]. For the last one we can mark that Gaussian fitting is more applicable and accurate for the central lobe of spectra [24].

2.2. Spectral characteristics of FBG reflection contours

FBG couple the fundamental mode of an optical fiber with the same mode propagating in the opposite direction. This means that radiation propagating in the fiber at a certain wavelength (9) is reflected from the grating completely or partially. Central or resonant frequency FBG λ_{BG} is defined by following expression [2]:

$$\lambda_{\mathrm{BG}} = 2n_{\mathrm{eff}}\Lambda, \tag{9}$$

where n_{eff} is effective refraction factor of the basic mode, Λ is the period of its modulation.

The characteristics of this reflection depend on the grating parameters. It is possible to describe the envelope R of FBG reflection spectra defined by detunes δ as:

$$R = \sinh^2\left[\kappa L\sqrt{1 - (\delta/\kappa)^2}\right] \Big/ \left\{\cosh^2\left[\kappa L\sqrt{1 - (\delta/\kappa)^2}\right] - (\delta/\kappa)^2\right\} \approx \tanh\left[\kappa L\sqrt{1 - (\delta/\kappa)^2}\right], \tag{10}$$

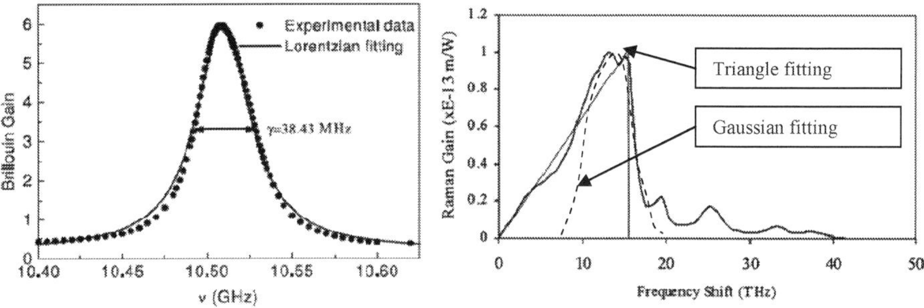

Figure 1. SMBS (a) and SRS (b) gain spectrums with the results of its fittings

where L – FBG length, κ – coupling factor of direct and return mode, (δ / κ) – relative detune. Detune of FBG with period Λ is equal to $\delta = \Omega - (\pi / \Lambda)$, where $\Omega = 2\pi n_{eff} / \lambda$ [2].

The spectral width of the resonance of a homogeneous FBG (fig. 2,a) measured between the first zeroes of its reflection spectrum is described by the expression

$$\Delta\lambda_{BG,0} = 2\lambda_{BG}\,\frac{\Lambda_{BG}}{L}\left[1+\left(\frac{\kappa L\sqrt{1-(\delta/\kappa)}}{\pi}\right)^{2}\right]^{1/2}. \tag{11}$$

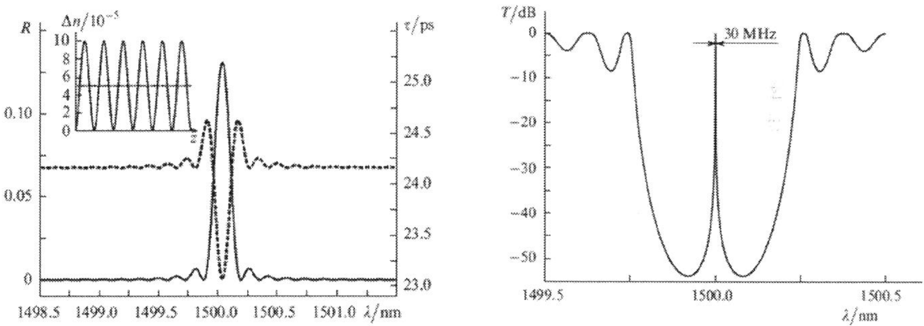

Figure 2. FBG reflection (a) and FBG-PS transmittance (b) calculated spectrums

The resonant FBG wavelength λ_{BG} depends on fiber temperature and from mechanical stretching or compressing pressure enclosed to it. This dependence is described by a following equation:

$$\Delta\lambda_{\text{ÂÐÁ}} = 2n_{\text{ýôô}}\Lambda\left(\left\{1-\left(\frac{\xi^2}{2}\right)\left[P_{12}-v\left(P_{11}+P_{12}\right)\right]\right\}\varepsilon+\left[\frac{1}{\Lambda}\frac{\partial\Lambda}{\partial T}+\frac{1}{n}\frac{\partial n}{\partial T}\right]\Delta T\right), \tag{12}$$

where ΔT is temperature change, ε is the enclosed pressure, the second item composed in a figure brace reflects photo elasticity factor. This parity gives typical values of λ_{BG} shift depending on temperature ~0,01 nm/°K from relative lengthening of a fiber ~ 10^3 (ΔL / L) (nm) [2].

The introduced phase shift (PS) [2] leads to the appearance of a narrow transmission band of width of a few tens of megahertz within the reflection band of FBG. Figure 2, b shows the calculated transmission spectrum of such FBG-PS grating. The phase shift in the grating can be introduced during the writing of the whole structure or later in the preliminary written grating. As the phase shift is increased (which is usually realized by writing two spatially separated gratings with the same FBG), the number of transmission regions in the reflection band increases, and such a structure is called, similarly to bulk optics, a Fabry-Perot interferometer (or filter). FBG-PS becomes the grate instrument in telecommunication and sensor nets [25-27].

The FBG reflective spectrum line shape can be approximated with a Gaussian profile [28]

$$R(\lambda)=R_B\exp\left[-4ln2\left[(\lambda-\lambda_B)/\Delta\lambda_B\right]^2\right]. \tag{13}$$

where λ is wavelength, λ_B is the center wavelength or peak wavelength of FBG, $\Delta\lambda_B$ is the full width at half maximum, and R_B is the maximum reflectivity.

As is known, the spectral dependence of the transmission band of FBG-PS has almost Lorentzian profile [2, 29]. If we assume that the spectral line width of the laser emission lines is negligibly small (~ several KHz), the spectral dependence of the transmission band of FBG-PS can be represented as follows:

$$T(\lambda)=\frac{T_B(\Delta\lambda_B/2)^2}{\left[(\lambda-\lambda_B)^2+(\Delta\lambda_B/2)^2\right]}. \tag{14}$$

where T_B is maximum transmittance on λ_B.

2.3. Discussion of results

Basics for the use of poly-harmonic probing methods of spectral characteristics for resonant circuits of arbitrary shape were described by us in a number of papers [4,9-10]. It is noted that their effective use (maximum slope of the measurement conversion) is possible at the location of equal amplitude symmetrical components of the probe radiation on the FWHM of contour with average frequency at the central (resonant) wavelength. Based on this requirement, the synthesis of the poly-harmonic (two-or four-frequency) probe radiation desired characteristics was carried out. The example results are shown in Tab. 1.

Effect	Fitting contour	Central peak	Bandwidth	FWHM
SRS	Gaussian	13,1 THz	25-45 THz	10-10 THz
SMBS	Lorentzian	10-11 GHz	20-100 MHz	10-50 MHz
FBG	Gaussian	1550 nm	0,1-1 nm	0,05-0,5 nm
FBG-PS	Gaussian/ Lorentzian	1550 nm	0,1-1/0,005 nm	0,05-0,5 /0,002 nm

Table 1. Spectral characteristics of Mandelstam-Brillouin gain contours, Raman scattering contours and FBG reflection spectra

We have show in [30-33] that poly-harmonic probing radiation can be forming by external electro-optic modulation of narrowband one-frequency laser one. For general estimations of requirements for electro-optic modulators we assume that 1 nm in wavelength band is 120 GHz in frequency bandwidth. Thus, the operating frequency of modulators, we need, must be equal from 10-20 MHz to 20-40 THz. Provision of the lower limit for the FBG-PS study causes no problems. To investigate the SMBS and FBG contours a wide range of modulators with a frequency range up to 100 GHz spacing are existed in $LiNbO_3$, GaAs, InP realizations. Direct solutions to achieve 20-40 THz frequency does not exist today. The use of electro-optic modulators in the mode of frequency multiplication (in 12-16 times) is possible to obtain bandwidth in units of THz. However, given that we are interested in the scattering of the central part of the SRS spectrum allocated by a Gaussian filter, you can use a tunable laser to deliver carrier laser radiation to 13.1 THz. Thus, the implementation of methods for poly-harmonic probing of Mandelstam-Brillouin gain, Raman scattering and FBG reflection contours spectral characteristics looks quite feasible.

3. Characterization of SMBS gain and FBG reflection spectra in telecom applications

3.1. Characterization of SMBS gain spectra

Characterization of SMBS gain spectrum in single-mode optical fiber is necessary in a number of applications. These are: an assessment of the distortions brought by SMBS in information, transferred on fiber-optical lines [34], processing and conversion of optical carriers and microwave sub-carriers in communication networks such as «radio-over-fiber» [35], failure or measuring conversion of temperature on ONU in phonon microwave spectroscopy of optical fiber in PONs [36,37]. For silica fibers shift of Mandelstam-Brillouin frequency is about 10-20 GHz, and Mandelstam-Brillouin gain is observed in a bandwidth of 20-100 MHz [7].

The main parameters are the central frequency of a gain spectrum, its Q-factor and gain coefficient.

The classical method for characterization of Mandelstam-Brillouin gain spectrum (MBGS) is based on use of two lasers: one – for SMBS pumping, and another – for probing of generated gain spectrum [5]. The disadvantage of this method is need in strong control of a frequencies difference of two sources. An advanced method gives the solution of this problem. The optical modulator generates the double-frequency signal. This signal is the sidebands of the pump laser, which are used then for probing [38]. But the disadvantage of this method is need in considering the input power in the gain spectrum and mechanisms of energy transfer between the pumping and probing components. The absence of these components can lead to saturation of contour and appearance of significant inaccuracies in characterization of MBGS. A certain progress in systems of characterization of MBGS was reached by generating the scanning double-band amplitude modulated probing radiation from pumping radiation [39]. However this method is characterized by the low sensitivity, caused by need of reception and processing of signals in a wide bandwidth (10-20 GHz), and also strong influence on measurement inaccuracy of the upper sideband existence. The solution of this problem also was in use of the double-frequency radiation generated unlike [38] for pumping radiation [40]. One frequency corresponded to pump frequency, and the second-to its Stokes component, thus frequency shifted absorption contour corresponded to Mandelstam-Brillouin gain spectrum. The absorption contour was used for suppression of the upper sideband. However, this system has a high complexity and need in strong control of positions of Stokes component and pump component, and also an absorption contour, especially when scanning of a probing signal within 20-100 MHz [7].

Not long ago the measurement system which is free from this drawback [6] was presented. It is based on MBGS conversion from optical to the electrical field by single-sideband amplitude modulated radiation, in which the upper sideband is suppressed. Despite advantages, realization of this method is not always effective; because relevant low sensitivity of measurements is remained, similar to measurements by double-band amplitude modulated probing radiation in a wide bandwidth. The new method for characterization of gain spectrum of SMBS in single-mode optical fiber is presented in this part. It is based on use of advantages of single-band modulation and double-frequency probing radiation, which gives possibility of transfer the data signal's spectrum in the low noise region of a photo-detector. Also this radiation is characterized by effective procedure of processing of received spectral information by envelope's characteristics of beats of two spectral components [7].

3.2. Two-frequency probing of MBGS

For conversion of the complex MBGS from optical to the electrical field the optical single-sideband modulation with scanning of frequency of a sideband component is used, including the information about the frequency shift and Q-factor of the SMBS gain spectrum. The measurement method offered by us is based on the double-frequency probing radiation of a MBGS, not on the single-frequency one. Experimental setup for measurements is shown on fig. 3 [7].

The optical signal from a 1550-nm laser diode with a bandwidth about 100 kHz is divided into two paths by a fiber-optic coupler. In the first path signal is modulated by an optical single-sideband modulator. Signal from the frequency combiner is applied to one of the modulator's inputs. The optical single-sideband modulator is based on a dual-drive Mach-Zehnder modulator design. The modulated signal is applied to the Fiber under test (FUT), where the

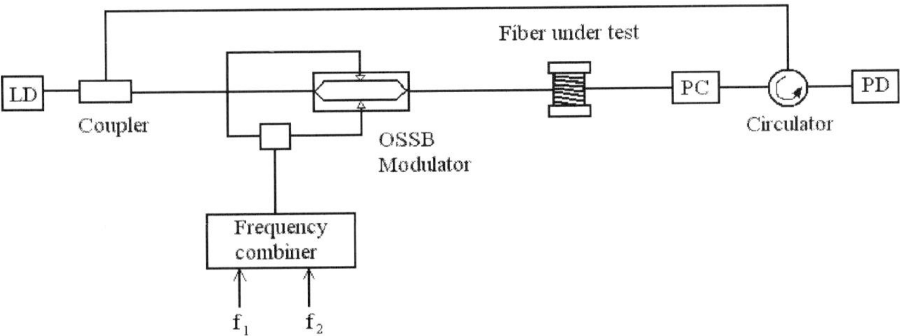

Figure 3. Experimental setup: LD – laser diode; PC – polarization controller; PD– photo-detector

optical radiation passed through the second path counter propagates. That non modulated radiation is the SMBS pump radiation in FUT [7].

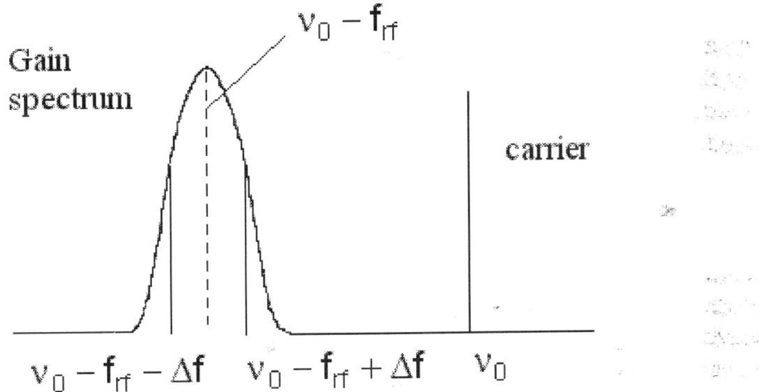

Figure 4. Probing of the gain spectrum by double-frequency signal

Thus, single-band double-frequency radiation with components $f_1 = f_{rf} - \Delta f$, $f_2 = f_{rf} + \Delta f$ probes the MBGS and the frequency $\nu_0 - f_{rf}$ tuned to the center of the gain spectrum conforms to its central frequency ν_{MB}, detuning Δf half of its FWHM, $\Delta \nu_{MB}$ and the carrier frequency ν_0 pump frequency $\nu_P = c / \lambda_P$. Double-frequency radiation, passed through the FUT, is received by photo-detector. Probing process is schematically shown in fig. 4 [57].

Radiation at the output of the optical single-sideband modulator is given by

$$E_{in}(t) = A_0 \exp(j 2\pi \nu_o t) + $$
$$+ A_{-1}\exp[\, j2\pi(\nu_0 - f_{rf} - \Delta f)t\,] + A_{-2}\exp[\, j2\pi(\nu_0 - f_{rf} + \Delta f)t\,], \tag{15}$$

where $A_0 = |A_0| \exp(j\varphi_0)$, $A_{-1} = |A_{-1}| \exp(j\varphi_{-1})$, $A_{-2} = |A_{-2}| \exp(j\varphi_{-2})$ complex amplitudes of the optical carrier and the double-frequency signal. This optical signal propagates through the FUT, which has an optical transfer function $H(v)$ characterizing the gain spectrum; therefore, the optical field at the output of the fiber is given by [57].

$$
\begin{aligned}
E_{out}(t) = &\ A_0 \left| H(v_0) \right| \exp\left[j \arg H(v_0) \right] \exp\left(j2\pi v_{0t} \right) + \\
&+ A_{-1} \left| H\left(v_0 - f_{rf} - \Delta f\right) \right| \exp\left[j \arg H\left(v_0 - f_{rf} - \Delta f\right) \right] \times \exp\left[j2\pi \left(v_0 - f_{rf} - \Delta f\right)t \right] + \\
&+ A_{-2} \left| H\left(v_0 - f_{rf} + \Delta f\right) \right| \exp\left[j \arg H\left(v_0 - f_{rf} + \Delta f\right) \right] \times \exp\left[j2\pi \left(v_0 - f_{rf} + \Delta f\right)t \right].
\end{aligned}
\tag{16}
$$

The output current on the beat frequency of the two probing components $2\Delta f$ is proportional to

$$
\begin{aligned}
\left| i_{out}(t) \right| \propto &\ \left| A_{-1} \right| \left| A_{-2} \right| \left| H(v_0 - f_{rf} - \Delta f) \right| \left| H(v_0 - f_{rf} + \Delta f) \right| \times \\
&\times \cos[4\pi t \Delta f + \varphi_{-1} - \varphi_{-2} + \arg H(v_0 - f_{rf} - \Delta f) - \arg H(v_0 - f_{rf} + \Delta f)].
\end{aligned}
\tag{17}
$$

Analysis of (17) shows that, we can get the image of the optical transfer function at the frequencies of the two probing signals from the electrical output signal of the photo-detector. The optical transfer function of the FUT is equivalent to concatenation of the fiber linear transfer function and the MBGS [57].

3.3. Four-frequency characterization of the MBGS

As we mentioned above, the main parameters of the MBGS are the central frequency of a gain spectrum, its Q-factor and gain coefficient. It is significant that at the moment when center frequency of a double-frequency signal $v_0 - f_{rf}$ gets to the resonance frequency of a gain spectrum v_{MB}, the envelope of the output signal is matched in phase with the envelope of the two-frequency signal at the FUT's input, and the modulation index of the output double-frequency signal's envelope is maximum and equal to 1 [57].

The measurement fractional inaccuracy of the central frequency can be 0,1% and determined by bandwidth of the laser radiation (in our case 0,1 MHz), and also by accuracy of keeping the difference frequency $2\Delta f$. Some part of the inaccuracy can be added by the appearance of not completely suppressed upper sideband of the double-frequency radiation in the spectrum. Among methods of its decreasing we can offer the usage of a chirp fiber Bragg grating, tuned on the suppression in the bandwidth of possible position change at scanning. We think that such solution is more effective, than offered in [40], as by efficiency of suppression, and also by ability to control the distortions, caused by chromatic dispersion [57].

Defining $v_0 - f_{rf} = v_{MB}$, we can find Q-factor of the MBGS. For this we offer the four-frequency method [58] or the method of the variation of difference frequency, which based on the dependence

$$Q_{1,2} = \frac{v_0 - f_{rf}}{f_1 - f_2} \sqrt{\frac{i_{\text{out}(v_0 - f_{rf})}}{i_{\text{out}1,2}} - 1}, \tag{18}$$

where $i_{\text{out}(v_0 - f_{rf})}$ and $i_{\text{out}1,2}$ amplitudes at center frequency and at components of the double-frequency signal at the output of the photo-detector when center frequency of the probing components at frequencies $f_1 = f_{rf} - \Delta f$ and $f_2 = f_{rf} + \Delta f$ is tuned on the center of gain spectrum. The value of $i_{\text{out}1,2}$ is determined by output signal of the photo-detector, and the value $i_{\text{out}(v_0 - f_{rf})}$ is unknown. If we change the Δf by a certain value $\Delta f\,'$, not changing the tuning on the center of gain spectrum, then we will get the new value of frequencies $f_3 = f_{rf} - \Delta f - \Delta f\,'$ and $f_4 = f_{rf} + \Delta f + \Delta f\,'$. For frequencies f_3 and f_4 we can rewrite the (18) as

$$Q_{3,4} = \frac{v_0 - f_{rf}}{f_3 - f_4} \sqrt{\frac{i_{v_0 - f_{rf}}}{i_{3,4}} - 1}. \tag{19}$$

Since $Q_{1,2} = Q_{3,4}$, from the combined solution of the equations (18) and (19) we get $i_{\text{out}(v_0 - f_{rf})}$ and then, inserting this value in one of the equations we find the Q-factor of the MBGS and half-width Δv_{MB}.

The advantage of the offered method is that in the measuring process the information signal is influenced by noises only of a bandwidth of the gain spectrum (20-100 MHz), not noises of all bandwidth from MBGS to the carrier (10-20 GHz). Therefore, SNR of the measurements in this case is 10^2-10^3 greater than similar ratio in previously offered methods. Inserting the known and determined by previously mentioned procedures frequency parameters v_P, v_{MB} and Δv_{MB} in (3), we get the Mandelstam-Brillouin maximum gain coefficient value [7].

The presented method was tested in the laboratory of R&D Institute of Applied Electrodynamics, Photonics and Living Systems on the basis of coil of optical fiber Corning SMF-28 6 km long. At the pump power LDI-DFB 1550 5 mW, power of probing sideband components 90 nW we found the SMBS frequency shift is 10,54 GHz, gain coefficient – 20 dB, half-width – 36 MHz. The optical single-sideband modulator is based on a Mach-Zehnder JDS Uniphase OC-192 design. The oscilloscope Agilent InfiniiVision 7000, stabilized power supply PSS-1, the spectrum analyzer FTB 5240-S and the photodiode LSIPD A75 were used [7].

So, a new method for characterization of gain spectrum of SMBS in single-mode optical fiber is presented. It is based on the usage of advantages of the scanning single-sideband modulation method and double-frequency probing method. For conversion of the complex SBS spectrum from optical to the electrical field single-sideband modulation is used. Detection of double-frequency components position in the gain spectrum occurs through the amplitude modulation index of the envelope and the phase difference between envelopes of probing and passing components. The method is characterized by high resolution, SNR of the measurements increased of a two order, simplicity of the offered algorithms for finding the central frequency, Q-factor and Mandelstam-Brillouin maximum gain coefficient. Measurement algorithm is realized on simple and stable experimental setup. Among the methods of measurement inaccuracy decreasing, caused by not completely suppressed upper sideband of the double-frequency radiation, the usage of a chirp fiber Bragg grating, tuned on the suppression of it in the bandwidth of possible position change at scanning, can be considered [7].

3.4. Two-frequency characterization of FBG reflection spectra

FBG represents a longitudinal, periodic variation in the refractive index in the core of an optical fiber [2]. The main parameters of the grating are the distributions of the amplitude and period of the refractive index modulation, as well as the average value of the induced refractive index along the fiber axis. These parameters specify the spectral and dispersion parameters of gratings and, thus, determine their use in different applications of the fiber optics. FBG are widely used fiber lasers and amplifiers, in the fiber systems for measuring physical quantities, optical communication lines, and etc.

One of possible ways to decide specified problems of FBG reflection spectra characterization is based on FBG probing by the two-frequency radiation which average frequency at calibration point is adjusted on the central wavelength of FGB spectrum, and its detune and-or amplitudes difference between components are used as informative factors for definition of enclosed physical field parameter [4, 41]. The two-frequency measurement technique finds more and more appendices in various problems, for example: research of atmospheric gases absorption contours [42,43], measurement of dielectric coverings thickness [44], the analysis of FBG spectrum contour [45], an estimation of communication lines selective devices temperature drift [46], etc. Distinctions consist in parameters of used two-frequency signal or radiation, the requirements shown to their stability and techniques of measuring transformation.

In the given part we will use the two-frequency radiation received by Il'in-Morozov technique in Mach-Zender modulator [4], differing as high spectral cleanliness and stability at admissible change of formation parameters, and possibility of differential frequency simple tuning for use with various FBG characteristics. The specified generalized characteristics meet requirements to construction of probing radiations sources for fiber-optical nets [3]. As a technique of measuring transformation we will choose an integrated technique of the passed through or reflected from FBG two-frequency radiation envelope characteristics analysis.

At FBG contour shift caused by the application of physical fields, there is inequality $R_1 \neq R_2$ and restoration phase opposition of two-frequency radiation components. The kind of an inequality and a phase sign is defined by a direction of FBG contour shift, i.e. increase or reduction of the enclosed field parameter. The analysis of amplitudes and phases of the received components can be spent separately after their allocation by optical filters or time division in the disperse environment; however these methods return us to problems of difficult spectral verification [47]. Therefore it was offered to spend processing of two-frequency radiation envelope [41].

Envelope amplitude U_R defined as:

$$U_R \approx \sqrt{R_1^2 + R_2^2 + 2R_1R_2\cos\left(k\Delta\delta t\right)}, \tag{20}$$

and an instant phase:

$$\phi_R \approx arctg\left\{\frac{\sin\left[\left(\phi_{R_2}-\phi_{R_1}\right)+k\Delta\delta t\right]}{R_1\big/R_2+\cos\left[\left(\phi_{R_2}-\phi_{R_1}\right)+k\Delta\delta t\right]}\right\}, \tag{21}$$

where ϕ_{R_1}, ϕ_{R_2} are accordingly the phases of output components R_1 and R_2.

For processing of the received values on amplitude we will enter modulation factor m:

$$m \approx \sqrt{1+\left(\delta_0+\Delta\delta/2\right)^2}\Big/\sqrt{1+\left(\delta_0-\left(\Delta\delta/2\right)^2\right)}, \tag{22}$$

and on phase – we will find a difference of phases between envelopes of input and output radiations $\Delta\phi$ [41].

The example of the received measuring characteristics of the temperature FBG sensor on amplitude and a phase is presented accordingly on fig. 5,a and fig. 5,b. Analysis of the envelope 2 Δf parameters (22) and (21) made it possible to depict the measurement characteristics for determination of the central frequency of the gain spectrum by its amplitude (Fig. 5, a) and phase difference or sign of the phase difference (Fig. 5, b) between the envelopes at the FBG's input and output, similar to [48]. If the amplitude characteristic of measurements (fig. 5,a) has symmetric character, phase (fig. 5,b) allows resolving shift sign. Advantages of the amplitude characteristic are shown at operation in the field of "zero" detune parameter where there is an area of small signals for the phase characteristic [41].

For practical realization a setup shown in fig. 6 was assembled. Setup consists of the laser LDI-DFB 1550-20/50-T2-SM3-FA-CWP, calibration source Superlum SLD Pilot-4, the oscilloscope Agilent InfiniVision 7000, random waveform signal generator AFG3000, multimeter, MZM JDS Uniphase OC-192 Modulator, stabilized power supply PSS-1, the spectrum analyzer FTB 5240-S, an optical splitter, a circulator, FBG, the photodiodes LSIPD-A75 [59]. The spectra of two-frequency laser radiation is shown in Fig. 7, carried out in a

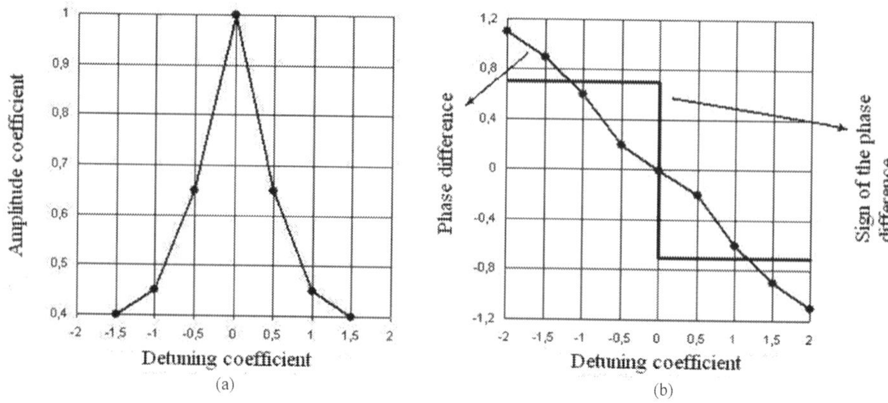

Figure 5. Amplitude (a), phase difference and sign of the phase difference (b) between the envelopes at the FBG's input and output as a function of detuning and of the central frequency of the FBG

Mach-Zehnder modulator by the Il'in-Morozov method for frequencies 2 GHz (fig. 7,*a*) and 8 GHz (fig. 7,*b*) at the output of FBG [60].

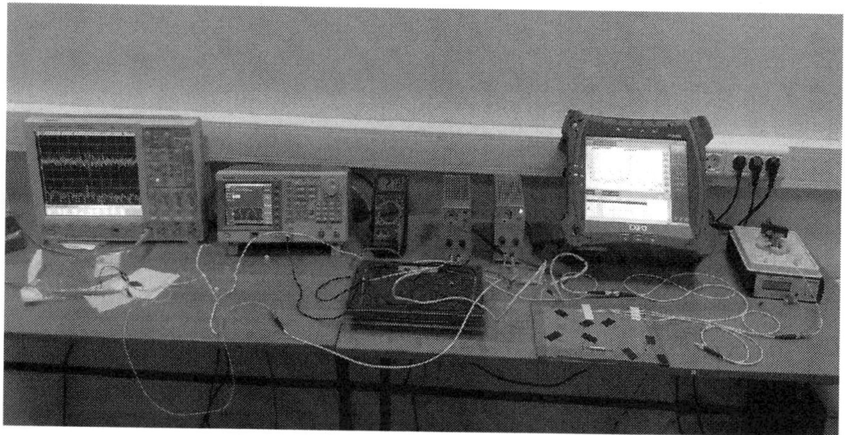

Figure 6. Setup for four-frequency FBG reflection spectra characterization

3.5. Four-frequency characterization of FBG reflection spectra

The analysis presented in [7], [59] shows that from the electrical output signal we can get the image of the optical transfer function at the beat frequency of the two components of the probing signal, which will be determined by the change of their amplitudes and phases. At the moment when center frequency of a probing signal gets to the resonance frequency of a sensor (FBG) amplitudes and modules of the phases of both components are equal, the modulation index of the output signal's envelope is maximum and equal to 1, and the envelope of the signal is matched in phase with the envelope of the probing signal at the input.

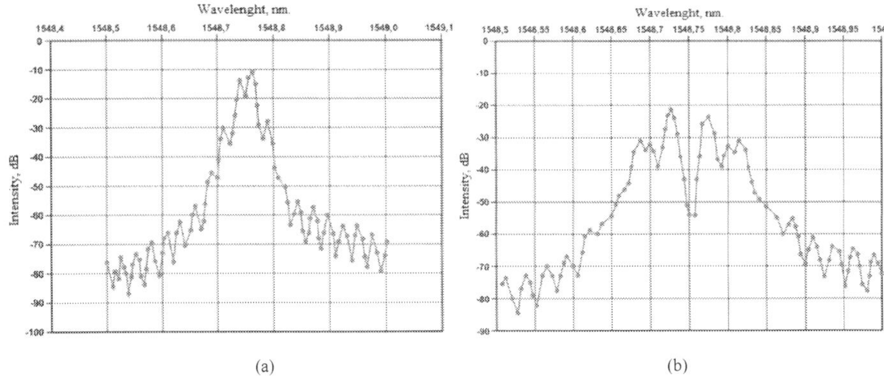

(a)   (b)

Figure 7. The spectra of the two-frequency laser radiation at frequencies 2 GHz (a) and 8 GHz (b)

When the two-frequency probing method is used the maximum measurement sensitivity is achieved by tuning its center frequency to the resonance frequency of a sensor (FBG), and the value of the detuning between the two components should be close to its band pass width at the half-maximum.

Inaccuracy of the measurements will depend on the correctness of maintaining the equality of the amplitudes and modules of the phases of probing signal's components, and on the signal/noise ratio of measurements. The phase measurements in the 30-100 GHz band are a complex task. Therefore, synthesis of poly-harmonic method that does not require processing of the phase information is an important problem. This method has been found and is a four-frequency method with two different average and difference frequencies. For reference, we note that in Section 2.3 we considered the four-frequency method with the overall average and different difference frequencies [7].

On fig.8 block diagram of device for precise four-frequency control method of FBG's resonant frequency is presented. In searching mode from driving generator of controller *10*, which is coupled with TFBG *3*, is coming retuning single frequency signal to the input of amplitude-phase four-frequency converter based on double port electro optic MZM.

Signal with frequency Ω, which corresponds the central peak half width of TFBG *3* or near it, is coming to the control input of amplitude-phase converter. Formed two two-frequency signals with two different average and difference frequencies is coming to the input of TFBG *3*. Output signal from TFBG *3* sends to the input of first and second photo-detectors *4, 7*. Output four-frequency signal from TFBG *3* comes to the first and second selective filters *5, 9*. Controller *10* receives information from first selective filter *5* on the frequency $\Omega1$ and second selective

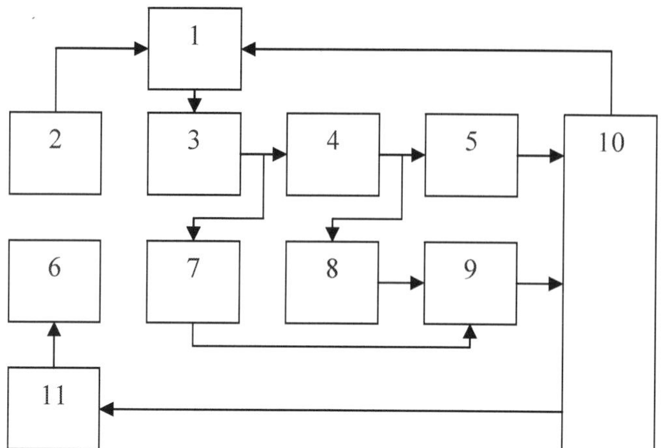

Figure 8. Device for precise four-frequency control method of FBG resonant frequency: 1 – four-frequency converter; 2 – DFB laser; 3 – FBG under test (TFBG); 4, 7 – first and second photo-detectors; 5, 9 – first and second selective filters; 6 – computer; 8 – amplifier; 10 – controller; 11-interface

filter 9 on the frequency $\Omega2$. Searching mode continuous till moment, when modulation index processing in controller detects $U_{\Omega1}U_{\Omega2}{=}0$. At the moment of resonance frequency adjustment output signal from controller 10 comes to the interface 11, and computer 6 performs frequency measuring and begins to monitoring FBG spectra characteristics. Second photo-detector 7 and amplifier 8 are used for calibration between two two-frequency channels.

Fig. 9 shows the dependence of the amplitudes of the envelopes of the signals' beats of the first $U_{\Omega1}$ and the second $U_{\Omega2}$ pair, passed through FBG (left vertical axis), and their difference $U_{\Omega1}$-$U_{\Omega2}$ (right vertical axis) on detuning of FBG pass band (horizontal axis) for the case of supplying a signals with equal amplitude and center frequency matched with central frequency of pass band. Difference frequencies of the pairs $\Omega1$ and $\Omega2$ are not identical and are in the range up to 300 MHz. This allows the use of a narrow-band photo-detector [60].

In the generated pairs of signals passing through the FBG, amplitudes of the several components change according to the direction and value of frequency shift of pass band. When there is a frequency shift of FBG pass band depending on temperature changes, position of generated pair of signals' components with respect to the pass band will change, amplitudes of the envelopes of the pairs' beats will change and the differences between amplitudes of the envelopes of the first and second pairs' beats will change (passed through FBG according to presented dependence $U_{\Omega1}$-$U_{\Omega2}$). In this case, the measurements are taken at frequencies of envelopes, which are in a region of minimal noise of the photo-detectors [60].

Tests of the skilled device have been spent on FBG, made in FORC of the Russian Academy of Sciences (Moscow), calibrated in laboratories PGUTI (Samara), and have shown, that use of two-frequency probing FGB has allowed to reach errors of measurement of temperature 0,01 °C in a range 50 °C [41]. Thus the measurement error was defined basically by error of analogue

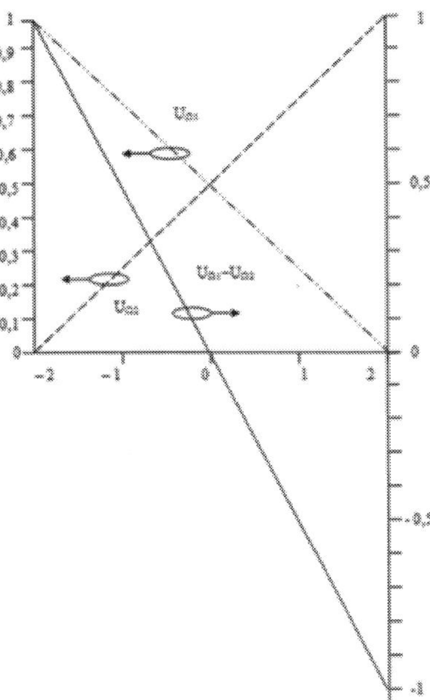

Figure 9. The dependence of the difference between the amplitudes of the envelopes of the signals' beats of the first and the second pair $U_{\Omega 1}U_{\Omega 2}$ on detuning of pass band

to digital coder of the controller *10* for the definition of temperature. The range of measured physical fields (temperature, pressure etc.) is defined by sensitivity of a grating to the measured parameter and value of differential frequencies of probing radiation, so in extreme points of frequency range displacement of a grating making radiations should not leave for level $(0{,}05\text{-}0{,}1)\,R_0$, where R_0 – factor of FGB reflection on the central frequency.

So, to reduce the measurement inaccuracy caused by phase fluctuations of system elements new poly-harmonic method was applied, based on a four-frequency narrow-band measurements without the use of phase analysis. The advantage of this approach is the ability of the measurement in a band up to 300 MHz with narrow-band low-noise photo-detector [59].

3.6. Discussion of results

This part of the chapter is devoted to poly-harmonic characterization of Mandelstam-Brillouin gain contour and FBG reflection spectra for telecommunication applications. Consideration of stimulated Raman scattering (SRS) in this part will be carried out only in general terms because of its negligible impact on the performance of telecommunication none-WDM lines [1]. We discussed the basics of MBGS and FBG spectrums characterization and proposed for the first time two new poly-harmonic methods. First is the four-frequency method with the overall average and different difference frequencies, discussed in section 2.3. Second is a four-frequency method with two different average and difference frequencies, discussed in section 2.5. The advantage of both methods is the absence of the need to measure the phase characteristics of the tested contours. The results of its practical realization proved the results of mathematical modelling.

We call the first as "the method of difference frequency variation" analogically to the methods of frequency and capacity variations for Q-factor measuring. This method can be widely used in different systems for Q-factor measuring as in optical, so in microwave range. For example, in [49] we applied this method to monitoring of cure processes in composite materials. In optical range it can be applied to the measuring of Q-factor of transmitting window of CFBG with phase shift, for example presented in [23,50], which isn't effected to shift of central wavelength. The second method with it simple realization can be widely used in precise sensor monitoring loops in laboratory conditions and special circuits for temperature control in the range of 5-10 °C with accuracy 0,01°C [11,51].

4. Poly-harmonic characterization of SMBS and SRS gain and FBG reflection spectra as the base of software defined down-hole telemetric systems

4.1. State-of-the-art in modern down-hole telemetric systems

In the last decade, fiber optic technologies (FOT) more intensively penetrate the oil and gas industry, especially in those interrelated topics in this industry as seismic, drilling, geophysical surveys in wells and oil and gas extraction. Different measuring system were developed based on FOT, which were characterized by high accuracy and better than electronic devices for the same purpose in terms of mechanical and thermal stability. Measuring elements (fiber or FBG) of such systems are not affected by magnetic and electric fields and are resistant to vibration and shock. Moreover, the measurements produced by the fiber optic wire line systems are not part of the wellbore which requires a power supply and operates only with the light sources.

Despite the fact that the installation of the pilot monitoring systems and production of oil wells, using FOT, began in 1980 and proved to be a number of positive factors, fiber telemetry means (FTM) installed today only in a small part of the hundreds of thousands of oil wells. On the one hand, this is due to the policy of preferential production of "easy" oil, on the other-a frequent haze fiber during hydrogen penetration into the ground and operating at high temperatures and pressures typical of oil production, and finally, the third one is the number of disadvantages of FTM structures themselves. Since stocks of "easy" oil is not unlimited, and emerging technologies allow to create a protected sealed fiber with the ability to work in the fields of temperatures up to 700 °C, the eyes of developers and operational organizations turned again to the possibility of installing FTM in wells by presenting to them the requirements to improve the metrological, feasibility, performance, which resulted in the appearance of the need for the optimization and upgrading of the structural construction of complexes of the specified class. This shows the relevance of the theme of alleged applied research.

The fiber-optic distributed temperature sensor DTS based on SRS is the most common and is used in almost the entire world's oil companies. However, a number of shortcomings associated with the complexity of instrumentation, instability of pulse parameters, the need for complex calibration procedures, highlights the use of its concurrent methods based on SMBS and the methods used pure system on FBG. If the methods based on SMBS can achieve a spatial resolution of 0.1 m, the ones using FBG – 0.01 m.

Another important factor is the cost of FTM installing, which is for SRS equal to 50-100, SBS – 100-150, and FBG – 15-25 thousand of US dollars. It would seem the choice is clear, however, for the complexes at SRS and SMBS simply fiber as sensor is need, but for complexes on FBG it is necessary to "write" gratings in fiber and then to "pack" each of them in order to bring the properties of sensors for temperature, strain, pressure, acoustic noise measuring flow parameters. The FBG requires a significant investment in the pipeline sensor cable and makes it significant costs during the operational phase.

FTM based on FBG are widely used for the construction of point sensors, such as temperature control of the submersible pump and quazi-distributed temperature sensors. It is known their use in wells with temperatures up to 374 °C and pressures up to 220 bar, when cable length is up to 10 km and the error of measurement of pressure – up to 1%. FTM based on SMBS with the analysis in the time domain allow you to measure the temperature and pressure at the same time. The possibility of measuring the pressure reached with an accuracy of 2 , the temperature with an accuracy of 0.1 °C with a spatial resolution of 0,1 m, which is comparable with the characteristics of a wide class of FMT on FBG (1, 0,2 °C) at discrete installation respectively. Discrete setting of FBG is determined by the presence of inter-grating distortion when you install more than 5 arrays in series or a significant appreciation of the interrogator with an increase in the number of gratings.

In recent years, SRS FTM with the analysis in the frequency domain of the company LIOS Technology, GMBH are appeared on the market. These systems are the main competitors of the systems developed by the authors since 2004. The main advantages of said system are lower cost as compared with systems with time analysis, using a more stable cw-laser compared to pulse, the application of heterodyning circuitry for increasing the sensitivity of measurements and thereby improve the metrological characteristics. However, the work of these systems is provided only up to a temperature of 90 °C, at what rate, made by heterodyning, is realized in the secondary electron receiver, so the noise of the photo-detector continue to play a significant role in reducing the metrological characteristics of this FTM type. Transfer of the heterodyning in the optical range is accompanied by significant cost of the system through the use of additional Mach-Zehnder modulators (up to $ 1,000 per channel).

4.2. Background of software defined down-hole telemetric systems

Following conclusions about combined methods were made on the base of Weatherford, Schlumberger, Halliburton companies patent and development analysis:

- Halliburton use the system of wells monitoring on SMBS, the benefits of which are based on modern technological solutions in the field of measurement and application of the reference temperature sensors or pressure to separate multiplicative Brillouin sensor readings, such as, FBG;

- Schlumberger has patents with the use of FBG, which relate only to point measurements. Particularly noteworthy is the patent for an integrated system of down-hole thermometry that uses backscattered signal of one/any of the species: Rayleigh, Raman, Mandelstam-Brillouin scattering;

- the patent portfolio of Weatherford differs from a similar Schlumberger one by presence of a large number of FBG patents, which are using to some extent of complementary Raman and Mandelstam-Brillouin reflectometry, and are also used as point sensors. Particularly noteworthy is the patent for an integrated system of down-hole thermometry that uses backscattered signal of one /any of the species: Rayleigh, Raman, Mandelstam-Brillouin obtained for 7 years before such Schlumberger one.

In addition, it should be noted kinds of systems combined in pairs – Raman and Mandelstam-Brillouin, Raman and Rayleigh. We did not find projects that would take advantage of all three types of measurement procedures simultaneously. Modern technology, including patent solutions of our research group lead to the formation of parallel procedures in the fiber response of different nature (Raman, Mandelstam-Brillouin, Rayleigh and reflection from FBG) on the temperature and pressure in the well and make a universal procedure of poly-harmonic sensing responses, given their similar quasi-resonance character. These factors determine the urgency of developing a fiber optic down-hole telemetry system based on a combined non-linear reflectometry.

Scientific novelty of the research is to create a scientifically based methodological basis for the construction and implementation of technical and algorithmic solutions for down-hole fiber optic telemetry systems based on a combined nonlinear reflectometry, including methods and means for:

- formation of quasi-resonant nonlinear response (Raman scattering, Mandelstam-Brillouin scattering and reflection from FBG) in the fiber, carrying information about the distribution of temperatures and pressures;

- probing of Raman, Mandelstam-Brillouin and Bragg structures based on poly-harmonic radiation and quasi-coherent Rayleigh scattering poly-harmonic registration of temperature and pressure profiles;

- design of algorithms and software procedures for measuring the temperature and pressure, and flow rate of the liquid component composition, including the presence of occluded gas, based on the information obtained from the distributed and point sensors.

4.3. Planned research directions

Planned methods and approaches of down-hole fiber optic telemetry development based on the software-defined combined nonlinear reflectometry based on unity structures formed feedback from optical fiber to an external influence – temperature, pressure, flow parameters of the liquid (crude oil). If Mandelstam-Brillouin and Raman scatterings are carrying distributed information of the measured parameters (temperature, pressure), the FBG allows to receive its point localization and flow velocity data. For more information OTDR with Rayleigh scattering can be used and characterized to the Mandelstam-Brillouin ones by Landau-Plyachek relation. So way it seems reasonable to use a single radiation source for getting fiber-forming response to external stimuli, forming its special signal shape or spectra, optimized for recording spectrally separated responses from various nonlinear effects and reflections, and monitoring their bound to the central wavelength of the radiation. Some pair wise effects of such an implementation are known. Comprehensive option to generate and use of responses from the three types of scattering and reflection from the Bragg gratings not yet been studied. Its implementation could put information redundancy in the process of measuring, the use of which would improve the metrological characteristics systems being developed.

The second approach is based on the poly-harmonic probing and resonance response obtaining. In recent years, significant progress in terms of accuracy and resolution of measurements, as well as practicality, was registered in the use of narrowband poly-harmonic technology for characterization of contour spectrums that makes them competitive for the above mentioned methods in metrological characteristics, ease and cost of implementation. Their main advantage is that no measurements in the resonance region of spectral characteristics are necessary, that allows eliminating the influence of power instability of probing laser radiation and to detect information on the difference inter-harmonic frequency in the region with small noise level of photo-detector. The above-mentioned circumstances determine the relevance of the topic and the scientific and technological objectives to develop poly-harmonic methods and tools for the analysis of the spectral characteristics intended for separate registration of physical fields of different nature (temperature, pressure, flow parameters of the three-phase) and the construction on the basis of their optoelectronic down-hole telemetry using complex effects of nonlinear reflectometry in their software-defined domain.

The third approach concerns the structure of the construction of down-hole flow-meter. Over the past five years, down-hole monitoring systems installations are significantly increased. More than 90% are deep and complex branched wells. Traditionally, continuous monitoring is primarily used to control the pressure and temperature in the borehole. Thanks to the development and establishment of fiber optic down-hole flow-meters it became possible to measure the productivity of wells in its branches. Sophisticated flow structure defines the requirements for the construction of the flow-meter, however, note that its primary purpose is the definition of a flow within the wellbore, and not on the surface. To construct the flow set point is used temperature and acoustic pressure sensors. We have proposed to use for such purposes FBG-PS, FBG with a phase shift, which is characterized by high resolution and the ability to check-in without a shift of the center wavelength, as shown in several studies [11,25,48,50].

4.4. Discussion of results

FOT systems and FTM for down-hole telemetry – a developing area of science and technology in Russia, which would create a competition international manufacturers of similar systems for the oil and gas industry and solve the problem of import substitution, significantly reduce the cost of the components used, displace traditional systems on electronic components. Based on the analysis of advanced domestic and foreign developments at the level of patents in the field of fiber optic systems of down-hole telemetry shows the relevance and scientific novelty of research areas, which determines the need to develop an integrated fiber optic down-hole telemetry system which is used to record the measured parameters for all kinds of scatterings assessments and distributed FBG – for point and the quasi ratings, including to resolve the multiplicative response to temperature and deformation (pressure) for FBG and Brillouin systems and error analysis in Raman systems. The studies will be established scientifically based methodological basis for building and technical and algorithmic solutions for the down-hole fiber optic telemetry systems based on a comprehensive nonlinear reflectometry and universal poly-harmonic probing of generated responses. These results allow to significantly improve the metrological characteristics of systems, including the reproducibility of results, because it will be used the measurement results with high redundancy on the basis of three or four feedback mechanisms of different nature of the optical fiber at the same environmental exposure. Scientific and methodological foundations and principles of systems can be used to monitor not only the down-hole structures, but any extended engineering structures and natural systems.

Proposed by us for the first time the concept, approaches and methods for its implementation allow reasonably formulate and solve the problem of creation of scientific and technical basis for designing software-defined down-hole fiber optic telemetry systems based on the combination of nonlinear reflectometry with improved metrological characteristics of poly-harmonic probing of sensing structures, including the removal of multiplicative and measurement errors caused by instability of the forming radiations in wide frequency range.

5. Elements of perspective reflectometeric systems with poly-harmonic probing

On the basis of two-frequency signal information structure investigation [4], [61], [62] with the purpose to find out main principles of combined interaction of its instantaneous values of amplitude, phase and frequency with arbitrary contour has been defined:

- instantaneous phase of two-frequency signal has a saw-tooth dependence. Speed of instantaneous phase changes is defined by components amplitude ratio. If $A_1/A_2=1$ the maximal speed of phase changing is observed. When two-frequency envelope has its minimum value, instantaneous phase has a shift, which depends on harmonic components amplitude ratio. If the amplitudes of two-frequency signal components are equal, phase shift is equal to π ;

- it has been shown, that instantaneous frequency of two-frequency signal changes with deviation of it components amplitude ratio in range limited by signal frequencies. In case of components amplitude equality, instantaneous frequency coincides with average frequency of two-frequency signal $\omega_a=(\omega_1+\omega_2)/2$. When the envelope of two-frequency signal has its minimum value, frequency overriding occurs, which depends on harmonic components amplitudes ratio. When amplitudes of two-frequency components are equal, value of frequency overriding is infinity;

- modulation coefficient has a linear dependence on amplitudes ratio of two-frequency components. When amplitudes of frequency components are equal, modulation coefficient is maximal and equals unit value.

All obtained dependences can be used in symmetric poly-harmonic reflectometeric systems as informative and directive functions and in order to get information from selective optical fiber structures, like SRS, SMBS and FBG contours and so on [61], [62].

A joint analysis of given information allows us to determine the range of individual tasks that must be addressed when constructing symmetrical poly-harmonic reflectometeric systems. These tasks include following problems: integral self heterodyning $(A_0=0,A_1=A_2)$; integral super heterodyning $(A_0>>A_1=A_2)$; differential self heterodyning $(A_0=0,A_1\neq A_2)$; differential super heterodyning $(A_0>>A_1\neq A_2)$; receiving and processing frequency information $(\omega_0\neq(\omega_1+\omega_2)/2)$; receiving and processing of phase information $(\phi_0\neq\phi_1=\phi_2,\ \phi_1\neq\phi_2)$; receiving and processing the polarization $(A_0\perp A_1\|\ A_2)$ and spatial information; information about instability (A0≠0). Given the diversity of the set measuring tasks, their decisions, are presented further in this part of chapter, describing the methods, tools and systems parameters of means to get poly-harmonic radiation on the base of dual-drive MZM, scanning and poly-harmonic (more than four harmonics) methods for SMBS characterization, designing of notch filters for blocking of elastic Rayleigh scattering in SRS and SMBS measurements [61], [62].

Additionally Raman and Mandelstam-Brillouin scattering in sensor nets so as a FBG reflection carry vast amount information about fiber conditions but sometime have low energy level. That's why it's one more cause to detect these types of scattering with high SNR and determine their properties. Applying of photo mixing allows significantly increase the reflectometeric system sensitivity under the condition of weak signals and receives information from fre-quency pushing of backscattered signal spectrum. We offered previously to use two-frequency heterodyne and the second nonlinear receiver in the structure of reflectometers and discuss these questions in [4] [56].

5.1. Numerical simulation of poly-harmonic conversion in dual-drive Mach-Zehnder modulator

The numerical simulation in dual-drive Mach-Zehnder modulator [31,52] was carried out for signals with following parameters: amplitude of RF modulating signals was equal $V_1=V_2=V_\pi$;

DC bias voltage applied to arm one and two was $V_{bias\ 1}=U_\pi/2$ and $V_{bias\ 2}=3U_\pi/2$, respectively; phase difference between RF signals of two arms was changing. Resulted spectrums of output MZM signal for different phase difference between RF signals of two arms are presented on fig. 10-14, [62].

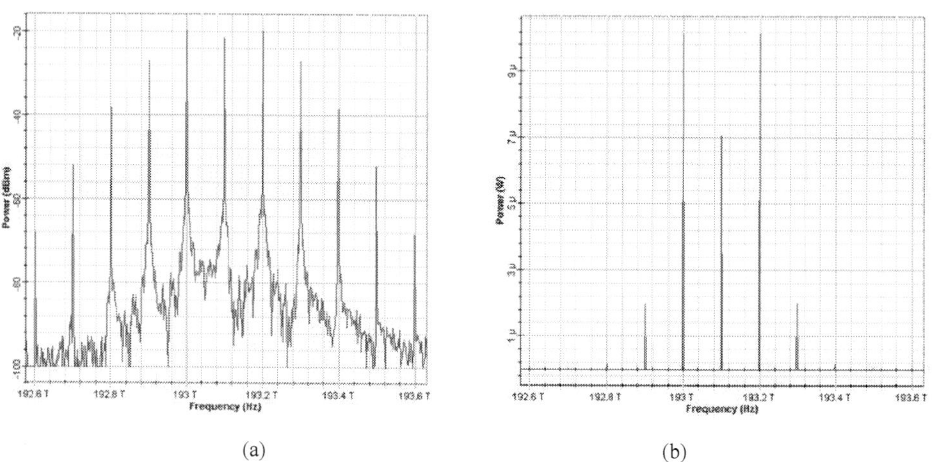

(a)   (b)

Figure 10. MZM output signal spectrum for the case of $\Delta\psi=0°$ (a – dBm, b – W)

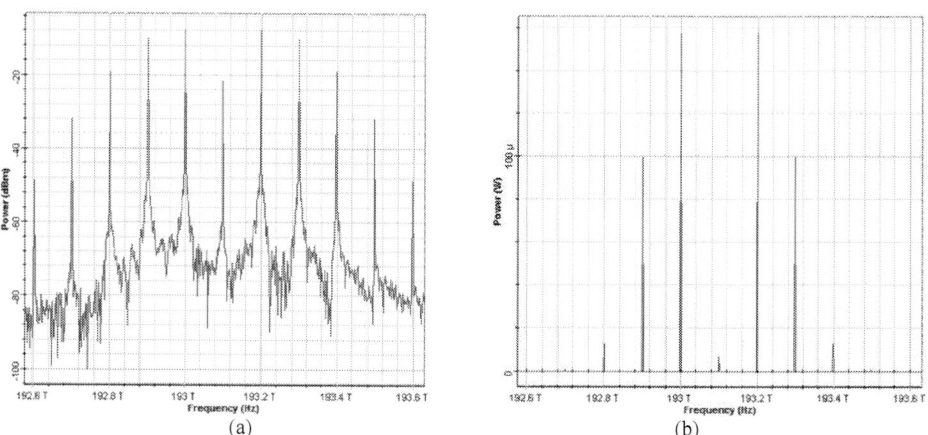

(a)   (b)

Figure 11. MZM output signal spectrum for the case of $\Delta\psi=45°$ (a – dBm, b – W)

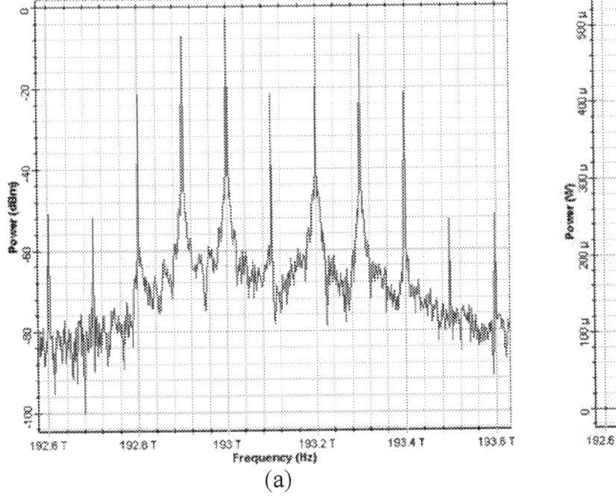 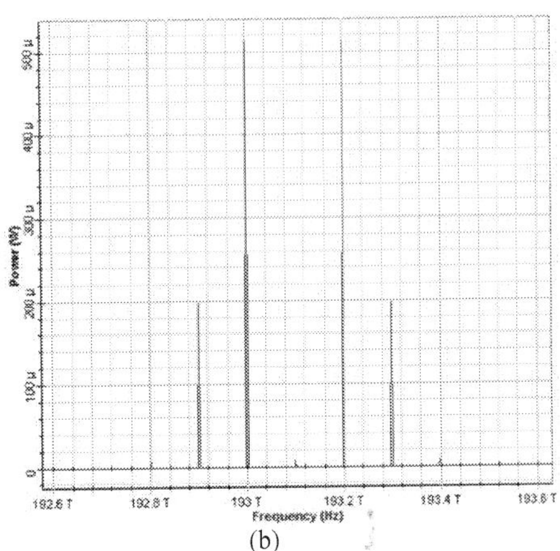

Figure 12. MZM output signal spectrum for the case of $\Delta\psi=90°$ (a – dBm, b – W)

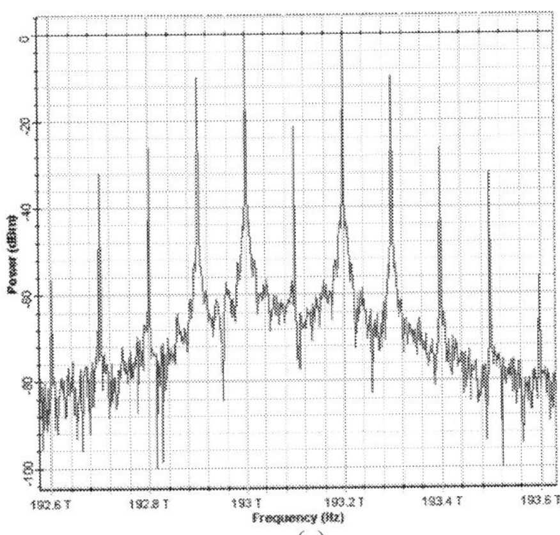 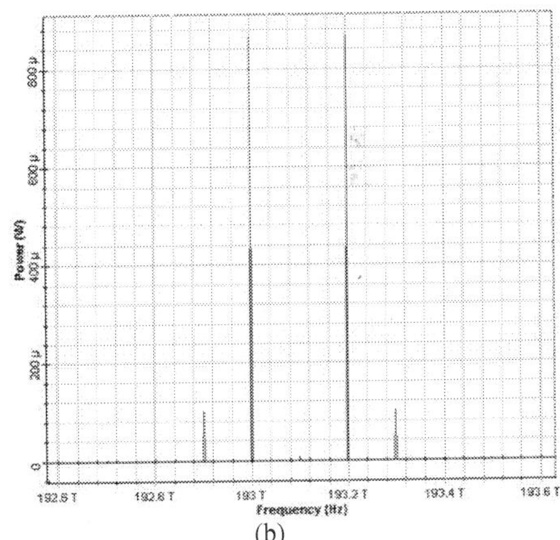

Figure 13. MZM output signal spectrum for the case of $\Delta\psi=135°$ (a – dBm, b – W)

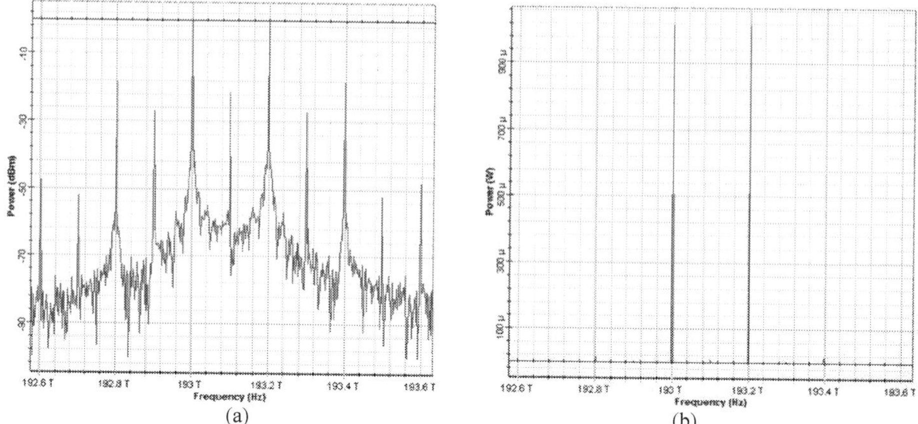

Figure 14. MZM output signal spectrum for the case of Δψ=180° (a – dBm, b – W)

Signal, containing only two spectral components, was obtained for a case of phase difference Δψ=180°. The career and the other side components were suppressed more than 20 dB. Signals, containing four spectral components, were obtained for a cases of phase difference Δψ=45° and Δψ=90°. In order to get four-frequency signal with equal components differential regime of dual-drive Mach-Zehnder modulator has to be used. But in this case problem of not full suppression of initial carrier (one-frequency) radiation may arise. Utilization of notch filter for purpose of carrier suppression will be discussed in 4.5.

5.2. Two-frequency scanning for SMBS gain spectra characterization

The questions of interaction between scanning two-frequency probing radiation and SMBS gain spectra are considered. If the laser wavelength is modulated by the frequency:

$$v(t) = v_0 + \Delta v \cos \Omega t, \tag{23}$$

where v_0 is initial laser frequency, Δv is modulation rate, than the intensity of SMBS gain spectra signal will could define from the equation:

$$I(t) = I_0 T(v_0) + I_0 T^{(1)}(v_0) \Delta v \cos \Omega t - 0,25 I_0 T^{(2)}(v_0) \Delta v^2 \cos 2\Omega t + ..., \tag{24}$$

where $T(v_0)$, $T^{(1)}(v_0)$, $T^{(2)}(v_0)$ are SMBS gain spectra transmittance (the reverse value of reflection) and its derivative by , correspondingly, I_0 is source intensity. Selective amplifier, which is tuned to the frequency Ω or 2Ω, allowed signal recovering from SMBS GAIN

SPECTRA even with low gain and eliminate high level constant component $I_0 T(v_0)$ of the signal [10].

Frequency modulation allows sensitivity increasing of spectral photometric method at least in two orders with good SNR. Spectral resolution for derivative method is defined by frequency modulation rate Δv. Time resolution defines by modulation frequency Ω. Spectrum recovering accurate within constant component is accomplished by sequenced signal integration [8].

During two-frequency probing SMBS gain spectra is exploring by two signals with central frequencies $v_0 - v_T$ and $v_0 + v_T$. When $v_T = \Delta v / 2$ modulation of sweeping frequencies is defined by:

$$v_1(t) = v_0 - \frac{\Delta v}{2}\left(1 + \cos\left(\Omega_0 t + \frac{\beta t^2}{2}\right)\right), v_2(t) = v_0 + \frac{\Delta v}{2}\left(1 + \cos\left(\Omega_0 t + \frac{\beta t^2}{2}\right)\right). \quad (25)$$

Interaction between frequency modulated laser radiation and SMBS gain spectra causes power amplitude modulation (AM) of receiving signal. Modulation rate of such signal is straight to Q-factor, and AM envelope follows linear-frequency modulation (LFM). When central laser wavelength is tuned to central peak of SMBS gain spectra and frequency deviation equals spectrum half width, initial frequency of amplitude modulation is Ω_0, and its frequency deviation is $2\Delta\Omega$ [10].

In work has been held modeling of signals processing for such system for direct and coherent detection. Usually in frequency ranging for extraction low frequency signal of telemetric frequencies which carries information of SMBS gain spectra properties within analyzing distance R in reception path received signal mixes with reference signal. In all cases measurement of subcarrier frequency increment $\Delta\Omega_R = 2\beta R / c$ is performed by means of registration diversity between subcarrier frequencies of transmitted and received signals. Simulation has shown that under matching the demands for scanning SMBS gain spectra and LFM chirp resolution of spatial measurements, more strict conditions are applied by SMBS gain spectra properties. That causes decrease of spatial resolution. In case of direct detection of double frequency reflected signal auto heterodyning is performed with spatial coincidence of mixing rays. In that case signal spectrum transfers to the zone with low noise level of photo detector. Range control is performed by measurement the reduplicated frequency changes of modulating LFM chirp signal, and output signal frequency from intermediate amplifier of receiving system. SMBS gain spectra parameters control is performed by measurement of signal power level at that frequency. For short fibers, which require bands of receiving paths in the range 20-100 MHz, sensitivity increasing may total two orders [8].

In some works has been shown the possibility of using additional signals, which were none selectively influenced by testing SMBS gain spectra for increasing metrological properties of a system. In case of two-frequency source frequency modulation could be accomplished by two signals. The first confirms the demands for SMBS gain spectra scanning and LFM chirping, the second represents modulating signal with constant frequency within the limits of SMBS gain spectra with amplitude two orders less than the scanning signal amplitude. In that case measuring signal will contain two items. The first $U_1 \approx k_1 f(I_0, T, R)\phi(\tau)$ depends on SMBS gain spectra optical properties, gauging equipment parameters and etc. The second item U_2, mostly is not under the influence of SMBS gain spectra, only from component $k_2 f(I_0, T, R)$. After normalization U_1 regarding U_2 for distance R (simultaneous signal selection) and assuming that $k_1=k_2$, we will get equation, which depends only on SMBS gain spectra properties $\phi(\tau)$.

Thus, two-frequency SMBS gain spectra scanning allows system increasing performance both in direct detection mode and normalizing mode, because reference signal parameters are determined and are not the signals of spurious modulation [8].

5.3. Multiple frequencies probing for SMBS gain spectra characterization

Analysis has shown that the main demand for SMBS gain spectra high precision measurements is high synchronization of phase, frequency, amplitude and polarization of probe signals. Sometimes it is impossible to fulfill those conditions even with the use of digital frequency synthesizer and locked-in lasers. Therefore the task of getting two-frequency and two-band oscillations with multiple frequencies and high synchronization is of current importance. Let us consider such method based on amplitude-phase conversion (Il'in-Morozov's method [4]).

Method is based on phase commutation by 180° of amplitude modulated signal at the moment when its envelope has minimum value. The basic task of research is defining the form and parameters of modulating signal $S(t)$ to receive an output two-frequency oscillation with suppressed carrier frequency ω_0. Total suppression of side components with $n \geq 3$ could be achieved with the use amplitude modulating oscillation $S(t)=S_0|\sin\Omega t|$. Than the resulting signal with amplitude modulation index b would have the following spectrum [8]:

$$e(t) = \frac{2E_0}{\pi}(1-b)\sum_n \frac{1}{n}\left\{\cos\left(\omega_0+n\Omega\right)t - \cos\left(\omega_0-n\Omega\right)t\right\} + \frac{\pi E_0 b}{4}\left\{\cos\left(\omega_0+\Omega\right)t - \cos\left(\omega_0-\Omega\right)t\right\}. \qquad (26)$$

Amplitude of spectral components will be defined by Fourier series indexes, and for $n=1$ $E_1=[2E_0/\pi][1b]+[\pi E_0 b/4]$, for $n\geq 3$ $E_n=[2E_0/\pi n][1b]$. When $b_{opt}=1$ spectrum contains two useful components with frequencies $\omega_0\pm\Omega$, and the spurious products are suppressed. When the modulation index varies between $(0,7-1)b_{opt}$ output signal nonlinear-distortions coefficient would be less than 1%.

Modulation signals given previously could be used for forming symmetric double band spectrum for multi frequency measurement systems. To realize them it is essential to solve equations set (25) for indexes, varying not only amplitude modulation index but also the value of phase commutation θ. Using the oscillation $S(t)=S_0|\sin\Omega t|$ we will get the following equations for amplitudes of spectral components [8]:

$$E_0 = \frac{E}{2}(1+\cos\theta); E_1 = E\left(\frac{1-b}{\pi} + \frac{\pi b}{8}\right)(1-\cos\theta); \qquad (27)$$

$$E_n = \frac{E(1-\cos\theta)}{\pi n}(1-b)\ \text{for n=3, 5, 7...;}\ E_n = \frac{bE(1+\cos\theta)}{1-n^2}\ \text{for n=2, 4, 6 ...} \qquad (28)$$

Equations (26) (27) allows to define spectral distribution of resulting signal with any b and θ. However, from point of view of simple technical realization we should look for forming oscillations without phase commutation using synthesis of oscillations with complex harmonic composition with k components, for example [8]:

$$S(t) = \sum_{k=1}^{\infty} S_k \cos\ (2k\ \Omega\ t + \pi), E_n = \frac{2E}{\pi}\left[\frac{1}{n} - \sum_{k=1}^{\infty}\frac{mk}{2}\left(\frac{1}{n+2k} + \frac{1}{n-2k}\right)\right], \qquad (29)$$

where S_k are the partial amplitudes, E_n are the equations for Fourier indexes of its spectrum, m_k are partial indexes of amplitude modulation. Such approach of looking for forming oscillations will also allowed to take into account influence of real devices modulation characteristics nonlinearities on spectral distribution of output radiation, which are used to realize amplitude-phase conversion.

Differential frequency of two-frequency oscillation (25) when b_{opt}=1 is defined by frequency Ω of phase commutation θ. Its stability unambiguously concerned with frequency stability of driving voltages and instabilities of commutations devices. Really achievable value of differential frequency's instability with thermo-stating of driving generators is 10^8. During differential frequency's retuning, which is required in some types of measurements and rather simple realized with the use of amplitude phase conversion, minimum frequency shift is determined by modulators gain slope, and maximum frequency shift is determined by higher cutoff frequency of modulator and correlation between modulator and modulated frequencies [8].

Energy equality of side lobes and the effectiveness of their conversion are of great importance in multiple frequencies systems. Using the derived equations for the spectrums of output oscillations and taking into consideration properties of amplitude-phase conversion, i.e. using the additional power of amplitude modulation and phase commutation for forming the side lobes, we could determine, that power of the last one is nearly 60% of initial single frequency oscillation, and conversion index equals unit value without taking into account the loss in real modulators. Moreover multi frequency radiation allows any pair of formed frequencies using for differential analysis without retuning central laser frequency [8].

As it has been mentioned above retuning both laser frequency and difference frequency between spectral components either is very complex or is carried out nonlinearly. Simple way of SMBS gain spectra analysis could be realized with its multiple frequencies probing (fig.15).

If initial SMBS gain spectra probing by two-frequency signal with components A_1 and A_2 did not allowed the measurement of the central wavelength of SMBS gain spectra we should change the conditions of amplitude phase forming of probe signal. Solving the equation (25) with E_3=0 and limiting to four components we will get multiple frequencies signal with $A_1, A_2,$

A_5 and A_6. At that components A_1, A_2, A_5 lie on the left slope and component A_6 lies on the right. It could be defined from phase characteristics. As far as $A_2>A_6$ what is evident from envelope analysis of that pair cutout by band pass filter we should change the parameters of amplitude phase conversion of probe signal one more time [10].

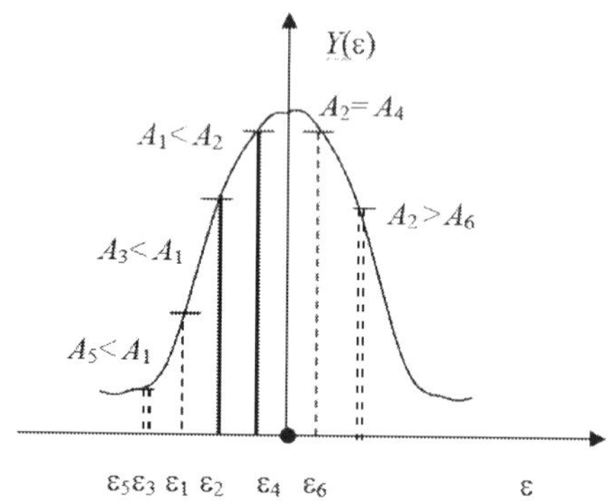

Figure 15. Multiple frequencies probing

At that components A_1, A_2, A_5 lie on the left slope and component A_6 lies on the right. It could be defined from phase characteristics. As far as $A_2>A_6$ what is evident from envelope analysis of that pair cutout by band pass filter we should change the parameters of amplitude phase conversion of probe signal one more time. From equation (25) with $E_5=0$ and limiting to four components will get multi frequency signal with A_1, A_2, A_3 and A_4, and with $A_2=A_4$, for example, what corresponds the tuning of that pair to the SMBS gain spectra central wavelength [10].

There are possible three types of analysis: analysis of each component alone (differential analysis), analysis of the envelope of each double frequency pair (integral-differential analysis), and the analysis of energy correlation of all components (integral analysis). All of these methods are possible and correspond to single, two-, and multiple frequencies probing of SMBS gain spectra with limited number of optical band pass filters which are conjugated with detuning frequency between spectral components [8].

In more complex way of searching the spectral center and half width we should solve an equation set with amplitude and phase coefficients. Process of multiple frequencies analysis of SMBS gain spectra consists on solving the following equations set [10]

$$[\mathbf{D}] = [\mathbf{A}] \times [\mathbf{E}] \times [\mathbf{E}]^*, \tag{30}$$

where [**D**] is matrix of output photo detector currents value with frequencies $k\Omega$; [**A**] is matrix describing required SMBS gain spectra components in band $\Delta\omega$; [**E**] is matrix describing spectrum of probe multiple frequencies oscillation with frequencies $\{\omega_0 \pm k\Omega\}\{\Delta\omega\}$, **[E]** is complex conjugate matrix with [E].

On the first step SMBS gain spectra is probing by two-frequency oscillation, amplitude of the components with $n \geq 3$ equals zero. At the same time hitting the SMBS gain spectra central peak is not required. From the analysis of photo detector components on frequency Ω we can define SMBS gain spectra slope, its steepness, frequency shift from central peak of reflection[8]. During the second step we use multi frequency oscillation, which component with $n=3$ does not equal zero. During the third step we use multi frequency oscillation, which component with $n=3$ equals zero, and component with $n=5$ does not equal zero. At the same time we analyze amplitudes of photo detector components with frequencies Ω, 2Ω, 3Ω. As far as the amplitude of probe signal components is known and stable, as it has been shown before, we could with given precision define SMBS gain spectra, such as central wavelength, steepness, symmetry of the spectrum curve, and etc. At the third step varying the frequency Ω, number of components n and working filters with frequencies $k\Omega$, we could optimize SMBS gain spectra analysis and tune the central frequency of probe laser to the reflection peak to take a tracking signal [8].

5.4. Some remarks for SNR of spectra characterization

Let's consider some different situations with single, two-, and multiple frequencies probing of real selective structures with limited number of optical band pass filters which are conjugated with detuning frequency between spectral components [8]. SMBS gain spectra characterization from optical to the electrical field by single-sideband amplitude modulated radiation, in which the upper sideband is suppressed and single frequency probing is used in [6]. SMBS gain spectra characterization in single-mode optical fiber was presented in [7]. It is based on the use of advantages of single-sideband modulation and two-frequency probing radiation, which gives possibility of transfer the data signal's spectrum in the low noise region of a photo detector. Two-frequency scanning and multiple frequencies probing of SMBS gain spectra are discussed in sections 4.2-4.3 and were partly considered in [8]. SMBS gain spectra characterization with two-frequency heterodyne and single frequency scanning was discussed in [4].

As we can see from [6,7] the needed bandwidth of photo detector is determined by Mandelstam-Brillouin frequency shift and equal to 10-20 GHz. One version of signal processing may be realized on the envelope of differential frequency $2\Delta f$ [7]. So the needed bandwidth of photo detector is determined by Mandelstam-Brillouin gain and equal to 20-100 MHz. The same value of bandwidth is a feature of methods which are presented now and discussed in [8]. Additional advantage of method [4] is heterodyning. Applying of photo heterodyning allows significantly increase the system sensitivity under the condition of weak signals and receives information from frequency pushing of counter propagated signal spectrum [8].

If we don't go into details of the physical nature phenomena, the basic noises level of the radiance receivers is more than background noises level and determine the detect ability of receiving signal. The gain in SNR relative to single frequency measurements can be calculated as [63]

$$G = \int_0^{\Delta f} S(f)\,df \Bigg/ \int_{f_0+\Delta f}^{f_0-\Delta f} S(f)\,df \qquad (31)$$

where $S(f)$ – spectral density of radiance receiver noises, Δf – bandwidth of photo receiver.

The gain will be determined by different nature and level of noise in different frequency regions (fig. 16)

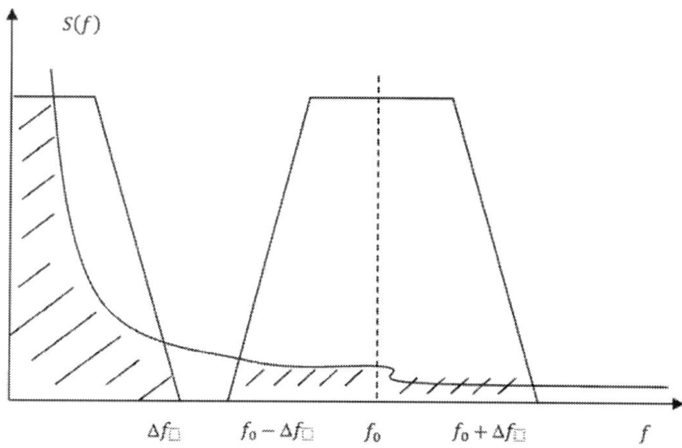

Figure 16. To SNR gain remarks

There are current noises with distribution type $1/f$ and other powerful noises of low frequency nature for region $\{0...\Delta f\}$ for variants 2,4. There are thermal agitations and shot noises with low intension for region $\{f_0\ \Delta f... f_0+\Delta f\}$ for present methods and heterodyning 9. For little distance routes the gain in SNR can be mount to 1–2 orders. All these summaries are correct and for multiple frequencies probing [8]. The results are much closed to [53].

5.5. Notch filters for suppressing of Rayleigh scattering

The Raman scattering of light is known to be accompanied by the emergence of spectral components shifted in terms of frequency [15]. The number and the spectral positions of these lines depend on the structural characteristics of the scattering medium. It is known that the intensity of the anti-Stokes line is very low (30 dB weaker than the amplitude of elastic Rayleigh scattering); therefore, registration of the ratio of intensities of the Stokes and anti-Stokes components is a difficult task. In addition, the power of the probing radiation should not exceed several watts to avoid such nonlinear effects as the stimulated Raman scattering and the stimulated Mandelstam-Brillouin scattering. The principle of the SMBS distributed temperature sensors is based on the Landau-Placzek ratio where the temperature-insensitive

Rayleigh signal provides a reference measurement of the fiber background losses and the temperature-sensitive Brillouin signal provides a measurement of the temperature [16]. It is estimated that the Rayleigh signal must be suppressed by at least 33 dB to reduce the effects of SRN to an acceptable level. In these circumstances, in order to achieve high accuracy of temperature measurement is required to choose the optimal method of Rayleigh scattering filtration and separation of the desired signal with minimal loss of information.

The separation of the relatively weak Brillouin signal from the Rayleigh has been reported using a bulk Fabry-Pérot interferometer (FPI) [54], which is lossy and expensive. A fiber Mach–Zehnder interferometer (MZI) has been demonstrated to achieve 27-dB Rayleigh rejection using a-switched fiber laser probe of 1.5-GHz bandwidth [55]. However, with a narrow-band source (20 MHz) and the resultant increased coherent Rayleigh noise, this is insufficient to achieve a temperature resolution of 1 °C. In [16] reports on the use of a narrow-band fiber Bragg grating filter (FBGF), which achieves recovery of the Brillouin signal by suppressing the Rayleigh signal. In this way, the Brillouin light path is subject to minimum attenuation and is frequency independent. Typically, the filter uses fiber Bragg gratings that produce narrow-band signals of the Stokes and anti-Stokes components [14], which are then sent on different channels registration. However, this method makes great demands on the quality of lattices: a spectral width (which should be as much as possible), the reflection coefficient, losses, and so on. Unfortunately, even under optimal conditions, a significant portion of the desired signal is lost during filtration. So, FBG is more effective to suppress Rayleigh scattering, then to detail Stokes and anti-Stokes components.

In [14,15] FBG with a spectral width of 0.5 nm reflection used to suppress the central Rayleigh line, which allowed to minimize the noise introduced by Rayleigh scattering without a substantial reduction in the integrated intensity of Raman shift lines. After filtering, the signal components of the SRS were separated by directional couplers with the appropriate wave-lengths and sent to pin-photodiodes. It was not possible in [16] to obtain a single grating that would provide rejection of Rayleigh line. Hence, the filter comprises two cascaded gratings separated with an isolator to prevent the formation of an étalon. The rating characteristics were all supplied to the following specifications: bandwidth 0.13 nm, reflectivity 0.98%. In this way the FBG offers minimum attenuation to the two Brillouin sidebands, which, therefore, pass relatively unaffected through the filter.

We developed a new method of FBG writing on the basis of phase mask and restrictive object positioned between mask and fiber. Such optical scheme was realized and we got FBG, which allowed measurements of intensity of the Raman or Mandelstam-Brillouin scattering components in a wide spectral range with the minimum losses; the sensitivity of a conventional InGaAs photo-detector turned out to be sufficient for these measurements. This filter suppressed the central area of the spectrum at the wavelength of 1552,6 nm and transmitted the anti-Stokes and Stokes lines of the Raman or Brillouin scattering. The spectra of FBG for Rayleigh line filtration in SMBS gain spectra characterization, obtained by the spectrum analyzer 5240 FTB-S with a resolution of 2 pm, is shown on fig. 17.

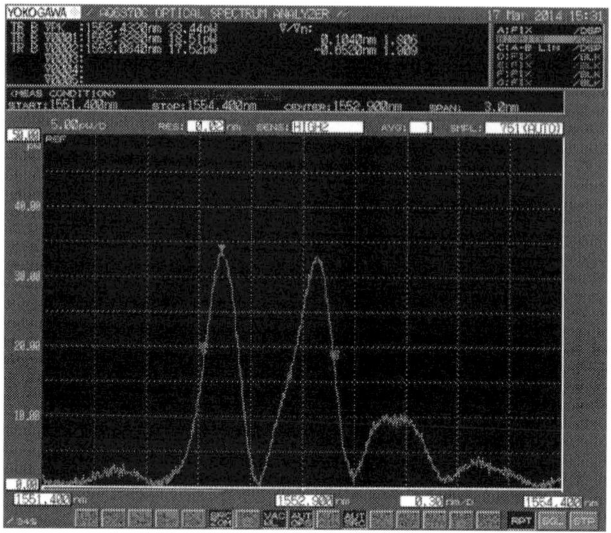

Figure 17. Spectra of FBG filter

Figure 18. Setup for FBG recording

FBG for the experiments was manufactured by a given method of continuous recording with restrictive object in the setup, shown in Fig. 18, at the R&D Institute of Applied Electrodynamics, Photonics and Living Systems. The setup for FBG recording was made in Novosibrsk State University on the base of amplitude-modulated ultraviolet laser with the conversion of the second harmonic on the Ar+. Laser was focused on the core of the aged in hydrogen-doped germanium-silicon fiber (SMF-28).

5.6. Discussion of results

On the basis of two-frequency signal information structure investigation main principles of combined interaction of its instantaneous values of amplitude, phase and frequency with arbitrary contour has been defined. The numerical simulation in dual-drive MZM was carried out and different poly-harmonic spectrums were realized. Novel methods for multiple frequency characterization of stimulated SMBS gain spectrum in single-mode optical fiber are presented. This method is based on the usage of multiple frequencies probing radiation [8]. For conversion of the complex SMBS gain spectra from optical to the electrical field single-sideband modulation, direct or heterodyne detection are used. Determining of a multiple frequencies positions over the gain spectrum occurs through the amplitude modulation index of the envelope or the phase difference between envelopes of probing and passing components. The methods are characterized by high resolution, SNR of the measurements increased of a two order, simplicity of the offered algorithms for finding the central frequency, Q-factor and SMBS maximum gain coefficient. Measurement algorithm is realized on simple and stable experimental setup. Applying of photo heterodyning allows significantly increase the system sensitivity under the condition of weak signals and receives information from frequency pushing of counter propagated signal spectrum. We developed a new method of FBG writing on the basis of phase mask and restrictive object positioned between mask and fiber. Such optical scheme was realized and we got FBG, which allowed measurements of intensity of the Raman or Mandelstam-Brillouin scattering components in a wide spectral range with the minimum losses and deep suppression of Rayleigh scattering line.

6. Conclusion

We reviewed the principle and the applications of the poly-harmonic (two-frequency or four-frequency) cw laser systems for characterization of Mandelstam-Brillouin gain contour, Raman scattering contours and FBG reflection spectra. Investigation methods and approaches are based on the unity of resonant structures of generated fiber responses on exciting and probing radiation or external physical fields for all given effects. The main decision is based on poly-harmonic probing of formed resonance responses. A variety of multiplexed measuring functions can be provided by the technique for telecom and sensing applications. As the examples, we introduced new four-frequency methods for Mandelstam-Brillouin gain contour and FBG spectra characterization classified by average and difference frequencies of probing radiation. These systems possess the advantages of the continuous-wave operation, high resolution, high accuracy, and absence of the need to measure the phase characteristics of the tested contours. We proposed for the first time the concept, approaches and methods for its implementation, which allowed reasonably formulate and solve the problem of creation of scientific and technical basis for designing software-defined down-hole fiber optic telemetry systems based on the combination of nonlinear reflectometry (inelastic Raman and Mandelstam-Brillouin, elastic Rayleigh and FBG) with improved metrological characteristics of poly-harmonic probing of sensing structures, including the removal of multiplicative and measurement errors caused by instability of the common source of forming radiations in wide frequency range. Universal poly-harmonic source of probing radiation, variants of scanning and multiple methods of probing, notch filters for Rayleigh scattering line suppression were demonstrated as elements of perspective systems of given class. Applications of designed methods and means in microwave range were also discussed.

Acknowledgements

This work was financially supported by the Ministry of Education and Science of the Russian Federation in the framework of the basic parts of the state and project assignments for works on the organization of scientific research conducted by Tupolev Kazan National Research Technical University (Kazan Aircraft Institute) at R&D Institute of Applied Electrodynamics, Photonics and Living Systems, and Department of Radiophotonics and Microwave Technologies (program "Photonics", 3.1962.2014K).

Author details

Oleg G. Morozov, Gennady A. Morozov*, Ilnur I. Nureev and Anvar A. Talipov

*Address all correspondence to: gmorozov-2010@mail.ru

Tupolev Kazan National Research Technical University (Kazan Aircraft Institute), R&D Institute of Applied Electrodynamics, Photonics and Living Systems, Department of Radiophotonics and Microwave Technologies, Kazan, Tatarstan, Russia

References

[1] Agrawal GP. Nonlinear Fiber Optics. Fifth Edition. Oxford: Academic Press; 2013.

[2] Vasil'ev SA et al. Fibre gratings and their applications. Quantum Electronics 2005; 35(12) 1085-1103.

[3] Udd E., Spillman WB, Jr. Fiber Optic Sensors: An Introduction for Engineers and Scientists. Hoboken, New Jersy: John Wiley & Sons, Inc.; 2011.

[4] Morozov O. et al. Synthesis of Two-Frequency Symmetrical Radiation and Its Application in Fiber Optical Structures Monitoring. In: Yasin M. (ed.) Fiber Optic Sensors. Rijeka: InTech; 2012. P137-164. DOI: 10.5772/27304. Available from: http://www.intechopen.com/books/fiber-optic-sensors/synthesis-of-two-frequency-symmetrical-radiation-and-its-application-in-fiber-optical-structures-mon (accessed 24 August 2014).

[5] Shibata N. et al. Identification of longitudinal acoustic modes guided in the core region of a single-mode optical fiber by Brillouin gain spectra measurements. Optics Letters 1988; 13(7) 595–597.

[6] Loayssa A. et al. Characterization of stimulated Brillouin scattering spectra by use of optical single-sideband modulation. Optics Letters 2004; 29(6) 638-640.

[7] Morozov O. et al. Characterization of stimulated Mandelstam-Brillouin scattering spectrum using a double-frequency probing radiation. Proceedings of SPIE – The International Society for Optical Engineering 2012; 8787 878709-6.

[8] Morozov O. et al. Principles of multiple frequencies characterization of stimulated Mandelstam-Brillouin gain spectrum. Proceedings of SPIE – The International Society for Optical Engineering 2014; 9156 91560K-7.

[9] Morozov O. et al. Methodology of symmetric double frequency reflectometry for selective fiber optic structures. Proceedings of SPIE – The International Society for Optical Engineering 2008; 7026 70260I-8.

[10] Morozov O. et al. Metrological aspects of symmetric double frequency and multi frequency reflectometry for fiber Bragg structures. Proceedings of SPIE – The International Society for Optical Engineering 2008; 7026 70260J-6.

[11] Morozov O. et al. Instantaneous frequency measurement of microwave signals in optical range using "frequency-amplitude" conversion in the π-phase-shifted fiber Bragg grating. Proceedings of SPIE – The International Society for Optical Engineering 2014; 9136 91361B-8.

[12] Long DA. Raman Spectroscopy. New York: McGraw-Hill; 1977.

[13] Grattan KTV and Sun T. Fiber Optic Sensor Technology: An Overview. Sensors Actuators 2000; A82(1–3) 40–61.

[14] Kuznetsov AG, Babin SA, and Shelemba IS. Distributed Fiber Sensor of Temperature with Spectral Filtration by Distributed Fiber Couplers. Kvant. Elektron. 2009; 39 (11) 1078–1081.

[15] Babin SA, Kuznetsov AG, and Shelemba IS. Comparison of Temperature Distribution Measurement Methods with the Use of the Bragg Gratings and Raman Scattering of Light in Optical Fibers. Optoelectronics, Instrumentation and Data Processing 2010; 46(4) 353–359.

[16] Wait PC and Hartog AH. Spontaneous Brillouin-Based Distributed Temperature Sensor Utilizing a Fiber Bragg Grating Notch Filter for the Separation of the Brillouin Signal. IEEE photonics technology letters 2001; 13(5) 508-510.

[17] Stolen RH, Ippen EP, and Tynes AR. Raman oscillaton in glass optical waveguide. Appl. Phys. Lett. 1972; 20 62-64.

[18] Ippen EP and Stolen RH. Stimulated Brillouin scattering in optical fibers. Appl. Phys. Lett. 1972; 21 539-540.

[19] Walrafen GE and Krishnan PN. Model analysis of the Raman spectrum from fused silica optical fibers. Appl. Opt. 1982; 21(3) 359-360.

[20] Rottwitt K. et al. Scaling of the Raman Gain Coefficient: Applications to Germanosilicate Fibers. J. Lightwave Techn. 2003; 21(7) 1652-1662.

[21] Felinskyi G. Nonlinear fitting of the complex Raman gain profile in singlemode optical fibers. In: 11 Int. Conf. on Mathematical Methods in Electromagnetic Theory, MMKT-06, June 26-29, 2006, Kharkiv, Ukraine. 378-380.

[22] Haijuan J. et al. Stability-improved slow light in polarization-maintaining fiber based on polarization-managed stimulated Brillouin scattering. J. Opt. 2013; 15 035404-5. DOI:10.1088/2040-8978/15/3/035404

[23] Al-Qazwini ZAT, Abdullah MK, Mokhtar MB. Measurements of stimulated-Raman-scattering induced tilt in spectral-amplitude-coding optical code-division multiple-access systems. Optical Engineering 2009; 48(1) 015001-6.

[24] André PS et al. Raman Gain characterization in Standard Single Mode Optical Fibres for Optical Simulation Purposes. http://www.researchgate.net/publication/228357255_Raman_gain_characterization_in_standard_single_mode_optical_fibres_for_optical_simulation_purposes (accessed 24 August 2014).

[25] Ming L. et al. Multichannel notch filter based on phase-shifted phase-only-sampled fiber Bragg grating. Optics Express 2008; 16(23) 19388-19394.

[26] Oliveira Silva SF. Fiber Bragg grating based structures for sensing and filtering. Porto: Porto University; 2007.

[27] Chehura E., James SW, and Tatam RP. A simple method for fabricating phase-shifted fibre Bragg gratings with flexible choice of centre wavelength. Proceedings of SPIE – The International Society for Optical Engineering 2009; 7503 750379-7.

[28] Junfeng J. et al. Investigation of peak wavelength detection of fiber Bragg grating with sparse spectral data. Optical Engineering 2012; 51(3) 034403-5.

[29] Bodendorfer T. et al. Comparison of different peak detection algorithms with regards to spectrometric fiber Bragg grating interrogation systems. In: International Symposium on Optomechatronic Technologies, ISOT-2009, 2009, Istanbul, Turkey. 122-126.

[30] Morozov OG, Aybatov DL. Spectrum conversion investigation in lithium niobate Mach-Zehnder modulator. Proceedings of SPIE – The International Society for Optical Engineering 2010; 7523 75230D-8.

[31] Aybatov DL, Morozov OG, and Sadeev TS. Dual port MZM based optical comb generator for all-optical microwave photonic devices. Proceedings of SPIE – The International Society for Optical Engineering 2011; 7992 799202-7.

[32] Sadeev TS, Morozov OG. Investigation and analysis of electro-optical devices in implementation of microwave photonic filters. Proceedings of SPIE – The International Society for Optical Engineering 2012; 8410 841007-6.

[33] Morozov OG. RZ, CS-RZ and soliton generation for access networks applications: problems and variants of decisions. Proceedings of SPIE – The International Society for Optical Engineering 2012; 8410 84100P-8.

[34] Mao XP, Tkach RW, Chraplyvy AR. Stimulated Brillouin threshold dependence on fiber type and uniformity. IEEE Photon. Technol. Lett. 1992; 4(1) 66–69.

[35] Loayssa A., Benito D., Garde MJ. Optical carrier Brillouin processing of microwave photonic signals. Opt. Lett. 2000; 25(17) 1234-1236.

[36] Honda N. In-Service Line Monitoring for Passive Optical Networks. In: Yasin M. et al. (ed.) Optical Fiber Communications and Devices. Rijeka: InTech; 2012. p203-218.

[37] Oh I., Yegnanarayanan S., Jalali B. High-resolution microwave phonon spectroscopy of dispersion shifted fiber. IEEE Photon. Technol. Lett. 2002; 14(3) 358–360.

[38] Nikles M., Thevenaz L., Robert PA. Brillouin gain spectrum characterization in single-mode optical fibers. J. Lightwave Technol. 1997; 15(10) 1842–1851.

[39] Loayssa A., Benito D., Garde MJ. Narrow-bandwidth technique for stimulated Brillouin scattering spectral characterization. Electron. Lett. 2001; 37(6) 367–368.

[40] Sagues M., Loayssa A. Swept optical single sideband modulation for spectral measurement applications using stimulated Brillouin scattering. Optics Express 2010; 18(16) 17555-17568.

[41] Morozov OG et al. Structural minimization of fiber optic sensor nets for monitoring of dangerous materials storage. Proceedings of SPIE – The International Society for Optical Engineering 2011; 7992 79920E-9.

[42] Petoukhov VM. Two-frequency IR CW LFM LIDAR for remote sensing of hydrocarbons and gas vapor. Proceedings of SPIE – The International Society for Optical Engineering 1997; 3122 339-346.

[43] Morozov OG. Two-frequency scanning LFM LIDARS: theory and applications. Proceedings of SPIE – The International Society for Optical Engineering 2002; 4539 158-168.

[44] Bogdanov NG, Plotnikov SN, Stchekotchihin SN. Monitoring of thickness of not magnetic coverings on a ferromagnetic basis. Factory laboratory. Materials Diagnostic 2007; 12 30-33.

[45] Aybatov DL. Distributed temperature fiber Bragg grating sensor. Proceedings of SPIE – The International Society for Optical Engineering 2009; 7374 73740B-6. DOI: 10.1117/12.829002.

[46] Weaver T. Thermal drift compensation system and method for optical network. Patent WO 020838, 2008.

[47] Jackson RG. Advanced sensors. Moscow: Technosphere; 2007.

[48] Sadykov IR, Morozov OG, Sadeev TS. The biosensor based on fiber Bragg grating to determine the composition of the fuel and biofuel. Proceedings of SPIE – The International Society for Optical Engineering 2012; 8410 84100F-8.

[49] Morozov GA. Application of microwave technologies for increase of efficiency of polymeric materials recycling. In:8th International Conference on Antenna Theory and Techniques, ICATT'11, 2011, Kiev, Ukraine. 321-323.

[50] Dong X. Bend measurement with chirp of fiber Bragg grating. Smart materials and structures 2001; 10 1111-1113.

[51] Denisenko PE, Sadeev TS, Morozov OG. Fiber optic monitoring system based on fiber Bragg gratings. Proceedings of SPIE – The International Society for Optical Engineering 2012; 8410 84100K-6.

[52] Morant M. et al. Dual-drive LiNbO3 interferometric Mach-Zehnder architecture with extended linear regime for high peak-to-average OFDM-based communication systems. Optics express 2011; 19(26) B450-B456.

[53] Xiao Y. et al. Multiple microwave frequencies measurement based on stimulated Brillouin scattering with improved measurement range. Opt. Express 2013; 21(26) 31740-31750.

[54] Wait PC and Newson TP. Landau–Placzek ratio applied to distributed fiber sensing. Opt. Commun. 1996; 122 141–146.

[55] Lees GP et al. Recent advances in distributed optical fiber temperature sensing using the Landau–Placzek ratio. Proceedings of SPIE – The International Society for Optical Engineering 1998; 3541 1–5. doi:10.1117/12.339104.

[56] Morozov O. et al. Two-frequency analysis of fiber-optic structures. Proceedings of SPIE – The International Society for Optical Engineering 2006; 6277 62770E-11.

[57] Morozov OG, Nurgazizov MR, Talipov AA. Double-frequency method for the instantaneous frequency and amplitude measurement. In: 9th International Conference on Antenna Theory and Techniques, ICATT-2013, Odessa, Ukraine, 2013. 381-383.

[58] Morozov OG. et al. Intellectual parachute and balloon systems based on fiber optic technologies. Proceedings of SPIE – The International Society for Optical Engineering 2014; 9156 91560B-8.

[59] Morozov O. et al. Instantaneous microwave frequency measurement with monitoring of system temperature. Proceedings of SPIE – The International Society for Optical Engineering 2014; 9156 91560N-7.

[60] Morozov O. et al. Instantaneous frequency measurement of microwave signals in optical range using "frequency-amplitude" conversion in the π-phase-shifted fiber-Bragg grating. Proceedings of SPIE – The International Society for Optical Engineering 2014; 9136 91361B-8.

[61] Morozov O. et al. Theory of symmetrical two-frequency signals and key aspects of its application. Proceedings of SPIE – The International Society for Optical Engineering 2014; 9156 91560M-11.

[62] Morozov O. et al. Training course and tutorial on optical two-frequency domain reflectometry. Proceedings of SPIE – The International Society for Optical Engineering 2012; 8410 84100Q-9.

[63] Natanson O. et al. Reflectometry in open and fiber mediums: technology transfer. Proceedings of SPIE – The International Society for Optical Engineering 2005; 5854 205-214.

Chapter 3

Dispersion Compensating Fibres for Fibre Optic Telecommunication Systems

Michal Lucki and Tomas Zeman

Additional information is available at the end of the chapter

http://dx.doi.org/10.5772/59152

1. Introduction

Chromatic dispersion (CD) of optical pulses in an optical fibre influences their width. It refers to changes in propagation of particular frequency components contained in optical pulses causing extension of optical pulses. The fibre reach is then significantly limited, unless signal regenerators are used. Because the pulse spread degrades optical systems, it is necessary to prevent its origination or to eliminate its results. CD ought to be suppressed or one shall prevent its origination by specialty optical fibres, including both conventional and micro-structured optical fibres (MOF). To keep the pulse length nearly constant, it is possible to use Dispersion Compensating Fibre (DCF) that shall be installed at the regenerators that provide amplification, renewal of timing and pulse duration. DCFs have negative dispersion parameter; CD accumulating between regenerators is then suppressed. Consequently, bit error rate could be improved and the fibre reach could be extended [1].

Photonic Crystal Fibres (PCF) could be a suitable solution for the problem of CD in high-speed transmission systems, especially those using Wavelength Division Multiplexing (WDM) to increase the symbol rate [2].

We systematize design approaches for specialty dispersion-tailored optical fibres in order to offer a guideline to flexibly design optical fibres used in telecommunications whose optimized CD is a key property. Last but not least, we present selected fibre designs prepared in last few years, including CD plots and information about structural parameters.

The knowledge about how to design fibres could be useful to design optical fibres, such as for example a submicron flat CD compensating fibre. In addition, fluoride compounds, as compared with silica glass, exhibit higher effective refractive index, the wider spectrum of working wavelengths (λ) or lower insertion losses. Additives using these materials are

promising for extending the application towards infrared region, where fluoride glasses are usually transparent.

2. Fibres with optimized dispersion [3]

DCFs could regenerate signals that are spread as a consequence of CD. This practically means that the bit error rate at the receiver's side could be improved or the spaces between adjacent symbols could be reduced. As a result, potential bit rate could be increased. One of the approaches how to deal with CD is to use zero CD fibres, offering near zero CD at the operating λs. (It shouldn't be exactly zero because of Four Wave Mixing problem occurring when propagation is with zero dispersion and the phases of all the frequency components are matched). Another approach is to use already mentioned DCFs at the signal regenerators. CD tailoring fibres could work at some λ, e.g. at 1.55 μm, where they could have negative value of CD being few hundred picoseconds (ps/nm/km), which expresses the delay between the slowest and fastest frequency component measured at the distance of 1 km, assuming the source of radiation emitting the spectrum of 1 nm. Sample compensating fibre was published in Refs. [4] and [5]. Very low CD parameter is observed whose properties could benefit in compensation of CD accumulating in some optical network. An unsolved problem seems to be relatively narrow window of λs. Such a fibre could then compensate CD in just one λ channel of a WDM system [3].

Resultant CD is a balance between waveguide and material component. In order to obtain flat CD, the evolution of waveguide dispersion has to be exactly opposite to the one of material dispersion, which means that both ought to be optimized at each channel step by step. The situation is simple when material dispersion is linear upon λ. Then, one could employ an algorithmic approach with iterations to calculate and precisely adjust the CD at each λ. A very important assumption is that the balance between both dispersion components could be a very useful tool in designing fibres for CD tailoring by optimizing material dispersion (through the material properties) and waveguide dispersion (as a result of optimization of the shape of the waveguide) [3] [6].

2.1. Dispersion compensating fibres [2]

When a compensating fibre works at some λ, optical symbols ought to be transmitted at small spectrum of light waves, without multiplying the bit rate in many channels. Such narrowband fibres can have very low value of CD that is possible without using additives (e.g. germanium dioxide). In Ref. [7], it could be found that CD of -18 ns/nm/km is possible. Other works concentrate rather on the optimization of the structural parameters of MOFs to obtain flat CD diagram. Microstructured fibres offer much greater flexibility because geometry could be optimized not only through the core and cladding diameter, but also by changing the air filling fraction, and the lattice pitch including its arrangements. In Ref. [8], flat CD over all the telecom bands is shown [2].

Index guiding MOFs (the one without the central inclusion, having solid core) offer high flexibility in CD compensating fibres design. For example microstructured based DCFs could have dual cores [9] or they could be doped [10]. Negative value of CD parameter in DCFs with dual core structure could also be combined with the idea of making some rings with smaller holes [7][10]. Such a DCF with low CD parameter in broad telecom band is shown in Ref. [11], in addition, a square lattice (with the defining angle of 90°) was proposed. The performance of DCFs with the concentric cores could be increased by using germanium dioxide in the core [12]. The relation between the amount of additives and resultant CD is shown in Ref. [13]. Some other substrates could be used to lower the effective index (n_{eff}), such as for example fluorides [14]. On the contrary, germanium dioxide raises n_{eff}. The key feature is not to raise or lower a certain index, but to create large index contrast between the fibre's core index and its n_{eff} of the cladding [2].

Fibres with flat CD, i.e. those having low value of CD in many λ channels concurrently are often considered as wideband fibres, but they can't be used to compensate CD in all λ channels at once, because accumulated CD in each WDM channel is different. They are rather suitable to compensate CD at one channel, which can be selected from the wide range of λs that are compatible with the used DCF. Real wideband fibres should compensate CD at every λ channel at the same time. They must have CD exactly opposite to dispersion in each channel. In Ref. [15], it is shown that larger lattice pitch in the 1st ring of holes is responsible for the CD slope that must be optimized for this purpose. In addition, the hole radius in the 1st ring should be larger to enhance dispersion. CD slope property is studied in Ref. [16] in the context of wavelength division multiplexing. Another slope compensating DCF is shown in Refs. [2] and [17].

2.2. Zero dispersion wavelengths in DCFs

Because a great part of MOFs have parabolic wavelength evolution of CD and the minimum CD at low values, those fibres can have two zero-CD wavelengths (ZDW). The 1st ZDW is the one utilized at visible wavelengths, whereas the longer ZDW is used at telecom wavelengths [18]. Such Highly Nonlinear PCF (HNPCF) could exhibit positive slope at 1st ZDW and negative slope at 2nd ZDW, as shown in Ref. [19]; the 1st ZDW was at 900 nm, the 2nd one at the λ of 1.6 µm.

The high index difference between the air-filled microstructure and pure or doped silica core enables tight mode confinement resulting in a low effective area and, thereby, a non-conventional behaviour of CD. Modifying the periodic cladding (i.e., hole-sizes, lattice pitch), the waveguide dispersion and the origination of ZDWs are influenced, keeping the fibre still in single-mode operation regime [20]. Then, the strong wavelength dependence in the characteristics of the fibre will be used to determine either huge CD with large slope or nearly-zero flat CD.

The bandwidth, at which MOFs are designed to have zero CD, could be divided into three categories: working in the region between 0.55 µm and 1 µm, where 1st zero-CD point could be determined; another region between 1 µm and 1.2 µm, where 1st or 2nd ZDW could be found. The 3rd region, starting at 1.2 µm and going up to longer λs, is used to create the 2nd zero CD and is not available for the 1st zero CD. An overview of the obtainable ZDWs as a function of MOF's structure was shown in Ref. [21].

Both core size and the radius of holes exhibit significant influence on the location of both ZDWs. Increasing core size tunes the ZDW to longer λ. The increase in air percentage in the cladding could extend the origination of the 1st ZDW at longer wavelengths and the 2nd ZDW at shorter λs (i.e. less negative CD is obtained for larger air-filling fraction).

2.3. Dispersion flattened fibres and wideband fibres [2][3]

Some compensating fibres could work at short spectrum of λs. When a compensating fibre works at a certain λ, optical pulses can be transmitted using one or – in general – low amount of WDM channels. As it has already been mentioned above, MOFs are suitable for designing the compensating fibres because they allow huge index contrast and offer many parameters to be optimized (core size, hole radius, lattice pitch, amount of rings of holes) in order to optimize waveguide dispersion [2].

Currently, flat CD fibres could for example be referred to as near zero CD transmission medium. In Ref. [22], a DCF made of pure (undoped) material is shown. Advanced DCFs with flat CD property over broad spectrum of working λs could be found in Ref. [23]. An interesting fibre with CD parameter close to zero with CD fluctuations less than 0.5 ps in S to L bands is shown in Ref. [24]. Results were obtained by careful optimization of holes in particular rings. To reduce losses, it is often required avoiding doping the MOF's core. Instead, one could consider using octagonal [25] and decagonal [26] lattice, or more sophisticated lattice arrangements [2].

Recent analyses concentrate on the low, flat CD over telecommunication λs for any telecom wavelength compensation. The value of CD and the width of the working range is a compromise. A DSF could have large negative CD value, but for medium-wide wavelength range, or, acceptable CD parameter, but designed for very broad spectrum of λ. Currently, the strongest demand is to design DCFs mainly for C and L bands, where modern WDM systems can work.

The idea how to obtain flat CD properties is to tune the 1st ZDW to shorter λs and the 2nd ZDW to longer λs, having little negative CD over the whole bandwidth at the same time. Practically, predicted CD diagram is a wide parabola. Considered properties are related to the diameter of the core and the radius of the holes in particular rings. Both were found to have different impact on the origination of each ZDW. There are requirements to locate the zero CD point at the λ of 1.55 µm or, in general, in the C-band [8]. Changing the air filling fraction and lattice pitch is not the only idea that could be used to optimize CD. For example elliptical holes instead of circular holes exhibit potential low and very stable CD properties, i.e. 0.6-1 ps/nm/km in the range from 1 µm to 1.9 µm [3].

Finally, there are a few examples that are worth to be noticed: a DCF in Ref. [27], with CD parameter being –1350 ps/nm/km at 1.55 µm; another one in Ref. [28] has CD of –440 to –480 ps/nm/km at the band of 1.5-1.62 µm [3].

3. State of the art

The employment of the considered fibres is mainly in high-speed transmission systems using wavelength division multiplexing and signal recovery, offering transmission rates of even more than 1 Tb/s, where it is proved that non-optimized CD could destroy the pulse spreading, but as well it could generate some nonlinear effects, such as four wave mixing, among others. Although the properties of the fibres significantly differ from one another, the techniques used for optimized CD could be systematized so that one could design a fibre with desired CD value and slope at a specific λ.

Understanding of the mechanisms governing CD tailoring is necessary not only permits for the fibre design, CD suppression and avoidance, but also to predict the potential manufacturing tolerances. There are a few techniques used to obtain the expected CD and we describe them in more details in the following sections.

3.1. Fibre designs [1]

One of the techniques applied is to dope the fibre's core. By inserting small doped inclusion in a MOF's core, the index contrast with result in much larger mode confinement. Because the waveguide dispersion is related to it and because the length of propagated light waves is of the range that is comparable to the core size, the inserted region will be responsible for very low waveguide dispersion component.

There is another technique that could be used to optimize dispersion, it assumes to inject some liquids into the microstructured holes. In Ref. [29], the 4th ring of holes is doped by using liquids for this purpose. But liquid's properties are strongly dependent on temperature, which means that constant temperature must be ensured. This means that such a fibre could for example be used for some experiments in laboratory conditions (e.g. sensors), but they can't be used in outdoor installations of telecommunication systems [1].

Germanium dioxide is very often used to raise the effective index. In Ref. [13] and [30], the idea how to use it to dope the microstructured fibre's core is described in details. But this solution is not perfect in terms of effective area of the mode which is rather small in the doped fibres, as well as from the perspective of increased attenuation caused by material absorption, which is high for germanium dioxide [1].

Another technique to be described and systematized is the one using air holes located close to the fibre centre. The 1st mode could be confined within the core by the index contrast, and the effective diameter of a mode is then usually small enough to control CD. Stronger confinement could be obtained in a MOF using other type of a lattice, like for example octagonal, decagonal or spiral.

An alternative way, how to control the effective index, is to have core defects. A "defect core" is usually understood as one allowing light coupling to some other regions by some imperfection that is introduced intentionally. This technique uses weak interactions between the core-guided mode and a mode localized in an intentionally introduced defect of the crystal.

Finally, the idea of concentric cores could be considered. What feature in dual core technique is responsible for low CD? It is large value of n_{eff} produced by significant asymmetry between the two cores where light is coupled. Light is propagated in both cores concurrently. Dual cores are responsible for the phase matching at a certain λ, and consequently at this wavelength large CD parameter is obtained.

Low value of CD parameter is possible in concentric cores. They could be obtained by removing [7],[10] or reducing [7],[31] the air fraction in some rings of holes [1].

There are many aspects that could influence resultant CD; it could be coupling modes and the phase matching wavelength, at which CD control is possible, could be obtained in many ways, for example by fibre bending. We found these conclusions as interesting, because many works assume that the most important consequence of bending a fibre is increased attenuation. It could be shown that large negative value of CD parameter could be the accompanying process.

3.2. Extension of transmission band towards infrared — Fluoride fibres [32]

Dealing with the structural parameters of MOFs is a solution to find optimal CD. They could be enhanced by very careful optimizing the material composition. For example fluorides could be used as additives, but they could also be used as a background material instead of silicon dioxide, which could be evaluated as significant progress in the field [32]. The fluorides have low effective index, broad range of λs and low attenuation. One could consider the use of BaF_2, CaF_2 and more advanced ZBGA (ZrF_4-BaF_2-GdF_3-AlF_3) or ZBLAN (ZrF_4-BaF_2-LaF_3-AlF_3-NaF) [32]. In addition, because those materials are composite, it is possible to change their composition in a compound in order to obtain ne_{ff} value or its slope in the investigated area of λ. Last but not least, these materials are transparent in infrared region (as well as in the telecommunication bands) which makes them interesting for some applications in sensors. The properties of fluorides allow using them in MOFs [33]. They would lower the effective index of the cladding or the index at the second, larger concentric core. It is usually combined with doping the main core by using germania. This idea could be considered as a significant progress in fibre optic technology. Then a fibre with a W-type profile or refractive index could be created. For example, the core doped by using germanium dioxide would exhibit the n_{eff} of about 1.48 and concurrently the cladding index doped by using F-SiO_2 would exhibit the index of about 1.43. Similar solution (in general modified W-type index profile fibre) could be employed in systems with WDM [32],[34].

Material and structural parameters are key features in order to obtain exact CD at each λ, but the accuracy and quality of production could become a critical issue, especially in submicron fibres whose small holes with the diameter being less than one micron is problematic for some manufacturing technologies. All this combined with the application of new materials that should be processed in different way compared to siliceous materials could result in both attenuation and CD properties far from the expectations [32].

At the telecommunication band both ZBLAN and ZBGA showed similar attenuation properties like the one of silicon dioxide, but their CD properties significantly differ [35], in addition fluorides have different mechanical properties and they have to be fused at different temperature than temperature for silica. The theoretical attenuation could even be of 0.01 dB/km, in the range from 0.3 μm to 4.3 μm. The n_{eff} in telecom band is 1.47 – 1.52 [32, 36].

4. Design approaches

To describe propagation of light in a given optical glass, we could consider the constant of propagation of a wave that is related to the phase variation [32]:

$$k = (\beta - j\alpha)$$

(1)

The real part β and the imaginary part ω could be specified [37]:

$$\beta = \omega \sqrt{\frac{\mu\varepsilon}{2}\left[\sqrt{1 + \left(\frac{\sigma}{\omega\varepsilon}\right)^2} + 1\right]}$$

(2)

$$\alpha = \omega \sqrt{\frac{\mu\varepsilon}{2}\left[\sqrt{1 + \left(\frac{\sigma}{\omega\varepsilon}\right)^2} - 1\right]}$$

(3)

Where ω is angular velocity, σ is conductivity, a μ is permeability and ε is permittivity. Because MOFs are dielectric it could be assumed that:

$$\left(\frac{\sigma}{\omega\varepsilon}\right) << 1$$

(4)

Both real and imaginary part of propagation constant in (3) could be simplified to a form:

$$\beta = \omega\sqrt{\mu\varepsilon}$$

(5)

$$\alpha = 1$$

(6)

MOFs could be described by an effective index n_{eff}, which expresses "the refractive index at the boundary of two entities", which in this particular case are the core and the cladding [32].

$$n_{eff} = \frac{\beta}{k_0}$$

(7)

β is real part of the constant of propagation of core, k_0 is the constant of propagation in vacuum. Propagation constant depends on λ, it could be written that:

$$k = \frac{2\pi}{\lambda} \tag{8}$$

Group velocity that integrates optical waves in an "envelope" could be considered to describe its velocity. It refers to CD and is less than the speed of electromagnetic wave in vacuum [32]:

$$v_g = \frac{\partial \omega}{\partial k} = \frac{c}{\sqrt{1 + \dfrac{\omega^2 \lambda^2}{4\pi^2 c^2}}} \tag{9}$$

Then group delay τ_g could be calculated as time necessary for the propagation of wave to certain distance with velocity of v_g (9):

$$\tau_g = \frac{1}{v_g} \tag{10}$$

In general, chromatic dispersion CD is known as the dependence of group delay τ_g on the wavelength λ, at which the signal is transmitted [38]:

$$CD(\lambda) = \frac{\partial \tau_g(\lambda)}{\partial(\lambda)} \tag{11}$$

In simulations it is more suitable to use material dispersion equation described by using Sellmeier approximation. CD equation suitable for Sellmeier approximations is following [39]:

$$CD(\lambda) = -\frac{\lambda}{c} \frac{\partial^2 Re\left[n_{eff} \right]}{\partial \lambda^2} \tag{12}$$

Designs of MOFs could be calculated by Finite-Difference Frequency Domain (FDFD) method. The distribution of light could be calculated by discretizing electric (13) and magnetic (14) field, it is described in Refs. [32, 40], and [41]:

$$\left(\nabla_t^2 + k_0^2 \varepsilon_r \right) \mathbf{E}_t + \nabla_t \left(\varepsilon_r^{-1} \nabla_r \varepsilon_r \cdot \mathbf{E}_t \right) = \beta^2 \mathbf{E}_t \tag{13}$$

$$\left(\nabla_t^2 + k_0^2 \varepsilon_r\right)\boldsymbol{H}_t + \varepsilon_r^{-1}\nabla_t \varepsilon_r \times \left(\nabla_t \times \boldsymbol{H}_t\right) = \beta^2 \boldsymbol{H}_t \tag{14}$$

k_0 is wavenumber in free space, β is real part of the propagation constant, ε_r is dielectric constant as in (5). The FDFD method uses the discretization scheme shown in Ref. [42][32].

FDFD is used in many commercial simulators of mode distribution and n_{eff} calculators. In most of them, it is possible to use user friendly graphical interface, but when a very accurate result is demanded, many iterations ought to be done by creating a loop using an appropriate scripting language. We use a mode solver from the Lumerical Inc. When the fibre's cross section is proposed and the simulation parameters and monitors are set, the simulation is run [3].

Creation of realistic models of sophisticated optical glasses is an interesting feature. To introduce materials to the simulator, we used the Sellmeier approximation [32, 43],[44]:

$$n^2\left(\lambda\right) = A + \sum_i \frac{B_i \lambda^2}{\lambda^2 - C_i^2} \tag{15}$$

where A, B_i, C_i are coefficients referring to index n. As a result, the wavelength evolution of refractive index could be plot by using the Sellmeier equation. The coefficients for the expanded version of Sellmeier equation (16) could found in [32][45]:

$$n^2 - 1 = \frac{B_1 \lambda^2}{\left(\lambda^2 - C_2^2\right)} + \frac{B_3 \lambda^2}{\left(\lambda^2 - C_4^2\right)} + \frac{B_5 \lambda^2}{\left(\lambda^2 - C_6^2\right)} \tag{16}$$

where $B_{1,3,5}$ $C_{2,4,6}$ are material constants and λ is wavelength.

5. Exemplary results

In this section, we present some selected fibre designs covering the wide area of telecom photonic fibres for CD tailoring, such as DCFs with large negative CD parameter, wideband fibres or slope compensating fibres, or fibre designs suitable for CD prevention, such as near zero CD fibres. We employ different design approaches and optimize them based on parametric investigation.

5.1. Dual core dispersion compensating fibre with large negative dispersion parameter [1]

A concentric core DCF based on microstructured optical fibre is proposed [1]. There is an assumption for the proposed structure to keep the 1st mode in a central core over the wide working spectrum of λ and in order to obtain low CD at 1.55 µm, and with low theoretical losses. The refractive index is optimized by making smaller the hole diameter in the 2nd ring. The design is specified in Table 1, the cross section is shown in Figure 1 [1].

Quantity [unit]	Value	Value	Value	Value	Value
Lattice pitch Λ [μm]	1.55	1.52	1.50	1.48	1.45
Minimum D at 1.55 μm [ps/nm/km]	-1010	-1180	-1300	-1460	-1690
Core index [-]	1.44	1.44	1.44	1.44	1.44
Theoretical loss [dB/cm]	$9 \cdot 10^{-10}$	$7 \cdot 10^{-10}$	$3.8 \cdot 10^{-7}$	$3.8 \cdot 10^{-4}$	$7 \cdot 10^{-2}$
Full width at half maximum [nm]	142	132	129	125	122
Hole diameter d_1 [μm]	1.33	1.34	1.314	1.27	1.23
Hole diameter d_2 [μm]	0.55	0.54	0.574	0.63	0.67
Hole diameter d_3 [μm]	1.35	1.35	1.2	1.04	0.92
Number of rings [-]	6	6	6	6	6

Table 1. Parameters of a dual core MOF for CD suppression [1].

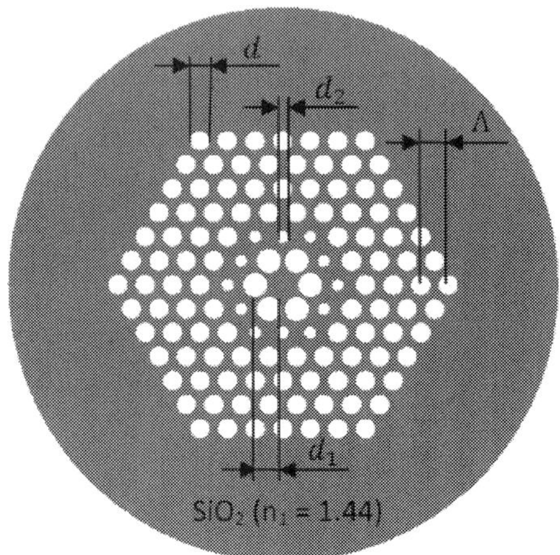

Figure 1. Structural parameters and the cross section of a dual core MOF for CD suppression purposes [1].

CD of the considered fibre is shown in Figure 2. Obtained CD parameter was -1460 ps/nm/km. The designed fibre work in the C-band, and the lowest CD is at 1.55 μm. Theoretical losses are 3.8×10^{-4} dB/cm. It could be concerned as one of the advantages of the MOF, when we have a look at losses presented for some other fibres [1].

In order to obtain very accurate results, a number of iterations were performed to find the most suitable structural parameters of a fibre. One could conclude that smaller holes in the 3rd ring, and concurrently larger radius of holes in the 2nd ring could result in improved CD properties (Figure 3). Optimized solution could be found in Table 2 [1].

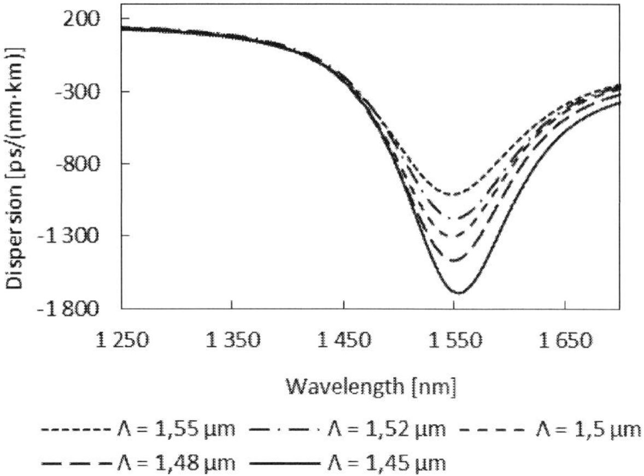

Figure 2. Chromatic dispersion in a dual core MOF for CD suppression at 1.55 µm [1].

Quantity [unit]	Value	Quantity [unit]	Value
Minimum CD [ps/nm/km]	-3290	Lattice pitch Λ [µm]	1.45
Core index [-]	1.44	Hole diam. d_1 [µm]	1.23
Theoretical losses [dB/cm]	7.50	Hole diam. d_2 [µm]	0.73
Number of rings [-]	6.00	Hole diam. d_3 [µm]	0.86

Table 2. Specification of parameters for dual core MOF for CD suppression [1].

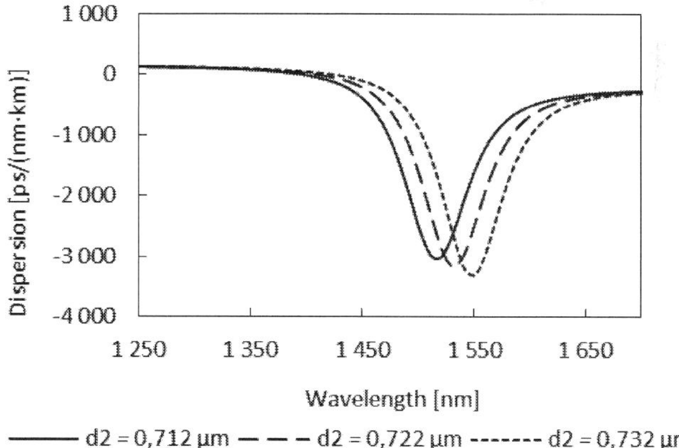

Figure 3. Dispersion in proposed dual core microstructured DCF [1].

In Figure 3, it could be observed that λ at which there is minimum CD is function of normalized hole diameter d/Λ of the holes; CD is mostly sensitive to the holes located in the 2nd ring. To move the operating λ towards 1.5 μm, one has to increase the size of the holes in this ring. Concurrently, the value of lowest CD is less. Summarizing, a one km long section of the designed DCF should be sufficient to tailor CD in a network that is created by using 75 km long conventional SMF with CD parameter being 17 ps/nm/km, (this value is in agreement with the recommendation of ITU-T for SMFs). The insertion losses of such a CD compensator are about 0.04 dB [1].

5.2. DCF with optimized dispersion slope [3]

Another design is done with the scope on fibres that could compensate CD in each channel of a system using wavelength multiplexing. It means that the fibre should be wideband, and its CD mustn't be flat, but it should have exactly opposite CD to CD of a fibre that is used in a WDM system. The proposal is shown in Figure 4 and in Table 3 [3].

n_{eff} [-]	1.47	Hole diameter [μm]	3.64
Core index [-]	1.48	Normalized hole diameter d/Λ [-]	0.58
Core size [μm]	4.46	Number of rings [-]	5

Table 3. Structural parameters of the HNPCF with dispersion evolution opposite to one in conventional MOF [3].

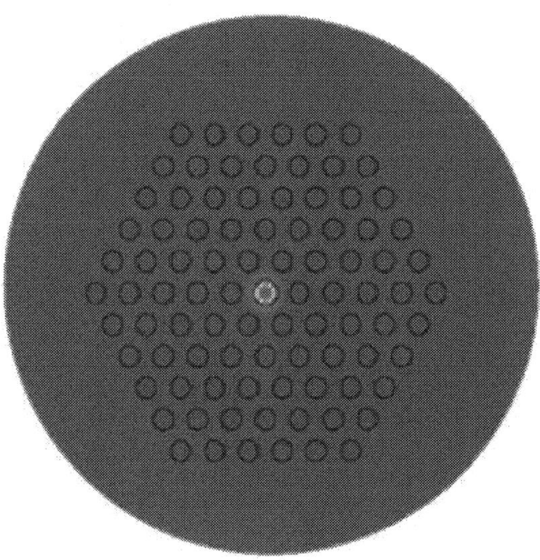

Figure 4. Cross section of a MOF with negative CD parameter [3].

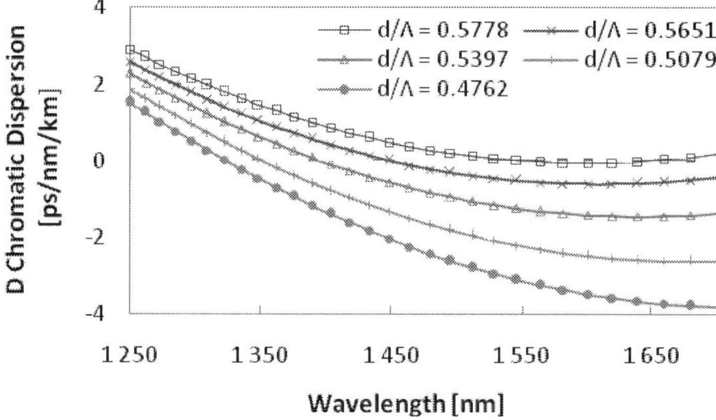

Figure 5. Reversed CD slope of the proposed DCF [3].

From Figure 5 it could be concluded that CD is reversed CD of conventional MOFs. The fibre is wideband. It could be optimized in terms of larger values of CD parameters, for example to exactly match CD evolution of conventional fibres. One shall also pay attention to the fact that larger d/Λ would result in multimode operation [3].

5.3. Dispersion flattened fibres and wideband dispersion compensating fibres [1][3]

A MOF with flat CD is demanded in design of transmission fibres (not suppression fibres) where it is d to have identical CD at each λ channel. Such a fibre could be doped in the core by using germania. The proposed core diameter could be 7.4 μm. In Figure 6, the 1st mode is kept in the core and the fibre is single-mode. Manufacturing of the MOF's core could be challenging, because of huge doping area; the fibre has large mode area. The proposed geometry is described in Table 4 [3].

Hole diameter d [μm]	1.32	Silica index [-]	1.46
Lattice pitch Λ [μm]	4.4	Propagating λ [μm]	1.55
Normalized hole diameter d/Λ [-]	0.3	Core size [μm]	7.4
Air index [-]	1	Effective cladding index at 1.55 μm	1.47
Core index [-]	1.48	Number of rings [-]	3

Table 4. Structure parameters for HNPCF flattened CD curve [3].

Obtained CD property could be evaluated as flat and oscillating around the value of -0.025 ps/nm/km. In Figure 7, comparison of CD in conventional MOF and the designed fibre is shown [3].

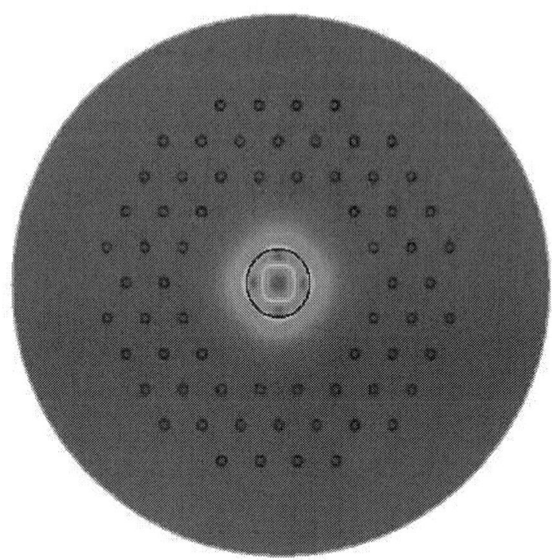

Figure 6. Wideband large mode area MOF with near zero flattened CD [3].

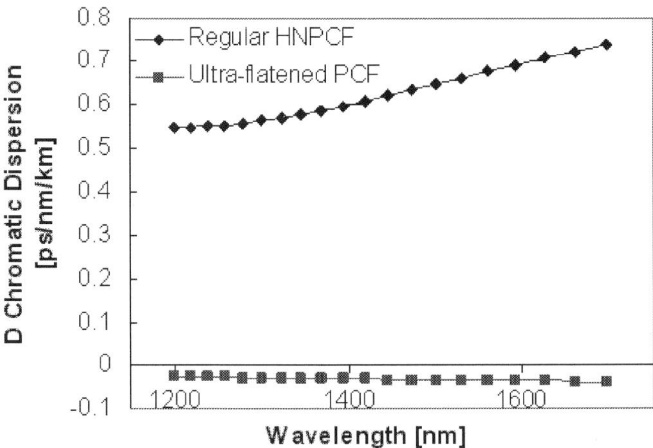

Figure 7. Dispersion in conventional microstructured fibre and proposed highly nonlinear MOF [3].

Wideband fibre could exhibit negative CD parameter, too. We propose a MOF for suppression in the band of 1.25-1.7 μm. We do not use any additives. Mode confinement is done by optimizing the geometry of the core and the normalized hole diameter d/Λ, the accepted view is that the core size should be small. In the 1st ring, there is $d_1/\Lambda=0.9$. The proposal could be found in Figure 8. Doping the core could additionally improve CD properties, but it would surely worsen attenuation. One should pay attention to the fact that optimization of CD shall be done in the context of attenuation properties. Optimizing one parameter and ignoring another is unacceptable in high-speed transmission system fibres [1].

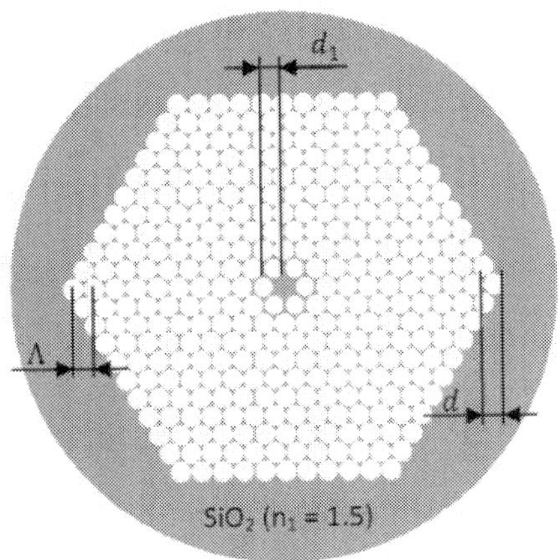

Figure 8. Cross section of designed microstructured fibre for wideband suppression of CD [1].

Having larger holes in the 1st t ring, d_1, is responsible for lower CD. On the contrary, by making smaller all the other holes (d) results in increasing dispersion. In Table 5, optimized structure is shown and its cross section is in Figure 9 [1].

Quantity [unit]	Value	Value	Value	Value	Value
CD [ps/nm/km]	-1580	-2040	-2259	-2094	-1930
Core index [-]	1.50	1.50	1.50	1.50	1.50
Theoretical losses [dB/cm]	$9.1 \cdot 10^{-6}$	$1.9 \cdot 10^{-3}$	$4.8 \cdot 10^{-2}$	$7 \cdot 10^{-2}$	$1.5 \cdot 10^{1}$
FWHM [nm]	Flat CD over 1.25-1.7 µm				
Lattice pitch [µm]	0.70	0.65	0.62	0.6	0.55
Hole diameter in 1st ring d_1 [µm]	0.70	0.65	0.62	0.6	0.55
Hole diameter d [µm]	0.70	0.65	0.62	0.6	0.55
Number of rings [-]	10	10	10	10	10

Table 5. Specification of parameters for wideband CD suppressing MOF [1].

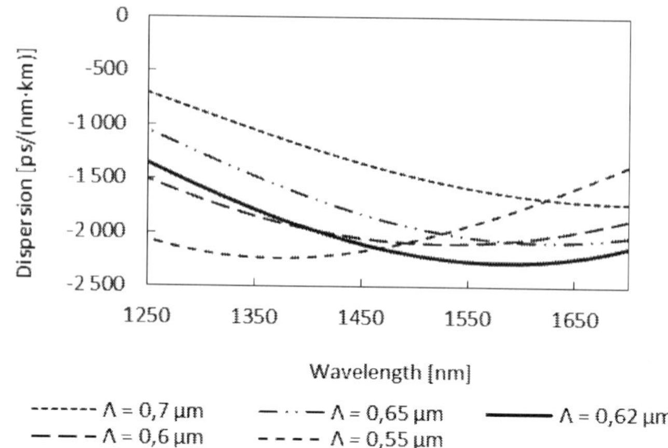

Figure 9. Resultant flattened negative value of CD over broad band of λs 1.25-1.7 μm (lattice pitch is a parameter) [1].

Concerning the values of CD, the best design is the one for Λ=0.62 μm, for which CD is -2259 ps/nm/km, obtained at 1.55 μm. Theoretical losses are 4.8 dB/cm. The optimized fibre is the one with Λ=0.70 μm, where CD parameter is -1580 ps/nm/km at 1.55 μm. Theoretical loss is lowered to 9.1 10-6 dB/cm. In this case it is possible to compensate dispersion of about 90 km of standard SMF.

5.4. Fluoride doped dispersion compensating fibres [32]

The considered fibre is a MOF employing the idea of a W-profile fibre with the core doped by using BaF_2 (n_{BaF2} = 1.468 to raise its refractive index) and containing three holes in the 1st ring doped by using CaF_2, n_{CaF2} = 1.426 to reduce effective cladding index. Increased index contrast is responsible for enhanced CD, as in eq. (12). As for example, models optical glasses could be expressed using coefficients shown in Table 6.

Material	B_1	C_2	B_3	C_4	B_5	C_6
Silica	0.6961663	0.0684043	0.4079426	0.11624	0.897479	9.89616
BaF_2	0.6433560	0.0577890	0.5067620	0.109680	3.82610	46.3864
CaF_2	0.5675888	0.0502636	0.4710914	0.10039	3.848472	34.6490

Table 6. Sellmeier coefficients for the fluoride additives used in investigated MOF [32].

The structural and material properties are in Table 7. Large radius of CaF_2 holes and the low core size are necessary for flat CD (Figure 10). CD is -413 ~ -415 ps/nm/km at 1.4 - 1.65 μm. Theoretical loss is 1.75 10^{-4} dB/cm. CaF_2 doped holes affect the CD (Figure 11). The larger are the doped holes, the lower is resultant CD [32].

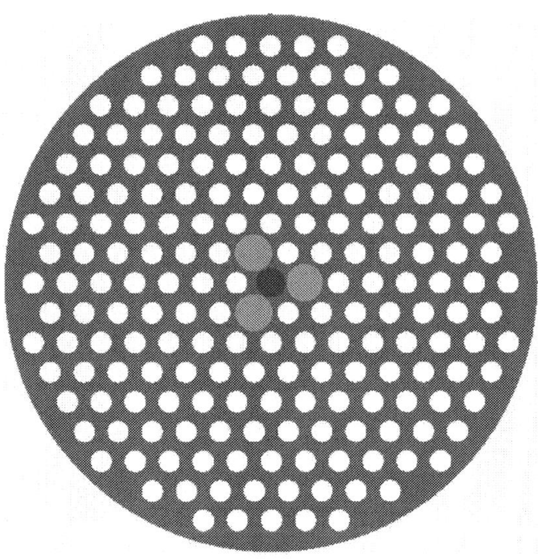

Figure 10. Fluoride-doped W-profile fibre's cross section for infrared and telecom band [32].

Proposed fibre:	
Air hole diameter d [μm]	0.4
Lattice pitch Λ [μm]	0.7
Core size [μm]	1.3
Doped hole diameter d_1 [μm]	0.76
Core dopant material	BaF_2 n = 1.468 at 1.55 μm
1st ring doping material	CaF_2 n = 1.426 at 1.55 μm
Normalized air-hole diameter d/Λ [-]	0.57

Table 7. Design specification of investigated W-profile MOF with doped holes in the 1st ring [32]

The air holes influence the dispersion slope. Larger d/Λ would limit the confinement losses. An interesting feature is that we use fluorides, ZBGA or ZBLAN, not just to dope the cladding, but we propose to use it as a background material instead of silicon dioxide (Figure 12). As a result, optimization of CD is more flexible. Last but not least, temperature and mechanical properties of such a fibre wouldn't be worse [32].

Right combination and the composition of additives is responsible for optimized properties of a compound. Let us consider zirconium (Zr), barium (Ba), gadolinium (Gd) or aluminium (Al) that are the compounds of ZBGA material: $ZrF_4 - BaF_2 - GdF_3 - AlF_3$ [32].The n_{eff} of ZBGA material could be represented by modified Sellmeier equation (17) [46], using the coefficients in Table 8 [45].

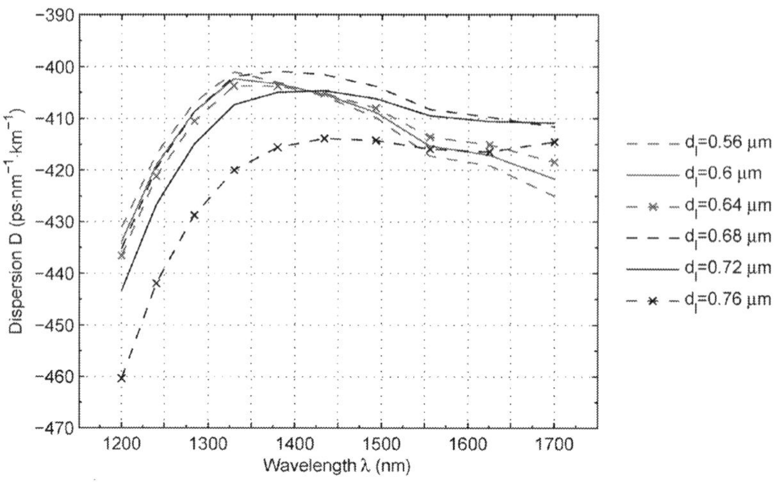

Figure 11. Wavelength evolution of CD with hole diameter of CaF_2 as a parameter [32].

$$n(\lambda) = \frac{A}{\lambda^4} + \frac{B}{\lambda^2} + C + D\lambda^2 + E\lambda^4 \tag{17}$$

Material	A	B	C	D	E
ZBGA at 25°C	2.98316e-6	3.39740e-3	6.81447e-3	-1.20276e-3	-5.48085e-6

Table 8. Sellmeier coefficients for the fluoride-background ZBGA fibre [45].

ZBGA has larger n_{eff} comparing with silicon dioxide. Tailored CD is expected [32].

Parametric sweep performed for the hole diameter showed that flatten CD is possible, which is shown in Figure 6. For hole diameter being equal to 0.44 μm, the evolution of CD is similar to one obtained for silicon dioxide. Larger n_{eff} of the ZBGA fibre requires larger radius of air-holes in the cladding to tailor CD [32].

BaF_2 could be used in the core and CaF_2 in the 1st ring, ZBGA as a background material (Figure 13). CD parameter is then possible within the range of -435 ~ -438 ps/nm/km and over the entire telecommunication band. At the same time, attenuation properties are not worse than those obtainable in fibres with silica as a background material [32].

5.5. Submicron dispersion compensating fibres with modified geometry [2]

One of the trends is to use a lattice with the pitch being less than one micrometre (so-called submicron lattice). Then, by careful adjustments of the diameter of holes (d_1-d_3 in Figure 14),

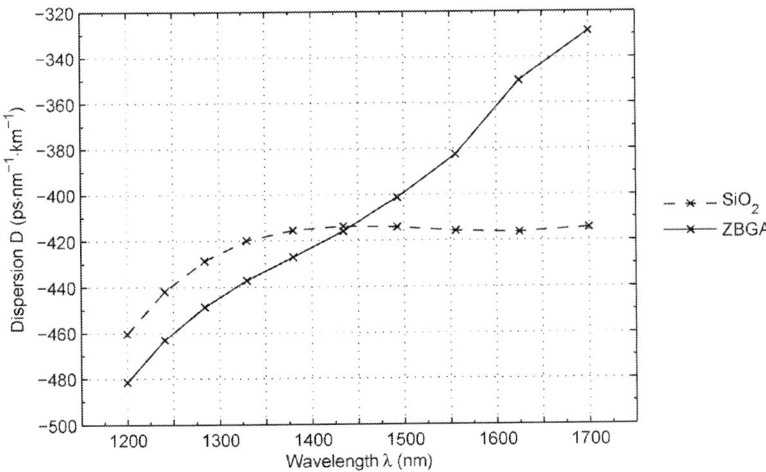

Figure 12. CD in SiO_2 and ZBGA used as a background material [32].

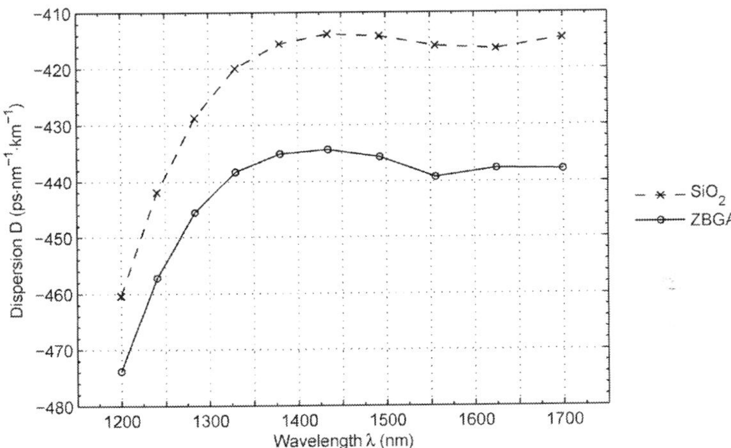

Figure 13. Proposed MOF with ZBGA material used as a background compared to a fibre with SiO_2 as a background [32].

where d_1 is hole diameter in the 1st ring of holes, one could obtain nearly zero CD for the λs from the window 1.25-1.7 μm. Hole diameter and lattice pitch are submicron. Then, high confinement of light waves and strong waveguide dispersion is possible. The geometrical parameters of the fibre are in Table 9 and in Figure 14 [2]. CD and loss are simulated for the 1st light mode. Confinement losses are 6.10^{-6} dB/km (Figure 15). Wavelength evolution of CD is shown in Figure 15 [2].

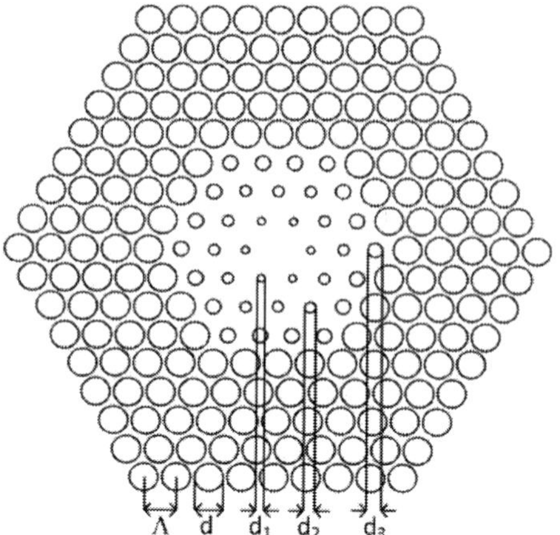

Figure 14. Cross section of a near zero CD flattened MOF with modified rings [2].

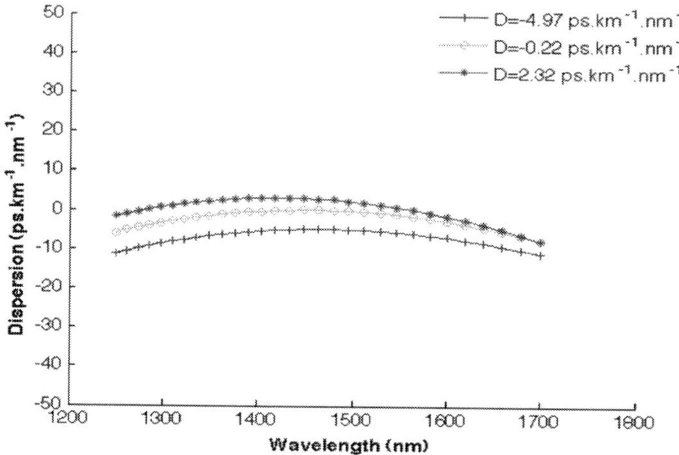

Figure 15. Near zero flattened CD for wideband utilization in telecommunications [2].

Obtained CD parameter is from -7.7 ps/nm/km to 3.1 ps/nm/km at the band of (1.25, 1.7 μm) and its average value is around 0.51 ps/nm/km. In the C-band, it is 1.35±0.46 ps/nm/km. For both C and L bands CD is 0.12±1.32ps/nm/km [2]. Results for parametric iterations for air fraction changed in three most internal rings are shown in Figures 15 and 16. From the results summarized in Table 10 it could be concluded that the fibre has CD slope of -0.09 ps.nm^{-2}.km^{-1} is obtained for little variation of d_1 from other holes and is suited to the slope of conventional ITU-T G.657 fibres. The slope is less than 0.09 ps.nm^{-2}.km^{-1} [2].

d_1	d_2	d_3	d = const.	Λ = const.	CD at 1.55 µm
µm	µm	µm	µm	µm	ps/nm/km
0.134	0.305	0.318	0.355	0.8	2.32
0.134	0.298	0.325	0.355	0.8	-0.22
0.134	0.290	0.325	0.355	0.8	-4.97

Table 9. Geometrical parameters of a fibre with nearly zero flattened CD [2].

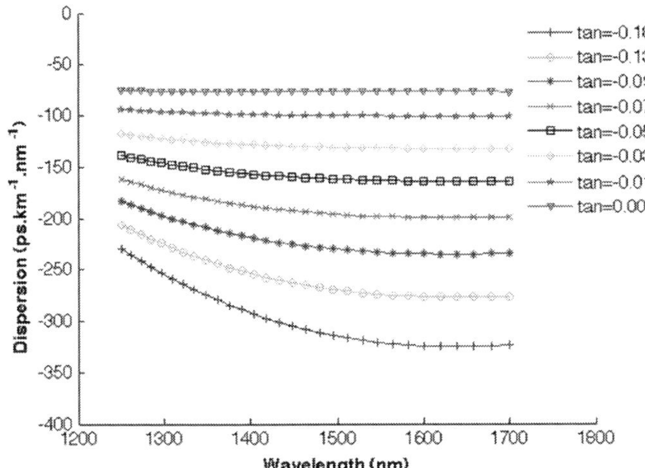

Figure 16. Optimized CD slope by varying the hole diameter in the 1st ring [2].

	d_1	d_2, constant	d_3, constant	d, constant	CD slope
	µm	µm	µm	µm	ps.nm⁻².km⁻¹
	0.18	0.31	0.355	0.355	0.01
hole diameter of	0.20	0.31	0.355	0.355	-0.03
the 1st ring	0.22	0.31	0.355	0.355	-0.07
	0.24	0.31	0.355	0.355	-0.13
	0.25	0.31	0.355	0.355	-0.18

Table 10. Optimized dispersion slope by varying diameter of holes in the 1st ring [2].

Larger CD slope is here obtained by adjusting the value of d_3 (see Table 11, Fig. 16). At d_1 being fixed, linear wavelength evolution of CD is possible. It becomes nonlinear when the slope exceeds -1 ps.nm⁻².km⁻¹. The obtained property could be applied in CD suppression performed on a fibre with the slope of 0.5 ps.nm⁻².km⁻¹ [2].

	d_1 = const.	d_2 = d, constant	d_3	d, constant	CD slope
	µm	µm	µm	µm	ps.nm^{-2}.km^{-1}
hole diameter of the 3rd ring	0.265	0.355	0.35	0.355	-0.39
	0.265	0.355	0.27	0.355	-0.63
	0.265	0.355	0.23	0.355	-0.85
	0.265	0.355	0.17	0.355	-1.32

Table 11. Large CD slope obtained by tuning the hole size in the 3rd ring [2].

6. Final conclusions

We described optical fibres from the perspective of their CD properties, which has to be solved in telecommunication. The mechanisms governing CD properties in telecom fibres are shown. The considered fibres are suitable for potential suppression of group velocity dispersion, utilized mainly in the C-band (dual core fibres with high negative CD parameter) and in wideband applications (CD flattened MOFs and exact slope suppression fibres).

A special family of fibres is so-called fluoride fibres that have CD optimized through material dispersion. This idea is combined with the proposal of a W-type fibre, where particular regions are doped by using different additives, including fluorides, among others. ZBLAN material offers broader range with low attenuation (telecom and infrared range). It makes them very attractive for getting eventually applied in spectrometry or in applications in fibre-optic sensors.

The use of fluorides assessed their flexibility in CD optimization and potential use in wideband CD suppression. Last but not least, it has been shown that optimization of fibre's CD slope is possible in submicron lattice pitch without additives, potentially for exact slope suppression of standard International Telecommunication Union fibres.

Acknowledgements

This work has been supported by the CTU foundation, SGS13-201-OHK3-3T-13.

Author details

Michal Lucki[*] and Tomas Zeman

*Address all correspondence to: lucki@fel.cvut.cz

Department of Telecommunication Engineering, Faculty of Electrical Engineering, Czech Technical University in Prague, Prague, Czech Republic

References

[1] Lucki, M., Zeleny, R. Broadband Dispersion Compensating Photonic Crystal Fibre: conference proceedings, SPIE Vol. 8306. Bellingham (Washington): SPIE, 83060Z-1-83060Z-6, 2011.

[2] Lucki, M., Jiruse, D., Kraus, S. Single-Material Submicron Microstructured Fibres for Broadband Applications in Exact Slope Compensation or Zero- Dispersion Propagation: conference proceedings, 14th International Conference on Transparent Optical Networks (ICTON). Piscataway: IEEE, Tu.P.28., 2012.

[3] Lucki, M. Optimization of Microstructured Fibre for Dispersion Compensation Purposes: conference proceedings, 2011 13th International Conference on Transparent Optical Networks (ICTON). Piscataway: IEEE, p. Tu.P.12, 2011.

[4] Veng M., et al. Dispersion compensating fibres. Opt. Fibre Technol. 2000; 6 164-80.

[5] Antos A., Smith D. Design and characterization of dispersion compensating fibre based on the LP_{01} mode. J. Lightwave Technol. 1994; 12(10) 1739-1745.

[6] Ferrando A., et al. Nearly zero ultraflattened dispersion in photonic crystal fibres. Opt. Lett.2000; 25 790-792.

[7] Ni Y. et al. Dual-Core Photonic Crystal Fibre for Dispersion Compensation. IEEE Photonics Technology Letters 2004; 16(6) 1516-1518.

[8] Liu Z. et al. A broadband ultra flattened CD microstructured fibre for optical communications. Optics Communications 2007; 272(1) 92-96.

[9] Fujisawa T. et al. Chromatic dispersion profile optimization of dual-concentric-core photonic crystal fibres for broadband dispersion compensation. Opt. Express 2006; 14(2) 893-900.

[10] Yang S. et al. Theoretical study and experimental production of high negative dispersion photonic crystal fibre with large area mode field. Opt. Express 2006; 14(7) 3015-3023.

[11] Nejad S. M., Ehtehsami N. A Novel Design to Compensate Dispersion for Square-lattice Photonic Crystal Fibre over E to L Wavelength Bands: conference proceedings, Communication Systems, Networks and Digital Signal Processing Symposium, 654-658, 2010.

[12] Hosaka T. et al. Dispersion of pure GeO2 glass core and F-doped GeO2 glass cladding single-mode opticalfibre. Eelectron. Letters 1987; 23(1) 24-26.

[13] Hoo Y. et al. Design of photonic crystal fibres with ultra-low, ultra-flattened chromatic dispersion. Optics Communications 2004; 242(4-6) 327-332.

[14] Ono-Kuwahara M. et al. Fluorine-doped silica fibre with high transparency and resistivity to deep ultra violet light, Lasers and Electro-Optics Lasers and Electro-Op-

tics: conference proceedings, Conference on Quantum Electronics and Laser Science. CLEO/QELS 1-2, 2008.

[15] Olyaee S., Taghipour F. Ultra-Flattened Dispersion Photonic Crystal Fibre with Low Confinement Loss: conference proceedings, Proceedings of the 11th International Conference on Telecommunications, pp. 531-534, 2011.

[16] Peckham D. W. et al. Reduced Dispersion Slope, Non-Zero Dispersion Fibre: conference proceedings, European Conferencce and Exhibition on Optical Communication, 139-140, 1998.

[17] Aikawa K. et al. High-performance Dispersion-slope and Dispersion Compensation Modules. Fujikura Technical Review 2002; 31.

[18] Andersen P. A., Paulsen H. N., and Larsen J. J. A photonic crystal fibre with zero dispersion at 1064 nm. Optical Communication 2002; 2 1-2.

[19] Hansen K. P. Introduction to nonlinear photonic crystal fibres. J. Opt. Fibre Commun. 2005; 2 226-254.

[20] Ferrando A., Silvestre E., and Andrés P. Designing the properties of dispersion flattened photonic crystal fibres. Opt. Express 2001; 9(13) 687-697.

[21] Lægsgaard J., Asger Mortensen N., and Bjarklev A. Mode areas and field energy distribution in honeycomb photonic bandgap fibres. J. Opt. Soc. Am. B 2003; 20(10) 2037-2045.

[22] Birks T. et al. Dispersion compensation using single-material fibres. IEEE Photonics Technology Letters 1999; 11 674-676.

[23] Haxha S., Ademgil H. Novel design of photonic crystal fibres with low confinement losses, nearly zero ultra-flatted chromatic dispersion, negative chromatic dispersion and improved effective mode area. Optics Communications 2008; 15(2) 278-286.

[24] Saitoh K. et al. Chromatic dispersion control in photonic crystal fibres: application to ultra-flattened dispersion. Opt. Express 2003; 11(8).

[25] Razzak S. M. A. Proposal for Highly Nonlinear Dispersion Flattened Octagonal Photonic Crystal Fibres. IEEE Photonics Technology Letters 2010; 20(4).

[26] Razzak S. M. A. et al. Chromatic Dispersion Properties of a Decagonal Photonic Crystal Fibre: conference proceedings, International Conference on Information and Communications Technology, 159-162, 2007.

[27] Ming W., et al. Broadband dispersion compensating fibre using index-guiding photonic crystal fibre with defected core. Chin. Opt. Lett 2008; 6 22-24.

[28] Zsigri B., Lægsgaard J., Bjarklev A. A novel photonic crystal fibre design for dispersion compensation. J. Opt. A: Pure Appl. Opt. 2004; 6 717-720.

[29] Yu Ch. et al. Tunable dual-core liquid-filled photonic crystal fibres for dispersion compensation. Optics Express 4443 2008; 16(17) 4443-4451.

[30] Zhong, Q., Inniss, D. Characterisation of lightguiding structure of optical fibres by atomic force microscopy. J. Lightwave. Technol. 1994; 12 1517-1523.

[31] Huttunen, A., Torma, P. Optimization of dual-core and microstructure fibre geometries for dispersion compensation and large mode area. Optics Express 2005; 13(2) 627-635.

[32] Lucki, M., Zeleny, R.: Broadband submicron flattened dispersion compensating fibre with asymmetrical fluoride doped core: conference proceedings, Micro-structured and Specialty Optical Fibres II. Bellingham: SPIE, 87750M-1-87750M-8, 2013.

[33] Tefelska, M. et al. Propagation Effects in Photonic Liquid Crystal Fibres with a Complex Structure. Optical and Acoustical Methods in Science and Technology, 2010; 118(6) 1259-1261.

[34] Petruzzi, P., Lowry, C., Goldhar, J., Sivanesan, P., Dispersion compensation using only Fibre Bragg Gratings: conference proceedings, Optical Fibre Communication Conference 4, 14-16, 1999.

[35] Saad, M. Fluoride glass fibres. Photonics Society Summer Topical Meeting Series 81-82; 2011.

[36] Tran, D., Sigel, G., Bendow, B. Heavy metal fluoride glasses and fibres: A review. Journal of Lighwave Technology 1984; 2 566-586.

[37] Poli, F., Cucinotta, A., Selleri, S. Photonic Crystal Fibres: Properties and Applications. Springer Series in Materials Science 102. The Nederlands; 2007.

[38] Hájek, M., Holomeček, P. Chromatická disperze jednovidových optických vláken a její měření Mikrokom; 2006.

[39] Saitoh, K., Koshiba, M. Chromatic dispersion control in photonic crystal fibres: application to ultra-flattened dispersion. Optics Express 2003; 11(8) 843-852.

[40] Stern, M. S. Semivectorial polarised finite difference method for optical waveguides with arbitary index profiles. Optoelectronics, IEE Proceedings 1988; 135 56-63.

[41] Huang, W. P., Xu, C. L. Simulation of three-dimensional optical waveguides by a full-vector beam propagation method. IEEE Journal of Quantum Electronics 1993; 29 2639-2649.

[42] Yee, K. Numerical solution of initial boundary value problems involving maxwell's equations in isotropic media. IEEE Transactions on Antennas and Propagation 1966; 14 302-307.

[43] Ghosh, G., Yajima, H. Pressure-dependent Sellmeier coefficients and material dispersions for silica fibre glass. Journal of Lightwave Technology 1998; 16 2002-2005.

[44] Iwasaki, T., Endo, M., Ghosh, G. Temperature-dependent Sellmeier coefficients and chromatic dispersions for some optical fibre glasses. Journal of Lightwave Technology 1994; 12 1338-1342.

[45] Weber. J. M. et al. Handbook of Optical Materials. The CRC Press Laser and Optical Science and Technology Series; 2003.

[46] Mitachi, S. Dispersion measurement on fluoride glasses and fibres. Lightwave Technology Journal 1989; 7(8) 1256-1263.

Chapter 4

Polarization Effects in Optical Fiber Links

Krzysztof Perlicki

Additional information is available at the end of the chapter

http://dx.doi.org/10.5772/59000

1. Introduction

Polarization effects were already observed in the first optical fiber transmission experiments. Initially, polarization effects in an optical fiber were a pure laboratory curiosity. During telecom expansion in 1990s these effects became the focus of many research groups. Optical fiber polarization effects and interaction between them become particular important as bit rate of a single optical channel increases. These effects must be overcome to implement more than 10 Gb/s transmission in a single wavelength over fiber plants in long haul optical systems. It can be a seriously limiting factor in systems in which the fiber plants were installed by 1998. Because, these old fibers are characterised by high internal birefringence such as core asymmetry and built in stress. The current fiber plants are characterised by low internal birefringence. However, external birefringence such as twists and external stress applied to optical fiber, significantly contributes to polarization effects.

Polarization effects are now a fundamental requirement to understand the signal propagation in modern long haul lighwave communication networks. The present chapter is designed to cover: description of polarized light, polarization phenomena in optical fiber links, modeling of polarization phenomena and polarizing component.

2. Description of polarized light

2.1. Polarization ellipse equation

All the important features of light wave follow from a detailed examination of Maxwell equations. Electromagnetic waves have two polarization along the x axis and along the y axis. The general form of polarized light wave propagating in z direction can be derived from two linear polarized components in the x and y directions [1]:

$$E_x(z,t) = E_{0x} \cos(\tau_\omega + \phi_x), \tag{1}$$

$$E_y(z,t) = E_{0y} \cos(\tau_\omega + \phi_y), \tag{2}$$

where: x and y refers to the components in the x and y directions, E_{0x} and E_{0y} are the real maximum amplitudes of electric field, ϕ_x and ϕ_y are the phases and τ_ω is so called propagator, which describes the propagation of the signal component in the z-direction.

Next, equations (1) and (2) can be written as: [1]:

$$\frac{E_x(z,t)}{E_{0x}} = \cos(\tau_\omega)\cos(\phi_x) - \sin(\tau_\omega)\sin(\phi_x), \tag{3}$$

$$\frac{E_y(z,t)}{E_{0y}} = \cos(\tau_\omega)\cos(\phi_y) - \sin(\tau_\omega)\sin(\phi_y). \tag{4}$$

Squaring and adding (3) and (4) then yields:

$$\frac{E_x^2(z,t)}{E_{0x}^2} + \frac{E_y^2(z,t)}{E_{0y}^2} - 2\frac{E_x(z,t)}{E_{0x}}\frac{E_y(z,t)}{E_{0y}}\cos(\phi) = \sin^2(\phi), \tag{5}$$

where: $\phi = \phi_y - \phi_x$.

Equation (5) is an ellipse equation. This equation is called the polarization ellipse.

Figure 1 shows the polarization ellipse for optical field.

The polarization ellipse presents some important parameters enabling the characterization of the state of light polarization (SOP) [2]:

1. Axis x and y are the initial, unrotated axes, ξ and η are a new set of axes along the rotated.

2. The area of polarization ellipse depends on the lengths of major and minor axes, amplitudes E_{0x} and E_{0y} and phase shift ϕ.

3. The angle $\beta_p = \text{arctg}(E_{0y}/E_{0x})$ is called the auxiliary angle ($0 \leq \psi \leq \pi/2$).

4. The rotation angle ψ ($-\beta_p \leq \psi \leq \beta_p$) is the angle between axis x and major axis ξ. This angle is called the azimuth angle.

5. The ellipticity is the major axis to minor axis ratio (b/a).

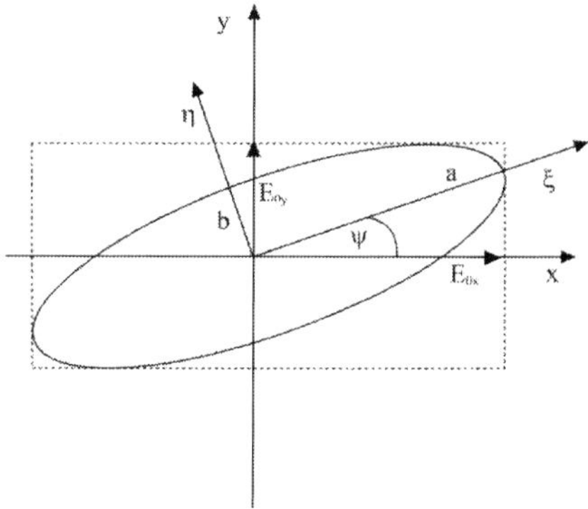

Figure 1. Polarization ellipse for optical field

6. The angle equals to $\upsilon=\mathrm{arctg}(b/a)$ is called ellipticity angle. For linearly polarized light $\upsilon=0^\circ$; for circularly polarized light $|\upsilon|=45^\circ$. In turn, for right polarized light: $0^\circ<\upsilon\leq45^\circ$ and for left polarized light: $-45^\circ\leq\upsilon<0^\circ$.

Figure 2 presents some polarization states; The phase shift ϕ is only changed.

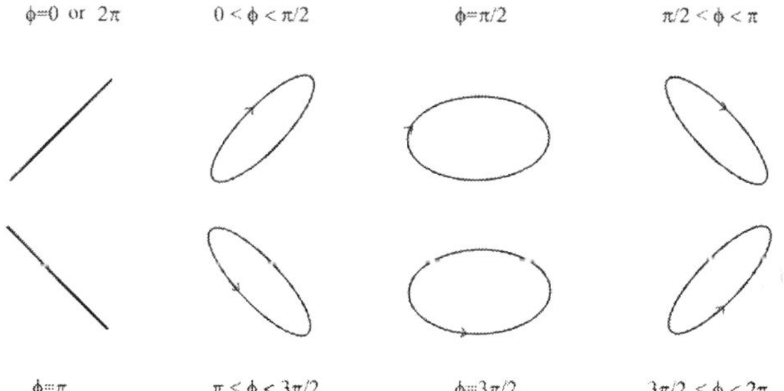

Figure 2. Different shapes of the polarization ellipse as a function of phase shift

2.2. Jones notation

The light wave components in terms of complex quantities can be expressed by means of the Jones vector [1]:

$$\begin{bmatrix} E_x \\ E_y \end{bmatrix} = \begin{bmatrix} E_{0x}e^{j\phi_x} \\ E_{0y}e^{j\phi_y} \end{bmatrix}. \tag{6}$$

The Jones vector representation is suited to all problems related to the totally polarized light.

Table 1 gives the Jones vectors corresponding to the fundamental SOPs.

State of polarization	Jones vector
Linear horizontal $E_{0y}=0;\ E_{0x}^{2}=1$	$\begin{bmatrix}1\\0\end{bmatrix}$
Linear vertical $E_{0x}=0;\ E_{0y}^{2}=1$	$\begin{bmatrix}0\\1\end{bmatrix}$
Linear 45° $E_{0x}=E_{0y};\ 2E_{0x}^{2}=1$	$\dfrac{1}{\sqrt{2}}\begin{bmatrix}1\\1\end{bmatrix}$
Linear -45° $E_{0x}=-E_{0y};\ 2E_{0x}^{2}=1$	$\dfrac{1}{\sqrt{2}}\begin{bmatrix}1\\-1\end{bmatrix}$
Right circular $E_{0x}=E_{0y},\ \phi=\pi/2;\ 2E_{0x}^{2}=1$	$\dfrac{1}{\sqrt{2}}\begin{bmatrix}1\\j\end{bmatrix}$
Left circular $E_{0x}=E_{0y},\ \phi=-\pi/2;\ 2E_{0x}^{2}=1$	$\dfrac{1}{\sqrt{2}}\begin{bmatrix}1\\-j\end{bmatrix}$

Table 1. Jones vectors of the fundamental SOPs

The Jones matrices for some polarization components are 2x2 matrices.

The relationship between the both output and input Jones vectors can be written as:

$$\begin{bmatrix}E_{x,out}\\E_{y,out}\end{bmatrix}=\begin{bmatrix}j_{xx}&j_{xy}\\j_{yx}&j_{yy}\end{bmatrix}\cdot\begin{bmatrix}E_{x,in}\\E_{y,in}\end{bmatrix}. \tag{7}$$

Where $\begin{bmatrix}j_{xx}&j_{xy}\\j_{yx}&j_{yy}\end{bmatrix}$ is the Jones matrix of a polarization component.

We now describe the matrix forms for the retarder (wave plate), rotator and polarizer (diatte-nuator), respectively.

1.   Retarder

The retarder causes a phase shift of $\phi/2$ along the fast (i.e. x) axis and a phase shift of $-\phi/2$ along slow (i.e. y) axis. This behavior is described by [3]:

$$\begin{bmatrix}E_{x,out}\\E_{y,out}\end{bmatrix}=\begin{bmatrix}e^{j\frac{\phi}{2}}&0\\0&e^{-j\frac{\phi}{2}}\end{bmatrix}\cdot\begin{bmatrix}E_{x,in}\\E_{y,in}\end{bmatrix}. \tag{8}$$

For quarter-wave plate ϕ is $\pi/2$ and for half-wave plate ϕ is π.

2.   Rotator

If the angle of rotation is Θ then the components of light emerging from rotation are written as [3]:

$$\begin{bmatrix} E_{x,out} \\ E_{y,out} \end{bmatrix} = \begin{bmatrix} \cos(\Theta) & \sin(\Theta) \\ -\sin(\Theta) & \cos(\Theta) \end{bmatrix} \cdot \begin{bmatrix} E_{x,in} \\ E_{y,in} \end{bmatrix}, \tag{9}$$

3.   Polarizer

The polarizer behavior is characterized by the transmission factor p_x and p_y. Here, for complete transmission $p_x=p_y=1$ and for complete attenuation $p_x=p_y=0$.

The output Jones vector for a polarizer is given by [3]:

$$\begin{bmatrix} E_{x,out} \\ E_{y,out} \end{bmatrix} = \begin{bmatrix} p_x & 0 \\ 0 & p_y \end{bmatrix} \cdot \begin{bmatrix} E_{x,in} \\ E_{y,in} \end{bmatrix}. \tag{10}$$

The Jones matrix (J_{ar}) a polarization component (J) rotated through an angle Θ is:

$$J_{ar} = J_R(-\Theta) \cdot J \cdot J_R(\Theta), \tag{11}$$

where $J_R(\Theta)$ is the rotation matrix.

2.3. Stokes parameters

Let us introduce the S_0, S_1, S_2 i S_3 real quantities defined by the following relations [4]:

$$S_0 = E_{0x}^2 + E_{0y}^2, \tag{12}$$

$$S_1 = E_{0x}^2 - E_{0y}^2, \tag{13}$$

$$S_2 = 2E_{0x}E_{0y}\cos(\phi), \tag{14}$$

$$S_3 = 2E_{0x}E_{0y}\sin(\phi), \tag{15}$$

where E_{0x}, E_{0y} are the real maximum amplitudes and ϕ is the phase difference.

These quantities are called the Stokes parameters. The Stokes parameters have a physical meaning in terms of intensity. The parameter S_0 represents the total intensity of light. The second parameter S_1 describes the difference in the intensities of the linearly horizontal polarized light and the linearly vertical polarized light. The third parameter S_2 represents the difference in the intensities of the linearly 45º polarized light and linearly-45º polarized light. The last parameter S_3 represents the difference in the intensities of the right circularly polarized light and the left circularly polarized light. The Stokes parameters are real values. The Stokes representation is the most adequate representation in treating partially polarized and unpolarized light problems. Moreover, Stokes representation is well suited to the definition of the Degree Of Polarization (DOP). This parameter is equal to [2]:

$$DOP = \frac{\sqrt{S_1^2 + S_2^2 + S_3^2}}{S_0},\tag{16}$$

with value between 0 (unpolarized light) and 1 (totally polarized light).

Often the normalized Stokes parameters are used to describe the light polarization: $\frac{S_1}{S_0}$, $\frac{S_2}{S_0}$, $\frac{S_3}{S_0}$; with value between-1 and 1.

Table 2 shows some Stokes vectors corresponding to the fundamental SOPs.

State of polarization	Stokes vector
Linear horizontal $E_{0y}=0,\ E_{0x}^2=1$	$\begin{bmatrix} 1 \\ 1 \\ 0 \\ 0 \end{bmatrix}$
Linear vertical $E_{0x}=0,\ E_{0y}^2=1$	$\begin{bmatrix} 1 \\ -1 \\ 0 \\ 0 \end{bmatrix}$
Linear 45º $E_{0x}=E_{0y}=E_0,\ \phi=0,\ 2E_0^2=1$	$\begin{bmatrix} 1 \\ 0 \\ 1 \\ 0 \end{bmatrix}$
Linear 45º $E_{0x}=E_{0y}=E_0,\ \phi=\pi,\ 2E_0^2=1$	$\begin{bmatrix} 1 \\ 0 \\ -1 \\ 0 \end{bmatrix}$

State of polarization	Stokes vector
Right circular $E_{0x}=E_{0y}=E_0,\ \phi=\pi/2,\ 2E_0^2=1$	$\begin{bmatrix} 1 \\ 0 \\ 0 \\ 1 \end{bmatrix}$
Left circular $E_{0x}=E_{0y}=E_0,\ \phi=3\pi/2,\ 2E_0^2=1$	$\begin{bmatrix} 1 \\ 0 \\ 0 \\ -1 \end{bmatrix}$

Table 2. Stokes vectors of the fundamental SOPs

The Poincaré sphere (Figure 3) is a very useful graphical tool representation of polarization in real three-dimensional space.

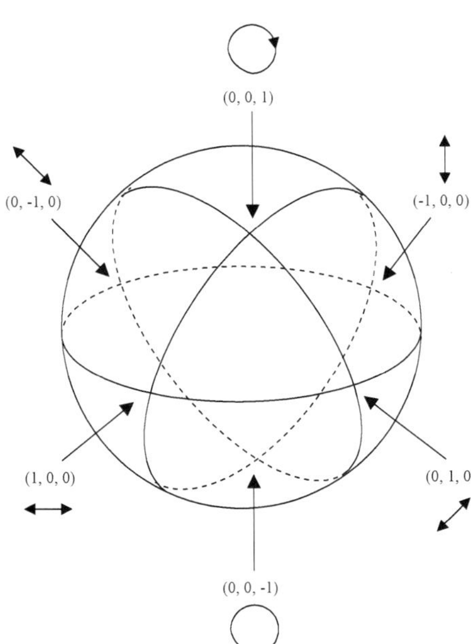

Figure 3. The Poincaré sphere and fundamental SOPs on this sphere

Each polarization is represented by a point on the Poincaré sphere (totally polarized light) or within the Poincaré sphere (partially polarized light) centered on rectangular coordinate system. Center of the Poincaré sphere represents unpolarized light. The coordinates of the point are normalized Stokes parameters. All linear SOPs lie on the equator. The right circular SOP and left one is located at the North and South Pole, respectively. Elliptically polarized states are represented everywhere else on the surface of the Poincaré sphere. The two orthogonal polarizations are located diametrically opposite on the Poincaré sphere. A continuous evolution of SOP is represented on the Poincaré sphere as a continuous path on this sphere (Figure 4).

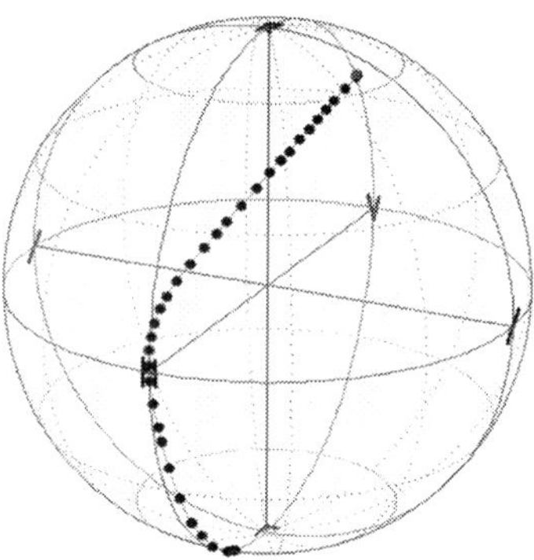

Figure 4. Example of a continuous evolution of SOP on the Poincaré sphere

Figure 5 presents changing the SOPs by means of the retarder and rotator.

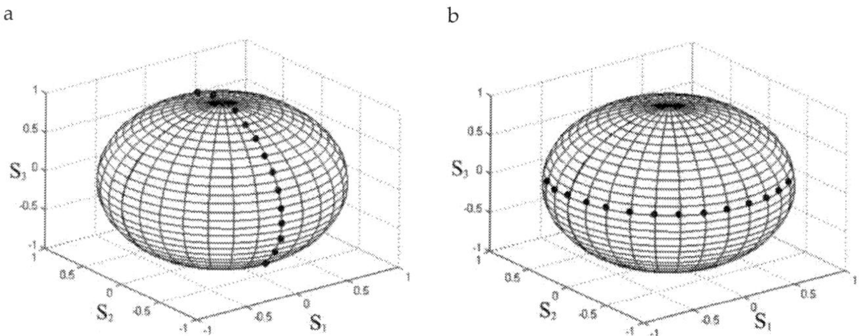

Figure 5. The effect of changing the SOPs by the retarder (a) and the rotator (b)

On the Poincaré sphere the phase shift causes that the intial SOP moves to a new SOP along the same longitude line (Figure 5a). The linear, elliptically and circular SOPs can be achieved by means of a single retarder. In turn, on the Poincaré sphere the rotation by a rotator causes that the intial SOP moves to a new SOP along the same latitude line (Figure 5b).

2.4. Mueller notation

Impact of an optical component (or optical system) properties on the polarization of light can be determined by constructing the Stokes vector for the input light and applying Mueller calculus, to obtain the Stokes vector of the light leaving the component:

$$\vec{S}_{out} = M \cdot \vec{S}_{in}, \tag{17}$$

where $\vec{S}_{in}$ and $\vec{S}_{out}$ is the input and output Stokes vector, respectively, M is the Mueller matrix of an optical component.

We now describe the Mueller matrix forms for the retarder, rotator and polarizer, respectively.

1.   Retarder

The retarder causes a total phase shift ϕ between fast (x) and slow (y) axis. The Mueller matrix of the retarder is seen to be [3]:

$$M_{Ret} = \begin{bmatrix} 1 & 0 & 0 & 0 \\ 0 & 1 & 0 & 0 \\ 0 & 0 & \cos(\phi) & -\sin(\phi) \\ 0 & 0 & \sin(\phi) & \cos(\phi) \end{bmatrix}. \tag{18}$$

2.   Rotator

The Mueller matrix of the rotator is given [3]:

$$M_{Rot} = \begin{bmatrix} 1 & 0 & 0 & 0 \\ 0 & \cos(2\Theta) & \sin(2\Theta) & 0 \\ 0 & -\sin(2\Theta) & \cos(2\Theta) & 0 \\ 0 & 0 & 0 & 1 \end{bmatrix}. \tag{19}$$

Because polarization effects are described in the intensity domain the physical rotation through an angle Θ leeds to the appearance of 2Θ.

3.   Polarizer

The Mueller matrix for polarizer is [3]:

$$M_{polar} = \frac{1}{2}\begin{bmatrix} p_x^2 + p_y^2 & p_x^2 - p_y^2 & 0 & 0 \\ p_x^2 - p_y^2 & p_x^2 + p_y^2 & 0 & 0 \\ 0 & 0 & 2p_x p_y & 0 \\ 0 & 0 & 0 & 2p_x p_y \end{bmatrix}, \tag{20}$$

where p_x and p_y are so called transmission factors.

Here, the transmission factors are $p_x=1$ and $p_y=0$ for the linear horizontal polarizer. In turn, the transmission factors $p_x=1$ and $p_y=0$ for the linear vertical polarizer.

The Mueller matrix (M_{ar}) for a polarization component (M) rotated through an angle Θ is:

$$M_{ar} = M_R(-2\Theta) \cdot M \cdot M_R(2\Theta), \tag{21}$$

where $M_R(\Theta)$ is the rotation matrix.

3. Polarization phenomena in optical fiber links

3.1. Polarization mode dispersion

The optical fiber transmission systems are exposed to some polarization effects. Changing of tranmsission quality (e.g. transmission capacity) of an optical fiber links during high bit rate transmission is caused by Polarization Mode Dispersion (PMD), Polarization Dependent Loss (PDL), Polarization Dependent Gain (PDG).

Polarization Mode Dispersion is impairment phenomenon that limits the transmission speed and distance in high bit rate optical fiber communication systems. The impairment results from PMD is similar to chromatic dispersion impairment.

According to [5]: There always exists an orthogonal pair of polarization states output a birefringent concatenation which are stationary to first order in frequency. These two states are called Principle States of Polarization (PSP).

A diffrential delay exists between signals launched along one PSP and its orthogonal comple-ment. This effect is quantified by Differential Group Delay (DGD). There are many ways in which an optical fiber can become birefringent. Birefringence can arise due to an asymmetric fiber core or asymmetric fiber refractive index or can be introduced through internal stresses during fiber manufacture or through external stresses during cabling and installation. Polarization Mode Dispersion of an optical fiber link is proportional to the square root of the fiber link length (strong coupling between the orthogonally polarized signal components) or to the fiber link length (weak coupling between ones). The frequency dependent evolution of SOPs in an optical fiber link (Figure 6) is described by the following equation [6]:

$$\frac{d\vec{S}}{d\omega} = \vec{\Omega} \times \vec{S},$$
(22)

where $\vec{S}$ is the Stokesa vector, ω is angular frequency and $\vec{\Omega}$ is the PMD vector.

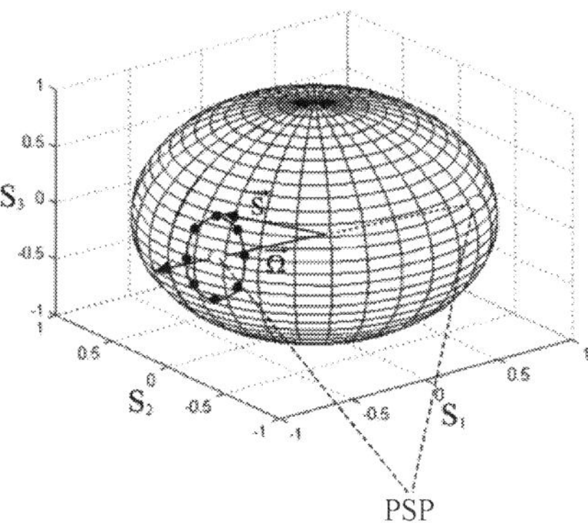

Figure 6. State of polarization transformation through the PMD vector

The pointing direction of the PMD vector is aligned to the slow PSP. The length of the PMD vector is the DGD value between the slow and fast PSP.

The Probability Density Function for DGD ($\tau_{g,r}$) is given by [7]:

$$\mathrm{PDF}\left(\tau_{g,r}\right) = \frac{8}{\pi^2 \langle \tau_{g,r} \rangle} \left(\frac{2\tau_{g,r}}{\langle \tau_{g,r} \rangle}\right)^2 e^{-\frac{\left(\frac{2\tau_{g,r}}{\langle \tau_{g,r} \rangle}\right)^2}{\pi}},$$
(23)

where $\langle \tau_{g,r} \rangle$ is average value of DGD.

Figure 7 shows Probability Density Function for DGD.

Differential Group Delay distribution is Maxwellian distribution. Differential Group Delay can be also expressed as [8]:

$$\langle \tau_{g,r} \rangle^2 = \frac{1}{3}\left(\frac{\lambda L_c}{c L_B}\right)^2 \left(\frac{L}{L_c} - 1 + e^{-\frac{L}{L_c}}\right),$$
(24)

where: λ is wavelength, c is light wave velocity in vacuum, L_B is beat length and L_c is correlation length, L is optical fiber length.

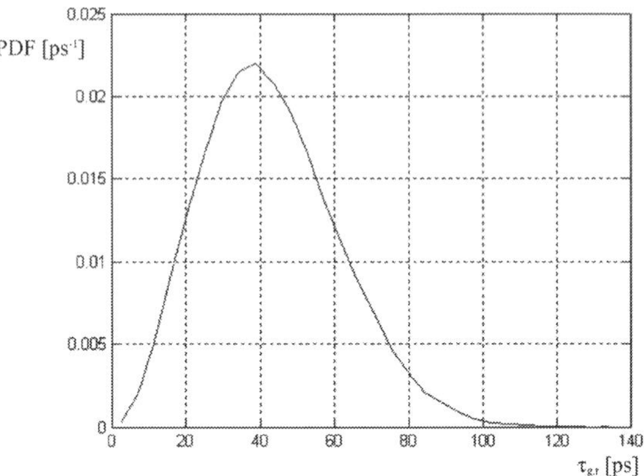

Figure 7. Probability Density Function (PDF) for DGD; average value of DGD is 40 ps

The beat length describes the length required for SOP to rotate 2π (360 degrees). In turn, the correlation length is defined to be length at which the difference between average power of orthogonally polarized signal components is within $1/e^2$.

Second order PMD is generated by a change of the PMD with frequency (wavelength) [9]:

$$\frac{d\vec{\Omega}}{d\omega} = \frac{d\tau_{g,r}}{d\omega}\vec{p} + \tau_{g,r}\frac{d\vec{p}}{d\omega} \ , \tag{25}$$

where $\vec{p}$ is the Stokes vector pointing in the direction of the fast PSP.

Differentiating the PMD vector with respect to frequency gives two components of second order PMD. The first term on the right side of equation (25) is so called polarization dependent chromatic dispersion, it is known to cause polarization-dependent pulse compression and broadening, while the second term causes depolarization. Figure 8 illustrates changing the SOP with frequency – second order PMD.

The Probability Density Function of second order PMD is given by [10]:

$$\mathrm{PDF}\left(\left|\vec{\Omega}_\omega\right|\right) = \frac{32\left|\vec{\Omega}_\omega\right|}{\pi\left\langle\left|\vec{\Omega}_\omega\right|\right\rangle^4}\tanh\left(\frac{4\left|\vec{\Omega}_\omega\right|}{\left\langle\left|\vec{\Omega}_\omega\right|\right\rangle^2}\right)\mathrm{sech}\left(\frac{4\left|\vec{\Omega}_\omega\right|}{\left\langle\left|\vec{\Omega}_\omega\right|\right\rangle^2}\right), \tag{26}$$

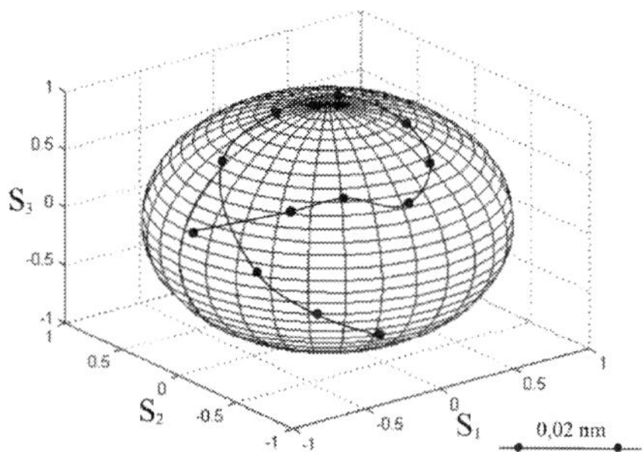

Figure 8. The effect of changing the SOP with frequency

Figure 9 shows Probability Density Function of second order PMD.

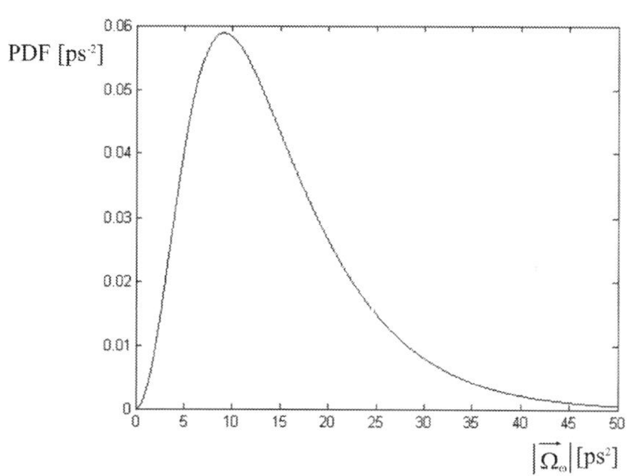

Figure 9. Probability Density Function (PDF) of second order PMD; average value of second order PMD is 10,0 ps²

It should be note that concatenation of two birefringent optical components (e.g. two sections of polarization maintaining optical fiber) generates only first order PMD (Figure 10a). These birefringent sections are orientated randomly relative to each other. In turn, concatenation of three (and more) birefringent optical components generates high order PMD (Figure 10b).

a   b

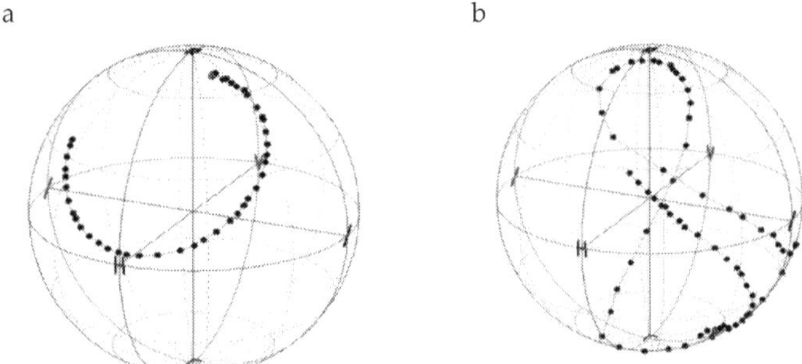

Figure 10. State of polarization evolution through rotating two birefringent optical components (a) and three birefringent optical components (b) which are orientated randomly relative to each other

3.2. Polarization Dependent Loss

Polarization Dependent Loss is defined as absolute value or the relative difference between an optical component maximium and minimum transmission loss given all possible input SOPs [3]. Dichroism phenomenon is responsible for the PDL effect. Dichroism can be achived by optical fiber bending or interaction between optical beam and tilted glass plate. The PDL value can be given by the following relationship [11]:

$$PDL[dB] = 10\log_{10}\left(\frac{T_{r,max}}{T_{r,min}}\right), \tag{27}$$

where $T_{r,max}$ and $T_{r,min}$ are the maximum and minimum transmission intensities through an optical component.

The PDL can be also written as [11]:

$$PDL[dB] = 10\log_{10}\left(\frac{1+\left|\vec{\Gamma}\right|}{1-\left|\vec{\Gamma}\right|}\right), \tag{28}$$

where $\vec{\Gamma}$ is the PDL vector.

This vector is equal to: $\dfrac{T_{r,max} - T_{r,min}}{T_{r,max} + T_{r,min}}$. The pointing direction of the PDL vector is aligned to maximum transmission direction. In other words, this vector is aligned to a polarization vector that imparts the least PDL value. The cumulative PDL vector over concatenate two optical components with the PDL vectors $(\vec{\Gamma_1}, \vec{\Gamma_2})$ is [12]:

$$\overrightarrow{\Gamma_{12}} = \frac{\sqrt{1-\Gamma_2^2}}{1+\overrightarrow{\Gamma_1}\overrightarrow{\Gamma_2}}\overrightarrow{\Gamma_1} + \frac{1+\overrightarrow{\Gamma_1}\overrightarrow{\Gamma_2}\left(\dfrac{1-\sqrt{1-\Gamma_2^2}}{\Gamma_2^2}\right)}{1+\overrightarrow{\Gamma_1}\overrightarrow{\Gamma_2}}\overrightarrow{\Gamma_2}. \tag{29}$$

If we want to calculate the resulting PDL value from PDL of each optical component we need to average over all possible orientation between $\overrightarrow{\Gamma_1}$ and $\overrightarrow{\Gamma_2}$ vectors [12]:

$$\langle\Gamma_{12}\rangle = \frac{1}{2}\int_{-1}^{1}\sqrt{\frac{\Gamma_1^2 + \Gamma_2^2 - \Gamma_1^2\Gamma_2^2 + 2\Gamma_1\Gamma_2\eta_a + \Gamma_1^2\Gamma_2^2\eta_a^2}{\left(1+\Gamma_1\Gamma_2\eta_a\right)^2}}\,d\eta_a, \tag{30}$$

where η_a is angle between $\overrightarrow{\Gamma_1}$ and $\overrightarrow{\Gamma_2}$ vectors.

Concatenation of N optical components with PDL gives the following result:

$$\overrightarrow{\Gamma} = \sum_{j=1}^{N}\overrightarrow{\Gamma_j} + 0\left(\Gamma^2\right). \tag{31}$$

Polarization Dependent Loss distribution is Rayleigh distribution (Figure 11).

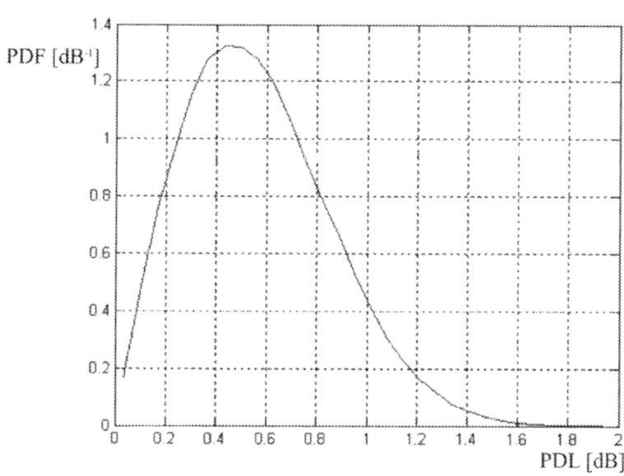

Figure 11. Probability density function (PDF) of PDL; average value of PDL is 0.5 dB

In the presence of PMD the PDL distribution is closed to Maxwellian distribution. It is important to note that in the case of single mode fiber the orthogonal SOPs pairs at the input lead to orthogonal output SOPs pairs, although the input SOP is not maintained in general. But, when the optical fiber link includes PDL the SOPs are no longer orthogonal. Moreover, polarization effects due to interaction between PMD and PDL can significantly impair optical fiber transmission systems. The accumulative PMD and PDL impairment is more dangerous for lightwave communication systems than a pure PMD or PDL impairment.

3.3. Polarization Dependent Gain

Another polarization effect which is closely related to PDL is PDG. This phenomenon is present in optical amplifiers (first of all in Semiconductor Optical Amplifier). Polarization dependent gain can be defined as absolute value or the relative difference between an optical amplifier maximimum gain (G_{max}) and minimum one (G_{min}):

$$PDG[dB] = 10\log_{10}\left(\frac{G_{max}}{G_{min}}\right), \tag{32}$$

Polarization Hole Burning phenomenon is responsible for the PDG effect. It is important to know that PDG effect is observed for linear polarized optical signals which are amplified. Polarization Dependent Gain for circular polarization can be neglected [13].

4. Modeling of polarization phenomena

The analysis of impact of optical fiber polarization properties on optical signal transmissions requires a detailed description of polarization effects. The most popular approaches are using an optical fiber links modeling based on homogeneous polarization segments and electromagnetic wave propagation equations.

In general, an optical fiber exhibits axially-varing birefringence and can be represented by a series of short and homogeneous polarization segments. Each polarization is described as the randomly rotated polarization elements. These polarization elements are characterised by birefringence (i.e. PMD) or dichroism (i.e. PDL).

The birefringent element can be represented by retarder (phase shifter). In terms of the Jones matrix this element is described by:

$$J_{DGD} = \begin{bmatrix} e^{j\frac{\phi}{2}} & 0 \\ 0 & e^{-j\frac{\phi}{2}} \end{bmatrix}, \tag{33}$$

where ϕ is is the total phase shift between the polarization signal components (two polarization modes).

In terms of the Mueller matrix the birefringent element is given by:

$$M_{DGD} = \begin{bmatrix} 1 & 0 & 0 & 0 \\ 0 & 1 & 0 & 0 \\ 0 & 0 & \cos(\phi) & -\sin(\phi) \\ 0 & 0 & \sin(\phi) & \cos(\phi) \end{bmatrix}. \tag{34}$$

The value of phase shift can be given by:

$$\phi = \tau_{g,r} \cdot \omega, \tag{35}$$

where $\tau_{g,r}$ is DGD and ω is angular frequency.

The phase shift between the two polarization signal components (two polarization modes) can be also expressed as:

$$\phi = b \cdot L_{el}, \tag{36}$$

where b is birefringence of birefringent element and L_{el} is birefringent element length.

Then PDL element is described by the following Jones matrix [14]:

$$J_{PDL} = \begin{bmatrix} e^{-\frac{\alpha_1}{2}} & 0 \\ 0 & e^{\frac{\alpha_1}{2}} \end{bmatrix}. \tag{37}$$

where α_1 is defined as: PDL [dB]$=10\log_{10}(\exp(2\alpha_1))$.

The Mueller matrix corresponding to equation (37) is equal to:

$$M_{PDL} = \begin{bmatrix} \dfrac{1+\alpha^2}{2} & \dfrac{1-\alpha^2}{2} & 0 & 0 \\ \dfrac{1-\alpha^2}{2} & \dfrac{1+\alpha^2}{2} & 0 & 0 \\ 0 & 0 & \alpha & 0 \\ 0 & 0 & 0 & \alpha \end{bmatrix}. \tag{38}$$

Here, value of α equals to: $\alpha = 10^{-\frac{PDL\,[dB]}{10}}$.

Figure 12 illustrates an optical fiber link which is split into some polarization segments (rotated polarization elements).

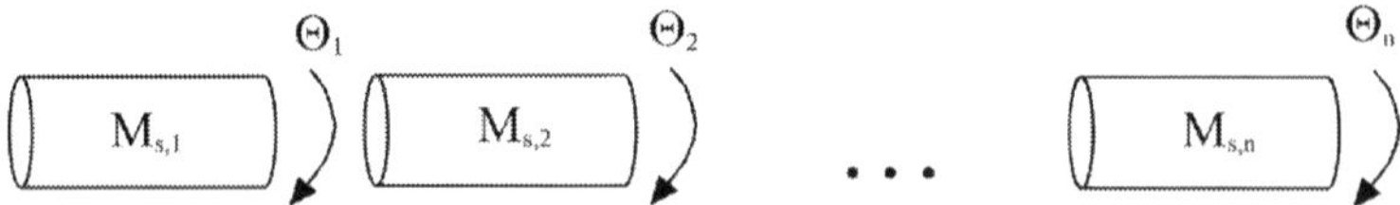

Figure 12. Optical fiber link model consists of N polarization segments; $M_{s,1}$, $M_{s,2}$, $M_{s,n}$ – matrix of polarization elements, Θ_1, Θ_2, Θ_n – angle of rotation

If we take into account only the PMD effect then the matrix of a single polarization segment $M_{s,n}$ is given by:

$$M_{s,n} = M_{R,n}(-2\Theta_n) \cdot M_{DGD,n} \cdot M_{R,n}(2\Theta_n).$$
(39)

When considering the PMD and PDL effect, matrix of a single polarization segment is can be writing as:

$$M_{s,n} = M_{R,n}(-2\Theta_n) \cdot M_{PDL,n} \cdot M_{DGD,n} \cdot M_{R,n}(2\Theta_n).$$
(40)

The matrix M_T of the whole optical fiber link which consists of N polarization segments is equal to:

$$M_T = M_{s,N} \cdot ... \cdot M_{s,3} \cdot M_{s,2} \cdot M_{s,1},$$
(41)

Furthermore, the PDG element matrix (M_{PDG}) can be described by the following equation:

$$\begin{bmatrix} \dfrac{g^2+1}{2} & \dfrac{g^2-1}{2} & 0 & 0 \\ \dfrac{g^2-1}{2} & \dfrac{g^2+1}{2} & 0 & 0 \\ 0 & 0 & g & 0 \\ 0 & 0 & 0 & g \end{bmatrix},$$
(42)

where g is the PDG coefficient equals to $g = 10^{\frac{PDG[dB]}{10}}$.

Moreover, to describe the backscattering process we treat an optical fiber link as a cascade of backscattering elements. We treat Rayleigh backscattering as many small reflections distributed over the optical fiber link. For a single element the matrix representing the round-trip propagation (fiber in forward direction, reflector, fiber in backward direction) is computed by:

$$M_{Rs,1} = M_{s,1}^{T} \cdot M_R \cdot M_{s,1},\tag{43}$$

where $M_{s,1}^{T}$ is the transpose of $M_{s,1}$, and M_R is the Mueller matrix of a reflection:

$$M_R = \begin{bmatrix} 1 & 0 & 0 & 0 \\ 0 & 1 & 0 & 0 \\ 0 & 0 & 1 & 0 \\ 0 & 0 & 0 & -1 \end{bmatrix}.\tag{44}$$

For light propagating to the end of the N-th element the round-trip Mueller matrix has the following form [4]:

$$M_{Rs,N} = M_{s,1}^{T} \cdot M_{s,2}^{T} \cdot \dots \cdot M_{s,N}^{T} \cdot M_R \cdot M_{s,N} \cdot \dots \cdot M_{s,2} \cdot M_{s,1}.\tag{45}$$

We can use polarization segments model for the DGD and PDL values calculation. We should take into account Jones and Muller notation.

4.1. Jones notation

Differential Group Delay value can be found by the following relationship [15]:

$$\tau_{g,r} = \frac{\left| \mathrm{Arg}\left(\dfrac{\lambda_{\tau,1}}{\lambda_{\tau,2}} \right) \right|}{d\omega},\tag{46}$$

where: Arg denotes the argument function, $\lambda_{\tau,1}$ and $\lambda_{\tau,2}$ are two eigenvalues of matrix $M_T(\omega + d\omega) \cdot M_T^{-1}(\omega)$; M_T^{-1} is inverse matrix.

Polarization Dependent Loss in the unit of dB at angular frequency ω is given by [16]:

$$PDL[dB] = 10\log_{10}\left(\frac{\lambda_{\alpha,1}}{\lambda_{\alpha,2}} \right),\tag{47}$$

where: $\lambda_{\alpha,1}$ and $\lambda_{\alpha,2}$ are two eigenvalues of matrix $M_T^{T}(\omega) \cdot M_T(\omega)$, where M_T^{T} is transpose matrix

4.2. Mueller notation

Differential Group Delay value can be expressed as the length of the PMD vector $\vec{\Omega}$ [6]:

$$\tau_{g,r} = \left|\vec{\Omega}\right| = \sqrt{\Omega_x^2 + \Omega_y^2 + \Omega_z^2}, \tag{48}$$

The PMD vector after (n+1)-th polarization segment may be written as [6]:

$$\vec{\Omega}_{n+1} = D\vec{\Omega}_{n+1} + MB\vec{\Omega}_n, \tag{49}$$

where $\vec{\Omega}_n$ is the PMD dispersion vector of the first n polarization segments, $\Delta\vec{\Omega}_{n+1}$ is the PMD vector of the (n+1)-th polarization segments, matrix MB represents a transformation of the PMD vector caused by the propagation through the (n+1)-th polarization segment.

The recursive relation for the PMD vector is given by:

$$\begin{bmatrix} \Omega_{x,n+1} \\ \Omega_{y,n+1} \\ \Omega_{z,n+1} \end{bmatrix} = \begin{bmatrix} \tau_{g,r,n+1} \\ 0 \\ 0 \end{bmatrix} + \begin{bmatrix} m_{11} & m_{12} & m_{13} \\ m_{21} & m_{22} & m_{23} \\ m_{31} & m_{32} & m_{33} \end{bmatrix} \cdot \begin{bmatrix} \Omega_{x,n} \\ \Omega_{y,n} \\ \Omega_{z,n} \end{bmatrix}. \tag{50}$$

We use 3x3 matrix in equation (50). Because we assume that SOP=1.

To calculation the PDL value of an optical component or optical fiber link, one must determine the minimum and maximum transmission. Because of this we should take into account 4x4 matrix:

$$\begin{bmatrix} m_{00} & m_{01} & m_{02} & m_{03} \\ m_{10} & m_{11} & m_{12} & m_{13} \\ m_{20} & m_{21} & m_{22} & m_{23} \\ m_{30} & m_{31} & m_{32} & m_{33} \end{bmatrix}, \tag{51}$$

Polarization Dependent Loss in the unit of dB is [17]:

$$PDL[dB] = 10\log_{10}\left(\frac{m_{00} + \sqrt{m_{01}^2 + m_{02}^2 + m_{03}^2}}{m_{00} - \sqrt{m_{01}^2 + m_{02}^2 + m_{03}^2}}\right). \tag{52}$$

For an understanding of linear and, first of all, nonlinear optical effects in optical fiber links it is necessary to consider the electromagnetic wave propagation. The linear and nonlinear optical effects in an optical fiber are described by so called nonlinear Schroedinger propagation equation. The nonlinear coupled Schroedinger propagation equations governing evolution of an optical pulse consisiting of the two polarization components along a fiber link (z) are given by [18]:

$$\frac{\partial E_x}{\partial z} = -\frac{\alpha_x}{2} E_x + j\frac{\beta_2}{2}\frac{\partial^2 E_x}{\partial t^2} + j\gamma\left(\left|E_x\right|^2 + \frac{2}{3}\left|E_y\right|^2\right)E_x + j\frac{\gamma}{3}E_x^*E_y^2,$$

(53)

$$\frac{\partial E_y}{\partial z} = -\frac{\alpha_y}{2} E_y + j\frac{\beta_2}{2}\frac{\partial^2 E_y}{\partial t^2} + j\gamma\left(\left|E_y\right|^2 + \frac{2}{3}\left|E_x\right|^2\right)E_y + j\frac{\gamma}{3}E_y^*E_x^2,$$

(54)

Where E_x, E_y are slowly varying amplitudes, α_x and α_y is attenuation coefficient for E_x and E_y, respectively. Moreover, β_2 is second-order term of the expansion of the propagation constant (the group velocity dispersion parameter), γ is the nonlinear parameter,* designates complex conjugation, j is imaginary unit.

A numerical approach is necessary for the polarization and nonlinear propagation equations solution.

The most popular numerical method is Split-Step Fourier Method. There is useful to write equations (53) and (54) formally in the following form [19]:

$$\frac{\partial \vec{E}(T,z)}{\partial z} = \left[\ell_1(T) + \ell_2(T) + \aleph(z)\right]\vec{E}(T,z),$$

(55)

where: $\vec{E} = \begin{bmatrix} E_x \\ E_y \end{bmatrix}$ and $T = t - \beta_1 z$; t is time, β_1 is first-order term of the expansion of the propagation constant (differential coefficient of the propagation constant with respect to optical frequency).

The operators on the right side of equation (55) are linear $\ell_1(T)$, $\ell_2(T)$ and nonlinear $\aleph(z)$. These operators have the following definitions [19]:

$$\ell_1(T) = \frac{1}{2}\begin{bmatrix} \beta_1 & 0 \\ 0 & -\beta_1 \end{bmatrix}\cdot\frac{\partial}{\partial T},$$

(56)

$$\ell_2(T) = -\frac{1}{2}I\left(j\beta_2\frac{\partial^2}{\partial T^2} + \frac{1}{3}\beta_3\frac{\partial^3}{\partial T^3}\right),$$

(57)

$$\aleph(z) = -j\frac{\gamma}{3}\begin{bmatrix} 3\left|E_x\right|^2 + 2\left|E_y\right|^2 & E_x^*E_y \\ E_y^*E_x & 3\left|E_y\right|^2 + 2\left|F_x\right|^2 \end{bmatrix} + $$
$$j\frac{1}{2}\begin{bmatrix} -j\alpha_1 + \beta_0 & 0 \\ 0 & j\alpha_1 - \beta_0 \end{bmatrix}$$

(58)

Where β_0 is zeroth-order term of the expansion of the propagation constant, β_3 is third-order term of the expansion of the propagation constant (third differential coefficient of the propagation constant with respect to optical frequency). The symbol I stands for the identity matrix.

The linear operators describe first-order and high-order PMD effect. It is a function of T alone. The nonlinear operator includes phenomena that do not depend on T i.e. PMD, nonlinear effects. It is a function of z alone.

The Split-Step Fourier Method obtains an aproximate solution by assuming that in propagating an optical pulse over a small distance h optical effects are independent [19].

$$E(T, z+h) \approx NL_1 L_2 E(T, z),\tag{59}$$

where:

$$L_1 = \exp(\ell_1 h),\tag{60}$$

$$L_2 = \exp(\ell_2 h),\tag{61}$$

$$N = \exp\left(\int_z^{z+h} \aleph(z')dz'\right) \approx \exp\left(h\frac{\aleph(z+h)+\aleph(z)}{2}\right).\tag{62}$$

Propagation from z to z+h is carried out in two steps. In the first step linear effects only (L1≠0, L2≠0, N=0) are taken into account. In the first step vice versa (L1=0, L2=0, N≠0).

Figure 13 shows schematic illustration of the Split-Step Fourier Method. Fiber length is split into a large number of small sygments of width h.

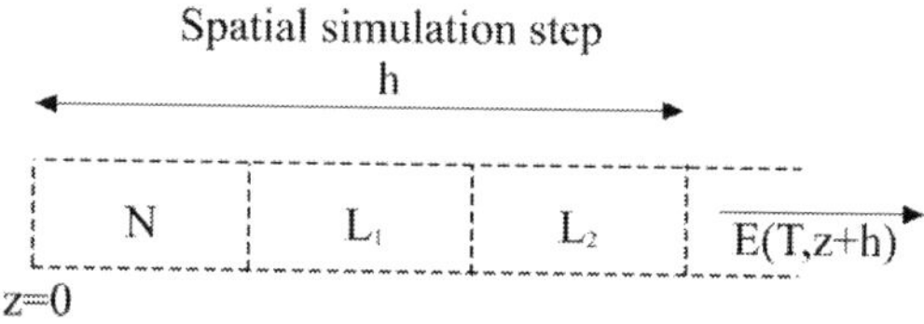

Figure 13. Schematic illustration of the Split-Step Fourier Method

It is important to know, that the linear operators are evaluated on the Fourier domain. In turn, nonlinear operator is evaluated on the time domain.

5. Polarizing components

Optical polarizing components belong to a class of optical components characterized by the modyfication of some polarization properties of light wave. Optical polarizing components are very useful for optical fiber communication technologies. Some of them are used for PMD and PDL compensating, Polarization Division Multiplexing transmission technique and measurement procedures. Ones of the most important optical polarizing components for modern, high capacity optical communication solutions are: polarization controller, polarization attractor, polarization scrambler and polarization effects emulator.

5.1. Polarization controller

The polarization controller is an optical component which allows one to modify the polarization state of light. The polarization controller is used to change polarized (or unpolarized) light into any well-defined SOP. Typically, the polarization controller consists of rotated retarders (wave plates). We can distingush the polarization controllers which are based on: two rotated quarter-wave plates, two rotated quarter-wave plates and one rotated half-wave plate or one rotated quarter-wave plate and one rotated half-wave plate. Figure 14 presents structure of polarization controller which is based on two rotated quarter-wave plates and distribution of SOPs at the polarization controller output port.

a   b

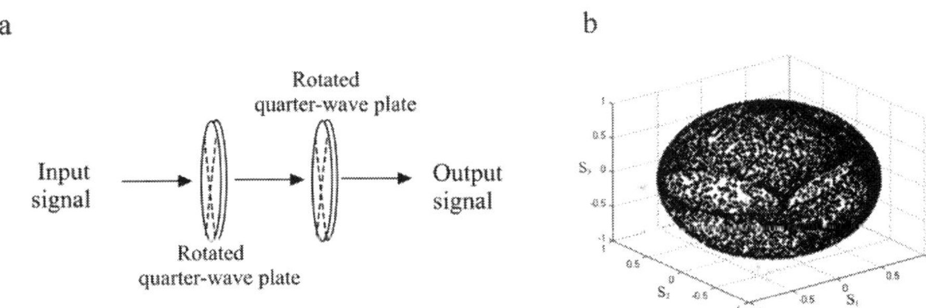

Figure 14. Polarization controller based on two rotated quarter-wave plates; structure of polarization controller (a), output SOPs distribution on the Poincaré sphere (b)

This polarization controller changes an arbitrary SOP into the other arbitrary SOP.

Figure 15 shows structure of polarization controller which is based on two rotated quarter-wave plates, one rotated half-wave plate and distribution of SOPs at the polarization controller output port.

This polarization controller is similar to above one. It transforms an arbitrary SOP into the other arbitrary SOP. Finally, Figure 16 presents structure of polarization controller which is based on one rotated quarter-wave plate, one rotated half-wave plate and distribution of SOPs at the polarization controller output port.

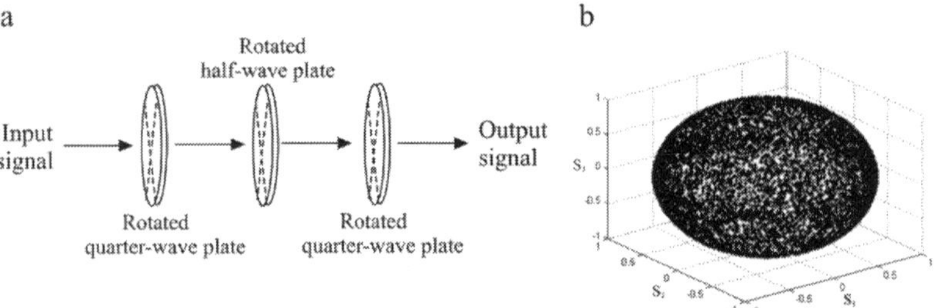

Figure 15. Polarization controller based on two rotated quarter-wave plates and one rotated half-wave plate; structure of polarization controller (a), output SOPs distribution on the Poincaré sphere (b)

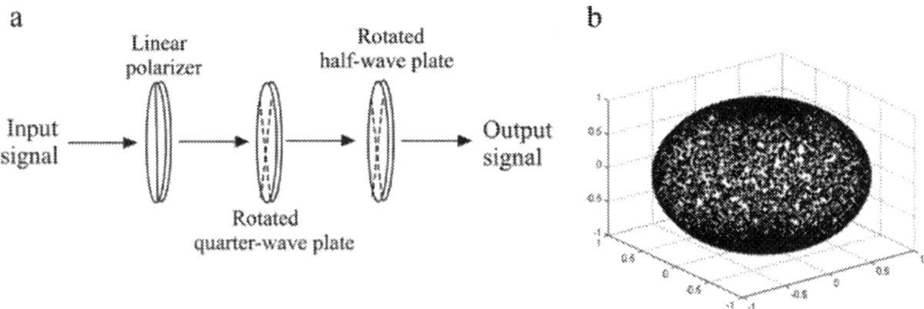

Figure 16. Polarization controller based on one rotated quarter-wave plate and one rotated half-wave plate; structure of polarization controller (a), output SOPs distribution on the Poincaré sphere (b)

This type of polarization controller only transforms linear polarization into an arbitrary SOP. We would expect flowed operation of this polarization controller with the other input SOP. This case is shown in Figure 17.

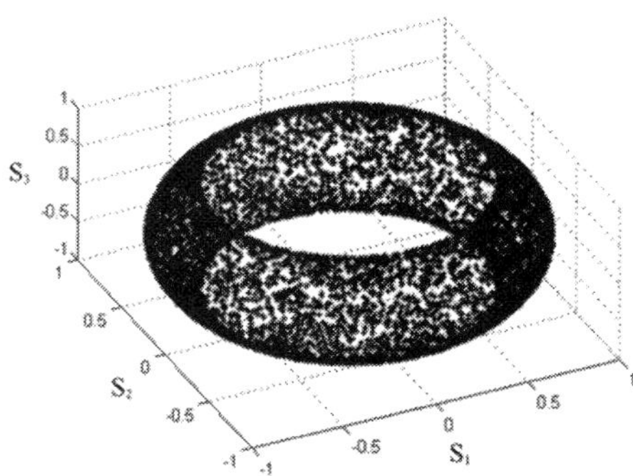

Figure 17. Distribution of SOPs at the polarization controller output port for circular input SOP

5.2. Polarization attractor

In real fibers the SOPs are not preserved because of the random birefringence. The uncontrolled SOPs variable can dramatically affect the performances of telecommunication systems. This phenomenon is very important, especially for demultiplexing process for Polarization Division Multiplexing transmission system. Possibility of polarization controlling is key issue for modern optical fiber communication technologies. The optical component which can stabilize an arbitrary polarized optical signal by lossless and instantaneous interaction is polarization attractor. This type of controlling the optical signal polarization can be based on: stimulated Brillouin scattering, stimulated Raman scattering or four wave mixing phenomenon. Here we focuse on the stimulated Raman scattering for polarization attraction effect [2, 3]. An arbitrary input SOP of the optical signal is pulled (attracted) by the SOP of the propagating pump (Raman pump), so that at the fiber output the signal SOP is matched the pump SOP. The power evolution of the pump ($\vec{P}$) and signal ($\vec{S}$) for copumped configuration along the optical fiber link can be modeled by means of coupled equations, respectively [20]:

$$\frac{d\vec{P}}{dz} = -\alpha_p \vec{P} - \frac{\omega_p}{2\omega_s} g_R \left(P_0 \vec{S} + S_0 \vec{P} \right) + \left(\omega_p \vec{b} + \overrightarrow{W_p^{NL}} \right) \times \vec{P}, \tag{63}$$

$$\frac{d\vec{S}}{dz} = -\alpha_s \vec{S} + \frac{1}{2} g_R \left(S_0 \vec{P} + P_0 \vec{S} \right) + \left(\omega_s \vec{b} + \overrightarrow{W_s^{NL}} \right) \times \vec{S}, \tag{64}$$

where ω_p and ω_s are pump and signal carrier angular frequencies, α_p and α_s are optical fiber attenuation coefficients for the pump and signal wavelengths, respectively. The g_R component is the Raman gain coefficient. The vector lengths $P_0 = |\vec{P}|$ and $S_0 = |\vec{S}|$ represent the pump and signal powers, respectively. The vector $\vec{b}$ is the local linear birefringence vector for optical fiber. The vectors $\overrightarrow{W_p^{NL}}$ and $\overrightarrow{W_s^{NL}}$ are given by [20].

$$\overrightarrow{W_p^{NL}} = \frac{2}{3} \gamma_p \left(-2S_{S,1}, -2S_{S,2}, S_{P,3} \right), \tag{65}$$

$$\overrightarrow{W_s^{NL}} = \frac{2}{3} \gamma_s \left(-2S_{P,1}, -2S_{P,2}, S_{S,3} \right), \tag{66}$$

where γ_p and γ_s are the nonlinear coefficients, $S_{P,1}$, $S_{P,2}$, $S_{P,3}$, $S_{S,1}$, $S_{S,2}$, $S_{S,3}$ are the Stokes parameters for the pump and signal, respectively.

The values of polarization attractor parameters (i.e.: pump power, pump SOP) should be accurately selected depending on expected polarization pulling.

Figure 18 shows scheme of polarization attractor based on stimulated Raman scattering.

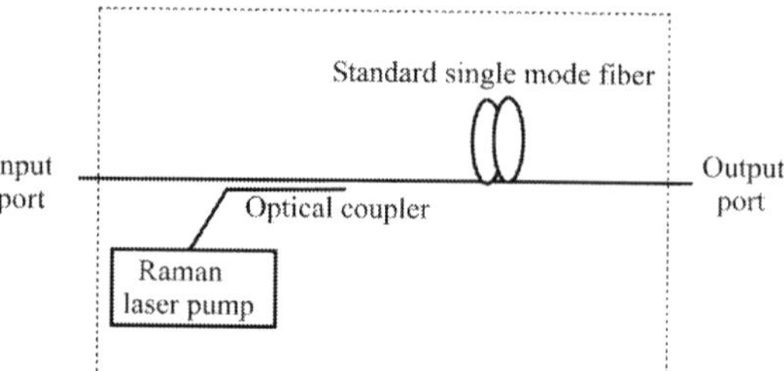

Figure 18. Polarization attractor based on Raman scattering

Figures 19 and 20 demonstrate simulated examples of polarization pulling effect for pump power equals to 1 W, 2W and 5 W. The simulated polarization attractor is based on standard single mode optical fiber [21].

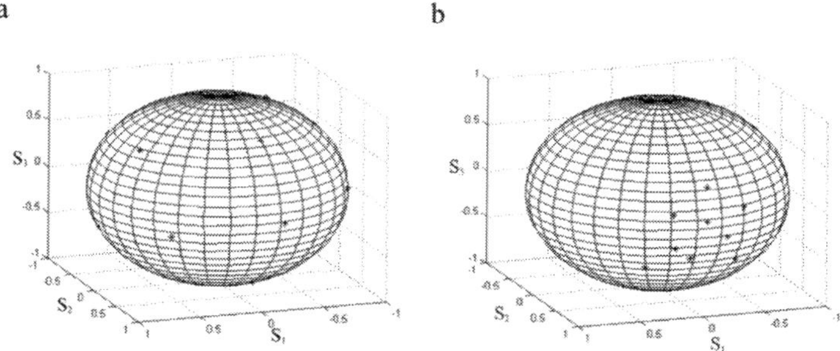

Figure 19. Simulated examples of polarization pulling; distribution of polarized signals at the attractor input port (a), distribution of output signal SOPs at the output attractor port for pump power 1 W (b)

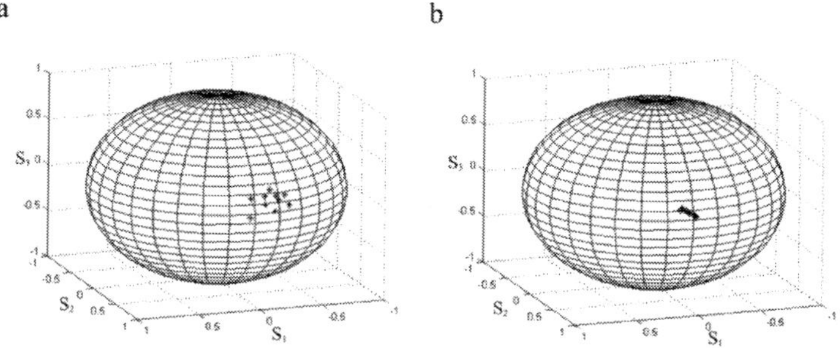

Figure 20. Simulated examples of polarization pulling; distribution of output signal SOPs at the output attractor port for pump power 2 W (a), distribution of output signal SOPs at the output attractor port for pump power 5 W (b)

It should be note that for stimulated Raman scattering the proper Raman polarization pulling and amplification for optical fiber communication systems may be simultaneously achieved.

5.3. Polarization scrambler

It is known that, d ue to the random nature of polarization mode coupling in an optical fiber several polarization effects (PMD, PDL, PDG) may occur that lead to impairments in long haul and high bit rate optical fiber transmission systems. Polarization scrambling the states of polarization has been shown to be technique that can reduction polarization impairments or the reduction of measurement uncertainly. A polarization scrambler actively changes the SOPs using polarization modulation method. In generally, the polarization scrambler configuration consists of rotating retarders (wave plates) or phase shifting elements. Furthermore, it is often necessary that the scrambler output SOPs are distributed uniformly on the entire Poincarè sphere. The spherical radial distribution function is very useful tool for the SOPs distribution analysis on the Poincarè sphere [22]. The spherical radial distribution function is the modified form of the well known plane radial distribution function. The spherical radial distribution function is defined as follows [22]:

$$g(d) = \frac{K(d)A_T}{A(d)K_T},\tag{67}$$

where $K(d)$ is the total number of pairs of the points separated by a given range of radial distances $(d, d+\Delta d)$, $A(d)$ is area of the sphere between two circles c_d and $c_{d+\Delta d}$, K_T is the total number of pairs of the points on the sphere; K_T is equal to N^2-N, where N is number of points on the sphere, A_T is area of the sphere (Figure 21).

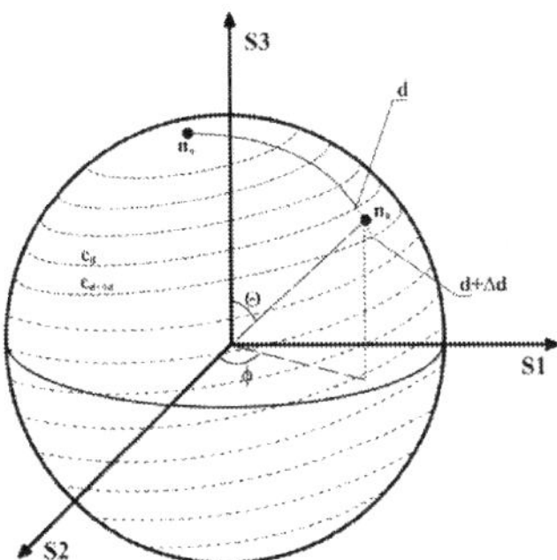

Figure 21. Example of spherical radial distribution function calculation for one reference point

The value of radial distance d changes from 0 to $\pi-\Delta d$, step is equal to Δd. The "great circle" distance between two points n_n and n_k (Figure 21), whose coordinates are (Θ_n, ϕ_n) and (Θ_k, ϕ_k), is given by so called Haversine formula:

$$d_{n,k} = 2R\arcsin\left(\sqrt{\sin^2\left(\frac{\Theta_n - \Theta_k}{2}\right) + \sin(\Theta_n)\sin(\Theta_k)\sin^2\left(\frac{\phi_k - \phi_n}{2}\right)}\right), \tag{68}$$

where R is the sphere radius.

We can distinguish three typical theoretical distributions: uniform (Figure 22a), random (Figure 23a) and clustered (Figure 24a). In turn, Figures 22b, 23b and 24b show the spherical radial distribution as a function of the distance d for the uniform, random and clustered distribution, respectively [22].

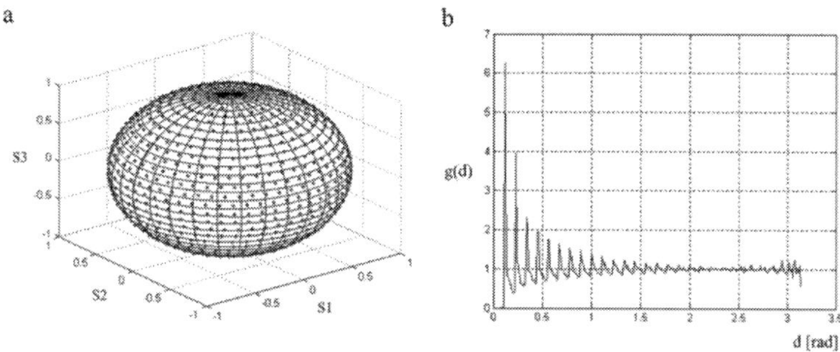

Figure 22. Theoretical uniform distribution; SOPs distribution on the Poincarè sphere (a), spherical radial distribution function versus radial distance (b)

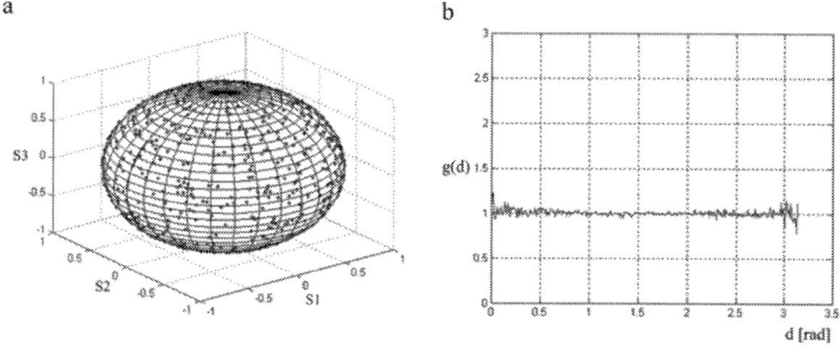

Figure 23. Theoretical random distribution; SOPs distribution on the Poincarè sphere (a), spherical radial distribution function versus radial distance (b)

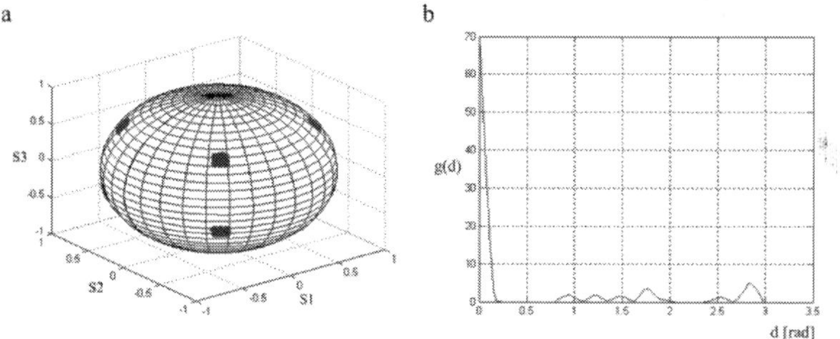

Figure 24. Theoretical clustered distribution; SOPs distribution on the Poincarè sphere (a), spherical radial distribution function versus radial distance (b)

For the uniform distribution (Figure 22b) the peaks on the g(d) curve provides information about the mean distance of the following neighbouring points (SOPs) on the Poincarè sphere. In the case of the random distribution (Figure 23b) the value of g(d) is close to 1. For the clustered distribution (Figure 24b) the localization of the first minimum on the g(d) curve provide information about the mean dimension (diameter) of the clusters. The location of the first lower peaks on the curve indicates the mean distance between clusters.

The analysis of the distribution of SOPs generated by polarization scramblers shows that SOPs distribution is clustered for three (and less) rotating retarders and for four (and less) phase shifting elements. For four and more rotating retarders and for five and more phase shifting elements random distribution is obtained [22].

5.4. Polarization effects emulator

You know well that polarization effects due to interaction between PMD and PDL can significantly impair optical fiber transmission systems. When PMD and PDL are both present they interact must be studied together. Emulating of PMD and PDL is one way to test and verify new transmission systems in the presence of PMD and PDL effects. Polarization effects emulation play a useful role, since it is possible to examine a large ensemble of system states far more rapidly than in a test bed with commercially available fiber optics. The polarization effects emulation devices can be split into two groups: statistical and deterministic emulators. Devices which are intended to mimic the random statistical behavior of a long single mode fiber links are termed as statistical emulators. Statistical polarization effect emulators should accurately reproduce the statistics of the polarization effect that a signal would see on a real link, as well as have good stability and repeatability. Devices that map the polarization effect space to the emulator settings and predictably generate the desired values are generally termed as deterministic emulators.

We typically have PMD and PDL statistical emulators and PDL deterministic emulators. Through pure statistical nature of the PMD effect, the PMD deterministic emulators should not be used for an optical communication systems testing.

Each statistical emulator that realistically simulates real optical fiber links should fulfil two criteria [6]:

1. Differential Group Delay should be Maxwellian distributed at any fixed optical frequency. This condition is also valid for the PDL effect. In the absence of the PMD the PDL distribution is Rayleigh distribution. But in the presence of PMD the PDL distribution is closed to Maxwellian distribution. Thus the PDL distribution in real optical fiber links can be also approximated by a Maxwellian function.

2. Frequency AutoCorrelation Function (ACF) of PMD and PDL vectors should tend toward zero as the frequency separation increases; so called Autocorrelation Function Background (BAC) should be lower than 10 %. Autocorrelation Function Background is defined as the mean absolute deviation of the ACF from the expected (mean) value of ACF for the frequencies larger than the autocorrelation bandwidth where the frequency autocorrelation bandwidth of the ACF is the frequency at the half of the variation of the ACF.

In [23] is demonstrated that 15 rotated polarization elements (e.g. sections of polarization maintaining fiber) realistically simulate the DGD and PDL distribution and BAC of real optical fiber links. Below, results for statistical emulator consisting of 15 rotated polarization elements. Figure 25 shows the histograms of DGD and PDL. For the statistical PMD emulator the DGD distribution is always indistinguishable from theoretical Maxwellian distribution. In turn, the PDL distribution (in the presence of PMD) is similar to Maxwellian distribution [24].

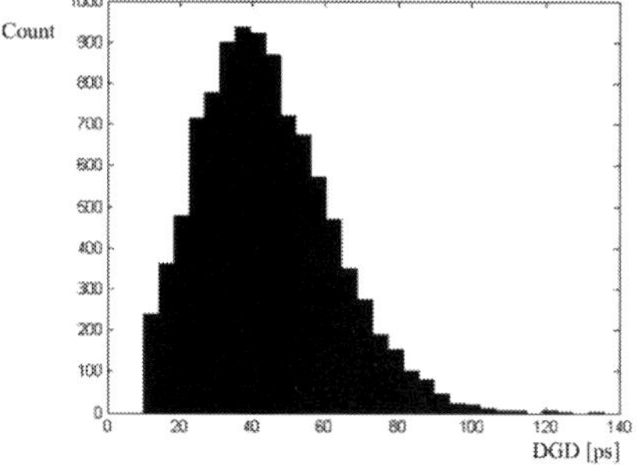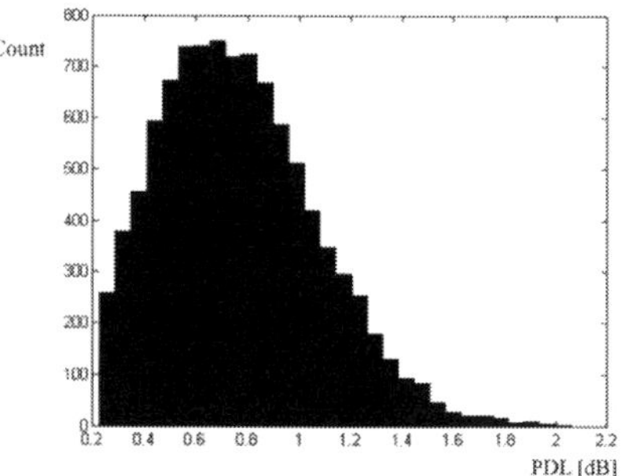

Figure 25. Statistical distribution of DGD values (a) and PDL values (b)

The theoretical and normalized ACF for both PMD and PDL vectors is shown in Figure 26.

Now, coming to to the PDL emulator, the simplest deterministic PDL emulator consists of tilted glass plate. The transmission coeffcients of a dielectric surface between two media were derived by Fresnel. They field the orthogonal component i.e. perpendicular coeffcient (T_s) and parallel coeffcient (T_p) to the plane of incidence [25]:

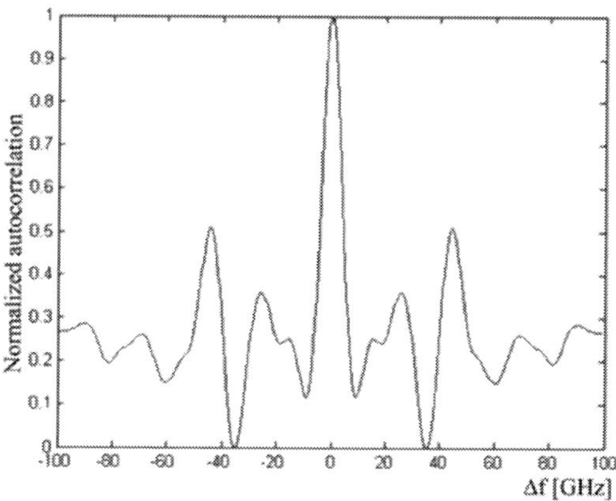

Figure 26. Theoretical normalized frequency autocorrelation function for PMD

$$T_s = \frac{n_s \cdot \cos(\xi_t)}{\cos(\xi_i)} \cdot |t_s|^2 \,,$$
(69)

$$T_p = \frac{n_s \cdot \cos(\xi_t)}{\cos(\xi_i)} \cdot |t_p|^2 \,,$$
(70)

where:

$$t_s = \frac{2\cos(\xi_i)}{\cos(\xi_i) + n_s \cdot \cos(\xi_t)} \,,$$
(71)

$$t_p = \frac{2\cos(\xi_i)}{\cos(\xi_t) + n_s \cdot \cos(\xi_i)} \,,$$
(72)

$$\xi_t = \arcsin\left(\frac{\sin(\xi_i)}{n_s}\right).$$
(73)

For above equations (69-73): angle ξ_i is the angle of incidence, angle ξ_t is the angle of refraction, n_s is the index of glass refraction. We assume that the index of air refraction is equal to 1.

Next, the PDL value is given by the following relation:

$$PDL[dB] = 10\log_{10}\left(\frac{T_s}{T_p}\right). \tag{74}$$

The PDL value is strong dependent on the angle of incidence. Figure 27 presents PDL in function of this angle. These PDL values are typically for some optical components which are used for optical fiber communication technologies.

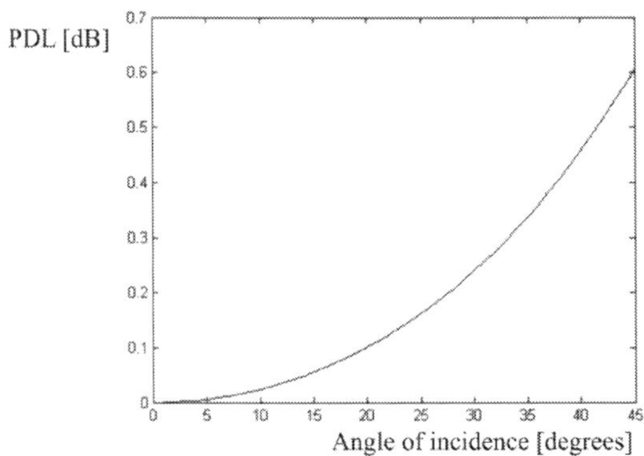

Figure 27. Polarization Dependent Loss value versus the angle of incidence; n_s=1.75

In conclusion, polarization issues become very important especially for long haul and high bit rate lightwave communication systems for which polarization effects, first of all, polarization mode dispersion and polarization dependent loss, become limiting factor. Optical fiber polarization phenomena must be taken into account during planning, installing and monitoring optical fiber communication systems. Additionally, the fast evolution of optical fiber transmission technologies requires powerful analysis and testing tools that must provide information about all relevant polarization phenomena in optical fiber links.

Author details

Krzysztof Perlicki[1,2*]

Address all correspondence to: perlicki@tele.pw.edu.pl

1 Institute of Telecommunications, Warsaw University of Technology, Warsaw, Poland

2 Orange Labs, Orange Polska, Warsaw, Poland

References

[1] Collet E. Polarization light. Fundamentals and applications. New York: Marcel Dekker; 1993.

[2] Huard S. Polarization of Light. Chichester: JohnWiley&Dons; 1997.

[3] Collet E. Polarized Light in Fiber Optics. Lincroft: The PolaWave Group; 2003.

[4] Mayers R. A. Encyclopedia of Physical Science and Technology Vol. 12. San Diego: Academic Press; 1992.

[5] Poole C. D, Wagner R. E. Phenomenological approach to polarisation dispersion in single mode fibres. Electronic Letters 1986;22(19) 1029-1030.

[6] Lima I. T, Khosravani R, Ebrahimi P. Comparison of Polarization Mode Dispersion Emulators. Journal of Lightwave Technology 2001;19(12) 1872-1881.

[7] . Curti F, Daino B, de Marchis G, Matera F. Statistical Treatment of the Evolution of the Principal States of Polarization in Single-Mode Fibers. Journal of Lightwave Technology 1990;8(8) 1162-1166.

[8] Galtarossa A, Palmieri L. Reflectometric measurement of PMD properties in long-single-mode fibers. Optical Fiber Technology 2003;9(3) 119-142.

[9] Kaminow I, Li T. Optical Fiber Telecommunications IVB Systems and Impairments. San Diego: Academic Press; 2002.

[10] Harris D. L, Kondamuri P. K. First and second order PMD statistical properties of intalled fiber. Proceeding of 17th Annual Meeting of the IEEE Lasers and Electro-Optics Society, LEOS2004, 7-11 November 2004, Rio Grande, Puerto Rico.

[11] Gisin N, Huttner B. Combined effects of polarization mode dispersion and polarization dependent losses in optical fibers. Optics Communications 1997;142(1-3) 119-125.

[12] El Amari A, Gisin N, Perny B. Statistical Prediction and Experimental Verification of Concatenations of Fiber Optic Components with Polarization Dependent Loss. Journal of Lightwave Technology 1998;16(3) 332-339.

[13] Wang L. J, Lin J. T. Analysis of Polarization-Dependent Gain in Fiber Amplifiers. Journal of Quantum Electronics 1998;34(3) 413-418.

[14] Fu X, O'Sullivan M, Goodwin J. Equivalent First-order Lumped-elements Model for Networks with both PMD and PDL. IEEE Photonics Technology Letters 2004;16(3) 939-941.

[15] Damask J. N. Polarization Optics in Telecommunications. Berlin: Springer Verlag; 2004.

[16] Phua P. B, Ippen E. P. A Deterministic Broad-Band Polarization-Dependent Loss Compensator. Journal of Lightwave Technology 2005;23(2) 771-780.

[17] Craig R. M. Visualizing the Limitations of Four-State Measurement of PDL and Results of a Six-State Alternative. Proceedings of the International Symposium on Optical Fiber Measurements, 24-26 September 2002, NIST Boulder, CO USA.

[18] Agrawal G. A. Nonlinear Fiber Optics, San Diego: Academic Press; 2001.

[19] Carena A, Curri V, Gaudino R, Poggiolini P, Benedetto S. A Time-Domain Optical Transmission System Simulation Package Accounting for Nonlinear and Polarization-Related Effects in Fiber. IEEE Journal on Selected Areas in Communications 1997;15(4) 751-765.

[20] Martinelli M, Cirigliano M. Evidence of Raman-induced polarization pulling. Optics Express 2009;17(2) 947-955.

[21] Perlicki K. Polarization pulling and signal amplification using Raman scattering for a Polarization Division Multiplexing transmission system. Photonics Letters of Poland 2011;3(4) 159-161.

[22] Perlicki K. Analysis of clusters and uniformity of distribution of states of polarization on the Poincare sphere. Applied Optics 2005;44(21) 4533-4537.

[23] M. C. Hauer M. C, Yu Q, Lyons E. R, Lin C.H, Au A. A, Lee H. P, Willner A. E. Electrically Controllable All-Fiber PMD Emulator Using a Compact Array of Thin-Film Microheaters. Journal of Lightwave Technology 2004;22(4) 1059-1065.

[24] Perlicki K. Statistical PMD and PDL effects emulator based on polarization maintaining optical fiber segments. Optical and Quantum Electronics 2009; 41(1) 1-10.

[25] Walraven R.L. Polarization by a tilted absorbing glass plate. Review of Scientific Instruments 1978;49(4) 537-540.

Simulation of Fiber Fuse Phenomenon in Single-Mode Optical Fibers

Yoshito Shuto

Additional information is available at the end of the chapter

http://dx.doi.org/10.5772/58959

1. Introduction

By utilizing dense wavelength-division-multiplexed (DWDM) transmission systems using an optical-fiber amplifier, large amount of data can be exchanged at a rate of over 60 Tbit/s [1]. In connection with this achievement, the danger of the "fiber fuse phenomenon" occurring has been pointed out. This occurs when high-power (W order) optical signals are transmitted in an optical-fiber cable [2].

The fiber fuse phenomenon was first observed in 1987 by British investigators [3]–[6]. A fiber fuse can be generated by bringing the end of a fiber into contact with an absorbing material, or melting a small region of a fiber by using an arc discharge of a fusion splice machine [3]. If a fiber fuse is generated, an intense blue-white flash occurs in the fiber core, and this flash propagates along the core in the direction of the optical power source at a velocity on the order of 1 m/s (see Figure 1). Fuses are terminated by gradually reducing the laser power to provide a termination threshold at which the energy balance at a fuse is broken.

When a fiber fuse is generated, the core layer in which the fuse propagates is seriously damaged, and the damaged fiber cannot be used in an optical communication system. The damage is made manifest by periodic or nonperiodic bullet-shaped cavities left in the core [7]–[13]. It was found that molecular oxygen was released and remained in the cavities while maintaining high pressure (about 4 atmospheres) at room temperature [4].

Several review articles [14], [15], [16] and a book [17] have been recently published, which cover many aspects of the current understanding of the fiber fuse.

Most experimental results of fiber fuse generation have focused on an intensity of 0.35–25 MW/cm^2 [3]–[13], [18]–[39]. This is many orders of magnitude below the intrinsic damage limit for silica, which exceeds 10 GW/cm^2 [3]. The threshold power for an SMF required to generate and/or terminate a fiber fuse was estimated at about 1.0 [7], 1.19 [37], 1.33 [37], and 1.4 W [30] at λ_0 = 1.064, 1.31, 1.48, and 1.56 μm, respectively.

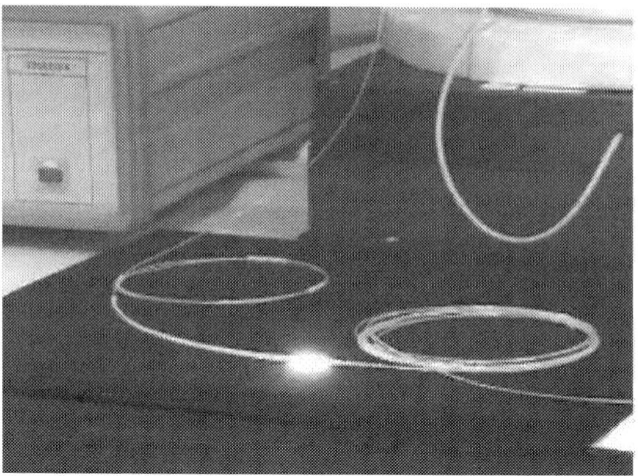

Figure 1. Fiber fuse phenomenon.

On the other hand, the fiber fuse effect in a microstructured optical fiber, in which thirty air holes are arranged around the fiber center, was reported by Dianov *et al.* [40]. No propagation of the fiber fuse was observed at the power of ≤ 3 W and $\lambda_0 = 0.5$ μm. This threshold power (3 W) was an order of magnitude higher than that of a conventional fiber. They also reported that the propagation of the fiber fuse was not observed when the laser light of up to 9 W and $\lambda_0 = 1.064$ μm entered in the microstructured optical fiber.

It has recently been reported that a hole-assisted fiber (HAF) [41], in which several air holes are arranged near the core of the optical fiber, exhibits high tolerance to fiber fuses [42]–[51]. The fiber fuse propagation in a HAF is affected by both the diameter of an inscribed circle linking the air holes (D_{hole}) and the diameter of the air hole (d_h) (see Figure 2).

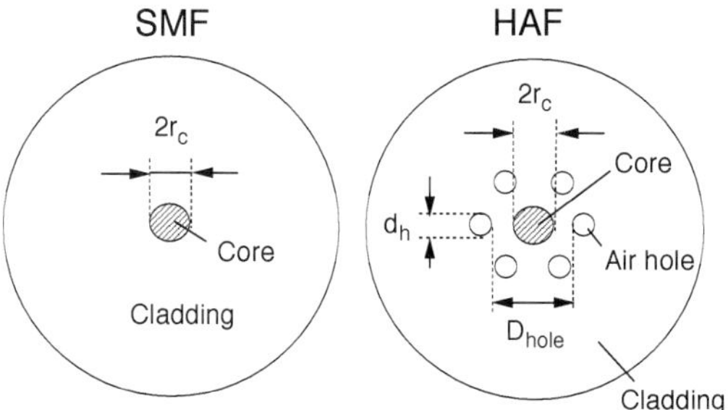

Figure 2. Schematic view of SMF and HAF.

No propagation of fiber fuse was observed in HAFs with $d_h = r_c = 4.5$ μm and a ratio R_h of the D_{hole} to the core diameter (2 r_c) of 2 or less when the laser power P_0 of 13.5 and/or 15.6 W at $\lambda_0 = 1.48 + 1.55$ μm was incident to HAFs [43], [45], [46].

Takenaga *et al.* investigated the power dependence of penetration length at a splice point of SMF and the HAF with d_h of about 16.3–16.9 μm and R_h of about 2.3, called "HAF2$_+$" [42],

[47]. The fiber fuse propagation immediately stopped in HAF2$_+$ for P_0 of 2.0–8.1 W at λ_0 = 1.55 μm, where the penetration length was maintained constant (about 120 μm). But the penetration length in HAF2$_+$ increased by nearly 600 μm when P_0 decreased from 2.0 W to 1.5 W.

They supposed that an inner ring area, which was observed around the cavities after fiber fuse propagation, was the trail of glass being melted. D_{melted} defined as the diameter of melting area (see Fugure 3) was considered as the radial size of plasma generated in the fiber fuse.

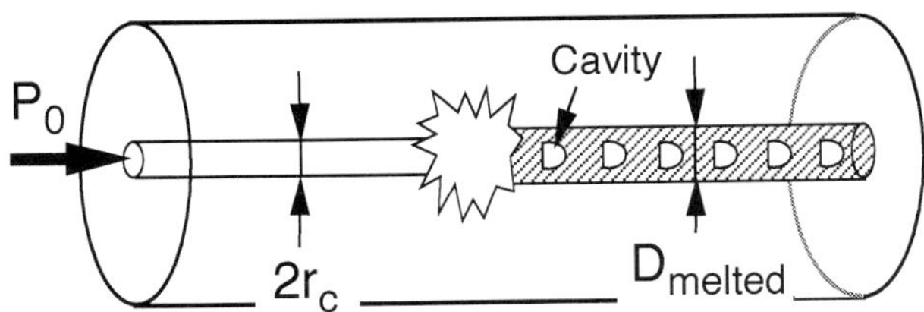

Figure 3. Schematic view of damaged SMF.

D_{melted} of HAF2$_+$ was maintained constant (about 20–22 μm) in the P_0 range of 2.0–8.1 W, and decreased with decreasing P_0 in the P_0 range of 1.33–2.0 W [42], [47]. The constant D_{melted} value (about 20–22 μm) was almost equal to the D_{hole} value (21.2 μm) of HAF2$_+$ in the P_0 range of 2.0–8.1 W. Therefore, an increase in penetration length observed in HAF2$_+$ for $P_0 \leq 2$ W was considered to be induced by the reduction of plasma size. On the other hand, D_{melted} of an SMF increased monotonously with increasing P_0 in the P_0 range of 4–14 W [46].

Several hypotheses have been put forward to explain the fiber fuse phenomenon. These include a chemical reaction involving the exothermal formation of germanium (Ge) defects [18], self-propelled self-focusing [3], thermal lensing of the light in the fiber via a solitary thermal shock wave [5], and the radiative collision of SiO and O complexes [52]–[55].

The similarities between the fiber fuse propagation and the combustion flame propagation were pointed out by Facão *et al.* [56], Todoroki [15], and Ankiewicz [57]. A fast detonation-like mode of fiber fuse propagation with a velocity of 3.2 km/s was observed under intense laser radiation intensity of 4,000 MW/cm^2 [58]. Combustion processes, including thermal self-ignition, can be mathematically expressed by the reaction-diffusion equations for temperature and fuel concentration [59], [60], [61]. If the reaction term for fuel concentration in these equations is replaced by the heat source term resulting from light absorption, the fiber fuse propagation can be described by solving the equations [56], [57], [62].

The optical absorption coefficient α of a fiber core at high temperatures is closely related to the generation of the fiber fuse. Kashyap reported a remarkable increase in the α value of a Ge-doped silica core above the critical temperature T_0 ($\sim$ 1,323 K), while the α value of about 0 dB/km at room temperature remained unchanged until the temperature (T) approached T_0 [4]. The α value increased by nearly 1,900 dB/km ($\sim$ 0.44 m^{-1}) at λ_0 = 1.064 μm when T

changed from T_0 to $T_0 + 50$ K. Furthermore, Kashyap *et al.* reported that the best fit between the experimental and theoretical fiber fuse velocities was obtained when the α value of the Ge-doped silica core at high temperatures of above T_0 was fixed to be 4.0×10^4 m^{-1} at 1.064 μm [19].

Hand and Russell found that this phenomenon was initiated by the generation of large numbers of Ge-related defects at high temperatures of above about 1,273 K, and the α values at $\lambda_0 = 0.5$ μm obtained at temperatures of below 1873 K were modeled quite accurately using an Arrhenius equation [5], [6]. By contrast, they reported that the best fit between the experimental and theoretical fiber fuse velocities was obtained when the α value of the Ge-doped silica core at 2,293 K was assumed to be 5.6×10^4 m^{-1} at 0.5 μm [5]. This large α value, however, could not be estimated using their Arrhenius equation [6]. As the focal length F of thermal lens effect is inversely as the α value [63], large α value of 5.6×10^4 m^{-1} is necessary to obtain small F value of 10 μm order, which is comparable with observed interval and/or large front size of the cavities (see Appendix).

Furthermore, Hand and Russell reported that the electrical conductivity σ of the fiber core increased with the temperature and that the hot spot at the fiber fuse center was plasmalike [5]. Kashyap considered that the large α values may be attributable to an increase in the σ value of the fiber core at high temperatures of above T_0 [4].

It is well known that silica glass is a good insulator at room temperature, and the electrical conductivity in silica glass below 1,073 K is due to positively charged alkali ions moving under the influence of an applied field [64], [65]. The ionic conduction in the glass is not related solely to optical absorption.

We previously investigated the optical absorption mechanism causing the increase in the σ value and reported the relationship between σ and α in silica glass at high temperatures of above 1,273 K [66]–[69]. It was found that the increase in loss observed at 1.064 μm can be well explained by the electronic conductivity induced by the thermal ionization of a Ge-doped silica core, and it is not directly related to the absorption of Ge E' centers.

However, the calculated α values resulting from the electronic conductivity at 1.064 μm were of 10^2 m^{-1} order at 2,873 K, about two orders smaller than the α values (1.0–4.0×10^4 m^{-1}) reported by Kashyap *et al.* [4], [19]. Therefore, we need another mechanism to explain the increase in loss at high temperatures of above 2,273 K to account for the large (10^4 m^{-1} order) α values.

To satisfy this requirement, we proposed a thermochemical SiO$_x$ production model in 2004 [68], [69]. Using this model, we theoretically estimated large α values of 10^4 m^{-1} order as a result of SiO$_x$ absorption at high temperatures of 2,800 K or above. This model was able to quantitatively explain the relation between the fiber fuse propagation velocity and the incident laser-power intensity previously reported by other research institutions.

On the other hand, since the parameters (particularly the light-absorbing parameter) used for the numerical simulation were not optimized, the shortcoming that the maximum temperature of the core center obtained by calculation became unusually high (10^5 K order) was observed.

We have improved this model by optimizing several parameters required for the numerical computation, and we proposed an improved model in 2014 [70].

In the first half of this chapter we describe the improved model in detail. That is, we explain the mechanism of the increased absorption in optical fibers at high temperatures due to the thermochemical production of SiO_x, and estimate high-temperature α values at $\lambda_0 = 1.064$ μm. Then, using these values, we theoretically study the non-steady-state thermal conduction process in a single-mode optical fiber using the explicit finite-difference technique.

Next we have analyzed the heat transfer of HAFs with $d_h = r_c$ on the basis of the improved model, and simulated the fiber fuse propagation behavior when a high optical power of 1–20 W at $\lambda_0 = 1.55$ μm is injected into an HAF. In the latter half of this chapter we describe the results of this analysis.

2. High-temperature optical absorption in optical fibers

2.1. Effect of SiO_x formation on absorption

It has been reported that, at elevated temperatures, silica glass is thermally decomposed by the reaction [71]

$$SiO_2 \rightleftarrows SiO_x + (x/2)O_2. \tag{1}$$

Among the reductants of the silica generated by this pyrolysis reaction with the formula SiO_x, the most thermally stable material is SiO ($x = 1$).

The internal core space heated at the elevated temperature is shielded from the external conditions. Thus, with increasing temperature, the internal pressure increases and the internal volume decreases. Dianov *et al.* reported that the internal core temperature is about 10,000 K and the internal pressure is about 10,000 atmospheres at the time of fiber fuse evolution [72]. It is thought that SiO_x generated under the high-temperature and high-pressure conditions is densely packed into the internal core space because it is not allowed to expand, and it exists in a liquidlike form. For this reason, it is thought that the optical absorption spectrum of the high-density SiO_x in the core space is similar to that of solid SiO_x.

Philipp reported that the optical absorption spectrum of a SiO_x film is similar to that of amorphous Si because of the many Si–Si bonds in a SiO_x film, and that the absorption coefficient α_{SiO} for SiO ($x = 1$) near the threshold energy should be about one-twentieth of that for amorphous Si α_{Si} [73]. Furthermore, Philipp estimated the theoretical concentration f_{Si} of Si–(Si$_4$) when the five possible tetrahedral conformations centering on Si, Si–(Si$_4$), Si–(Si$_3$O), Si–(Si$_2$O$_2$), Si–(SiO$_3$), and Si–(O$_4$), were completely distributed at random within amorphous SiO_x [74]. The relationship between f_{Si} and x for SiO_x is illustrated in Figure 4. According to Figure 4, f_{Si} for a SiO ($x = 1$) film is 6.25% (= 1/16). This is very close to the factor by which the α value was reduced (about 1/20) as reported by Philipp [73].

The optical absorption near the absorption edge of a SiO_x film is dominated by that of amorphous Si with the Si–(Si$_4$) tetrahedral conformation, and the optical absorption coefficient α_{SiO_x} near the absorption edge of a SiO_x film can be calculated by multiplying the value of α_{Si} for amorphous Si by f_{Si}. That is, if the production rate of SiO_x given by Eq.

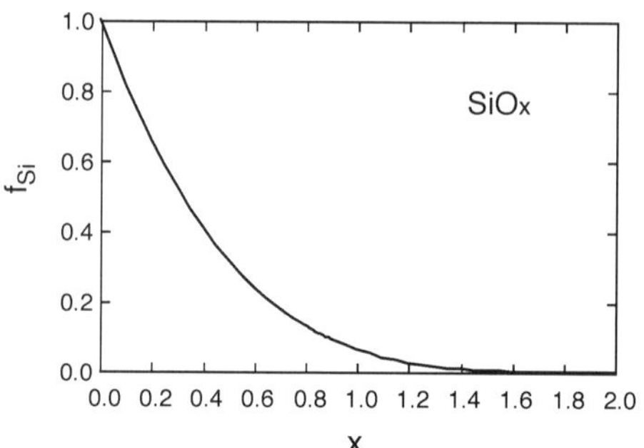

Figure 4. Relative content of Si–(Si$_4$) in SiO$_x$.

(1) is denoted by g_{SiO_x}, then α_{SiO_x} at temperature T is given by

$$\alpha_{SiO_x}(T) = g_{SiO_x}(T)f_{Si}(x)\alpha_{Si}(T).\tag{2}$$

2.2. Production rate of SiO$_x$

The most thermally stable compound in SiO$_x$ generated via Eq. (1) is SiO ($x = 1$). Therefore, the production rate g_{SiO_x} of SiO$_x$ in Eq. (2) is assumed to be almost equal to the production rate g_{SiO} of SiO.

g_{SiO} at temperature T can be calculated as the ratio of the molar concentration c_{SiO} of SiO at temperature T to the maximum value c_{SiO}^0. This c_{SiO}^0 is the molar concentration when SiO$_2$ changes to SiO via Eq. (1) with a yield of about 100%.

We assumed that the pyrolysis reaction system given by Eq. (1) reaches its equilibrium state during fiber fuse propagation. The equilibrium constant for Eq. (1) is denoted as K_c. The value of c_{SiO} at temperature T was calculated using K_c as described below.

First, the initial molar concentration $c_{SiO_2}^s$ (= 0.0366 mol cm^{-3}) is denoted by a. We consider the case that the SiO$_2$ concentration decreases to $a - y$ via the pyrolysis reaction of Eq. (1). In this case, the molar concentration of SiO and the molar concentration c_{O_2} of O$_2$ are expressed as y and $y/2$, respectively. K_c is given in terms of a and y as follows:

$$K_c = \frac{c_{SiO}c_{O_2}^{1/2}}{c_{SiO_2}} = \frac{y(y/2)^{1/2}}{a-y}.\tag{3}$$

Rearranging Eq. (3), we obtain a cubic equation in y. The solution of this equation is given by

$$y = \sqrt[3]{C} + \frac{4K_c^2}{3\sqrt[3]{C}}\left(\frac{K_c^2}{3} - a\right) + \frac{2K_c^2}{3},\tag{4}$$

where

$$C = a^2 K_c^2 - \frac{4}{3} a K_c^4 + \frac{8}{27} K_c^6$$
$$+ a^2 K_c^2 \sqrt{1 - \frac{8 K_c^2}{27 a}}. \tag{5}$$

It is well known that the equilibrium constant K_c is related to the standard Gibbs energy change $\Delta_r G^0$ for Eq. (1). The relationship between K_c and $\Delta_r G^0$ is given by [75]

$$\ln K_c = \frac{-\Delta_r G^0}{RT}, \tag{6}$$

where R is the gas constant. The $\Delta_r G^0$ value for Eq. (1) is given by

$$\Delta_r G^0 = \Delta_f G^0_{SiO} + (1/2)\Delta_f G^0_{O_2}$$
$$- \Delta_f G^0_{SiO_2}, \tag{7}$$

where $\Delta_f G^0$ is the standard production Gibbs energy of a reactant and/or a product.

Vitreous silica (SiO_2) is a solid at the standard temperature (298.15 K). It melts at high temeperatures of above 1,996 K, and becomes a liquid. It also becomes a vapor at temperatures of above 3,000 K.

The standard production Gibbs energies $\Delta_f G^0_{SiO_2}$, $\Delta_f G^0_{SiO}$, and $\Delta_f G^0_{O_2}$ in each phase have been published [76]. Thus, using these $\Delta_f G^0$ values, we first calculated the standard Gibbs energy change $\Delta_r G^0$ for Eq. (1). Next, we calculated K_c by substituting the $\Delta_r G^0$ value into Eq. (6). In this way, we computed c_{SiO} (= y) at temperature T by substituting K_c and a (= 0.0366 mol cm^{-3}) into Eqs. (4) and (5). The relationship between c_{SiO} and T is shown in Figure 5. c_{SiO} increases with increasing T and gradually approaches its maximum value ($c_{SiO}^0 \cong 0.0366$ mol cm^{-3}) at T of about 3,200 K.

Therefore, g_{SiO_x} at temperature T was estimated by dividing the value of c_{SiO} calculated above by c_{SiO}^0 ($\cong 0.0366$ mol cm^{-3}).

2.3. Absorption coefficient of amorphous Si

The optical absorption spectrum of amorphous Si was reported by Brodsky et al. [77]. The absorption coefficient α_{Si} of amorphous Si in Eq. (2) was estimated as follows.

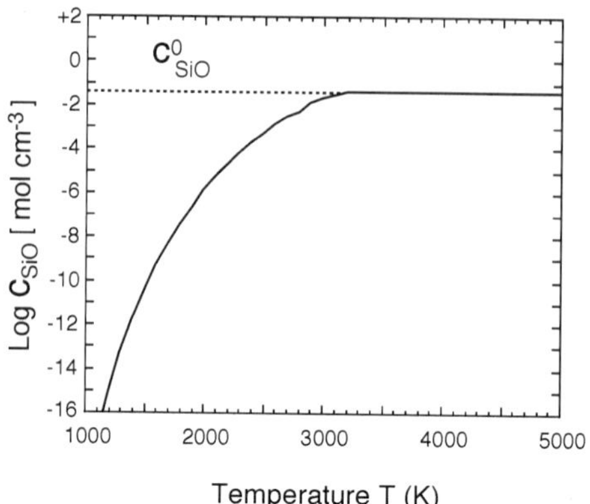

Figure 5. Molar concentration of SiO vs. temperature.

First, we consider the absorption coefficient α_{Si} of an interband transition region, where the photon energy $\hbar\omega$ is larger than the energy gap E_g (= 1.26 eV) of amorphous Si. In this case, the values of α_{Si} (cm^{-1} unit) near E_g is given by

$$\alpha_{Si}(\omega) = B\frac{(\hbar\omega - E_g)^2}{\hbar\omega},\tag{8}$$

where the parameter B is 5.06×10^5 cm^{-1} eV^{-1}. The α_{Si} values calculated by Eq. (8) are in good agreement with the experimental values reported by Brodsky *et al.* [77] for the case of $\hbar\omega > 1.5$ eV.

Next, we consider the values of α_{Si} in the low-energy region, where $\hbar\omega < E_g$. In this region, an absorption edge spectrum is broadened as a result of the interaction between optical phonons and electrons (or excitons). α_{Si} in this region exhibits exponential behavior (referred to as an"exponential tail" or "Urbach tail") as follows [78]:

$$\alpha_{Si}(\omega) = \alpha_0 \exp\left[\frac{\gamma(\hbar\omega - E_g)}{kT^*}\right],\tag{9}$$

where α_0 and γ are parameters and k is the Boltzmann constant. T^* is the effective temperature. If the characteristic temperature of phonons is denoted by θ, then T^* is given by [79]

$$T^* = \frac{\theta}{2} \coth\left(\frac{\theta}{2T}\right).\tag{10}$$

For Si, θ = 600 K [80]. α_0 and γ are estimated to be 9.016×10^3 cm^{-1} and 0.14, respectively, by simulation using the experimental α_{Si} values reported by Brodsky *et al.* [77].

It is known that the energy gap E_g of Si decreases linearly with increasing T at $T > 200$ K
[81]. The temperature dependence of E_g (eV unit) at $T > 300$ K is given by

$$E_g = 1.26 - \beta T, \tag{11}$$

where β is the temperature coefficient of the energy gap and takes a value of 4.2×10^{-4}
eV/K [82].

2.4. Absorption coefficient of SiO_x

On the basis of the above results, the temperature dependence of α_{SiO_x} at 1.064 μm ($\hbar\omega$ =
1.17 eV) was calculated using Eq. (2) and the c_{SiO} values shown in Figure 5. f_{Si} = 0.0625 (x
= 1) was used in the calculation. The results are shown in Figure 6.

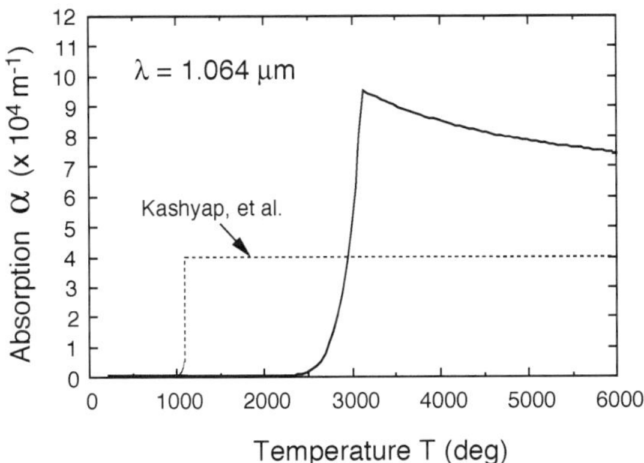

Figure 6. Absorption coefficient of SiO_x at 1.064 μm vs. temperature. The thick solid line was calculated using Eq. (2). The dotted line shows the values estimated by Kashyap *et al.*

The α_{SiO_x} values (shown as the thick solid line in Figure 6) are about 1.5×10^4 m^{-1} at 2,800
K and about 3.5×10^4 m^{-1} at 2,950 K. These values are close to 4×10^4 m^{-1} (shown as the
dotted line in Figure 6), which was estimated by simulation using the finite element method
[19].

Moreover, it turns out that α_{SiO_x} reaches its maxium value (about 9.47×10^4 m^{-1}) at about
3,150 K, then decreases gradually with increasing temperature.

Compared with the stepwise shape (dotted line in Figure 6) of the absorption coefficient
assumed by Kashyap *et al.*, the maximum value (about 9.47×10^4 m^{-1}) of α_{SiO_x} is about 2.4
times the assumed value (4×10^4 m^{-1}), and the graph has the shape of a somewhat distorted
echelon form.

As mentioned above, SiO_x is produced in the pyrolysis of vitreous silica, and it induces a very
large amount of optical absorption with a large absorption coefficient of 10^4 m^{-1} order at

high temperatures of above 2,800 K. It is thought that the large amount of optical absorption at high temperatures is the cause of the genesis of the fiber fuse phenomenon.

We investigated the thermal conduction behavior within an SMF by numerical computation using the thermochemical SiO_x production model. In the next section, we describe the results of some numerical calculations related to the thermal conduction process in an SMF.

3. Simulation of fiber fuse in SMF

3.1. Heat conduction in SMF

We assume the SMF to have a radius of r_f and to be in an atmosphere with temperature $T = T_a$. We also assume that part of the core layer of length ΔL is heated to a temperature of $T_c^0 (> T_a)$ (see Figure 7). Such a region, called the "hot zone" in Figure 7, can be generated by heating the fiber end faces using the arc discharge of a fusion splice machine.

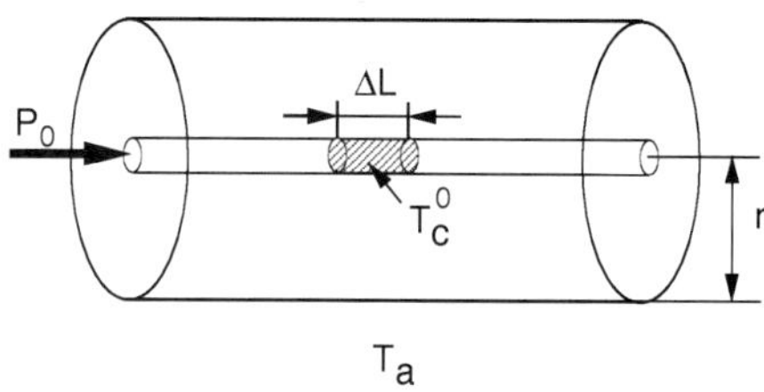

Figure 7. Hot zone in the core layer.

As explained above, the optical absorption coefficient α of the core layer in an optical fiber is a function of temperature T, and α increases with increasing T. In the hot zone in Figure 7, the α values are larger than those in the remainder of the core region because of its elevated temperature. Thus, as light propagates along the positive direction (away from the light source) in this zone, considerable heat is produced by light absorption.

In the case of a heat source in part of the core layer, the nonsteady heat conduction equation for the temperature field $T(r, z, t)$ in the SMF is given by [83]

$$\rho C_p \frac{\partial T}{\partial t} = \lambda \left(\frac{\partial^2 T}{\partial r^2} + \frac{1}{r} \frac{\partial T}{\partial r} + \frac{\partial^2 T}{\partial z^2} \right) + \dot{Q}, \tag{12}$$

where ρ, C_p, and λ are the density, specific heat, and thermal conductivity of the fiber, respectively. The values of ρ, C_p, and λ used for the calculation are described in the next subsection.

The last term $\dot{Q}$ in Eq. (12) represents the heat source resulting from light absorption, which is only required for the hot zone in the fiber core. $\dot{Q}$ can be expressed by

$$\dot{Q} = \alpha I, \tag{13}$$

where I is the optical power intensity in the core layer, which can be estimated by dividing the incident optical power P by the effective area A_{eff} of the fiber.

3.2. Heat conduction parameters

In the heat conduction calculation for SMFs, we used the following values of λ (W m^{-1} K^{-1}), ρ (kg m^{-3}), and C_p (J kg^{-1} K^{-1}) in each temperature range. The unit of T is K (Kelvin).

(1) Parameters in the temperature range from room temperature (298 K) to 1,996 K [84]:

$$C_p = 1194.564 + 31.541 \times 10^{-3} T$$
$$- 651.396 \times 10^5 T^{-2}$$
$$\lambda = 9.2$$
$$\rho = 2,200.0.$$

(2) Parameters in the temperature range from 1,996 to 3,000 K [84]:

$$C_p = 1,430.490$$
$$\lambda = 9.2$$
$$\rho = 2,200.0.$$

(3) Parameters for $T > 3,000$ K:

$$C_p = 844.4$$
$$\lambda - 0.0025247\sqrt{T}$$
$$+ 1.84275 \times 10^{-13} T^{5/2}$$
$$\rho = 2,024.0.$$

In (3), the first term in the expression for λ was estimated by substituting the parameters for SiO in the following equation, which represents the thermal conductivity of diatomic molecules derived from the kinetic theory of gases [85]:

$$\lambda = \frac{5}{3\sigma^2} \sqrt{\left(\frac{k}{\pi}\right)^3 \frac{N_A}{M}} \sqrt{T}, \tag{14}$$

where N_A is Avogadro's number, M (=44.0854) is the molecular weight of SiO, and σ (= 1.5 Å) is half of the collision diameter.

Moreover, in (3), the second term in the expression for λ was estimated by mutiplying the following thermal conductivity equation for weakly ionized gas plasma [86] by the correction factor of $1/20$:

$$\lambda = \frac{5N_e k^2 T}{2m_e \nu_c},$$
(15)

where m_e is the electron mass, N_e is the number density of electrons in the plasma, and ν_c is the collision frequency. If we assume that electrons mainly collide with ions with a charge of $+1$ in the plasma, then ν_c is given by [87]

$$\nu_c = \sqrt{\frac{2}{9\pi}} \frac{N_e e^4}{\varepsilon_0^2 m_{fe}^2} \left(\frac{m_e}{3kT}\right)^{3/2} \ln \Lambda,$$
(16)

where e is the electron charge and ε_0 is the permittivity of free space. $\ln \Lambda$ is the so-called Coulomb logarithm, and it takes values of 4–34 in the electron temperature range of 10^2–10^8 K and the N_e range of 1–10^{24} cm^{-3} [88]. In our calculation, we used $\ln \Lambda = 16.155$, which corresponds to the case of $T \sim 10^4$ K and $N_e \sim 10^6$ cm^{-3}.

3.3. Boundary and initial conditions for heat conduction

We solved Eq. (12) using the explicit finite-difference method (FDM) [89] under the boundary and initial conditions described below.

The area in the numerical calculation had a length of $2L$ ($= 4$ cm) in the axial (z) direction and a width of $2r_f$ ($= 125$ μm) in the radial (r) direction. There were 24 and 2,000 divisions in the r and z directions, respectively, and we set the calculation time interval to 1 μs. We assumed that the hot zone was located at the center of the fiber (length $2L$) and that the length ΔL of the hot zone was 40 μm.

The boundary conditions are as follows:

(1) Since the temperature distribution of the optical fiber is axisymmetric, the amount of heat conducted per unit area (heat flux) in the r direction is set to 0 at the fiber center ($r = 0$) as follows:

$$-\lambda \frac{\partial T}{\partial r}\bigg|_{r=0} = 0.$$
(17)

(2) At the outer fiber surface ($r = r_f$), the amount of heat conducted per unit area (heat flux) is dissipated by radiative transfer or heat transfer to the open air and/or the jacketing layer ($T = T_a$) as follows:

$$-\lambda \frac{\partial T}{\partial r}\bigg|_{r=r_f} = \sigma_S \epsilon_e \left(T^4 - T_a^4\right)$$

$$+ \frac{\lambda}{\delta r_t}\left(T - T_a\right),$$
(18)

where σ_S is the Stefan-Boltzmann constant and ϵ_e (~ 0.9) is the emissivity of the surface. δr_t is the thickness of the thermal boundary layer and $\delta r_t = \delta r$ is assumed in our calculation, where δr is the step size along the r axis.

(3) At the center ($z = 0$) of the hot zone, the temperature of the fiber core center is T_c^0. Also, there is heat inflow along the z axis only at the core center ($r = 0$), which is attributable to light absorption. Moreover, when $r \neq 0$ and $z = 0$, there is neither heat inflow nor heat outflow along the z axis. These conditions are given by

$$-\lambda \frac{\partial T}{\partial z}\bigg|_{z=0} = \begin{cases} \alpha I, & \text{if } r = 0 \\ 0 & \text{if } r \neq 0. \end{cases} \tag{19}$$

(4) At both fiber ends ($z = \pm L$), the amount of heat conducted per unit area (heat flux) is dissipated by radiation transfer to the open air ($T = T_a$) as follows:

$$-\lambda \frac{\partial T}{\partial z}\bigg|_{z=\pm L} = \sigma_S \epsilon_e \left(T^4 - T_a^4 \right). \tag{20}$$

In contrast, as initial conditions, $T = T_a$ in the optical fiber at $t = 0$, except in the hot zone, and the core-center temperature in the hot zone is equal to $T_c^0 (> T_a)$ as follows:

$$T(0, z, 0) = \begin{cases} T_a, & \text{if } -L \leq z < -\Delta L/2 \\ T_c^0, & \text{if } -\Delta L/2 \leq z \leq \Delta L/2 \\ T_a & \text{if } \Delta L/2 < z \leq L. \end{cases} \tag{21}$$

When light propagates through the fiber core ($r = 0$) along the z direction (away from the light source), the incident laser power P decreases because of the nonzero optical absorption coefficient α. When the laser light propagates from z to $z + \delta z$ along the z axis at $r = 0$, the P value is given by

$$P = P_0 \exp\left(-\alpha \delta z - \int_{-L}^{z} \alpha(T) dz \right), \tag{22}$$

where P_0 is the initial laser power. The second term on the right-hand side expresses the optical absorption loss when the light propagates through a distance of $z - L$.

The results described above assume that the laser light propagates through the fiber core along the positive z direction (away from the light source).

When the core layer is heated to above the vaporization point of silica (~ 3273 K), an enclosed hollow cavity is produced in the core center. This cavity contains oxygen, which is produced by the pyrolysis reaction of Eq. (1). The heat conductivity κ of the oxygen (0.03 W m^{-1} K^{-1}) is two orders smaller than that of the silicate glass. Therefore, the heat transferred in the silica core is stopped at the cavity.

Moreover, as the cavity has a refractive index of $n \sim 1$, which is smaller than that of the silica core ($n_1 = 1.46$), the light propagating in the core layer is reflected at the cavity wall. When the light direction is reversed at the cavity, the heat source term αI_r resulting from the optical absorption of the reflected light is added to αI in Eq. (13), where I_r is the optical power intensity of the reflected laser light.

We consider the P value at a z position located near the cavity wall. The length of this position is assumed to be δz. The laser light reaches z, propagates through a distance of δz, and then reaches the cavity wall, whose coordinate is z_v. Next, the light is reflected at the cavity wall and propagates in the negative z direction, and then reaches z again. In such a case, the P value at z is given by

$$P = P_0 \exp\left(-\alpha\delta z - \int_{-L}^{z} \alpha(T)dz\right)$$
$$\times R\exp\left(-2\int_{z+\delta z}^{z_v} \alpha(T)dz\right), \tag{23}$$

where R is the reflectivity at the boundary of the silica core and the cavity. R is given by

$$R = \left(\frac{n_1 - 1}{n_1 + 1}\right)^2. \tag{24}$$

The second integral on the right-hand side of Eq. (23), which is related to the reflected light, slightly affects the occurrence of the fiber fuse. However, by taking this term into consideration, the calculated fiber fuse velocities fit the experimental values. For this reason, in the present work, we decided to take into account the effect of the reflected light.

In the following section, we describe the calculated time (t) dependence of $T(r,z)$ in an SMF.

3.4. Propagation of fiber fuse in SMF

In the calculation, we used $T_c^0 = 2923$ K and $T_a = 298$ K. It was assumed that laser light of wavelength $\lambda_0 = 1.064$ μm and $P_0 = 2$ W was incident to an SMF-28 optical fiber, which has a core diameter of $2\,r_c = 8.2$ μm, a refractive index difference of $\Delta = 0.36\%$, and $A_{eff} = 49.4091 \times 10^{-12}$ m^2.

We calculated the $T(r,z)$ values at $t = 1$, 11, and 21 ms after the incidence of the 2 W laser light. The calculated results are shown in Figures 8 $\sim$ 10, respectively.

As shown in Figure 8, the core center temperature near the end of the hot zone ($z = -0.7$ mm) changes abruptly to a large value of about 3.4×10^4 K after 1 ms. This rapid rise in the temperature initiates the fiber fuse phenomenon as shown in Figures 9 and 10. After 11 and 21 ms, the high-temperature front in the core layer reached z values of -6.1 and -11.5 mm, respectively. The average propagation velocity v_f was estimated to be 0.54 m/s using these data.

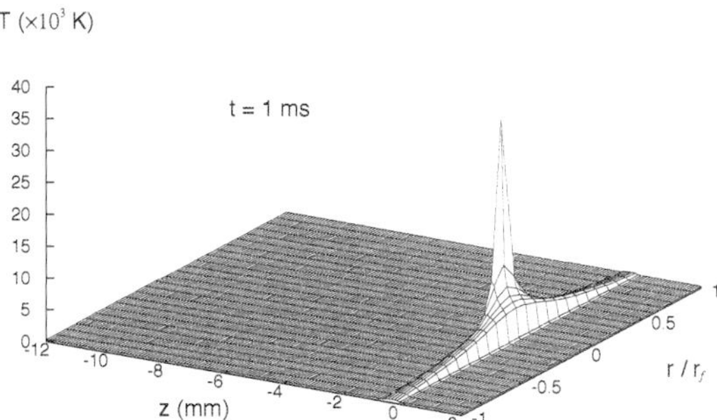

Figure 8. Temperature field in SMF-28 after 1 ms when $P_0 = 2$ W at $\lambda_0 = 1.064$ μm.

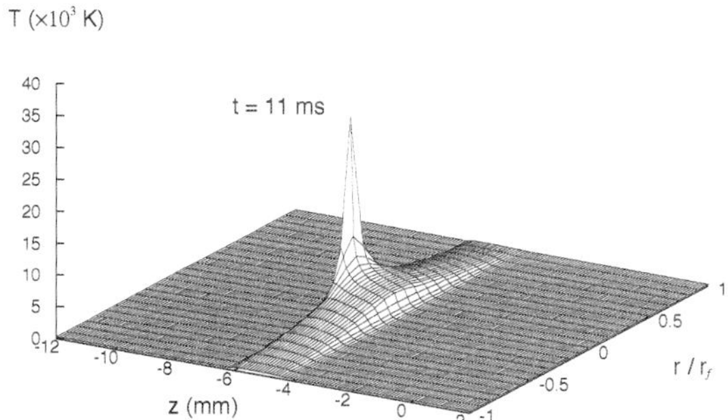

Figure 9. Temperature field in SMF-28 after 11 ms when $P_0 = 2$ W at $\lambda_0 = 1.064$ μm.

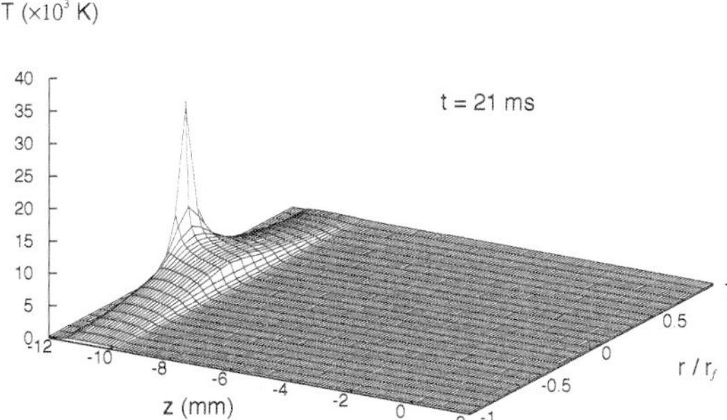

Figure 10. Temperature field in SMF-28 after 21 ms when $P_0 = 2$ W at $\lambda_0 = 1.064$ μm.

When the laser light of $\lambda_0 = 1.064$ μm and $P_0 = 2$ W is incident to the SMF-28 optical fiber, the optical power intensity is $I = 4.048$ MW/cm^2. The fiber fuse velocity at this value of I is estimated to be about 0.55 m/s (see Figure 10 in [69]). This value is in very good agreement with the upper value obtained by calculation (0.54 m/s).

On the other hand, Hand and Russell measured the fiber fuse temperature to be 5,400 K [5], and Dianov *et al.* obtained a temperature of 4,700–10,500 K [28],[90] by measurement. They estimated the temperatures from precisely measured spectral data in the 600–1,400 nm [5] and 500–800 nm [28],[90] regions, while assuming blackbody radiation.

In our calculation, the temperature distribution of the fiber fuse in the core center is shown in Figure 11. Similar shapes of temperature distribution were reported by Kashyap *et al.* [19]

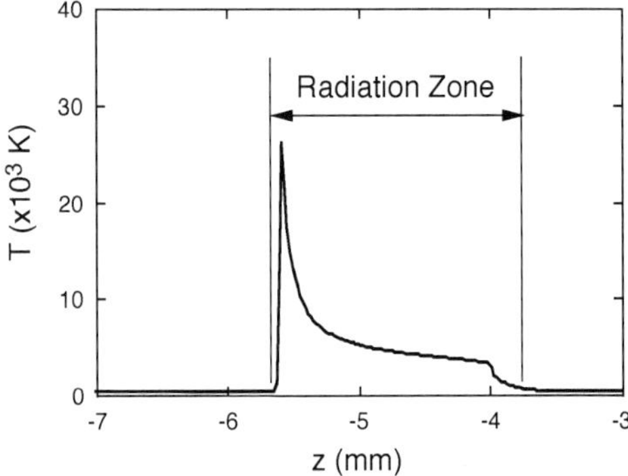

Figure 11. Temperature distribution of the SMF-28 core center vs. length along the z direction when $P_0 = 2$ W at $\lambda_0 = 1.064$ μm. The time after the laser incidence is 10 ms.

and Facão *et al.* [56].

As shown in Figure 11, it is clear that the sharp temperature peak is located near the light source, and a relatively high temperature plateau of about 5,000 K extends over about 1.5 mm behind the sharp peak. This region, called the "radiation zone" in Figure 11, exhibits high temperatures of above 323 K.

When gaseous SiO and/or SiO$_2$ molecules are heated to high temperatures above 5,000 K, they decompose to form Si and O atoms, and finally becomes Si$^+$ and O$^+$ ions and electrons in the ionized gas plasma state [91]. The numbers of electrons and ions in the plasma front, which exhibits sharp temperature peak, are larger than those in the plateau region. However, as the plasma tends to restore electrical neutrality, the motions of the electrons and ions will not produce any change on the initial temperature distribution shown in Figure 11, except for the peak temperature reduction of the plasma front due to energy loss induced by electron-ion collisions.

In the ionized gas plasma, electron-ion collisions generate electromagnatic radiation because of the deceleration during the collisions. This bremsstrahlung emission [92], [93] is a universal and irreducible process of energy loss. If we assume that electrons mainly collide with ions with a charge of +1 in the plasma, the spectral radiance function I_p for the bremsstrahlung emission is given by [94]

$$I_p = \frac{N_e^2 v^2}{\sqrt{T} c^2} \exp\left(-\frac{h\nu}{kT}\right),$$ (25)

where N_e is the number density of electrons in the plasma, h is Planck's constant, and ν is the optical frequency. This functin is directly proportionate to ν^2 in the case of $h\nu \ll kT$.

On the other hand, in the case of $h\nu \ll kT$, the spectral radiance function I_b for blackbody radiation is given by [95]

$$I_b = \frac{2\pi v^2}{c^2} kT.$$ (26)

This is well-known as the Rayleigh-Jeans formula, and this function is proportional to ν^2, too. Therefore, we assumed that the radiation zone, in which the bremsstrahlung emission of the plasma is liberated, can be treated as a blackbody because of its similar dependence on ν.

If we consider the radiation zone as a blackbody, that is isolated from the surrounding nonheated regions, it is expected that the radiation zone will exhibit a radiation spectrum with a broad range of optical frequencies ranging from ultraviolet to infrared. The blackbody temerature T_b of the zone is related to the frequency ν_m of the spectral peak as follows [95]:

$$\nu_m = \frac{2.82 k T_b}{h}.$$ (27)

If T_{av} is defined as the average temperature of the radiation zone, the T_b will be close to the T_{av}.

The relationship between the T_{av} value and the time after fiber fuse generation is shown in Figure 12. T_{av} exceeds 10000 K immediately after fiber fuse generation but is less than 7000 K after 4 ms. Then it gradually approaches about 5,700 K. The value of 5,760 K shown in Figure 12 is the average T_{av} value from 4 to 30 ms. This temperature (5,760 K) is close to the reported temperatures of 5,400 K [5] and about 5,800 K [90].

Thus, it was found that if a fiber fuse, which exhibits a sharp temperature peak located near the light source, is approximated as a blackbody isolated from the surrounding nonheated regions, its average temperature from 4 ms after the generation of the fiber fuse approaches the experimentally estimated radiation temperatures.

4. Modeling of Hole-Assisted Fiber

In a Hole-Assisted Fiber (HAF), some holes exist around the main core as shown in Figure 1. In this HAF, the minimum distance of the holes from the main core is only $(D_{hole} - 2r_c)/2$. These holes are filled with air at the same temperature as the surrounding air (T_a). Therefore, at the inner surfaces of the holes, it can be expected that heat transfer or radiative transfer

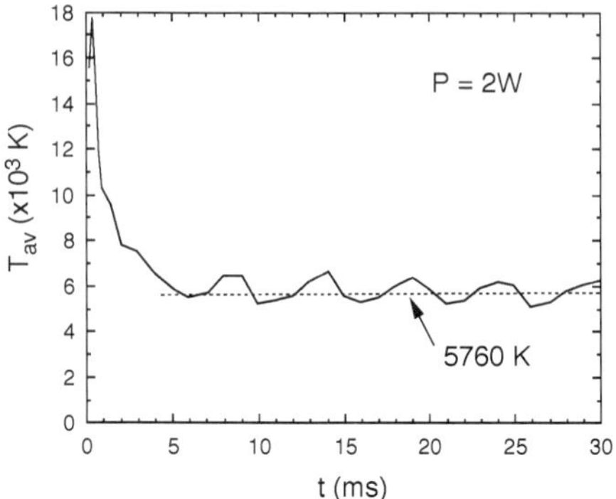

Figure 12. Average temperature of radiation zone vs. time after fiber fuse generation.

will occur between the heated solid inner surfaces of the holes and the gaseous fluid (air) with the low temperature of $T = T_a$. This is expected to affect the heat conduction behavior in the core center of the HAF.

Thus, we used the model proposed by Takara *et al.* [44] for heat conduction analysis, which includes heat transfer or radiative transfer between the inner surfaces of the holes and the gaseous air in the HAF. The proposed model for the HAF is shown in Figure 13.

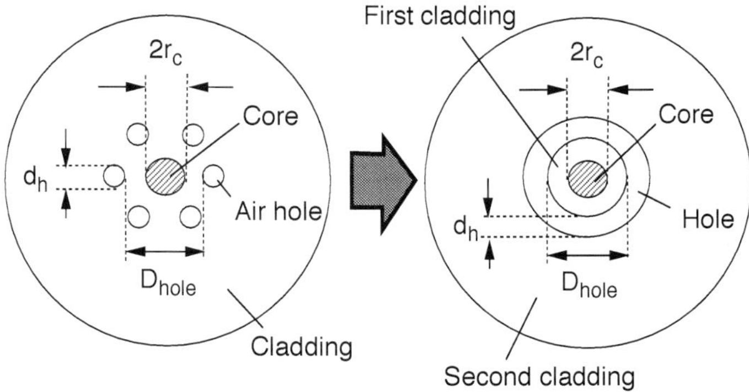

Figure 13. Heat conduction model of HAF [44].

Although bridges of silica glass exist between the holes in an actual HAF, to simplify the calculation, the first cladding layer was assumed to be a cylindrical layer of D_{hole} diameter, and the hole layer was treated as a gap of d_h width, which was inserted between the first and second cladding layers (see Figure 13). The hole layer was assumed to be fulfilled with a silica-air mixture. In the hole layer, the volume ratio of a silica glass to an air was $(1 - \gamma)$: γ, where γ_1 is the occupancy of the 6-air holes in the cross section of HAF. Furthermore, it was assumed that heat transfer and radiative transfer occurred at the inner surface of the hole layer in addition to heat conduction between the silica glass and the silica-air mixture.

When a certain quantity of heat per unit area (heat flux) is conducted from the heated core center to the end of the first cladding layer ($r = D_{hole}/2$) of an optical fiber, a considerable amount of the heat flux is transmitted through the hole layer, and the residual heat flux stagnates in the first cladding layer because the thermal conductivity κ of the hole layer is lower than that of silica glass.

This heat flux stagnating in the first cladding layer can be dissipated by radiative transfer and heat transfer between the outer surface of the first cladding layer and the air in the hole layer with a temperature of T_a.

It is noteworthy here that the total surface area of the 6-air holes existing in an actual HAF is larger than the outer surface area of the first cladding layer when $R_h = 2$ (see Table 1).

The heat transfer results from the convection of air, which is generated by warming the air near the outer surface of the first cladding layer. In this heat transfer, there is a region (called the "thermal boundary layer") near the surface of the first cladding layer where the temperature of the air changes rapidly from a high value to T_a.

5. Heat conduction modeling of HAF

We assume the HAF to have a radius of r_f and to be in an atmosphere with temperature $T = T_a$. We also assume that part of the core layer of ΔL length is heated to a temperature of $T_c^0 (> T_a)$ (see Figure 14).

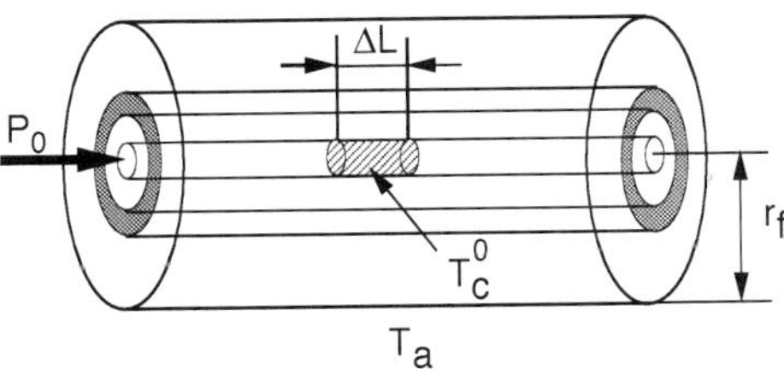

Figure 14. Hot zone in core layer of HAF.

The optical absorption coefficient α of the core layer in a HAF is a function of temperature (T), and it increases with increasing T [70]. In the heating zone (called the "hot zone") shown in Figure 14, α is larger than in other parts of the core because of its high temperature $T_c^0 (> T_a)$. Thus, as light propagates along the positive direction (away from the light source) in this zone, considerable amount of heat is produced by light absorption.

In the case of a heat source in part of the core layer, the nonsteady heat conduction equation for the temperature field $T(r, z, t)$ in the HAF is given by Eq. (12). We can solve Eq. (12) using the explicit finite-difference method (FDM) under the boundary and initial conditions, described in section 3.3, and the additional boundary conditions as follows:

(1) At the outer surface ($r = D_{hole}/2$) of the first cladding layer, the amount of heat conducted per unit area (heat flux) is dissipated by radiative transfer or heat transfer to the air ($T = T_a$) in the hole layer as follows:

$$-\lambda \frac{\partial T}{\partial r}\bigg|_{r=D_{hole}/2} = \gamma_1 \sigma_S \epsilon_e \left(T^4 - T_a^4 \right)$$

$$+ \gamma_1 \frac{\lambda}{\delta r_t} (T - T_a), \tag{28}$$

where γ_1 is the ratio of the total surface area of the 6-air holes existing in an actual HAF to the outer surface area of the first cladding layer, δr_t is the thickness of the thermal boundary layer, and $\delta r_t = x \, \delta r$ is assumed in the HAF calculation, where $x = 0.5$.

(2) At the inner surface ($r = D_{hole}/2 + d_h$) of the second cladding layer, the amount of heat conducted per unit area (heat flux) is dissipated by radiative transfer or heat transfer to the air ($T = T_a$) in the hole layer as follows:

$$-\lambda \frac{\partial T}{\partial r}\bigg|_{r=D_{hole}/2+d_h} = \gamma_2 \sigma_S \epsilon_e \left(T^4 - T_a^4 \right)$$

$$+ \gamma_2 \frac{\lambda}{\delta r_t} (T - T_a), \tag{29}$$

where γ_2 is the ratio of the total surface area of the 6-air holes existing in an actual HAF to the inner surface area of the second cladding layer.

The fiber parameters of the HAFs used for calculation were $d_h = r_c = 4.5~\mu m$, which were used in the experiments reported by Hanzawa *et al.* [45]. R_h was set to 2, 3, and 4. In the following, HAFs with R_h = 2, 3, and 4 are called HAF2, HAF3, and HAF4, respectively.

The parameters γ, γ_1, γ_2 for HAF2, HAF3, HAF4 are shown in Table 1.

Type	γ	γ_1	γ_2
HAF2	0.300	1.50	1.00
HAF3	0.214	1.00	0.75
HAF4	0.167	0.75	0.60

Table 1. Parameters for HAFs.

The area in the numerical calculation has a length of $2L$ (= 4 cm) in the axial (z) direction and a width of $2r_f$ (= 125 μm) in the radial (r) direction. There were 28 and 2,000 divisions in the r and z directions, respectively, and we set the calculation time interval to 1 μs. We assume that the hot zone is located at the center of the fiber (length $2L$) and that the length ΔL of the hot zone is 40 μm.

6. Simulation of fiber fuse in HAF

On the basis of the model described above, we calculated the generation and propagation behavior of a fiber fuse in HAFs with $d_h = r_c = 4.5$ μm using the FDM.

For reference, we also carried out the calculation for an SMF which has $r_c = 4.5$ μm, a refractive index difference $\Delta = 0.3\%$, and $A_{eff} = 98.1318 \times 10^{-12}$ m^2.

The calculation using the FDM was conducted in accordance with the procedure in [70]. In the calculation, we set the time interval δt to 1 μs, the step size along the r axis δr to $r_f/14$, the step size along the z axis δz to 20 μm, respectively, and assumed that $T_c^0 = 2{,}923$ K and $T_a = 298$ K.

We estimated the temperature field $T(r,z)$ of HAFs and/or SMF at $t = 10$ ms after the incidence of laser light with wavelength $\lambda_0 = 1.55$ μm and an initial power P_0 of 4 W. The calculation results for HAF2, HAF3, and HAF4 are shown in Figures 15–17, respectively, and the result for the SMF is shown in Figure 18.

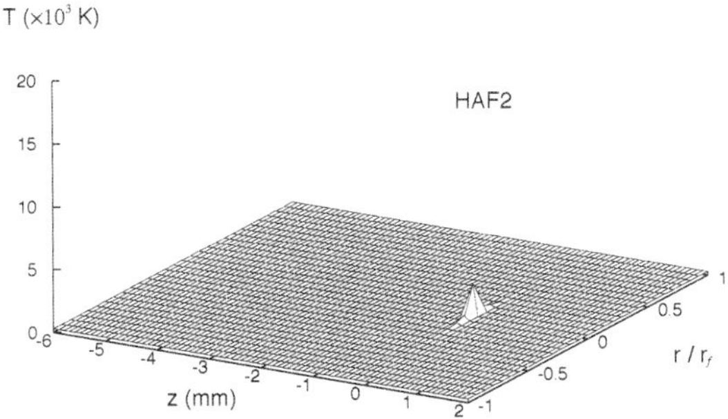

Figure 15. Temperature field in HAF2 after 10 ms after when $P_0 = 4$ W at $\lambda_0 = 1.55$ μm.

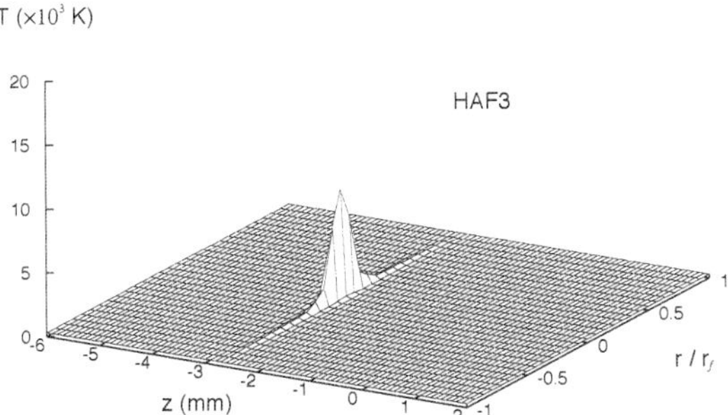

Figure 16. Temperature field in HAF3 after 10 ms when $P_0 = 4$ W at $\lambda_0 = 1.55$ μm.

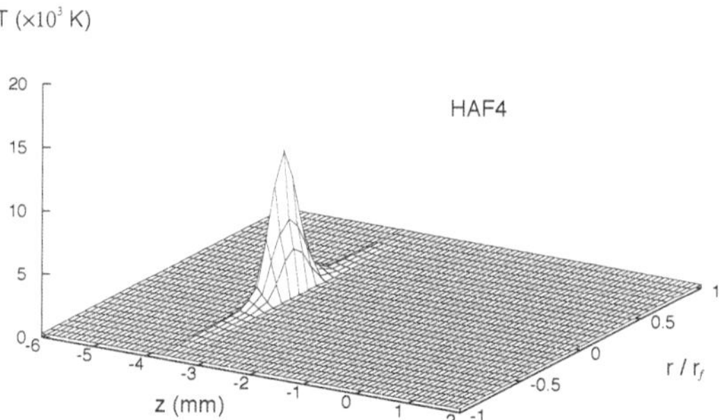

Figure 17. Temperature field in HAF4 after 10 ms when $P_0 = 4$ W at $\lambda_0 = 1.55$ μm.

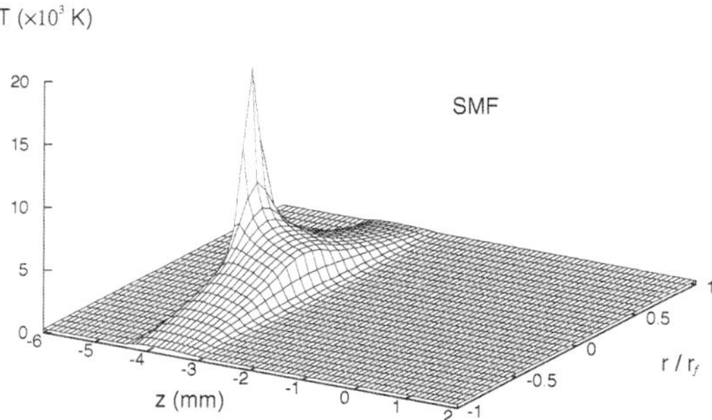

Figure 18. Temperature field in SMF after 10 ms when $P_0 = 4$ W at $\lambda_0 = 1.55$ μm.

As shown in Figures 15–18, the core center temperature at the thermal peak position of HAF2, HAF3, HAF4, and SMF are 2,923 K, 8,756 K, 11,603 K, and 17,528 K, respectively. The papid rise in the temperature shown in HAF3, HAF4, and SMF initiates the fiber fuse phenomenon. In SMF, the propagation velocity v_f of the fiber fuse was estimated to about 0.43 m/s from the peak shift distance (4.3 mm) per 10ms of the core center temperature shown in Figure 18. In HAF3 and HAF4, the propagation velocities of the fiber fuse were estimated to about 0.26 and 0.36 m/s, respectively, by using the same procedure as above.

On the other hand, in HAF2, there is no temperature increase in the core layer, as shown in Figure 15, and it means that a fiber fuse was not generated when $P_0 = 4$ W.

We investigated the temperature field $T(r, 0)$ at the end ($z = 0$ mm) of the hot zone in HAF2 after 10 ms when $P_0 = 4$–10 W at $\lambda_0 = 1.55$ μm. The calculated temperature fields are shown in Figure 19.

When the power of the light entering HAF2 increases from 4 W to 5 W, a thermal wave with a peak temperature of higher than 60,000 K was generated at the end of the hot zone, as

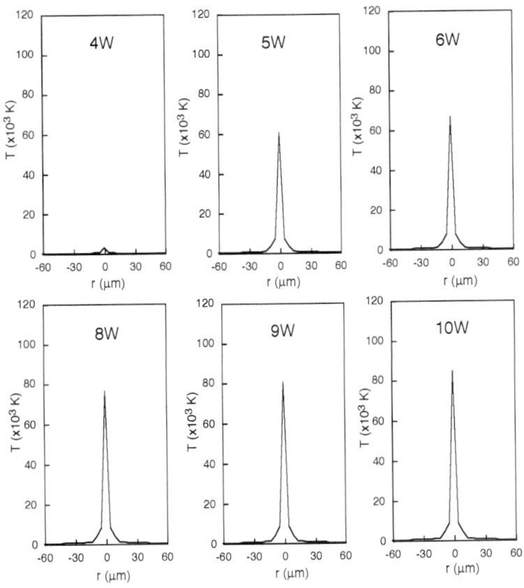

Figure 19. Temperature field at the end of the hot zone in HAF2 after 10 ms when P_0 = 4–10 W at λ_0 = 1.55 μm.

shown in Figure 19. Figure 20 shows the peak temperature change at the core center ($r = 0$ μm) with passage of time after the incidence of laser light with P_0 = 5 W at λ_0 = 1.55 μm.

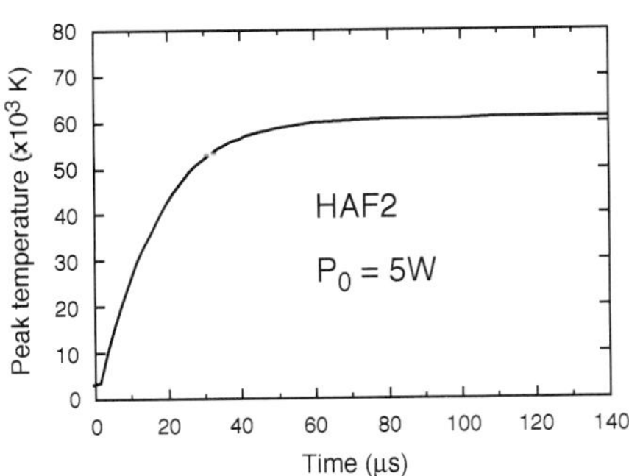

Figure 20. Peak temperature at the core center vs. time after the incidence of laser light with P_0 = 5 W at λ_0 = 1.55 μm.

The core center temperature at the end of the hot zone ($z = 0$ mm) changes abruptly to a large value of $\geq$ 50,000 K after 28 μs.

Although a thermal wave with a peak temperature of higher than 60,000 K was generated at the end of the hot zone, it was found that this thermal wave did not increase in size or propagate in the negative z direction when P_0 increased from 5 W to 10 W, as shown in Figure 19. Such a "stationary" thermal wave was reported by Kashyap [14].

If P_0 is further increased to 11 W and above, the thermal wave increases in size and propagates in the negative z direction toward the light source, as shown in Figure 21.

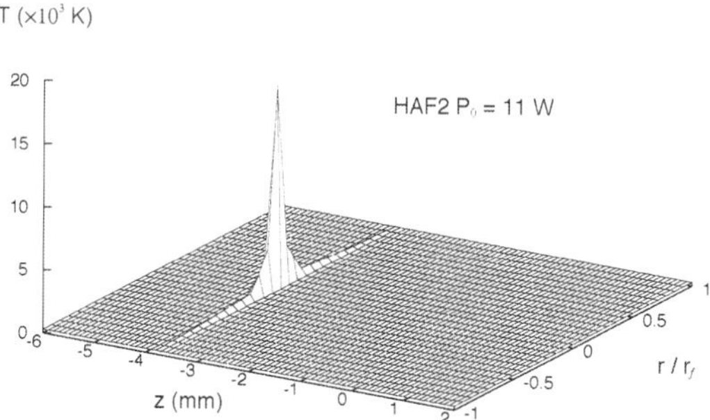

Figure 21. Temperature field in HAF2 after 10 ms after when $P_0 = 11$ W at $\lambda_0 = 1.55$ μm.

When $P_0 = 11$ W, the propagation velocities of the fiber fuse were estimated to about 0.38 m/s by using the same procedure as above.

We calculated the power intensity dependence of the propagation velocity v_p for each HAF and the SMF at $\lambda_0 = 1.55$ μm. The calculation results for the SMF, HAF2, HAF3, and HAF4 are shown in Figure 22.

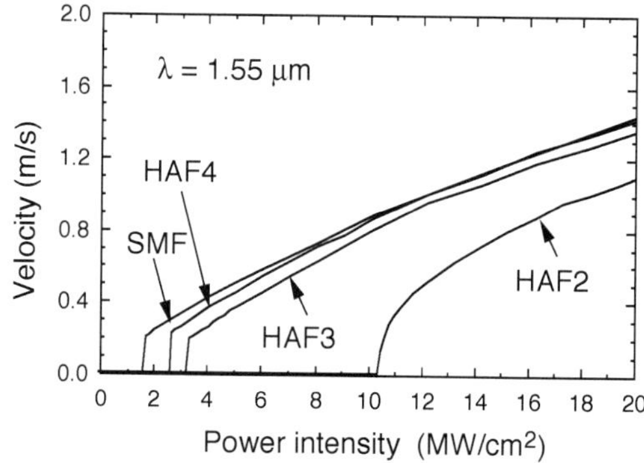

Figure 22. Power intensity vs fiber-fuse propagation velocity for SMF, HAF2, HAF3, and HAF4.

It is clear that the propagation velocity of the HAF4 and/or HAF3 approaches that of the SMF with increasing power intensity.

The threshold power intensity I_{th} for HAF2, HAF3, HAF4, and SMF are shown in Table 2.

Type	I_{th} (MW/cm^2)
HAF2	10.41
HAF3	3.27
HAF4	2.65
SMF	1.63

Table 2. Threshold power intensity for HAFs and SMF.

On the other hand, in HAF2, fiber fuse propagation begins at a threshold power intensity I_{th} of about 10.4 MW/cm^2 (P_0 = 10.2 W), which is much higher than that in HAF3 (about 3.3 MW/cm^2) or HAF4 (about 2.7 MW/cm^2), as shown in Table 2. The I_{th} of HAF2 is six times that of SMF.

When laser light enters the hot zone of the core layer, heat is produced in the zone by optical absorption of the incident light. In HAF2, the heat generated by optical absorption is effectively dissipated by the heat transfer and radiative transfer between the inner surfaces of the hole layer and the gaseous air. The dissipation of heat in HAF2 is more effective than that in HAF3 or HAF4 because the first cladding layer in HAF2 is thinner than that in HAF3 or HAF4.

Figure 23 shows a schematic view of the relationship between P_0 and the accumulated heat in the hot zone of the core layer for HAF2 and SMF.

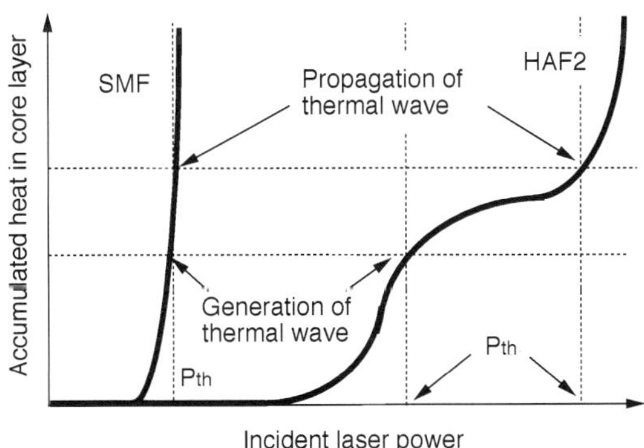

Figure 23. Schematic view of incident laser power vs accumulated heat in hot zone for HAF2 and SMF.

When the light power is small, the heat generated in the hot zone flows into the adjacent cladding layer, and the rise in the temperature of the zone is prevented. However, if light with a threshold power P_{th} and above enters the hot zone, it becomes difficult for all the generated heat to escape into the cladding layer, and part of the heat accumulates in the hot zone.

In the SMF, the accumulated heat raises the temperature in the hot zone and a thermal wave, *i.e.*, fiber fuse, is generated at the center of the zone. The thermal wave then rapidly increases in size and starts to propagate toward the light source as shown in Figure 23.

On the other hand, in HAF2, a large P_0 value is required to obtain sufficient accumulated heat for both the generation and propagation of a thermal wave, compared with that in the SMF.

Several researchers observed the dynamics of fiber fuse termination near a splice point between a HAF and an SMF by using a high-speed camera [42], [46], [48], [51].

Takenaga *et al.* reported that when a laser light of $P_0 = 8.1$ W and $\lambda_0 = 1.55$ μm was input into HAF2$_+$, a fiber fuse was generated at the splice point between the HAF and the SMF, propagated about 100 μm in the direction of the light source, and then stopped [42]. Kurokawa and Hanzawa [46] defined the length between the splice point and the termination point of fiber fuse as a penetration length L_p. In this case, $L_p \simeq 100$ μm.

Similarly, Kurokawa and Hanzawa investigated the power dependence of both the propagation velocity v_f for the SMF and L_p for HAF2 at $\lambda_0 = 1.48$ μm [46]. When a laser light of $P_0 = 8.1$ W was input into HAF2, the observed v_f for the SMF was 1.1 m/s and L_p for HAF2 was $\simeq 80$ μm.

As shown in Figure 19, even if a high power of 9 W was input into HAF2, a fiber fuse with a high peak temperature of about 80,000 K was generated at the center of the heated core, but it did not propagate forward the light source. This phenomenon, in which the propagation of a fiber fuse is controlled, is in good agreement with the experimental results observed by Takenaga *et al.* and Kurokawa and Hanzawa [46] for HAF2.

On the other hand, Kurokawa and Hanzawa reported that $v_f = 1.3$ m/s and $L_p = 110$ μm when a laser light of $P_0 = 12.0$ W (6.0 W at both 1.48 μm and 1.55 μm) was input into HAF2 [46]. This behavior for HAF2 cannot be explained by our calculation described above. As shown in Figure 22, a fiber fuse, generated at the splice point between the HAF2 and the SMF, was expected to maintain propagation in the direction of the light source when $I_{th} = 12.2$ MW/cm^2 ($P_0 = 12.0$ W). However, it was reported that the fiber fuse propagated only about 120 μm, and then stopped.

Furthermore, they reported that the hole part in the HAF disappeared in the domain in which the fiber fuse penetrated [46]. They considered that the plasma density of the core decreased in connection with the disappearance of the hole part and that the propagation of a fiber fuse can be controlled even if a high power of 10 W order is input into an HAF. In practice, they observed the dynamics of fiber fuse termination near the splice point between HAF2 and the SMF and found that the termination was accompanied by the evolution of a gas jet in the case of $P_0 = 12$ W [48] and 18.1 W (6.6 W at 1.48 μm and 11.5 W at 1.55 μm) [51].

In order for the hole part to disappear with the gas jet, it is necessary for the (first) cladding layer inscribed in this hole part to be destroyed by the incident high power.

We considered the destruction of the cladding layer using our heat conduction model, the results of which are described below.

7. Destruction mechanism of inner cladding layer of HAF

The internal core space of an HAF heated to an elevated temperature is shielded from the external conditions. Thus, with increasing temperature, the internal pressure increases and the internal volume decreases. Dianov *et al.* reported an internal core temperature of about 10,000 K and an internal pressure of about 10,000 atmospheres (1 GPa) at the time of fiber fuse evolution [72].

In HAF2, $D_{hole}/2 = 9.0$ μm and $r_c = 4.5$ μm. Therefore, the thickness h of the first cladding layer was 4.5 μm before the core was heated to a high temperature. When the core is heated to a high temperature and the internal pressure p reaches 1 GPa, the first cladding layer is partially melted by heat transmitted from the core and h becomes smaller than its initial value (4.5 μm).

We consider a tensile stress σ_θ acting on the inner wall of the first cladding layer with thickness h, as shown in Figure 24.

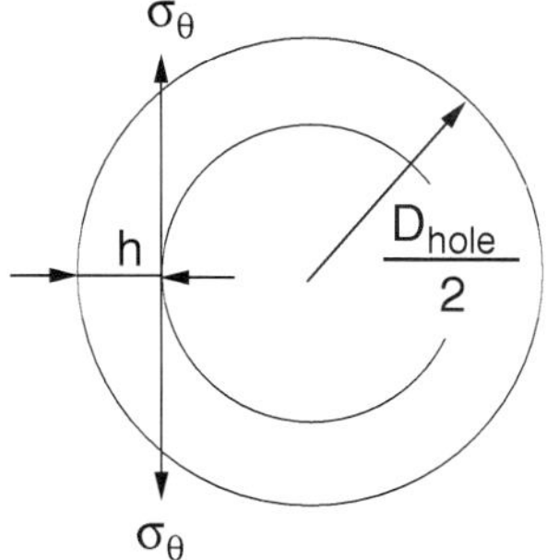

Figure 24. Tensile stress acting on inner wall of first cladding layer in HAF.

σ_θ is related to p by the following expression [96]:

$$\sigma_\theta = \frac{D_{hole}^2 - 2D_{hole}h + 2h^2}{2D_{hole}h - 2h^2} \cdot p. \tag{30}$$

σ_θ increases with increasing p. If σ_θ exceeds the ideal fracture strength σ_0 of the silica glass, a crack will generate on the inner wall of the first cladding layer.

On the other hand, it is well known for various solid materials that the σ_0 value is related to the Young's modulus E of the material by the following equation [97]:

$$\sigma_0 \approx E/10. \tag{31}$$

By using Eq. (31) and $E = 73$ GPa for silica glass, we can estimate σ_0 to be approximately 7.3 GPa.

For the HAF2, we estimated the relationship between h and the tensile stress σ_θ acting on the inner wall of the first cladding layer for $p = 1$ GPa using Eq. (30). The calculation result is shown in Figure 25.

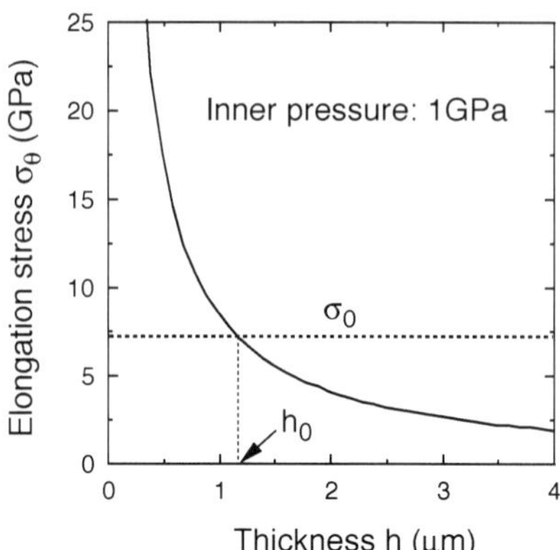

Figure 25. Thickness vs tensile stress acting on inner wall of cladding layer in HAF2.

σ_θ increases with decreasing h. It was found that when h decreases to a critical thickness h_0 (~ 1.2 μm), σ_θ becomes to σ_0 (~ 7.3 GPa). That is, σ_θ becomes larger than σ_0 when $h < h_0$. This means that the inner wall of the first cladding layer will crack and be destroyed when $h < h_0$.

Silica glass has a melting temperature of $T_m = 1{,}973$ K. If solid silica glass is heated above T_m, it will become a liquid "melt" and its mechanical properties, such as tensile tolerance, will be lost.

If the heat conduction model discussed above is used, the temperature distribution along the internal radial direction can be estimated for the HAF2.

We can estimate the radial distance (equivalent to h in Figure 24) from the outer wall of the first cladding layer with temperature T_a to an inner point at which T reaches T_m using the heat conduction model. Then, the minimum power of incident laser light necessary for $h < h_0$ to be satisfied can be determined by estimating the h value when laser light with various powers enters the HAF2.

For incident laser powers P_0 of 4–10 W with $\lambda_0 = 1.55$ μm, the temperature distribution in the first cladding layer as a function of the radial distance h' was calculated for the HAF2. The result is shown in Figure 26.

The horizontal axis h' in Figure 26 represents the distance between the outer wall of the first cladding layer and an inner point located closer to the core center.

When P_0 is 4 W or less, the temperature T in the cladding layer is lower than T_m, and the cladding layer is not destroyed.

On the other hand, when P_0 is 5 W and above, T increases with increasing h'. When $P_0 = 10$ W, T reaches $\sim T_m$ at $h' \sim 1.2$ μm. This value of h' (~ 1.2 μm) is the same as the threshold value of h_0 (~ 1.2 μm) for crack generation on the inner wall of the cladding layer. Therefore, the destruction of the cladding layer is predicted when laser light with $P_0 \geq 10$ W enters the HAF2.

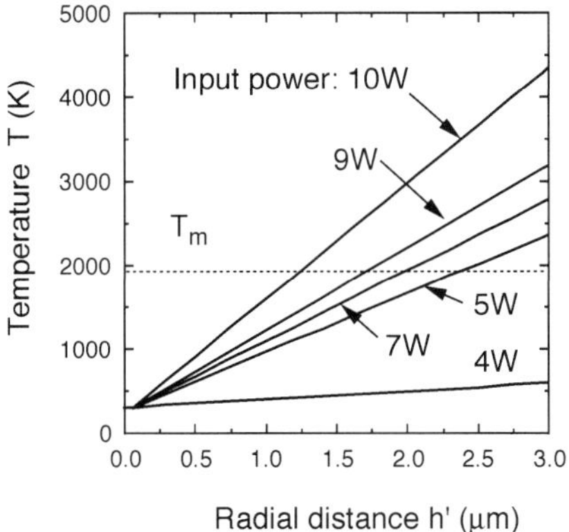

Figure 26. Temperature field in first cladding layer of HAF2 when $P_0 = 4$–10 W at $\lambda_0 = 1.55$ µm.

If a thermal wave propagates along the fiber axis, because of the pressure difference of about 10,000 atmospheres between the internal core and the air hole layer, the destruction of the cladding layer will change the propagation direction of the thermal wave from the axial direction to the radial direction. As a result, the thermal wave will propagate toward the air hole layer with a low pressure (about 1 atmosphere). This is the reason for the observation of a gas jet reported by Kurokawa *et al.* [48].

When the thermal wave propagates into the air hole layer, the hole layer disappears as a result of thermal heating and the core temperature decreases owing to the departure of the thermal wave. Such cooling of the core will prevent the generation of a new thermal wave in the core, and the propagation of the thermal wave will stop at this time.

Kurokawa and Hanzawa [46] reported that when laser light with a high power entered HAF2, the increase in penetration length started at an incident power of 8 W.

As described above, if the incident power increases from 9 W to 10 W, the cladding layer is destroyed and the direction of propagation of the thermal wave changes to the radial direction, and the air hole layer vanishes owing to the melting of the cladding layer.

There is a slight time delay τ for an crack propagating from the inner surface to the outer surface of the first cladding layer. The crack generated on the inner surface of the first cladding layer grows and propagates to the outer surface of the cladding layer. When the crack reaches the outer surface, the first cladding layer is destroyed because of the pressure difference of about 10,000 atmospheres between the internal core and the air hole layer.

The crack propagation rate V_c of the silica glass is related to the sonic rate V_s of the glass as follows [98]:

$$0 \leq V_c \leq 0.38 V_s. \tag{32}$$

The value of V_s for silica glass is 5,570 m/s. For a crack propagating through a small length of ~ 1.2 µm ($= h_0$), the minimum time delay τ_m was estimated to be

$$\tau_m \equiv \frac{h_0}{0.38V_s} = 5.7 \times 10^{-10}\text{s}. \tag{33}$$

This time delay is not dependent on the incident power, as indicated by Eq. (32), and the propagation velocity v_f of the thermal wave increases with increasing incident power.

We assumed that τ for the HAF2 corresponded to the time (50–90 μs) for rapid rise in the core temperature after the incidence of fiber fuse, which propagated in the SMF. If we assume that $\tau \sim 80$ μs and the penetration length L_p is the product of v_f and τ, the L_p values at a incident light of $P_0 = 8.1$ W and 12.0 W are 88 μm and 104 μm. These values are fairly good agreement with the experimental results (~ 80 μm and 110 μm) reported by Kurokawa and Hanzawa [46].

In closing, we should comment on the fiber fuse propagation with a long-period damage track, which was observed in the HAF2 with d_h of smaller than r_c [46], [48]–[50]. The most striking characteristic of this phenomenon is the long period (several 100 μm order) of the cavities, which were generated by entering of a relatively low power (about 1.5–4.5 W) into HAF2.

The long period damage tracks such as those in HAF2 were reported by Bufetov *et al.* [99]. They found that such a phenomenon was observed as a result of the interference of LP_{01} and LP_{02} modes excited in an optical fiber with a W-index profile. However HAF2 is a step-index optical fiber, and only LP_{01} (or TE_{11}) mode can be excited in HAF2 at 1.48 and/or 1.55 μm. Therefore, it is very difficult to consider that the mode interference is responsible for the long period damage tracks observed in HAF2.

Instead of the mode interference, we consider that this phenomenon may relate to a thermal lens effect [63], [101]. The focal length F of thermal lens effect is given by (see Appendix)

$$F = \frac{n_0\pi\kappa\omega_0^2}{\alpha Pl(\partial n/\partial T)}, \tag{34}$$

where n_0 (= 1.46) is the characteristic refractive index of the core layer, P is the incident power of the light reflecting from the cavity wall, which is estimated by Eq. (24). ω_0 (~ 4.5 μm) is the spot size radius of the laser beam when optical power in the optical fiber was assumed to take on Gaussian distribution, $\partial n/\partial T$ (= 1.23 $\times 10^{-5}$ K^{-1} [5]) is the thermal coefficient of refractive index for silica glass, and l is the length of the heating core, where α exhibits large value.

As F is inversely as the product of α and P, large F value may be obtained by small α and/or P value, which is comparable with observed long period of the cavities. If we assume $\lambda = 9.2$ W/mK [5], $l = 20$ μm, and $\alpha = 5 \times 10^4$ m^{-1}, the F value at $P = 0.158$ W ($P_0 = 4.5$ W) can be estimated to about 440 μm by using Eq. (34). This value is fairly good agreement with the observed period (460 μm) of the cavities [46], [48]–[50].

The cause of mechanism of the long-period damage track has yet to be sufficiently clarified. It largely depends upon future multilateral studies.

8. Conclusions

We investigated the unsteady thermal conduction status in a single-mode optical fiber by numerical computation in order to visualize the mode of fiber fuse propagation. We assumed that the vitreous silica optical fiber underwent pyrolysis at elevated temperatures to form SiO_x ($x \sim 1$). We also proposed a model in which the optical absorption coefficient of the core layer increased with increasing molar concentration of SiO_x. By using the model, we calculated the temperature distribution in the fiber with the explicit finite-difference method. It was found that when a short core with 40 μm length was heated to 2,923 K and a 2 W laser light (wavelength of 1.064 μm) entered the core layer of an SMF-28 optical fiber, a thermal wave, *i.e.*, a fiber fuse, with a peak temperature of about 34,000 K was generated at the boundary of the heating region near the light source. The fiber fuse was enlarged and propagated toward the light source at a rate of about 0.54 m/s. The calculated propagation velocity of the fiber fuse was in agreement with the experimental value. Moreover, the average temperature of the radiated region of the core layer was less than 7,000 K at a time of 4 ms after the generation of the fiber fuse and gradually approached a temperature of about 5,700 K. The final average temperature was close to the experimentally reported values.

We evaluated the threshold power of fiber fuse propagation in hole-assisted fibers (HAFs) using the finite-difference method and the model proposed by Takara *et al.* The HAFs with ratios of hole-space distance to core diameter of 3 and 4 exhibited fiber fuse propagation when a 1.55 μm laser with a power of 4 W was input into the core layer, as observed for an SM optical fiber.

On the other hand, the HAF with a ratio of hole-space distance to core diameter of 2 exhibited no fiber fuse generation or propagation when a 1.55 μm laser with a power of 10 W was input into the core layer.

Furthermore, when the incident power was 5 W and above, the temperature of the central core increased owing to the absorption of a large amount of power, causing the melting of the first cladding layer adjacent to the heated core. Thus, the thickness of the first cladding layer decreased below the value at which the solid cladding layer could be maintained, and the cladding layer was destroyed following the disappearance of the air hole layer. The destruction of the cladding layer caused the direction of the thermal wave to change from the axial direction to the radial direction, stopping the propagation of the thermal wave. By using this phenomenon, Kurokawa and Hanzawa proposed a novel fiber fuse terminotor composed of a short length of HAF [100].

Appendix: Focal length of thermal lens effect

If part of the core layer is heated by light absorption, the refractive-index gradient in the core is induced by the thermal coefficient of the refractive index $\partial n/\partial T$ of the silica glass and the temperature distribution ΔT in the core layer. The refractive index n of the heated core is expressed as follows:

$$n(r,t) = n_0 + \frac{\partial n}{\partial T}\Delta T(r,t) \tag{35}$$

where n_0 (= 1.46) is the characteristic n value of the core layer, r is the radial distance from the center of the optical fiber, and t is time, respectively.

If we assume that optical power in the optical fiber takes on Gaussian distribution, ΔT in the core layer is given by [63], [101]

$$\Delta T(r,t) = \frac{\alpha P}{4\pi\lambda}\left[\ln\left(1+\frac{8Dt}{\omega_0{}^2}\right) - \frac{16Dt}{\omega_0{}^2+8Dt}\frac{r^2}{\omega_0{}^2}\right] \tag{36}$$

where α is the absorption coefficient, P is the incident optical power, ω_0 is the spot size radius of the laser beam, and λ is the thermal conductivity of the silica glass, respectively. Parameter $D = \lambda/C_p\rho$, where C_p and ρ are the specific heat and dnsity of the silica glass.

Substitution of Eq. (36) into Eq. (35) gives

$$n \cong n_0\left[1+\delta\left(\frac{r}{\omega_0}\right)^2\right], \tag{37}$$

$$\delta = -\frac{2\alpha P}{4n_0\pi\lambda}\left(\frac{\partial n}{\partial T}\right)\frac{8Dt}{\omega_0{}^2+8Dt}. \tag{38}$$

When refractive index of the core layer takes a radial distribution as shown in Eq. (37), propagating laser beam will be focussed as a result of the thermal lens effect. In this case, the focal length F is given by [63], [101]

$$F(t) = \frac{n_0\pi\lambda\omega_0{}^2(\omega_0{}^2+8Dt)}{\alpha Pl(\partial n/\partial T)(8Dt)} = F_\infty\left[1+\frac{t_c}{2t}\right], \tag{39}$$

$$F_\infty = \frac{n_0\pi\lambda\omega_0{}^2}{\alpha Pl(\partial n/\partial T)}, \tag{40}$$

$$t_c = \frac{\omega_0{}^2}{4D} \tag{41}$$

where l is the length of the heating core, where α exhibits large value. In typical SM fiber, $\lambda = 9.2$ W/mK, $C_p = 788$ J/kgK, $\rho = 2{,}200$ kg/m^3 [5], and $\omega_0 \sim 4.5\ \mu$m. If we insert these values into Eqs. (39)–(41), we obtain $t_c \sim 0.95\ \mu$s. This means that $F \cong F_\infty$ when t is 10 μs or above. If we assume $P = 2$W, $l \sim 40\ \mu$m, $\alpha = 5.6\times10^4$ m^{-1} [5], and $\partial n/\partial T = 1.23\times10^{-5}$ K^{-1} [5], F is given by

$$F \cong F_\infty \sim 15.5\mu\text{m}$$

This F value is of the same order of the observed interval and/or large front size of the cavities [11], [23].

Author details

Yoshito Shuto

Ofra Project, Iruma, Japan

References

[1] A. Asano, T. Kobayashi, E. Yoshida, and Y. Miyamoto. Ultra-high capacity optical transmission technologies for 100 Tbit/s optical transport networks, *IEICE Trans. Commun.* 2011; E94-B(2): 400–408.

[2] T. Morioka. New generation optical infrastructure technologies: "EXAT initiative" toward 2020 and beyond, *OptoElectron. Commun. Conf.* 2009 (OECC 2009): FT4.

[3] R. Kashyap and K. J. Blow. Observation of catastrophic self-propelled self-focusing in optical fibres, *Electron. Lett.* 1988; 24(1): 47–49.

[4] R. Kashyap. Self-propelled self-focusing damage in optical fibres, *Proc. Xth Int. Conf. Lasers* 1988: 859–866.

[5] D. P. Hand and P. St. J. Russell. Solitary thermal shock waves and optical damage in optical fibers: the fiber fuse, *Opt. Lett.* 1988; 13(9): 767–769.

[6] D. P. Hand and P. St. J. Russell. Soliton-like thermal shock-waves in optical fibers: Origin of periodic damage tracks, *Eur. Conf. Opt. Commun.* 1988: 111–114.

[7] D. D. Davis, S. C. Mettler, and D. J. DiGiovanni. Experimental data on the fiber fuse, in H. E. Bennett, A. H. Guenther, M. R. Kozlowski, B. E.. Newnam, and M. J.Soileau (eds), *Proc. Soc. Photo-Opt. Instrum. Eng.* 1995; Vol. 2714: 202–210.

[8] D. D. Davis, S. C. Mettler, and D. J. DiGiovanni. A comparative evaluation of fiber fuse models, in H. E. Bennett, A. H. Guenther, M. R. Kozlowski, B. E.. Newnam, and M. J.Soileau (eds), *Proc. Soc. Photo-Opt. Instrum. Eng.* 1996; Vol. 2966: 592–606.

[9] S. Todoroki. Animation of fiber fuse damage, demonstrating periodic void formation, *Opt. Lett.* 2005; 30(19): 2551–2553.

[10] S. Todoroki. In-situ observation of fiber fuse propagation, *Jpn. J. Appl. Phys.* 2005; 4(6A): 4022–4024.

[11] S. Todoroki. Origin of periodic void formation during fiber fuse, *Opt. Express* 2005; 13(17): 6381–6389.

[12] S. Todoroki. Transient propagation mode of fiber fuse leaving no voids, *Opt. Express* 2005; 13(23): 9248–9256.

[13] S. Todoroki. In situ observation of modulated light emmission of fiber fuse synchronized with void train over hetero-core splice point, *PLos ONE* 2008; 3(9): e3276-1–4.

[14] R. Kashyap. The fiber fuse - from a curious effect to a critical issue: a 25th year retrospective, *Opt. Express* 2013; 21(5): 6422–6441

[15] S. Todoroki. Fiber fuse propagation behavior, in Y. Moh, S. W. Harun, and H. Arof (eds), *Selected Topics on Optical Fiber Technology*, InTech, Croatia 2012: 551–570.

[16] P. André, A. Rocha, F. Domingues, and M. Facão. Thermal effects in optical fibres, in M. A. D. Bernardes (eds), *Developments in Heat Transfer*, InTech, Croatia 2011: 1–20.

[17] S. Todoroki. Fiber Fuse: Light-Induced Continuous Breakdown of Silica Glass Optical Fiber. NIMS Monographs. Springer: Tokyo; 2014.

[18] T. J. Driscoll, J. M. Calo, and N. M. Lawandy. Explaining the optical fuse, *Opt. Lett.* 1991; 16(13): 1046–1048.

[19] R. Kashyap, A. Sayles, and G. F. Cornwell. Heat flow modeling and visualization of catastrophic self-propagating damage in single-mode optical fibres at low powers, in H. E. Bennett, A. H. Guenther, M. R. Kozlowski, B. E.. Newnam, and M. J.Soileau (eds), *Proc. Soc. Photo-Opt. Instrum. Eng.* 1996; Vol. 2966: 586–591.

[20] E. M. Dianov, I. A. Bufetov, A. A. Frolov, V. G. Plotnichenko, V. M. Mashinskii, M. F. Churbanov, and G. E. Snopatin. Catastrophic destruction of optical fibres of various composition caused by laser radiation, *Quantum Electron.* 2002; 32(6): 476–478.

[21] E. M. Dianov, I. A. Bufetov, A. A. Frolov, V. M. Mashinskii, V. G. Plotnichenko, M. F. Churbanov, and G. E. Snopatin. Catastrophic destruction of fluoride and chalcogenide optical fibres, *Electron. Lett.* 2002; 38(15): 783–784.

[22] R. Kashyap. High average power effects in optical fibres and devices, in H. G. Limberger, and M. J. Matthewson (eds), *Proc. Soc. Photo-Opt. Instrum. Eng.* 2003; Vol. 4940: 108–117.

[23] R. M. Atkins, P. G. Simpkins, and A. D. Yabon. Track of a fiber fuse: a Rayleigh instability in optical waveguides, *Opt. Lett.* 2003; 28(12): 974–976.

[24] K. Seo, N. Nishimura, M. Shiino, R. Yuguchi, and H. Sasaki. Evaluation of high-power endurance in optical fiber links, *Furukawa Rev.* 2003; (24): 17–22.

[25] M. M. Lee, J. M. Roth, T. G. Ulmer, and C. V. Cryan. The fiber fuse phenomenon in polarization-maintaining fibers at 1.55 μm, *Proc. Conf. on Lasers and Electro-Optics* 2006 (CLEO): JWB66.

[26] E. D. Bumarin and S. I. Yakovlenko. Temperature distribution in the bright spot of the optical discharge in an optical fiber, *Laser Phys.* 2006; 16(8): 1235–1241.

[27] I. A. Bufetov, A. A. Frolov, E. M. Dianov, V. E. Frotov, and V. P. Efremov. Dynamics of fiber fuse propagation, *Optical Fiber Commun./Nat. Fiber Optic Engineers Conf.* 2005 (OFC/NFOEC 2005): OThQ7.

[28] E. M. Dianov, V. E. Frotov, I. A. Bufetov, V. P. Efremov, A. E. Rakitin, M. A. Melkumov, M. I. Kulish, and A. A. Frolov. High-speed photography, spectra, and temperature of optical discharge in silica-based fibers, *IEEE Photon. Technol. Lett.* 2006; 18(6): 752–754.

[29] J. Wang, S. Gray, D. Walton, and L. Zentero. Fiber fuse in high power optical fiber, in M.-J. Li, P. Shum, I. H. White, and X. Wu (eds), *Proc. Soc. Photo-Opt. Instrum. Eng.* 2008; Vol. 7134: 71342E-1–9.

[30] K. S. Abedin and M. Nakazawa. Real time monitoring of a fiber fuse using an optical time-domain reflectometer, *Opt. Express* 2010; 18(20): 21315–21321.

[31] P. S. André, M. Facão, A. M. Rocha, P. Antunes, and A. Martins. Evaluation of the fuse effect propagation in networks infrastructures with different types of fibers, *Optical Fiber Commun./Nat. Fiber Optic Engineers Conf.* 2010 (OFC/NFOEC 2010): JWA10.

[32] P. S. André, A. M. Rocha, F. Domingues, and A. Martins. Improved thermal model for optical fibre coating owing to small bending diameter and high power signals, *Electron. Lett.* 2010; 46(10): 695–696.

[33] M. Yamada, O. Koyama, Y. Katsuyama, and T. Shibuya. Heating and burning of optical fiber by light scattered from bubble train formed by optical fiber fuse, *Optical Fiber Commun./Nat. Fiber Optic Engineers Conf.* 2011 (OFC/NFOEC 2011): JThA1.

[34] A. M. Rocha, P. Antunes, F. Domingues, M. Facão, and P. S. André. Detection of fiber fuse using FBG sensors, *IEEE Sensors J.* 2011; 11(6): 1390–1394.

[35] A. M. Rocha, F. Domingues, M. Facão, and P. S. André. Threshold power of fiber fuse effect for diffent types of optical fiber, *Int. Conf. on Transparent Optical Networks* 2011 (ICTON 2011): Tu.P.13.

[36] W. Ha, Y. Jeong, and K. Oh. Fiber fuse in hollow optical fibers, *Opt. Lett.* 2011; 36(9): 1536–1538.

[37] S. Todoroki. Threshold power reduction of fiber fuse propagation through a white tight-buffered single-mode optical fiber, *IEICE Electron. Express* 2011; 8(23): 1978–1982.

[38] F. Domingues, A. R. Frias, P. Antunes, A. O. P. Sousa, R. A. S. Ferreira, and P. S. André. Observaion of fuse effect discharge zone nonlinear velocity regime in erbium-doped fibres, *Electron. Lett.* 2012; 48(20): 1295–1296.

[39] S. Todoroki. Fiber fuse propagation modes in typical single-mode fibers, *Optical Fiber Commun./Nat. Fiber Optic Engineers Conf.* 2013 (OFC/NFOEC 2013): JW2A.

[40] E. M. Dianov, I. A. Bufetov, A. A. Frolov, Y. K.Chamorovsky, G. A. Ivanov, and I. L. Vorobjev. Fiber fuse effect in microstructured fibers, *IEEE Photon. Technol. Lett.* 2004; 16(1): 180–181.

[41] K. Nakajima, K. Hogari, J. Zhou, K. Tajima, and I. Sankawa. Hole-assisted fiber design for small bending and splice losses, *IEEE Photon. Technol. Lett.* 2003; 15(12): 1737–1739.

[42] K. Takenaga, S. Tanigawa, S. Matsuo, M. Fujimaki, and H. Tsuchiya. Fiber fuse phenomenon in hole-assisted fibers, in *Technical Digest of European Conf. Opt. Commun.* 2008 (ECOC 2008): P.1.14.

[43] N. Hanzawa, K. Kurokawa, K. Tsujikawa, T. Matsui, and S. Tomita. Suppression of fiber fuse propagation in photonic crystal fiber (PCF) and hole assisted fiber (HAF), *Proc. 15th Microoptics Conf.* 2009 (MOC'09): M7.

[44] H. Takara, H. Masuda, H. Kanbara, Y. Abe, Y. Miyamoto, R. Nagase, T. Morioka, S. Matsuoka, M. Shimizu, and K. Hagimoto. Evaluation of fiber fuse characteristics of hole-assisted fiber for high power optical transmission systems, in *Technical Digest of European Conf. Opt. Commun.* 2009 (ECOC 2009): 918–919.

[45] N. Hanzawa, K. Kurokawa, K. Tsujikawa, T. Matsui, K. Nakajima, S. Tomita, and M. Tsubokawa. Suppression of fiber fuse propagation in hole assisted fiber and photonic crystal fiber, *IEEE J. Lightwave Technol.* 2010; 28(15): 2115–2120.

[46] K. Kurokawa and N. Hanzawa. Fiber fuse propagation and its suppression in hole assisted fibers, *IEICE Trans. Commun.* 2011; E94-B(2): 384–391.

[47] K. Takenaga, S. Tanigawa, S. Matsuo, and M. Fujimaki. Fiber fuse phenomenon in hole-assisted fibers, *Fujikura Technical Rev.* 2011: 12–15.

[48] K. Kurokawa, N. Hanzawa, K. Tsujikawa, and S. Tomita. Hole-size dependence of fiber fuse propagation in hole-assisted fiber (HAF), *Proc. 17th Microoptics Conf.* 2011 (MOC'11): H-30.

[49] N. Hanzawa, K. Kurokawa, K. Tsujikawa, K. Takenaga, S. Tanigawa, S. Matsuo, and S. Tomita. Observation of a propagation mode of a fiber fuse with a long-period damage track in hole-assisted fiber, *Opt. Lett.* 2010; 35(12): 2004–2006.

[50] K. Kurokawa, N. Hanzawa, K. Tsujikawa, and S. Tomita. Power dependence of fiber fuse propagation with a long-period damage track in hole-assisted fiber, *IEICE Electron. Express* 2011; 8(11): 802–807.

[51] K. Kurokawa. Optical fiber for high-power optical communication, *Crystals* 2012; 2: 1382–1392.

[52] S. I. Yakovkenko. On reasons for strong absorption of light in an optical fibre at high temperature, *Quantum Electron.* 2004; 34(9): 787–789.

[53] S. I. Yakovkenko. Plasma behind the front of a damage wave and the mechanism of laser-induced production of a chain of caverns in an optical fibre, *Quantum Electron.* 2004; 34(8): 765–770.

[54] S. I. Yakovkenko. Mechanism for the void formation in the bright spot of a fiber fuse, *Laser Phys.* 2006; 16(3): 474–476.

[55] S. I. Yakovkenko. Physical processes upon the optical discharge propagating in optical fiber, *Laser Phys.* 2006; 16(9): 1273–1290.

[56] M. Facão, A. Rocha, and P. André. Traveling solution of the fuse effect in optical fibers, *IEEE/OSA J. Lightwave Technol.* 2011; 29(1): 109–114.

[57] A. Ankiewicz, W. Chen, P. St. J. Russell, M. Taki, and N. Akhmediev. Velocity of heat dissipative solitons in optical fibers, *Opt. Lett.* 2008; 33(19): 2176–2178.

[58] E. M. Dianov, V. E. Frotov, I. A. Bufetov, V. P. Efremov, A. A. Frolov, M. Y. Schelev, and V. I. Lozovoi. Detonation-like mode of the destruction of optical fibers under intense laser radiation, *JETP Lett.* 2006; 83(2): 75–78.

[59] B. J. Matkowsky and G. I. Sivashinsky. Propagation of a pulsating reaction front in solid fuel combustion, *SIAM J. Appl. Math.* 1978; 35(3): 465–478.

[60] S. Ei and M. Mimura. Relaxation oscillations in combustion models of thermal self-ignition, *J. Dynamics and Differential Equations* 1992; 4(1): 191–229.

[61] R. O. Weber, G. N. Mercer, H. S. Sidhu, and B. F. Gray. Combustion waves for gases ($Le = 1$) and solids ($Le \to \infty$), *Proc. R. Soc. Lond. A* 1997; 453: 1105–1118.

[62] N. Akhmediev, P. St. J. Russell, M. Taki, and J. M. Soto-Crespo. Heat dissipative solitons in optical fibers, *Phys. Lett. A* 2008; 372: 1531–1534.

[63] M. Katayama. Laser Chemistry; Nonlinear Spectroscopy and Laser Induced Chemical Processes. Chap. 11. Syokabo, Inc.; Tokyo: 1985.

[64] A. E. Owen and R. W. Douglas. The electrical properties of vitreous silica, *J. Soc. Glass Technol.* 1959; 43: 159–178.

[65] R. H. Doremus. Ionic transport in amorphous oxides, *J. Electrochem. Soc.* 1968; 115: 181–186.

[66] Y. Shuto, S. Yanagi, S. Asakawa, M. Kobayashi, and R. Nagase. Simulation of fiber fuse phenomenon in single-mode optical fibers, *IEEE/OSA J. Lightwave Technol.* 2003; 21(11): 2511–2517.

[67] Y. Shuto, S. Yanagi, S. Asakawa, and R. Nagase. Generation mechanism on fiber fuse phenomenon in single-mode optical fibers, in *Electronics and Communications in Japan, Part 2* 2003; 86(11): 11–20.

[68] Y. Shuto, S. Yanagi, S. Asakawa, M. Kobayashi, and R. Nagase. Evaluation of high-temperature absorption coefficients of optical fibers, *IEEE Photon. Technol. Lett.* 2004; 16(4): 1008–1010.

[69] Y. Shuto, S. Yanagi, S. Asakawa, M. Kobayashi, and R. Nagase. Fiber fuse phenomenon in step-index single-mode optical fibers, *IEEE J. Quantum Electron.* 2004; 40(8): 1113–1121.

[70] Y. Shuto. Heat conduction modeling of fiber fuse in single-mode optical fibers, *J. Photon.* 2014; Vol. 2014, Article ID 645207: 1–21.

[71] H. L. Schick. A thermodynamic analysis of the high-temperature vaporization properties of silica, *Chem. Rev.* 1960; 60: 331–362.

[72] E. M. Dianov, I. A. Bufetov, and A. A. Frolov. Destruction of silica fiber cladding by the fuse effect, *Opt. Lett.* 2004; 29(16): 1852–1854.

[73] H. R. Philipp. Optical properties of non-crystalline Si, SiO, SiO_x and SiO_2, *J. Phys. Chem. Solids* 1971; 32: 1935–1945.

[74] H. R. Philipp. Optical and bonding model for non-crystalline SiO_x and SiO_xN_y materials, *J. Non-Cryst. Solids* 1972; 8–10: 627–632.

[75] W. J. Moore. Physical Chemistry. 4th Ed. Chap. 8. Prentice-Hall, Inc.; New York: 1972.

[76] *JANAF Thermochemical Tables*. 2nd Ed. U.S. Department of Commerce and National Bureau of Standards: 1971.

[77] M. H. Brodsky, R. S. Title, K. Weiser, and G. D. Pettit. Structural, optical, and electrical properties of amorphous silicon films, *Phys. Rev. B* 1970; 1(6): 2632–2641.

[78] D. L. Dexter. Interpretation of Urbach's rule, *Phys. Rev. Lett.* 1967; 19(24): 1383–1385.

[79] H. Mahr. Ultraviolet absorption of KI diluted in KCl crystals, *Phys. Rev.* 1962; 125(5): 1510–1516.

[80] G. G. Macfarlane and V. Roberts. Infrared absorption of silicon near the lattice edge, *Phys. Rev.* 1955; 98(6): 1865–1866.

[81] G. G. Macfarlane, T. P. McLean, J. E. Quarrington, and V. Roberts. Fine structure in the absorption-edge spectrum of Si, *Phys. Rev.* 1958; 111(5): 1245–1254.

[82] E. Burstein, G. Picus, and N. Sclar. Optical and photoconductive properties of silicon and germanium, in *Photoconductivity Conference.* pp. 353–413. John Wiley & Sons, Inc.; New York: 1956.

[83] H. S. Carslaw and J. C. Jaeger. Conduction of Heat in Solids. 2nd Ed. Chap. 13. Oxford University Press; Oxford: 1959.

[84] I. Barin and O. Knacke. Thermochemical Properties of Inorganic Substances. p. 690. Springer-Verlag; Berlin: 1973.

[85] M. Shoji. Heat Transfer Textbook. Chap. 2. Univ. Tokyo Press; Tokyo: 1995.

[86] T. Okuda. Plasma Engineering. Chap. 3. Corona-Sha; Tokyo: 1975.

[87] T. Okuda. Plasma Engineering. Chap. 1. Corona-Sha; Tokyo: 1975.

[88] L. Spitzer, Jr. Physics of Fully Ionized Gases. 2nd Ed. p. 128. John Wiley & Sons, Inc.; New York: 1962.

[89] G. D. Smith. Numerical Solution of Partial Differential Equations: Finite Difference Methods. 3rd Ed. Chap. 2. Clarendon Press; Oxford: 1985.

[90] E. M. Dianov, V. E. Frotov, I. A. Bufetov, V. P. Efremov, A. E. Rakitin, M. A. Melkumov, M. I. Kulish, and A. A. Frolov. Temperature of optical discharge under action of laser radiation in silica-based fibers, *Proc. Eur. Conf. Opt. Commun.* 2005: Vol. 3: 469–470 (We3.4.4).

[91] Y. Shuto. Evaluation of high-temperature absorption coefficients of ionized gas plasmas in optical fibers, *IEEE Photon. Technol. Lett.* 2010; 22(3): 134–136.

[92] L. Spitzer, Jr. Physics of Fully Ionized Gases. 2nd Ed. Chap. 5. John Wiley & Sons, Inc.; New York: 1962.

[93] Y. B. Zel'dovich and Y. P. Raizer. Physics of Shock Waves and High-Temperature Hydrodynamic Phenomena. Chap. 5. Dover; New York: 2002.

[94] T. Sekiguchi. Plasma Engineering. Chap. 1. Ohm-Sha; Tokyo: 1997.

[95] G. R. Fowles. Introduction to Modern Optics. 2nd Ed. Chap. 7. Dover; New York: 1975.

[96] S. P. Timoshenko and J. N. Goodier. Theory of Elasticity. 3rd Ed. p. 71. McGraw-Hill Book Co.; New York: 1970.

[97] T. Yokobori. An Interdisciplinary Approach to Fracture and Strength of Solids. Chap. 1. Gordon & Breach Publishers; Amsterdam: 1968.

[98] H. Kobayashi. Fracture Mechanics. Chap. 3. Kyoritsu Press Ltd.; Tokyo: 1993.

[99] I. A. Bufetov, A. A. Frolov, A. V. Shubin, M. E. Likhachev, S. V. Lavrishchev, and E. M. Dianov. Propagation of an optical discharge through optical fibres upon interference of modes, *Quantum Electron.* 2008; 38(5): 441–444.

[100] K. Kurokawa and N. Hanzawa. Suppression of fiber fuse propagation and its break in compact fiber fuse terminator, *OptoElectron. Commun. Conf. held jointly with 2013 Int. Conf. on Photonics in Switching* 2013 (OECC/PS 2013): WS4-5.

[101] J. P. Gordon, R. C. C. Leite, R. S. Moore, S. P. S. Porto, and J. R. Whinnery. Long-transient effects in lasers with inserted liquid samples. *J. Appl. Phys.* 1965; 36(1): 3–8.

Chapter 6

Fiber-Based Cylindrical Vector Beams and Its Applications to Optical Manipulation

Renxian Li, Lixin Guo, Bing Wei, Chunying Ding and Zhensen Wu

Additional information is available at the end of the chapter

http://dx.doi.org/10.5772/59151

1. Introduction

Radiation pressure force (RPF) indeced by a focused laser beam has bean widely utlized for the manipulation of small particles, and has found more and more applications in various fields including physics [1], biology [2], and optofludics [3, 4]. Accurate prediction of optical force exerted on particles enables better understanding of the physical mechanicsm, and is of great help for the design and improvement of optical tweezers.

Many researches have been devoted to the prediction of radiation pressure force (RPF), and different approaches have been developed for the theoretical calculation of RPF exerted on a homogeneous sphere. The geometrical optics [5-7] and Rayleigh theory [8] are respectively considered for the particles much larger and smaller than the wavelength of incident beam. Since geometrical optics and Rayleigh theory are both approximation theories, rigorous theories based on Maxwell's theory have been considered [9-13]. Generalized Lorenz-Mie Theory (GLMT) [14] has been used to investigate the RPFs exerted on some regular parti-cles[10-13, 15, 16] induced by a Gaussian beam. GLMT can rigorously calculate RPF induced by any beam. To isolate the contribution of various scattering process to RPF, Debye series is introduced [17, 18].

Traditional optical tweezers use Gaussian beams as trapping light sources. This approach works well for the manipulation of microscopic spheres. However, the deveopment of science and technology brings new challenges to optical tweezers, and several approaches have been developed. Holographic methods have been used to increase the strength and dexterity of optical trap [19]. Another approch is the employment of non-Gaussian beam including Laguerre-Gaussian beams [20] and Bessel beams [28]. Laguerre-Gaussian beams have zero on-

axis intensity, and can increase the strength of optical trap. Bessel beams consist of a series of concentric rings of decreasing intensity, and have characteristics of non-diffraction and self-reconstruction. A single Bessel beam can be used to simultaneously trap and manipulate, accelerate, rotate, or guide many particles. Bessel beams can trap and manipulate both high-index and low-index particles.

In addition to Laguerre-Gaussian and Bessel beams, there is a speical class of beams which have cylindrical symmetry in both amplitude and polarization, hence the name Cylindrical Vector Beams (CVBs) [29-32]. CVBs are solutions of vector wave equation in the paraxial limit. The special features of CVBs have attracted considerable interest for a variety of novel applications, including lithography, particle acceleration, material processing, high-resolution metrology, atom guiding, optical trapping and manipulation. The most interesting features for optical trapping arise from the focusing properties of CVBs. A radially polarized beam focused by a high numerical aperture objective has a peak at the focus, and can trap a high-index particle. On the contrary, an azimuthally polarized beam has null central intensity, and can trap low-index particle. These two kinds of beams can be experimentally switched.

CVBs can be generated by many methods, which are categorized as active or passive depending on whether amplifying media is used. The simplest mothod is to convert an incident Gaussian beam to a radially polarized beam using a radial polariser. However this method does not produce very high purity tansverse modes. Moer efficient methods use interferometry. Since a CVB can be expressed as the linear superpostion of two Hermite-Gaussian or Laguerre-Gaussian beams with different orientations of polarization. Another efficient method is based on optical fiber [33]. This technique takes advantage of the similarity between the poarization propeties of the modes that propagate inside a step-index optical fiber and CVBs. When TE_{01} or TM_{01} is excited in the fibre, it excites a CVB in free space. Fiber-generated CVB, taking Bessel-Gaussian as example, and its applicaions top optical manipulations will be discussed in this chapter.

2. Mathematical description of cylindrical vector beams

Cylindrical vector beams are solutions of vector wave equation

$$\nabla \times \nabla \times \vec{E} + k^2 \vec{E} = 0, \tag{1}$$

where $k=2\pi/\lambda$ is wavenumber with λ being the wavelength. In the paraxial approximation, the radially and azimuthally polarized vector Bessel-Gaussian beams, two kinds of typical CVBs, can be expressed as

$$\vec{E}_{rad} = E_0 \frac{\rho}{w_0} e^{-\frac{\rho^2}{w_0^2}} e^{i(\omega t - kz)} \hat{e}_\rho \tag{2}$$

$$\vec{E}_{azi} = E_0 \frac{\rho}{w_0} e^{-\frac{\rho^2}{w_0^2}} e^{i(\omega t - kz)} \hat{e}_\phi \tag{3}$$

where r and ϕ are respectively the radial and azimuthal coordinates, $\hat{e}_\rho$ and $\hat{e}_\phi$ are unit vectors in ρ and ϕ directions, and the subscripts *rad* and *azi* denote the polarization state. w_0 is the width of beam waist, and E_0 is a constant. Fig. 1(a) and (b) respectively give the intensity distribution of radially and azimuthally polarized Bessel-Gaussian beam in the plane $z=0$. Note that the longitudial component of CVB is negligible under the condition of paraxial approximation. A general CVB can be considered as a linear superposition of a radially polarized CVB and an azimuthally polarized one.

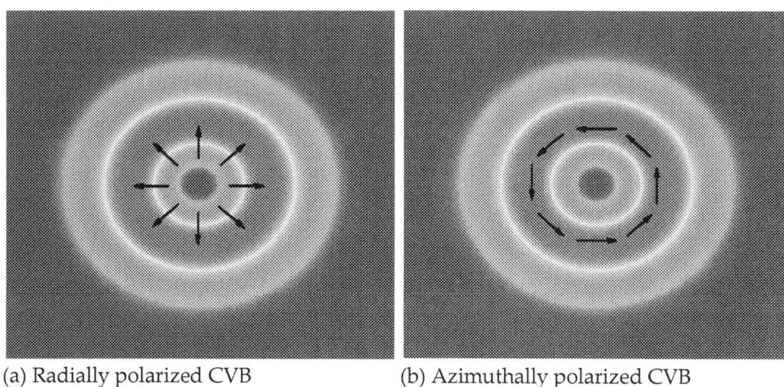

(a) Radially polarized CVB   (b) Azimuthally polarized CVB

Figure 1. Intensity distribution of CVB. The arrows indicate the direction of polarization

3. Radiation force induced by CVB

3.1. Optical force on Rayleigh particles

In the Rayleigh regime, particles can be considered as infinitesimal induced dipoles which interact with incident beam. Here we assume that the particle is a microsphere. RPF will be decomposed into scattering force and gradient force.

The oscillating dipole, which is induced by time-harmonic fields, can be considered as an antenna. The antenna will radiate energy. The difference between energy removed from incident beam and energy radiated by the antenna accounts for the change of momentum flux, and hence rusults in a scattering force. The scattering force can be expressed as

$$\vec{F}_{scat} = \hat{e}_z C_{pr} n_1^2 \varepsilon_0 |E|^2 \tag{4}$$

with

$$C_{pr} = C_{scat} = \frac{8}{3}\pi(ka)^4 a^2 \left(\frac{m^2-1}{m^2+2}\right)^2 \tag{5}$$

and

$$m = n_2 / n_1 \tag{6}$$

where n_1 is the refractive index of surrounding media, and n_2 is the refractive index of the particle. a is the radius of microsphere. ε_0 is the dielectric constant in the vacumm. Note that the scattering force always points in the direction of incident beam.

When a particle is illuminated by a non-uniform electric field, it will experience a gradient force.

$$\vec{F}_{grad} = \pi n_1^2 \varepsilon_0 a^3 \left(\frac{m^2-1}{m^2+2}\right) \nabla |E|^2 \tag{7}$$

For a time-harmonic field, the gradient force can also be expressed in terms of the intensity I of incident beam:

$$\vec{F}_{grad} = \frac{2\pi n_1 a^3}{c}\left(\frac{m^2-1}{m^2+2}\right)\nabla I \tag{8}$$

where c is the speed of light in the vacumm. It is obvious that the gradient force depends on the gradient of the intensity. By sbustituting Eqs. (2) and (3) into Eqs. (4) and (7), we can obtain the scattering and gradient force of vector Bessel-Gaussian beams exerted on a microsphere. For radially polarized Bessel-Gaussian beam, they can be expressed as

$$\vec{F}_{scat} = \hat{e}_z C_{pr} n_1^2 \varepsilon_0 E_0^2 \frac{\rho^2}{w_0^2}\left(e^{-\frac{\rho^2}{w_0^2}}\right)^2 \tag{9}$$

$$\vec{F}_{grad} = \pi n_1^2 \varepsilon_0 a^3 \left(\frac{m^2-1}{m^2+2}\right)\left[\frac{2E_0^2\rho}{w_0^2}\left(e^{-\frac{\rho^2}{w_0^2}}\right)^2 - \frac{4E_0^2\rho^3}{w_0^4}\left(e^{-\frac{\rho^2}{w_0^2}}\right)^2\right]\hat{e}_\rho \tag{10}$$

From Eq. (10), we can find that the gradient force has only ϱ component. This is because $|E|^2$ is only dependent on ϱ. Here we give only the force for radially polarized beam incidence, and that for azimuthally polarized beam incidence can be derived in the same way.

3.2. Radiation force exerted on Mie particles

Many practical particles manipulated with optical tweezers, such as bioloical cells, are Mie particles, whose size is in the order of the wavelength of trapping beam. To calculate the radiation force exerted on such particles, a rigorous electromagnetic theory based on the Maxwell equations must be considered. Generalized Lorenz-Mie Thoery (GLMT) developed by Gouesbet et al. can solve the interaction between homogeneous spheres and focused beams with any shape, and has been entended to solve the scattering of shaped beam by multilayered spheres, homogeneous and multilayered cylinders, and homogeneous and multilayered spheroids. GLMT has been applied to the rigorous calculation of radiation pressure and optical torque. In GLMT, the incident beam is described by a set of beam shape coefficients(BSCs), which can be evaluated by integral localized approximation (ILA) [34].

This section is devoted to the GLMT for radiation force exerted on a sphere illuminated by a vetor Bessel-Gaussian beam. The general theory for radiation force based on electromagnetic scattering theory is followed by BSCs for CVB. To clarify the physical interpretation of various features of RPF that are implicit in the GLMT, Debye Series Expansion (DSE) is introduced.

3.2.1. Generization Lorenz-Mie theory

Consider a sphere with radius a and refractive index m_1 illuminated by a CVB of wavelength λ in the surrounding media. The center of the sphere is located at O_P, origin of the Cartesian coordinate system $O_{P\text{-}xyz}$. The beam center is at O_G, origin of coordinate system $O_{G\text{-}uvw}$, with u axis parallel to x and similarly for the others. The coordinates of O_G in the system $O_{P\text{-}xyz}$ is (x_0, y_0, z_0). The refractive index of surrounding media is m_2. The other parameters are defined in Fig. 2.

When a sphere is illuminated by focused beam, the RPF is proportional to the net momentum removed from the incident beam, and can be expressed in terms of the surface integration of Maxwell stress tensor

$$< \mathbf{F} >=< \oint_s \mathbf{n}\cdot\overset{\frown}{\vec{T}}dS >$$

(11)

where $< >$ represents a time average, $\overset{\wedge}{\mathbf{n}}$ the outward normal unit vector, and S a surface enclosing the particle. The Maxwell stress tensor $\vec{T}$ is given by

$$< \vec{T} >= \frac{1}{4\pi}\left(\varepsilon\mathbf{EE} + \mathbf{HH} - \frac{1}{2}\left(\varepsilon E^2 + H^2\right)\vec{I} \right)$$

(12)

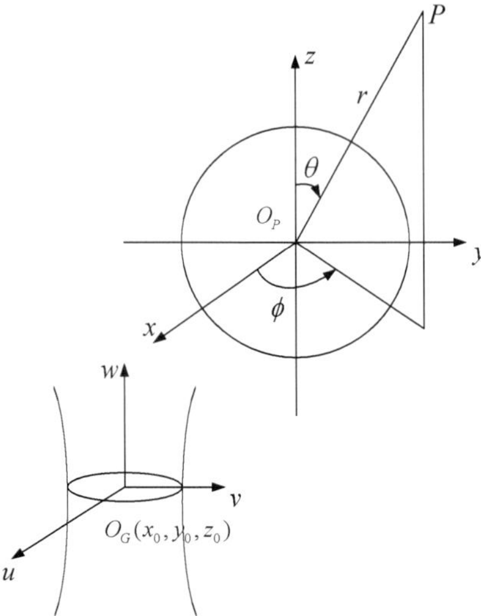

Figure 2. Coordinate systems in GLMT. $O_{P\text{-}xyz}$ is attached to the sphere, and $O_{G\text{-}uvw}$ to the incident beam.

Where the electromagnetic fields E and H are the total fields, namely the sum of the incident and scattered fields, given by

$$\mathbf{E} = \mathbf{E}_i + \mathbf{E}_s, \mathbf{H} = \mathbf{H}_i + \mathbf{H}_s \tag{13}$$

E_i and H_i are the incident electromagnetic wave, and can be expaned as:

$$E^i_r = \frac{E_0}{k_0^2 r^2} \sum_{n=1}^{\infty} \sum_{m=-n}^{+n} c_n^{pw} g_{n,TM}^m n(n+1) \psi_n(k_0 r) P_n^{|m|}(\cos\theta) e^{(im\varphi)} \tag{14}$$

$$E^i_\theta = \frac{E_0}{k_0 r} \sum_{n=1}^{\infty} \sum_{m=-n}^{+n} c_n^{pw} \left[g_{n,TM}^m \psi_n'(k_0 r)\tau_n^{|m|}(\cos\theta) + m g_{n,TE}^m \psi_n(k_0 r)\pi_n^{|m|}(\cos\theta) \right] e^{(im\varphi)} \tag{15}$$

$$E^i_\varphi = i\frac{E_0}{k_0 r} \sum_{n=1}^{\infty} \sum_{m=-n}^{+n} c_n^{pw} \left[m g_{n,TM}^m \psi_n'(k_0 r)\pi_n^{|m|}(\cos\theta) + g_{n,TE}^m \psi_n(k_0 r)\tau_n^{|m|}(\cos\theta) \right] e^{(im\varphi)} \tag{16}$$

$$H^i_r = \frac{H_0}{k_0^2 r^2} \sum_{n=1}^{\infty} \sum_{m=-n}^{+n} c_n^{pw} g_{n,TE}^m n(n+1) \psi_n(k_0 r) P_n^{|m|}(\cos\theta) e^{(im\varphi)} \tag{17}$$

$$H_\theta^i = -\frac{H_0}{k_0 r} \sum_{n=1}^{\infty} \sum_{m=-n}^{+n} c_n^{pw} \left[m g_{n,TM}^m \psi_n(k_0 r) \pi_n^{|m|}(\cos\theta) - g_{n,TE}^m \psi_n'(k_0 r) \tau_n^{|m|}(\cos\theta) \right] e^{(im\varphi)} \tag{18}$$

$$H_\varphi^i = -\frac{iH_0}{k_0 r} \sum_{n=1}^{\infty} \sum_{m=-n}^{+n} c_n^{pw} \left[g_{n,TM}^m \psi_n(k_0 r) \tau_n^{|m|}(\cos\theta) - m g_{n,TE}^m \psi_n'(k_0 r) \pi_n^{|m|}(\cos\theta) \right] e^{(im\varphi)} \tag{19}$$

with

$$\pi_n^m(\cos\theta) = \frac{dP_n^m(\cos\theta)}{d\theta} \tag{20}$$

$$\tau_n^m(\cos\theta) = m \frac{P_n^m(\cos\theta)}{\sin\theta} \tag{21}$$

$$c_n^{pw} = (-i)^{n+1} \frac{2n+1}{n(n+1)} \tag{22}$$

where $P_n^m(\cos\theta)$ represents the associated Legendre polynomials of degree n and order m, $\psi(\cdot)$ is the spherical Ricatti-Bessel functions of first kind, and the prime indicates the derivative of the function with respect to its argument. $g_{n,TM}^m$ and $g_{n,TE}^m$ are so-called BSCs and will be discussed in next subsection.

Similarly, the scattered fields E_s and H_s have the expression :

$$E_r^s = -\frac{E_0}{k_0^2 r^2} \sum_{n=1}^{\infty} \sum_{m=-n}^{+n} c_n^{pw} A_n^m n(n+1) \xi_n(k_0 r) P_n^{|m|}(\cos\theta) e^{(im\varphi)} \tag{23}$$

$$E_\theta^s = -\frac{E_0}{k_0 r} \sum_{n=1}^{\infty} \sum_{m=-n}^{+n} c_n^{pw} \left[A_n^m \xi_n'(k_0 r) \tau_n^{|m|}(\cos\theta) + m B_n^m \xi_n(k_0 r) \pi_n^{|m|}(\cos\theta) \right] e^{(im\varphi)} \tag{24}$$

$$E_\varphi^s = -i\frac{E_0}{k_0 r} \sum_{n=1}^{\infty} \sum_{m=-n}^{+n} c_n^{pw} \left[m A_n^m \xi_n'(k_0 r) \pi_n^{|m|}(\cos\theta) + B_n^m \xi_n(k_0 r) \tau_n^{|m|}(\cos\theta) \right] e^{(im\varphi)} \tag{25}$$

$$H_r^s = -\frac{H_0}{k_0^2 r^2} \sum_{n=1}^{\infty} \sum_{m=-n}^{+n} c_n^{pw} B_n^m n(n+1) \xi_n(k_0 r) P_n^{|m|}(\cos\theta) e^{(im\varphi)} \tag{26}$$

$$H_\theta^s = \frac{H_0}{k_0 r} \sum_{n=1}^{\infty} \sum_{m=-n}^{+n} c_n^{pw} \left[m A_n^m \xi_n(k_0 r) \pi_n^{|m|}(\cos\theta) - B_n^m \xi_n'(k_0 r) \tau_n^{|m|}(\cos\theta) \right] e^{(im\varphi)} \tag{27}$$

$$H_\varphi^s = \frac{iH_0}{k_0 r} \sum_{n=1}^{\infty} \sum_{m=-n}^{+n} c_n^{pw} \left[A_n^m \xi_n(k_0 r) \tau_n^{'m}(\cos\theta) - m B_n^m \xi_{nn}^{'}(k_0 r) \pi_n^{'m}(\cos\theta) \right] e^{(im\varphi)} \tag{28}$$

Where $\xi_n(k_0 r)$ is Ricatti-Hankel functions, and scattering coefficients A_n^m and B_n^m can be expressed by traditional Mie scattering coefficients a_n, b_n and BSCs $g_{n,TM}^m$, $g_{n,TE}^m$:

$$A_n^m = a_n g_{n,TM}^m, \qquad B_n^m = b_n g_{n,TE}^m \tag{29}$$

with

$$a_n = \frac{-m_1 \psi_n^{'}(x)\psi_n(y) + m_2 \psi(x)\psi_n^{'}(y)}{-m_1 \xi_n^{(1)'}(x)\psi_n(y) + m_2 \xi_n^{(1)}(x)\psi_n^{'}(y)} \tag{30}$$

$$b_n = \frac{-m_2 \psi_n^{'}(x)\psi_n(y) + m_1 \psi(x)\psi_n^{'}(y)}{-m_2 \xi_n^{(1)'}(x)\psi_n(y) + m_1 \xi_n^{(1)}(x)\psi_n^{'}(y)} \tag{31}$$

$$x = m_2 k_0 a, \qquad y = m_1 k_0 a \tag{32}$$

Substituting Eqs. (14) - (19) and (23) - (28) into Eqs. (11) - (12), and after some algebra, we can get the formula for RPFs which can be characterized by radiation pressure cross section (RPCS):

$$\mathbf{F(r)} = \frac{2_z I_0}{c} \left[\hat{\mathbf{e}}_\mathbf{x} C_{pr,x}(\mathbf{r}) + \hat{\mathbf{e}}_\mathbf{y} C_{pr,y}(\mathbf{r}) + \hat{\mathbf{e}}_\mathbf{z} C_{pr,z}(\mathbf{r}) \right] \tag{33}$$

where RPCS $C_{pr,i}$ ($i = x,\ y,\ z$) has a longitudinal cross section $C_{pr,z}$

$$\begin{aligned}
C_{pr,z} = \frac{\lambda^2}{\pi} \sum_{n=1}^{\infty} Re \Bigg\{ &\frac{1}{n+1}(A_n g_{n,TM}^0 g_{n+1,TM}^{0*} + B_n g_{n,TE}^0 g_{n+1,TE}^{0*}) + \sum_{m=1}^{n} \Bigg[\frac{1}{(n+1)^2} \frac{(n+m+1)!}{(n-m)!} \\
&\times (A_n g_{n,TM}^m g_{n+1,TM}^{m*} + A_n g_{n,TM}^{-m} g_{n+1,TM}^{-m*} + B_n g_{n,TE}^m g_{n+1,TE}^{m*} + B_n g_{n,TE}^{-m} g_{n+1,TE}^{-m*}) \\
&+ m \frac{2n+1}{n^2(n+1)^2} \frac{(n+m)!}{(n-m)!} C_n (g_{n,TM}^m g_{n,TE}^{m*} - g_{n,TM}^{-m} g_{n,TE}^{-m*}) \Bigg] \Bigg\}
\end{aligned} \tag{34}$$

and two transverse cross section $C_{pr,x}$ and $C_{pr,y}$

$$C_{pr,x} = \mathrm{Re}(C) \qquad C_{pr,y} = \mathrm{Im}(C) \tag{35}$$

where

$$C = \frac{\lambda^2}{2\pi} \sum_{n=1}^{\infty} \left\{ -\frac{(2n+2)!}{(n+1)^2} F_n^{n+1} + \sum_{m=1}^{n} \frac{(n+m)!}{(n-m)!} \frac{1}{(n+1)^2} \right.$$
$$\times \left[F_n^{m+1} - \frac{n+m+1}{n-m+1} F_n^m + \frac{2n+1}{n^2} (C_n g_{n,TM}^{m-1} g_{n,TE}^{m*} \right.$$
$$\left. \left. - C_n g_{n,TM}^{-m} g_{n+1,TE}^{-m+1*} + C_n^* g_{n,TE}^{m-1} g_{n,TM}^{m*} - C_n^* g_{n,TE}^{-m} g_{n,TM}^{-m+1*}) \right] \right\} \tag{36}$$

with

$$F_n^m = A_n g_{n,TM}^{m-1} g_{n+1,TM}^{m*} + B_n g_{n,TE}^{m-1} g_{n+1,TE}^{m*} + A_n^{m*} g_{n+1,TM}^{-m} g_{n,TM}^{-m+1*} + B_n^{m*} g_{n+1,TE}^{-m} g_{n,TE}^{-m+1*} \tag{37}$$

$$A_n = a_n + a_{n+1}^* - 2a_n a_{n+1}^* \tag{38}$$

$$B_n = b_n + b_{n+1}^* - 2b_n b_{n+1}^* \tag{39}$$

$$C_n = -i(a_n + b_{n+1}^* - 2a_n b_{n+1}^*) \tag{40}$$

Note that substituting Eq. (13) into Eqs. (11) - (12) shows that the total RPF can be devided into thress parts:

$$< \mathbf{F} >=< \mathbf{F}_i > + < \mathbf{F}_{mix} > + < \mathbf{F}_s > \tag{41}$$

where $<F_i>$ depends only on the incident fields, $<F_s>$ is associated with the scattered fields, and $<F_{mix}>$ involves the interactions of the incident beam with the scattered field. After a great deal of algebra, we can get that $<F_i>=0$, which can be understood by the momentum conservation law for monochromatic fields in free space. The RPCS for $<F_{mix}>$ and $<F_s>$ can be directly given using Eqs. (34) - (40) by changing Eqs. (38) - (40) using

$$\begin{aligned} A_n &= a_n + a_{n+1}^* \\ B_n &= b_n + b_{n+1}^* \\ C_n &= -i(a_n + b_{n+1}^*) \end{aligned} \tag{42}$$

for $<F_{mix}>$, and

$$\begin{aligned} A_n &= -2a_n a_{n+1}^* \\ B_n &= -2b_n b_{n+1}^* \\ C_n &= 2i a_n b_{n+1}^* \end{aligned} \tag{43}$$

for $<F_s>$.

3.2.2. Beam shape coefficients for CVB

This section is devoted the derivation of BSCs for CVB using ILA. In the ILA, the beam shape coefficients $g_{n,TM}^m$ and $g_{n,TE}^m$ are obtained respectively from the radial component of electric and magnetic field E_r and H_r according to [34]

$$g_{n,TE}^m = \frac{Z_n^m}{2\pi H_{B0}} \int_0^{2\pi} \overline{H}_r(r,\theta,\phi) e^{-im\phi} d\phi \tag{44}$$

$$g_{n,TM}^m = \frac{Z_n^m}{2\pi E_{B0}} \int_0^{2\pi} \overline{E}_r(r,\theta,\phi) e^{-im\phi} d\phi \tag{45}$$

with

$$Z_n^m = \begin{cases} \dfrac{2n(n+1)i}{2n+1} & m = 0 \\[2ex] \left(\dfrac{-2i}{2n+1}\right)^{|m|-1} & m \neq 0 \end{cases} \tag{46}$$

$\overline{E}_r$ and $\overline{H}_r$ are respectively the localized fields of E_r and H_r, and they are obtained by changing kr to $(n+1/2)$ and θ to $\pi/2$ in their expression. For a radially polarized Bessel-Gaussian beam, the localized radial component of electric field are derived from Eq. (2):

$$\begin{aligned} \overline{E}_{rad}^r &= E_0 \overline{\Omega}_0 \left[-(\rho_n \cos\phi - \xi_0)\cos\phi + (\rho_n \sin\phi - \eta_0)\sin\phi\right] \\ &= E_0 \overline{\Omega}_0 \left[\rho_n - \rho_0 \sin(\phi + \phi_0)\right] \end{aligned} \tag{47}$$

with

$$\overline{\Omega}_0 = -\sqrt{2} \exp\left[-(\rho_n^2 + \xi_0^2 + \eta_0^2)\right] \exp\left[2\rho_n(\xi_0 \cos\phi + \eta_0 \sin\phi)\right] \tag{48}$$

$$\rho_n = \frac{kr}{kw_0} = \frac{1}{kw_0}\left(n + \frac{1}{2}\right) \tag{49}$$

$$\xi_0 = \rho_0 \sin\phi_0, \quad \eta_0 = \rho_0 \cos\phi_0 \tag{50}$$

Substituting Eq. (47) into Eq. (45), and considering the formula of Bessel function

$$J_n(x) = \frac{1}{2\pi}\int_{-\pi}^{\pi} e^{i(x\sin\theta - n\theta)}d\theta = \frac{1}{2\pi}\int_0^{2\pi} e^{i(x\sin\theta - n\theta)}d\theta \tag{51}$$

we can obtain the final expression of BSCs

$$g_{n,TM}^{m,rad} = \frac{1}{2}Z_n^m\overline{\Omega}_n e^{im\phi_0}\left[2\rho_n J_m(-2i\rho_n\rho_0) + i\rho_0(J_{m-1}(-2i\rho_n\rho_0) - J_{m+1}(-2i\rho_n\rho_0))\right] \tag{52}$$

Here we only derive $g_{n,TM}^{m,rad}$, and $g_{n,TE}^{m,rad}$ can be derived in the similar way.

3.2.3. Debye series expansion

GLMT is a rigorous electromagnetic theory, and can exactly predict the RPF exerted on a sphere by focused beam. Whereas the solution is complicated combinations of Bessel functions, and the mathematical complexity obscures the physical interpretation of various features of RPF. The DSE, which is a rigorous electromagnetic theory, expresses the Mie scattering coefficients into a series of Fresnel coefficients and gives physical interpretation of different scattering processes. The DSE is an efficient technique to make explicit the physical interpretation of various features of RPF which are implicit in the GLMT. The DSE is firstly presented by Debye in 1908 for the interaction between electromagnetic waves and cylinders. Since then, the DSE for electromagnetic scattering by homogeneous, coated, multilayered spheres, multilayered cylinders at normal incidence, homogeneous, multilayered cylinder at oblique incidence, and spherical gratings are studied. In our previous work, DSE has been employed to the analysis of RPF exerted on a sphere induced by a Gaussian and Bessel beam.

As shown in Fig. 3, when an incoming spherical multipole wave, which is

$$\Psi = \xi_n^{(1)}(m_2 kr)P_n^m(\cos\theta)\begin{Bmatrix}\cos m\phi\\\sin m\phi\end{Bmatrix}, \tag{53}$$

encounters the interface of the sphere at $r=a$, portion of it will be transmitted into the sphere, and another portion will reflected back. The transmitted and reflected waves are respectively:

$$\Psi_1 = T_n^{21}\xi_n^{(1)}(m_1 kr)P_n^m(\cos\theta)\begin{Bmatrix}\cos m\phi\\\sin m\phi\end{Bmatrix} \qquad r \leq a \tag{54}$$

$$\Psi_2 = \left[\xi_n^{(1)}(m_2 kr) + R_n^{212}\xi_n^{(2)}(m_2 kr)\right]P_n^m(\cos\theta)\begin{Bmatrix}\cos m\phi\\\sin m\phi\end{Bmatrix} \qquad r \geq a \tag{55}$$

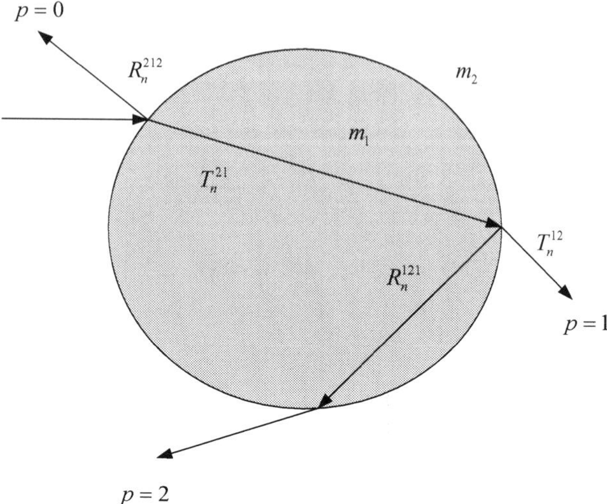

Figure 3. Debye model of light scattering by a sphere

Applying the boundary conditions, which reqires continuity of the tangential components of filds at the interface, to the incident, transmitted and reflected waves,we can obtain:

$$T_n^{21} = \frac{m_1}{m_2}\frac{2i}{D_n} \tag{56}$$

$$R_n^{212} = \frac{\alpha\xi_n^{(1)'}(\kappa_2)\xi_n^{(1)}(\kappa_1) - \beta\xi_n^{(1)}(\kappa_2)\xi_n^{(1)'}(\kappa_1)}{D_n} \tag{57}$$

with

$$D_n = -\alpha\xi_n^{(2)'}(\kappa_2)\xi_n^{(1)}(\kappa_1) + \beta\xi_n^{(2)}(\kappa_2)\xi_n^{(1)'}(\kappa_1) \tag{58}$$

$$\kappa_j = m_j ka \tag{59}$$

$$\alpha = \begin{cases} 1, & for\ TE \\ \dfrac{m_1}{m_2}, & for\ TM \end{cases}, \qquad \beta = \begin{cases} \dfrac{m_1}{m_2}, & for\ TE \\ 1, & for\ TM \end{cases} \tag{60}$$

Similarly, the consideration of outgoing multipole waves can get

$$T_n^{12} = \frac{2i}{D_n}$$
(61)

$$R_n^{121} = \frac{\alpha \xi_n^{(2)'}(\kappa_2)\xi_n^{(2)}(\kappa_1) - \beta \xi_n^{(2)}(\kappa_2)\xi_n^{(2)'}(\kappa_1)}{D_n}$$
(62)

Substituting all Fresnel coefficients into

$$\left(1 - R_n^{121}\right)\left(1 - R_n^{212}\right) - T_n^{21}T_n^{12}$$
(63)

and after much algebra, we get

$$\begin{aligned} \left.\begin{array}{c} a_n \\ b_n \end{array}\right\} &= \frac{1}{2}\left[1 - R_n^{212} - \frac{T_n^{21}T_n^{12}}{1 - R_n^{121}}\right] \\ &= \frac{1}{2}\left[1 - R_n^{212} - \sum_{p=1}^{\infty} T_n^{21}\left(R_n^{121}\right)^{p-1}T_n^{12}\right] \end{aligned}$$
(64)

where the prime indicates the derivative of the function with respect to its argument. $\xi_n^{(1)}(\cdot)$ and $\xi_n^{(2)}(\cdot)$ are respectively the spherical Ricatti-Hankel functions of first and second kinds. The definition of all Fresnel coefficients and Debye term p are given in Fig. 3. For convenience, we note $p = -1$ and $p = 0$ respectively for the diffraction and direct reflection. In our previous work, we have theoretically and numerically proved that when p ranges from 1 to , the Eq. (64) is identical to the traditional Mie scattering coefficients. Here we provide the DSE for homogeneous spheres, and the DSE for multilayered spheres can be found in our previous work.

4. Numerical results and discussions

In this section, the GLMT and DSE will be employed to analyze the RPF exerted on a homogeneous sphere induced by a radially polarized vector Bessel-Gaussian beam. Xu et al. used GLMT to analyze the RPF exerted on a slightly volatile silocone oil of refractive index , which can be levitated in the air by a beam of wavelength . We first use GLMT to analyze the RPF exerted on such oil induced by vector Bessel-Gaussian beam, and DSE will be employed to the study of the contribution of various scattering process to RPF. In our calculation, the radius of the particle is .

We first explore the influence of beam center location on the RPF. In our calculation, we assume the beam center is located on the x axis so that . Fig. 4 gives the transverse RPCS ∞ versus $m_1 = 1.5$ for various beam-waist radius $\lambda = 0.5\mu m$. Here we consider $a = 2.5\mu m$ and $y_0 = z_0 = 0$, which are respectively larger, equal and smaller than the radius of the particle. We can find that the particle can not be trapped at the beam center $C_{pr,x}$ for all beams. This results from the fact all beams have null central intensity. It is worth pointing out that a stable trap corresponds to a particle position where the RPF is zero and its slope is positive. All curves have two equilibrium points, which are symmetric with respect to beam axis (x_0). This is decided by the intensity maxima of beams. So a vector Bessel-Gaussian beam can simultaneously trap more than one particles. We can also find that the interval of equilibrium points increases with the increasing of w_0. This can be easily explained from the fact that the interval of intensity peaks increases with the increasing of $w_0 = 5\mu m$, $2.5\mu m$.

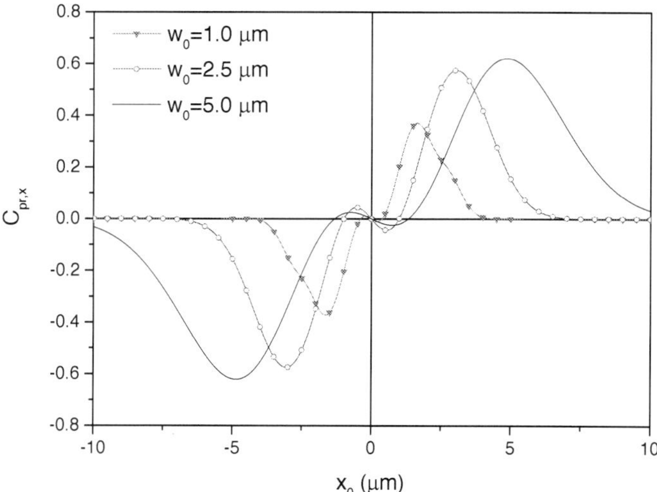

Figure 4. Transverse cross-section $1\mu m$ versus $x_0 = 0$ with parameter $x_0 = 0$. w_0, w_0, $C_{pr,x}$, x_0, w_0 and $\lambda = 0.5\mu m$

To clarify the physical explanation of some features of RPCS, it is necessary for us to consider the contribution of each mode p to RPCS. The contribution of each p mode to RPCS can be computed separately by considering a single term in Eq. (64). Now we consider the contribution of a single p mode to transverse RPCS $m_1 = 1.5$. Here we set beam-waist radius $m_2 = 1$.

It is shown in Fig. 5 the transverse RPCS $y_0 = 0$ versus $a = 2.5\mu m$ with parameter $p_{max} = \infty$. In the calculation, the beam-waist radius is assumed $C_{pr,x}$. Comparison of Fig. 5 with Fig. 4 shows that when $w_0 = 5\mu m$ the results obtained by DSE are identical to those by GLMT. In fact, when

$C_{pr,x}$ is large enough, the difference between two theories should be negilible. For example, if x_0, the results of DSE is very close to GLMT results. Special attention should be paid to the case of p_{max}. Fig. 5 shows that when $w_0 = 5\mu m$, the agreement between the results of GLMT and DSE is already good. This concludes that main contribution of RPF comes from the scattering processes of diffraction ($p_{max} \to \infty$), specular reflection (p_{max}) and direct transmission ($p_{max} = 100$).

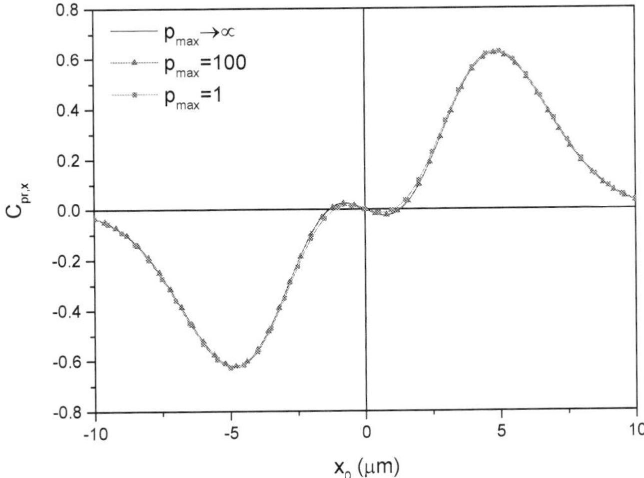

Figure 5. Transverse cross-section $p_{max} = 1$ versus $p_{max} = 1$ with parameter $p = -1$. $p = 0$, $p = 1$, $C_{pr,x}$, x_0, p_{max} and $\lambda = 0.5\mu m$.

To clarify the physical explanation of some phenomena of RPF, it is necessary to consider the contribution of each mode p to RPCS, which can be computed separately by considering a single term in Eq. (64). Now we consider the contribution of a single p mode to transverse RPCS $m_1 = 1.5$. Fig.6 depicts the transverse RPCS $m_2 = 1$ versus $y_0 = 0$ for $a = 2.5\mu m$ and $w_0 = 5\mu m$. Generally, the RPF at $C_{pr,x}$ is zero for any mode p because of the symmetry. The magnitude of $C_{pr,x}$ for x_0 is much greater than that for $p = 1$. This validates that the transverse RPCS $p = 2$ is dominated by the contributions of direct transmission ($x_0 = 0$). Note that the curve for $C_{pr,x}$ has two equilibrium points at about $p = 1$, while the curve for $p = 2$ has only one points at $C_{pr,x}$. Near the beam axis ($p = 1$), the curvesfor $p = 2$ has positive slope, while that for $x_0 - \pm 1.3\mu m$ has negative one. To explain such phenomena, we must consider the integral effect of all intensity peaks.

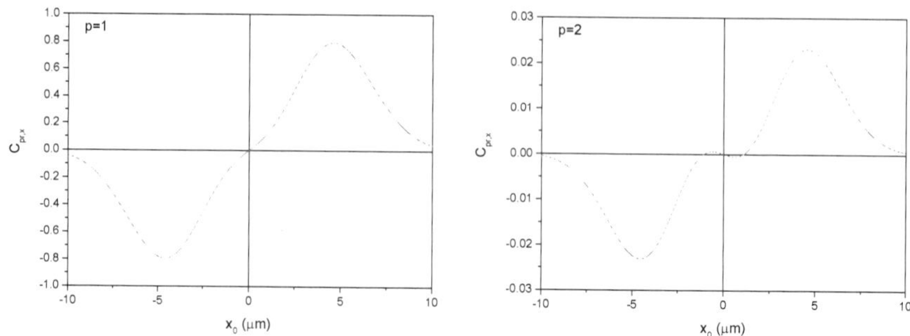

Figure 6. Transverse cross-section $p=1$ versus $x_0=0$ corresponding to single mode p. $x_0=0$, $p=1$, $p=2$, $C_{pr,x}$ x_0 and $\lambda=0.5\mu m$.

5. Conclusions

Rigorous theories including GLMT and DSE for RPF exerted on spheres induced by CVB is derived. The incident beam is described by a set of BSCs which is calculated by integral localized approximation, and the scattering coefficients are expanded using Debye series. For very small particles, namely Rayleigh particles, an approximation model is also given. Such thoery can be easily extended to the RPF exerted on multilayered sphere, and also to the RPF induced by other beams. Debye series is used to isolate the contribution of various scattering process to the RPF. The results are of special importance for the improvement of optical tweezers system.

Acknowledgements

The authors acknowledge support from the Natural Science Foundation of China (Grant No. 61101068), the National Science Foundation for Distinguished Young Scholars of China (Grant No.61225002), and the Fundamental Research Funds for the Central Universities.

Author details

Renxian Li*, Lixin Guo, Bing Wei, Chunying Ding and Zhensen Wu

*Address all correspondence to: rxli@mail.xidian.edu.cn

School of Physics and Optoelectronic Engineering, Xidian University, China

References

[1] Dholakia,, Dholakia, K.K., Reece, P.. Optical micromanipulation takes hold. Nano Today 2006;1(1):18–27.

[2] Neuman,, K.C.,, Block, S.. Optical trapping. Rev Sci Instrum 2004;75:2787.

[3] Enger, J., Goksor, M., Ramser, K., Hagberg, P., Hanstorp, D.. Optical tweezers applied to a microfluidic system. Lab Chip 2004;4(3):196–200.

[4] Domachuk, P., Cronin-Golomb, M., Eggleton, B., Mutzenich, S., Rosengarten, G.,
Mitchell, A.. Application of optical trapping to beam manipulation in optofluidics.
Opt Express 2005;13(19):7265–7275.

[5] Ashkin, A.. Forces of a single-beam gradient laser trap on a dielectric sphere in the
ray optics regime. Biophys J 1992;61(2):569–582.

[6] Roosen, G.. A theoretical and experimental study of the stable equilibrium positions
of spheres levitated by two horizontal laser beams. Opt Commun 1977;21(1):189–194.

[7] Yao, X., Li, Z., Cheng, B., Zhang, D.. Analysis and calculation of the optical force on a
double-layer dielectric sphere. Acta Optica Si 2000;10:13051310.

[8] Chaumet, P.C., Rahmani, A., Nieto-Vesperinas, M.. Optical trapping and manipulation of nano-objects with an apertureless probe. Phys Rev Lett 2002;88(12):123601.

[9] Barton, J. P., Alexander, D.R., Schaub, S.A.. Theoretical determination of net radiation force and torque for a spherical particle illuminated by a focused laser beam. J
Appl Phys 1989;66(10):4594–4602.

[10] Ren, K.F., Gréhan, G., Gouesbet, G.. Prediction of reverse radiation pressure by generalized Lorenz-Mie theory. Appl Opt 1996;35(15):2702–2710.

[11] Ren, K., Gréhan, G., Gouesbet, G.. Radiation pressure forces exerted on a particle arbitrarily located in a Gaussian beam by using the generalized Lorenz-Mie theory,
and associated resonance effects. Opt Commun 1994;108(4-6):343–354.

[12] Lock, J.A.. Calculation of the radiation trapping force for laser tweezers by use of
generalized Lorenz-Mie theory. i. localized model description of an on-axis tightly
focused laser beam with spherical aberration. Appl Opt 2004;43(12):2532–2544.

[13] Lock, J.A.. Calculation of the radiation trapping force for laser tweezers by use of
generalized Lorenz-Mie theory. ii. on axis trapping force. Appl Opt 2004;43(12):2545–
2554.

[14] Gouesbet, G., Maheu, B., Gréhan, G.. Light scattering from a sphere arbitrarily located in a Gaussian beam, using a bromwich formulation. J Opt Soc Am A 1988;5(9):
1427–1443.

[15] Polaert, H., Gréhan, G., Gouesbet, G.. Forces and torques exerted on a multilayered spherical particle by a focused Gaussian beam. Opt Commun 1998;155(1-3):169–179.

[16] Xu, F., Ren, K.F., Gouesbet, G., Cai, X., Gréhan, G.. Theoretical prediction of radiation pressure force exerted on a spheroid by an arbitrarily shaped beam. Phys Rev E 2007;75(2):026613.

[17] Li, R., Han, X., Shi, L., Ren, K., Jiang, H.. Debye series for Gaussian beam scattering by a multilayered sphere. Appl Opt 2007;46(21):4804–4812.

[18] Li, R., Han, X., Ren, K.F.. Debye series analysis of radiation pressure force exerted on a multilayered sphere. Appl Opt 2010;49(6):955–963.

[19] Bo Sun, Yohai Roichman, and David G. Grier, Theory of holographic optical trapping, Optics Express 2008, 16: 15765-15776

[20] Anna T. O'Neil, Miles J. Padgetta, Axial and lateral trapping efficiency of Laguerre–Gaussian modes in inverted optical tweezers, Optics Communications 2001, 193: 45–50

[21] Garces-Chavez, V., McGloin, D., Melville, H., Sibbett, W., Dholakia, K., et al. Simultaneous micromanipulation in multiple planes using a self-reconstructing light beam. Nature 2002;419(6903):145–147.

[22] Arlt, J., Garces-Chavez, V., Sibbett, W., Dholakia, K.. Optical micromanipulation using a Bessel light beam. Opt Commun 2001;197(4-6):239–245.

[23] Garc'es-Ch'avez, V., Roskey, D., Summers, M., Melville, H., McGloin, D., Wright, E., Dholakia, K.. Optical levitation in a Bessel light beam. Appl Phys Lett 2004;85:4001.

[24] Cizm'ar, T.. Optical traps generated by non-traditional beams. Ph. D. thesis, Masaryk University in Brno; 2006.

[25] Ambrosio, L., Hern'andez-Figueroa, H., et al. Gradient forces on double-negative particles in optical tweezers using Bessel beams in the ray optics regime. Opt Express 2010;18(23):24287– 24292.

[26] Ambrosio, L., Hern'andez-Figueroa, H., et al. Radiation pressure cross sections and optical forces over negative refractive index spherical particles by ordinary Bessel beams. Appl Opt 2011;50(22):4489–4498.

[27] Milne, G., Dholakia, K., McGloin, D., Volke-Sepulveda, K., Zem'anek, P.. Transverse particle dynamics in a Bessel beam. Opt Express 2007;15(21):13972–13987.

[28] Preston, T.C., Mason, B.J., Reid, J.P., Luckhaus, D., Signorell, R.. Size-dependent position of a single aerosol droplet in a Bessel beam trap. Journal of Optics 2014;16(2): 025702.

[29] Q. Zhan, Cylindrical vector beams: from mathematical concepts to applications, Advances in Optics and Photonics 2008, 1:1-57.

[30] Q. Zhan, Trapping metallic Rrayleigh particles with radial polarization, Opt. Express 2004, 12: 3377-3382.

[31] Y. Kozawa and S. Sato, Optical trapping of micrometer-sized dielectric particles by cylindrical vector beams, Opt. Express 2010, 18: 10828-10833.

[32] L. Huang, H. Guo, J. Li, L. Ling, B. Feng, and Z.-Y. Li, Optical trapping of gold nano-particles by cylindrical vector beam, Opt. Lett. 2012, 37:1694-1696.

[33] G. Volpe, D. Petrov, Generation of cylindrical vector beams with few-mode fibers excited by Laguerre–Gaussian beams 2004, Opt. Commun. 237: 89–95.

[34] Ren, K.F., Gouesbet, G., Gr´ehan, G.. Integral localized approximation in generalized lorenz-mie theory. Appl Opt 1998;37(19):4218–4225.

Fiber Laser and Sensor

Chapter 7

Advanced Optical Fibers for High Power Fiber Lasers

Liang Dong

Additional information is available at the end of the chapter

http://dx.doi.org/10.5772/58958

1. Introduction

The nonlinear limits of conventional single-mode fibers were well recognized in the nineties during the earlier efforts to scale the peak powers of pulsed fiber lasers. These earlier efforts focused on lowering the NA and optimizing the refractive index profile of optical fibers to increase the effective mode area [1, 2]. The emergence of photonic crystal fibers in the late nineties gave new impetus to the mode area scaling of single-mode optical fibers. The observation of the "endlessly single-mode" nature of *photonic crystal fibers* (PCF) at small hole sizes in 1996 [3] led to an early realization of the dispersive nature of a photonic crystal cladding [4], which limits the increase of normalized frequency V at short wavelengths. It was quickly realized that the scalability of Maxwell's equation allows for single-mode operation at a very short wavelengths in a small core to be directly translated into single-mode operation in a large core at longer wavelengths [5]. This led to a rapid progress in scaling of core diameters of single-mode PCF, culminating in the 100μm-core diameter demonstrated in 2006 [6].

The design and fabrication techniques developed for photonic crystal fibers also led to a realization of the potential of optical fibers with more open cladding structures than the closed concentric circles found in conventional optical fibers. It then became possible to design and fabricate leaky waveguides where differential mode losses can be used to control the number of propagation modes. As a consequence, *leakage channel fibers* (LCF) were proposed and first demonstrated in 2005 [7]. Two years later, single-mode operation in a record core diameter of ~180μm was demonstrated in a LCF [8].

Many other approaches based on conventional fibers were also pioneered during the last decade. One notable example is the *chirally coupled core* (CCC) fiber [9, 10], which relies on out-coupling of *higher-order modes* (HOM) to side cores adjacent to the main core. Phase-matching is achieved with the help of angular momentum from the helical side cores, which are formed by spinning the preform during fiber drawing. Another notable example is based on the

propagation of a higher-order mode in a specially designed multimode fiber [11, 12]. It is argued that perturbations mostly have anti-symmetry in optical fibers and promote mode coupling mostly between modes of opposite symmetries. The mode spacing between a radially symmetric LP0n mode and its nearest neighbor modes with opposite symmetry is, in fact, larger for higher order modes. These higher-order modes are also more resistant to bend-induced mode compression. A special fiber design facilitates the ease of mode conversion to and from the LP0n mode using a *long period grating* (LPG).

In this chapter, we will give a brief introduction to the key approaches to effective mode area scaling which have shown great promise for future high power fiber lasers. Basic concepts are introduced and the latest developments are also discussed.

2. Photonic crystal fibers

The authors of the paper in 1996 [3] were looking for photonic bandgap guidance in the defect core of the PCF with a core size of ~4.6 μm. The reason for the absence of photonic bandgap guidance in a solid-core PCF with air holes in the cladding only became clear a few years later. To their surprise, they observed very broadband single-mode guidance from 458–1550 nm (extended to 337-1550 nm in [4]). A conventional single-mode step-index optical fiber would become multimode at shorter wavelength. Even more surprising, the diffraction angle of the output beam was smaller at shorter wavelengths, indicating a smaller fiber numerical aperture at short wavelength.

2.1. Fundamental space-filling mode of photonic crystal cladding

In a follow-on paper [4] the initial observation of the "endlessly single-mode" nature of photonic crystal fibers was explained by the dispersive nature of the cladding, which can be viewed as a composite of two materials, i.e. silica and air. The effective refractive index of the composite cladding was identified for the first time as the effective index of the fundamental mode of the composite cladding, referred to as the *fundamental space-filling mode (FSM)*. In the extreme case of conventional optical fibers with an infinite cladding, the effective index of the fundamental space-filling mode becomes the refractive index of the cladding. In PCFs with air holes, the cladding index n_{cl}, i.e. the effective index of the fundamental space-filling mode, is lower than the refractive index of the background glass n_b due to the existence of the air holes, i.e. $n_{cl} < n_b$. The core index n_{co} is the same as the index of the background glass, $n_{co} = n_b$. Fundamental optical guidance in PCFs can, therefore, in principle, be understood in a similar way to that in conventional optical fibers (see Figure 1).

This paper [4] established the principle for understanding the basic guidance properties of PCFs. In conventional optical fibers, the refractive indexes of core and cladding are only weakly dependent on wavelength due to material dispersion. This is also expected of the core refractive index of a PCF. The refractive index of the composite photonic crystal cladding, however, behaves very differently. At the long wavelength extreme, i.e. $\lambda \rightarrow \infty$, the wavelength is much larger than the air holes and the fundamental space-filling mode will occupy all areas

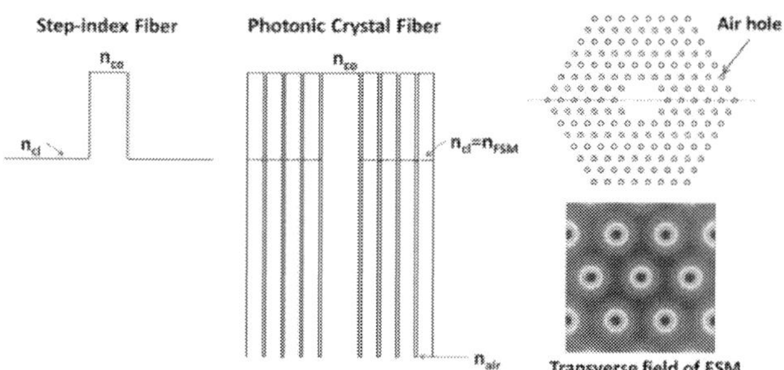

Figure 1. Refractive index of photonic crystal fiber cladding.

of the cladding equally. In this case, the refractive index of the FSM can be obtained by averaging the square of the refractive index of the composite photonic crystal cladding, i.e. $n_{cl}^2 = n_b^2(1-F) + n_{air}^2 F = n_b^2 - F(n_b^2 - n_{air}^2)$ where F is the area fraction of air; and n_{air} is the refractive index of air. The effective NA of the PCF for $\lambda \to \infty$ is, therefore, $F^{1/2}(n_b^2 - n_{air}^2)^{1/2}$. At the short wavelength extreme, i.e. $\lambda \to 0$, the wavelength is much smaller than the pitch of the air holes and the fundamental space-filling mode is mostly in the region with high refractive index, i.e. the background glass. In this case, the cladding index is the refractive index of the background glass n_b. The effective NA of a PCF at $\lambda \to 0$ trends to zero. The guidance of a PCF still gets weaker at large λ, as in a conventional optical fiber, due to the inverse wavelength dependence of the normalized frequency V. In addition, its guidance also gets weaker at short wavelengths due to the diminishing NA, which also limits the growth of V and leads to the "endlessly single-mode" nature of the PCFs with small air holes. Guidance of PCF will, therefore, diminish at both long and short wavelength ends, possessing two bend-induced cut-off edges.

2.2. Single mode and multimode regime of photonic crystal fibers

In conventional fibers, a mode is guided when the effective mode index n_{eff} is between the core and cladding index, i.e. $n_{co} > n_{eff} > n_{cl}$. The mode cut-off can be identified when $n_{eff} = n_{cl}$. The second-order mode cuts off in a step-index fiber at V=2.405. In the first reported work on determining the single-mode regime of PCFs with a 1-cell core, i.e. one hole missing at the defect core, by Birks et al [4], a somewhat arbitrary equivalent step-index core radius equaling *pitch*Λ, i.e. center-to-center hole separation, was used to calculate the V value. Work by Saitoh [13] using a *finite element model* (FEM) to determine n_{FSM} and the condition $n_{eff} = n_{cl}$ to determine the second-order mode cut-off, put the effective step-index core radius to be $\Lambda/3^{1/2}$ for a one-cell core. Later work by Saitoh [14] determined the effective step-index core radius to be Λ for a 3-cell core and $2^{1/2}\Lambda$ for a 7-cell core.

The optical waveguide equation allows the scaling of all parameters measured in the length scale by the same factor. For PCF, the most convenient scaling factor is $1/\Lambda$. The second-order mode cut-off is typically plotted as a normalized wavelength λ/Λ versus normalized hole diameter d/Λ plot. This is shown for PCFs with 1-cell, 3-cell and 7-cell core in Figure 2 [14]. For each of the curves, the area above the curve is in the single-mode regime, i.e. the wavelengths above the second-order mode cut-off wavelength, and, below it, multimode regime. The "endlessly single-mode" nature of PCFs can be easily identified in Figure 2. When d/Λ<0.424, 0.165 and 0.046 respectively for 1-cell, 3-cell and 7-cell core PCFs, the PCFs will remain in the single-mode regime for all wavelengths. It is worth noting that Figure 2 is for PCFs with infinite cladding. For PCFs with finite cladding, guidance is weaker and the second-order-mode cut-off is expected to happen at slightly shorter wavelengths. The curves in Figure 2 are expected to move downwards slightly. It needs to be noted as discussed earlier, that the PCFs are too weak to guide any light at long and short extremes of wavelength. The critical bend radius at the short wavelength edge is determined to be dependent on pitch Λ and wavelength λ such that [4]

$$R_c \propto \frac{\Lambda^3}{\lambda^2} \tag{1}$$

The critical bend radius at the short length edge increases in proportion to $1/\lambda^2$ as wavelength decreases. This relation was verified experimentally for the critical bend radius at 3dB excessive bend loss [4]. It comes directly from the dispersive nature of the photonic crystal cladding.

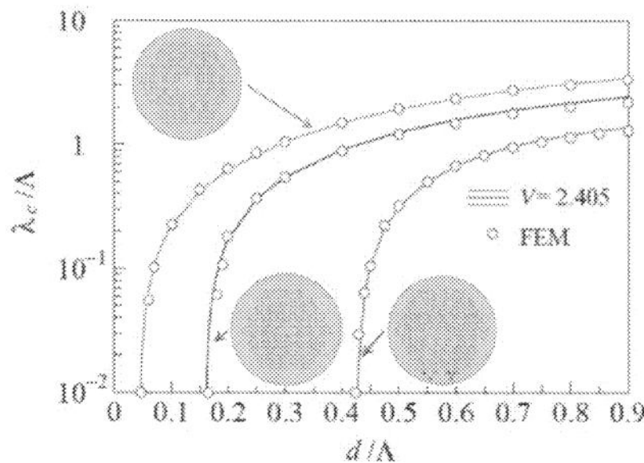

Figure 2. The second-order mode cut-off determined by $n_{eff}=n_{FSM}$ (dots) and V=2.405 (solid lines) using core radius ρ=Λ/3¹ᐟ², Λ and 2¹ᐟ²Λ respectively for 1-cell, 3-cell and 7-cell cores [14].

2.3. Waveguide loss of photonic crystal fibers

PCFs are intrinsically leaky, i.e. there is always a finite waveguide loss for each mode in a PCF with a finite cladding. The waveguide loss can be found by calculating the imaginary part of the effective mode index using a numerical mode solver. When plotting the waveguide loss versus wavelength, the slope is expected to change around cut-off. This can also be used to determine the second-order-mode cut-off. This was done for a 1-cell core PCF [15]. The results are consistent with those in [13].

In conventional optical fibers, the fully enclosed core and cladding boundary ensures that waveguide loss is zero for all guided modes, i.e. those that satisfy the conditions for total internal reflection at the boundary. For the open structure of PCFs with a finite cladding, all modes are leaky with finite waveguide losses. At the extreme of an infinite number of layers, the waveguide losses are zero. The waveguide losses are lower for larger air holes and can be substantially lowered by increasing the number of air-hole layers. In practice, the waveguide loss can be made below other material and process related losses. By employing appropriate polishing, etching and dehydration processes, a PCF with a loss 0.28dB/km at 1550nm has been demonstrated [16]. For applications in fiber lasers where a length of not more than a few tens of meters is used, the waveguide loss does not present any problem if appropriate designs are used.

2.4. Modeling of photonic crystal fibers

Due to the complexity of the geometrical structures of PCFs, numerical models are typically used to find mode properties including the effective mode index, waveguide loss and effective mode area. For any numerical mode solvers, electric and magnetic fields are described by an expansion series. Eigenvalue equations are then established by requiring the fields to satisfy all boundary conditions. These equations are typically expressed as a set of linear equations, which can be solved for complex effective mode index. The more complex a waveguide design is, the larger is the number of linear equations. The waveguide dispersion can be calculated from the real part of the complex effective mode index. The waveguide loss can be obtained from the imaginary part of the complex effective mode index. Electric and magnetic fields can be calculated once the effective mode index is determined. The most common and flexible numerical mode solver is a *finite element mode solver* (FEM). This is commercially available from COMSOL (*http://www.comsol.com/*). A FEM is capable of handling practically any design. It can, however, be computationally very demanding. If only circular boundaries are involved, a Multipole mode solver is a good option [17, 18]. It is based on the decomposition of fields into circular Bessel series, which are the most accurate and efficient method for modeling circular boundaries. For non-commercial research and teaching purposes, it can be downloaded from the University of Sydney website (*http://sydney.edu.au/science/physics/cudos/research/mofsoft-ware.shtml*). A plane wave expansion method can also be used. This, however, assumes an infinite cladding and, therefore, cannot determine the waveguide loss.

2.5. Mode area scaling with photonic crystal fibers

The first demonstration of a large-core PCF was performed by Knight el at [5]. The 1-cell PCF with relative hole diameter $d/\Lambda \approx 0.12$ and a core diameter of $2\varrho = 22.5\mu m$, provided robust single-mode guidance at 458nm. According to Figure 2, a 1-cell PCF with $d/\Lambda < 0.424$ is expected to be single-mode over its entire wavelength range. It is, therefore not surprising that the PCF was single mode at 458nm. The critical bend radius was measured to be 50cm at 458nm and 4cm at $1.55\mu m$ [5]. This fiber is expected to have a critical bend radius of ~10cm at $1\mu m$ using equation 1.

Considering the wavelength scalability of the waveguide equation, this 1-cell PCF with $2\varrho/\lambda \approx 50$ can be scaled by a factor of ~2.2 to a ~$50\mu m$ core diameter to operate at $1\mu m$. The critical bend radius for this single-mode PCF with $50\mu m$ core diameter is expected to be ~1.1m at $1\mu m$

according to equation 1! A similarly scaled 1-cell PCF with a core diameter of 30μm will have a critical bending radius of ~24cm. This is probably close to the practical limit of coiled single-mode 1-cell PCFs. Above a core diameter of ~40μm, 1-cell single-mode PCF can only be used in a straight configuration in practice. The single-mode operation of PCFs with large core diameters comes at the cost of weak guidance as a result of the diminishing effective NA. This fundamentally limits the use of single-mode PCFs with large cores in coiled configurations. If PCFs are allowed to operate in the few-moded regime, coiled PCFs with slightly larger core diameters are possible. For high average power fiber lasers with outputs exceeding kW, effective thermal management becomes increasingly critical. Long fiber lengths of many meters are required. The constraint of not being able to coil the fibers can become a major issue considering the additional space constraints.

2.6. Rare-earth-doped glass for large-core photonic crystal fibers

The diminishing NA of PCFs with large cores also has additional implications, as realized by the authors of [5]. To fabricate rare-earth-doped large-core PCFs, the core refractive index needs to be accurately controlled at levels far below the very small effective index difference between the core and cladding in order not to disturb the guidance properties of the PCFs. This requires much improved techniques for the fabrication of the active core in large-core PCFs.

The first ytterbium-doped large-core PCF with 15μm core diameter was reported in 2001 at Bath University [19]. The effective mode area at 1μm was ~100μm². The laser operated in single mode with poor efficiency. The key advance was the use of a repeated stack-and-draw process to achieve a uniformly doped core with a refractive index close to silica. Conventional fabrication techniques for rare-earth doped silica fibers result in significant non-uniformity in refractive index across the core as well as a raised refractive index above silica. Due to the weak guidance in large-core PCFs, any index non-uniformity across the core can lead to mode distortion. It also requires the refractive index of the core to be very close to the silica background in PCFs. The authors of [19] fabricated ytterbium-doped glass with low doping levels. The doped glass is surrounded by some undoped silica glass. The glass is stacked and drawn twice before being finally incorporated into the core of a PCF. The resulting PCF has an ytterbium-doped core which consists of 425 original doped glass sections. The dimension of each of the original doped glass sections is much smaller than the wavelength of light and the core of the PCF, therefore, appears to be homogeneous to light at the operating optical wavelength. The doped core is also heavily diluted by the addition of silica glass (90%), leading to an average refractive index close to that of silica [20]. In the same paper, it was pointed out that rare-earth-doped glass with a higher index can be stacked and drawn with undoped glass with a low index to achieve a better match to that of silica background [20].

2.7. Double-clad photonic crystal fibers with high NA air-clad for pump guidance

The concept of a composite air-glass clad with a high air-filling fraction to provide a high NA pump guide in a double-clad fiber was first proposed in 1999 at what was then then Lucent Technologies [21]. The high pump NA can enable a significant improvement in pump coupling especially from low brightness multimode diode lasers for a given pump waveguide dimension. In conventional double-clad fibers, low-index polymer coatings are typically used to form the pump cladding. The air-clad pump isolates high pump powers from the polymer coatings, leading to potentially improved reliability.

The first such fiber was demonstrated in 2000, where the pump cladding was mostly air except for a single connecting element [22]. The measured pump NA over 1m was below 0.2. The fiber was passive and there was no laser demonstration in this first attempt. The first cladding-pumping demonstration in an active fiber with an air-clad pump guide and a conventional core design was in 2001 at Southampton University [23]. The measured NA of the pump guide with air-cladding was 0.4-0.5. In a separate demonstration from Southampton University later in the same year, a conventional core with a very low NA together with a photonic crystal cladding was used in combination with a pump guide with air-cladding [24]. The core guidance came from a combination of the raised core index and the photonic crystal cladding. Cladding pumping with a low brightness laser diode at 915nm (100μm core with a NA=0.22) was demonstrated with a slope efficiency of 70% relative to the absorbed pump power. The measured pump NA over a short length (~10cm) was 0.3-0.4. NA over much longer fiber length as in a fiber laser is expected to be lower. The first true active double-clad photonic crystal fibers with a pump air-cladding was demonstrated in 2003 with a measured pump NA of 0.8 at 1μm [20].

The theoretical basis for the high NA of optical waveguides formed by air-cladding was established in [25]. It had been understood previously that the effective index of a glass and air composite can be obtained by the effective index of the fundamental space filling mode. An example of high NA air-clad used in double-clad fibers is shown in Figure 3 from [25]. In such air-glass composites, the glass webs can be considered as slab waveguides. The separations between the webs are typically far enough that the webs can be considered as isolated. In this case, the effective index of the glass-air composite cladding can be well approximated by that of the fundamental slab waveguide mode. This can be easily calculated. The resulting pump NA is only dependent on the ratio of the web width w and wavelengthλ. This is plotted in Figure 3(c), showing that a small w/λ is critical for high NA.

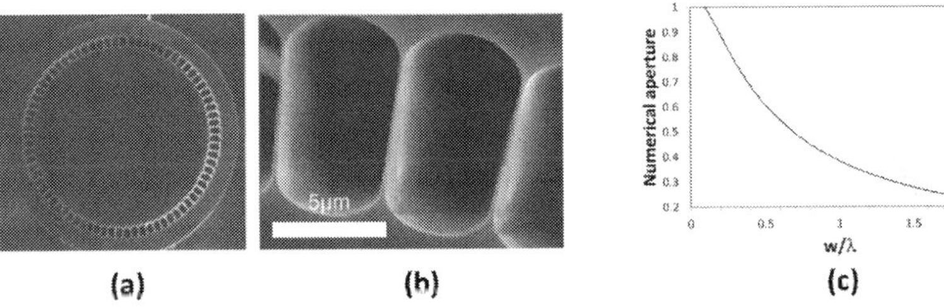

Figure 3. (a) High NA air-clad from [25], (b) close-up of the high NA air-clad, and (c) NA of air-clad waveguide based on slab waveguide model plotted against the ratio of waveguide width w and wavelength λ.

2.8. Progress of active large-core photonic crystal fibers

Since 2003, the group at Friedrich-Schiller-University Jena has been playing a significant role in the development large-mode-area photonic crystal fibers with pump air-clad. In their first work in collaboration with then Crystal Fibre A/S, now NKT Photonics, an ytterbium-doped PCF with effective mode area of ~350μm^2 and mode field diameter (MFD) of 21μm was demonstrated [26] (see Figure 4). The 3-cell PCF had a hole diameter d=2μm and pitch Λ=11.5, giving an d/Λ=0.18. The core diameter was 28μm. A relatively small circular area of 9μm was doped in the 3-cell core. The doped area had a high ytterbium-doping level of ~0.6 at%. It was further co-doped with aluminum and fluorine to provide an index merely 2×10^{-4} above the silica background. The PCF with the raised index in the doped area was simulated, showing that the fiber guides the second-order mode, which is, however, close to cut-off. The fiber had a 150μm pump guide with an air clad. The webs in the air clad were ~50μm long with a thickness of ~390nm, giving w/λ=~0.4 at 976nm. The measured pump NA was 0.55, slightly below the NA=0.68 predicated by the slab model in Figure 3. The fiber had an outer glass diameter of 450μm and was coated with standard acrylic coating. Due to the high doping level, the fiber has a pump absorption of ~9.6dB/m at 976nm. An impressive slope efficiency of 78% was demonstrated with respect to the launched pump power.

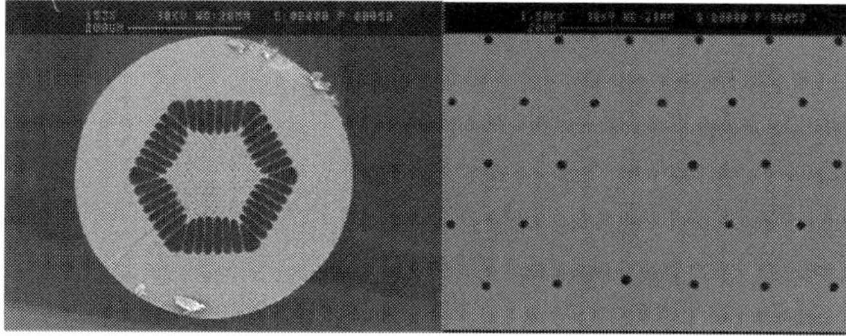

Figure 4. SEM image of the air-clad ytterbium-doped large-mode-area photonic crystal fiber in [26].

In a subsequent paper in 2003 [27], the fiber in [26] and a new fiber with narrower pump air-clad were studied with the FEM for temperature distributions in the fiber at various thermal loads in the core, considering both convective and radiative heat transportation by air at the fiber surface. The results show that, in the case of natural cooling, the thermal transportation is mainly limited by the heat transfer at the fiber surface. The air-clad impedes heat flow, especially when the width of the air clad is large. A narrower pump air-clad is advantageous, especially in actively cooled fiber lasers.

In 2004, the bar was raised in a collaborative work between Friedrich Schiller University Jena and Crystal Fibre A/S [28]. Their 7-cell fiber has an effective area of ~1000μm^2 and MFD of ~35μm (see Figure 5). Core diameter is ~40μm. Hole diameter d is 1.1μm and the pitch Λ is 12.3μm, giving a d/Λ=0.09. The fiber has a pump guide diameter of 170μm and a measured pump NA of 0.62 at 950nm. The pump absorption is ~13dB/m at 976nm. The pump air-clad is

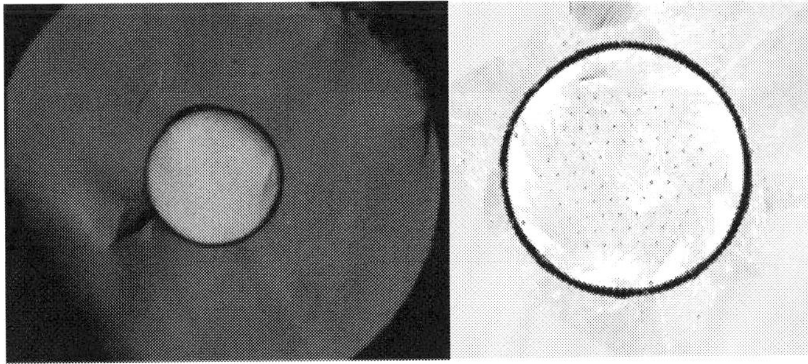

Figure 5. Microscope image of the air-clad ytterbium-doped single-mode PCF and close-up to the 40-µm core formed by seven missing air holes [28].

much narrower for more efficient heat diffusion. It is worth noting that the fiber is in the multimode regime for a 7-cell PCF (see Figure 5). This helps to ease bend loss in the weakly guided fiber.

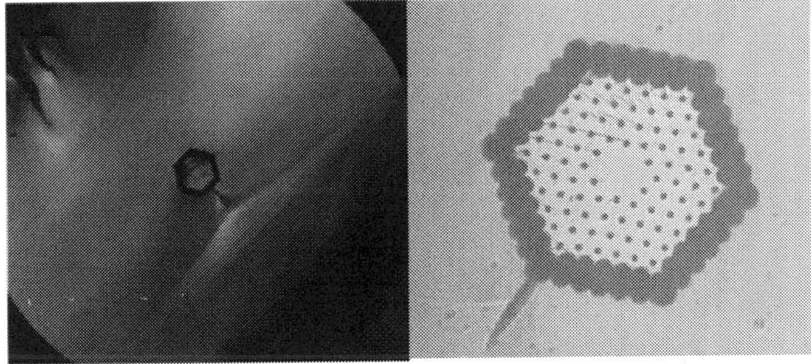

Figure 6. Microscope image of a rod-type photonic crystal fiber and close-up view of the inner cladding and core regions.

To further mitigate bend loss, a rod-type PCF was developed in 2005 [29] (see Figure 6). The rod had an outer diameter of 1.7mm so that it cannot be bent. A 48cm long length was used in the demonstration. The relative hole diameter was increased to d/Λ=0.33. This reduces bend sensitivity. This larger d/Λ in a 7-cell core PCF, however, puts this fiber firmly in the multimode regime (see Figure 2). The pump guide was reduced to a hexagon with flat-to-flat dimension of 117µm and corner-to-corner dimension of 141µm, similar to earlier fibers. The ytterbium doping level was also increased compared with earlier fibers. The increased doping level and reduced pump guide led to a high pump absorption of ~30dB/m at 976nm. The fiber was used to demonstrate power extraction of ~250W/m.

The rod-type PCF was further developed with the demonstration of a 19-cell PCF with a core diameter of 60µm, effective mode area of ~2000µm² and MFD of 75µm in 2006 [30] (see Figure

7). The fiber had a d/Λ=0.19, again firmly in the multimode regime (19-cell PCF not shown in Figure 2). The pump guide was 175μm in diameter. The pump air-clad had a web thickness of 400nm, which was 10μm long. The measured pump NA was 0.6 at 975nm, fairly close to the 0.67 NA predicted by the slab model (see Figure 3). The pump absorption is ~30dB/m at 976nm. The rod diameter is ~1.5mm. A 58cm long fiber was used to demonstrate 55W/m power extraction. In the same paper, a passive 100μm core fiber was also demonstrated. The 19-cell fiber had a d/Λ=0.2, again in the multimode regime. The effective mode area was ~4500μm^2 and MFD was ~75μm.

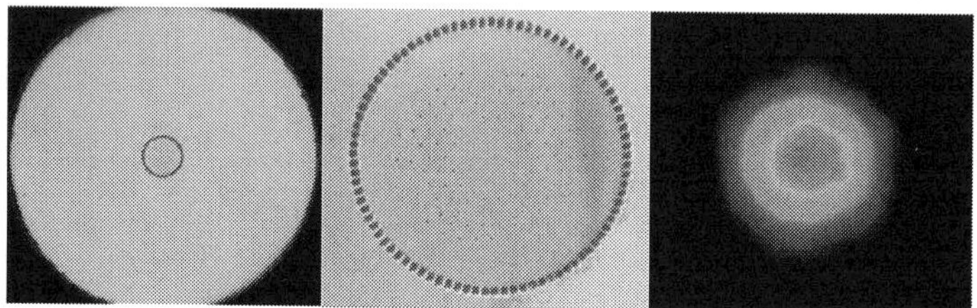

Figure 7. Microscope-image of the extended-mode-area rod-type photonic crystal fiber, SEM-picture of the microstructured region and measured near-field intensity profile of the 60μm core fiber [30].

The demonstration of a 100μm active 19-cell PCF finally came in 2006 [6]. The pump guide had a diameter of 290μm. The rod diameter was 1.5mm. A 90cm long fiber was used to amplify 1ns pulses at 9.6kHz to record peak power of 4.5MW and pulse energy of 4.3mJ with M^2=1.3.

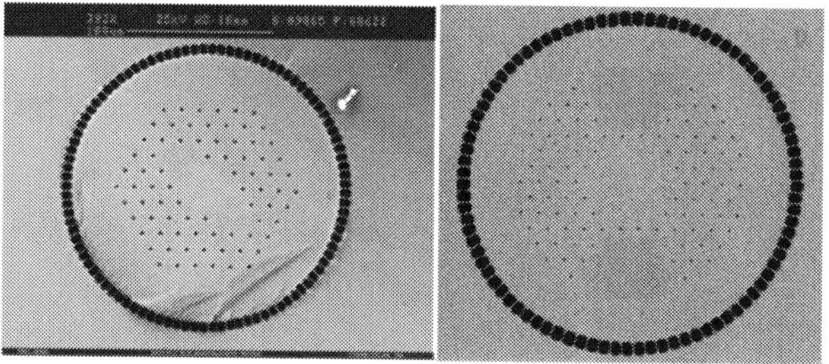

Figure 8. SEM image of the single polarization PCF with an effective area of ~700μm^2 [31] (left) and 2300μm^2 (right) [32].

Polarization-maintaining (PM) PCFs have also been developed by introducing stress elements. A PM 7-cell PCF with a mode area of ~700μm^2 was demonstrated in 2005 [31] (see left figure in Figure 8). The pitch Λ was 12.3μm with a d/Λ of 0.2. In the weakly guided PCFs, the polarization modes on slow and fast axis have different bend losses. This enables single-polarization operation where the polarization mode on the slow axis is still guided while polarization mode on the fast axis suffers high bend loss. Another single-polarization 19-cell

PCF was demonstrated in 2008 with a mode area of ~2300μm² (right figure in Figure 8) [32]. The corner-to-corner distance of the core was 70μm. The pitch Λ was 11μm with a d/Λ of 0.1.

3. Leakage channel fibers

A 2D micro-structured cladding, which is made possible by the stack-and-draw technique developed for photonic crystal fibers, enables new designs which do not possess the closed core-and-clad boundaries of conventional optical fibers. When a mode is guided in a conventional optical fiber, total internal reflection everywhere around the closed core-and-clad boundary, traps light entirely in the core, leading to zero waveguide loss. In designs with an open cladding, light can leak out, leading to finite waveguide loss associated with each mode. The waveguide loss is mode-dependent, providing opportunities for mode control by minimizing loss of the desired mode while maximizing loss of the unwanted modes. *Leakage channel fibers* (LCF) takes advantages of these new opportunities made possible by open cladding designs. A LCF can be precisely engineered to have high confinement loss for all higher order modes and low confinement loss for the fundamental mode and can, therefore, significantly extend the effective mode area of the fundamental mode. LCFs essentially exploit the increased ability of higher order modes to leak through small gaps in the cladding while maintaining good fundamental mode confinement.

3.1. Leakage channel fibers with air holes

The first LCF was demonstrated in 2005 [7]. It has a simple cladding design with air holes in the cladding [7]. The LCF is shown in in the left figure in Figure 9. The LCF had an outer diameter of ~270μm. The two smaller holes had a diameter of d=39.6μm and pitch Λ=51.2μm. The four larger holes had a diameter of d=46.0μm and pitch Λ=51.1μm. The passive LCF provided robust single-mode operation with a measured mode area of ~1417μm² (MFD=42.5μm). The most significant aspect of this work is that the LCF can be coiled down to 15cm with negligible loss penalty. This is a significant improvement over PCFs.

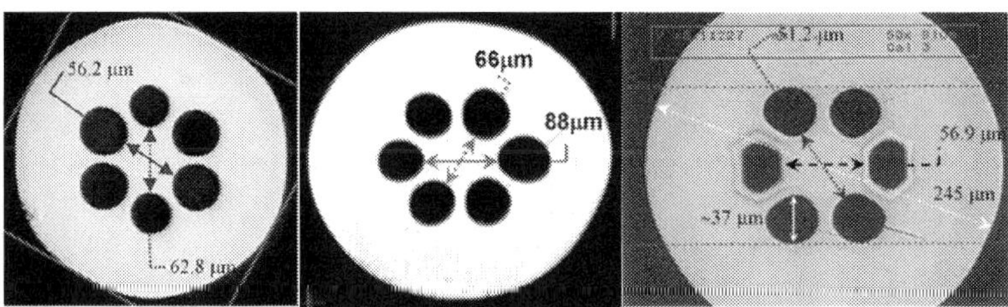

Figure 9. The LCF used in the first demonstration [7] (left), the first ytterbium-doped LCF (center) [33] and the first ytterbium-doped PM LCF (right) [34].

The first Ytterbium-doped LCF was demonstrated in 2006 [33] (see the middle figure in Figure 9). The fiber outer diameter was ~350μm and it was coated with a low index polymer to give a pump NA of 0.46. The average hole diameter was ~55μm and average pitch was ~67μm, giving an average d/Λ=0.82. The effective mode area was $3160\mu m^2$ (MFD=63.4μm). Pump absorption was measured to be ~3.6dB/m at 976nm. Slope efficiency versus launched power was measured to be ~60% in a 5m long amplifier coiled at 40cm diameter. M^2 was measured to be ~1.3.

The first ytterbium-doped PM LCF was also demonstrated in 2006 [34] (see the right figure in Figure 9). A pair of boron-doped silica stress rods was used to replace two opposing air holes. The hole diameter was ~37μm. The effective mode area was ~$1400\mu m^2$ (MFD=42.2μm). The birefringence was measured to be ~2.1×10^{-4} over 1010-1080nm. The LCF had an outer diameter of ~245μm and was coated with a low index polymer to give a pump NA of ~0.46. The pump absorption was ~2.6dB/m at 976nm. Slope efficiency versus launched power was measured to be ~60%. M^2 was measured to be ~1.2. The PM LCF could be coiled to 12cm diameter with negligible bend loss.

3.2. All-glass leakage channel fibers

High refractive index contrast is not necessary for large core fiber designs. Low refractive index contrast is sufficient and often advantageous for further limiting higher order mode propagation. Fluorine-doped silica can be used to replace air holes in the LCFs described in the last section. The all-glass LCFs can provide much improved ease of fabrication and use, compared with fibers with air holes.

Despite the fact that air is a readily available ingredient, there are a number of drawbacks related to the use of air holes in fibers. The first one is the difficulty in precisely controlling the dimension of air holes in fiber fabrication. This is an intrinsic problem of a holey structure due to the air hole's tendency to collapse during fiber drawing. This is usually countered by a precise control of pressurization of the air holes, a process dependent on drawing conditions such as furnace temperature, feed rate, and drawing speed. When small air holes are desirable as in endless single-mode PCFs, higher pressure is required to maintain air hole dimensions due to the significantly increased tendency for the air holes to collapse by surface tension in this regime. This can make air hole dimensions to become highly sensitive to drawing conditions. This delicate balance of pressurization and collapse can lead to issues of controllability and repeatability in PCF fabrication. Air holes can also disturb smooth fracture wave propagation during fiber cleaving, leading to a poor cleaved surface due to the appearance of deep fractures behind the air holes, a problem often aggravated by large air holes and high cleaving tensions. In addition, air holes often have to be thermally sealed at the fiber ends to minimize environmental contamination. Mode distortion can occur from the air holes collapsing during splicing. This is especially true for large-mode-area fibers, which are much more susceptible to small perturbations.

A detailed analysis of all-glass LCFs was reported in [35]. For a LCF formed by one layer of features as shown in the inset of Figure 10, the core of diameter 2ϱ is formed by six features with diameter d and refractive index n_f. Center-to-center feature spacing is Λ. The refractive

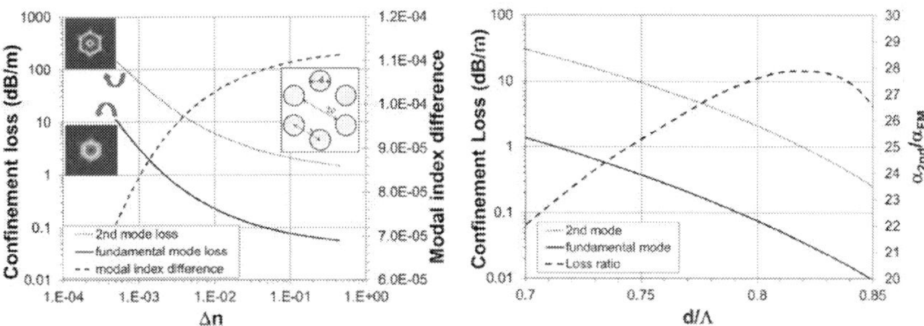

Figure 10. Effect of index contrast on confinement loss and modal index difference for one-cladding-layer LCFs with d/Λ=0.675 and 2ϱ=50µm (left) and effect of d/Λ on confinement loss and the loss ratio between the second-order mode and fundamental mode for LCFs with Δn=1.2×10^{-3} and 2ϱ=50µm (right) [35].

index of the background glass, usually silica, is n_b. The fiber was studied by a multipole mode solver for the effect of index contrast $\Delta n = n_b - n_f$. For the simulation in Figure 10, the following parameters were used: 2ϱ=50 µm, d/Λ=0.675, and n_b=1.444. The wavelength of the simulations was at 1.05µm. It can be seen from the left figure in Figure 10 that confinement loss for both fundamental, α_{FM}, and second-order modes, α_{2nd}, increases with a reduction of index contrast Δn, with the loss of the second-order mode, α_{2nd}, remaining over an order of magnitude higher than the loss of the fundamental mode, α_{FM}. The modal index difference, the difference between the effective mode indices of the fundamental and second modes, decreases toward low index with Δn by just ~40% over three orders of magnitude change in Δn.

The effect of normalized hole diameter d/Λ was also studied in [35] and is shown in the right figure in Figure 10 for confinement losses and the ratio of the second-mode loss to the fundamental-mode loss. The confinement loss for both the fundamental and second modes increases toward small d/Λ with the loss ratio changing very little over the entire range of d/Λ, from 22 to 28. The normalized hole diameter d/Λ is typically chosen to give an acceptable fundamental-mode loss. For LCFs with one layer of features as shown in Figure 10, the loss ratio of all-glass LCFs is very similar to that of an LCF with air holes. Slightly larger d/Λ is, however, required for achieving a similar confinement loss.

LCFs with two layers of features can be used to further improve the differential confinement loss between the fundamental and second-order modes at the expense of bending perform-ance. Acceptable fundamental-mode loss at smaller feature sizes can be realized in LCFs with two layers of features, while leakage of higher order modes is substantially increased by a reduction of feature size despite the additional layer of features. Higher differential loss between modes can therefore be realized. Since bending loss of the fundamental mode is very strongly dependent on feature size, a reduction of feature size increases the bend loss of the fundamental mode in LCFs. An LCF with two layers of features was studied in [35] and the results are shown in Figure 11. Both the fundamental mode and second-order-mode loss shows the characteristic increase at small d/Λ, while the loss ratio α_{2nd}/α_{FM} is increased by over an order of magnitude compared to the one-layer designs in Figure 10. At d/Λ ≈ 0.548, the fundamental-mode loss $\alpha_{FM} \approx 0.1$ dB/m, while the second-order-mode loss $\alpha_{2nd} \approx 48$ dB/m, a loss ratio α_{2nd}/α_{FM} of ~480. A very high loss ratio α_{2nd}/α_{FM} of ~700 is possible at d/Λ=0.62.

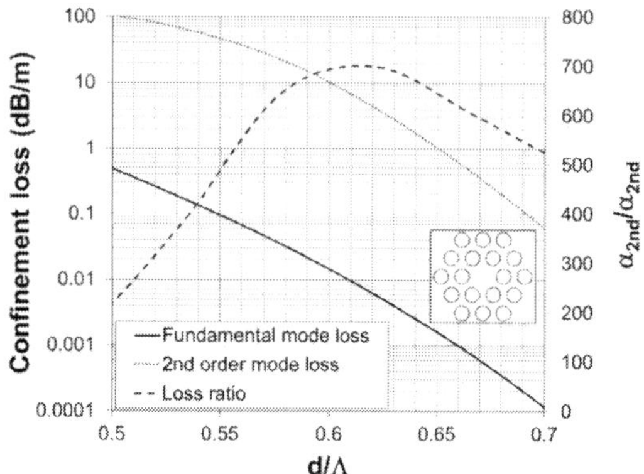

Figure 11. Effect of d/Λ on confinement loss and the loss ratio between the second-order mode and fundamental mode for an LCF with two layers of features, Δn=1.2×10⁻³ and 2ϱ=50µm. [35].

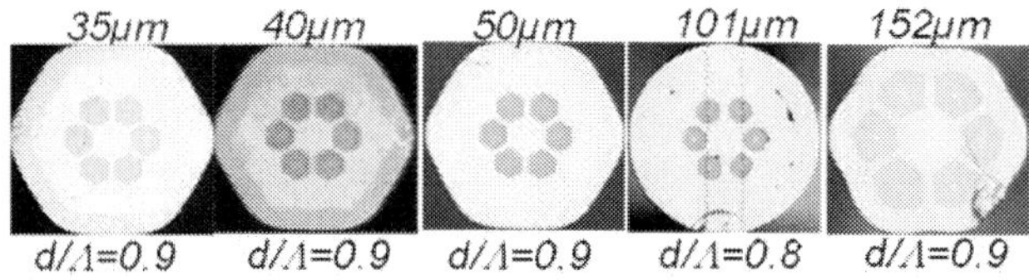

Figure 12. Some examples of fabricated all Glass LCFs. Core diameter is given above the fiber [35].

A wide range of all-glass LCFs were fabricated from core diameters from 35µm to well over 100µm [35] (see Figure 12). All fibers were made with silica glass as the background glass and slightly fluorine-doped silica glass as the cladding features and coated with standard coating with index of 1.54. The refractive index difference between the background and the low index feature wasΔ n=1.2×10⁻³. LCFs with both circular and hexagonal features were fabricated and tested. The conditions for the fabrication of LCFs with hexagonal features also created LCFs with a rounded hexagonal outline. Such a shape is known to be preferred for the pump mode mixing in a double clad fiber where the pump light propagates in a much larger cladding guide. All the fabricated LCFs in Figure 12 operated in the fundamental mode with a varying degree of bend loss performance. In general, bend loss increases rapidly with core diameter increase (see Figure 13). This effect is fundamentally related to the fact that the ability of guided modes to navigate a bend is related to how rapidly a mode can change its spatial pattern without breaking up while propagating, i.e. maintain adiabatic transition. As the mode gets larger, this ability to change diminishes very quickly due to larger Rayleigh range.

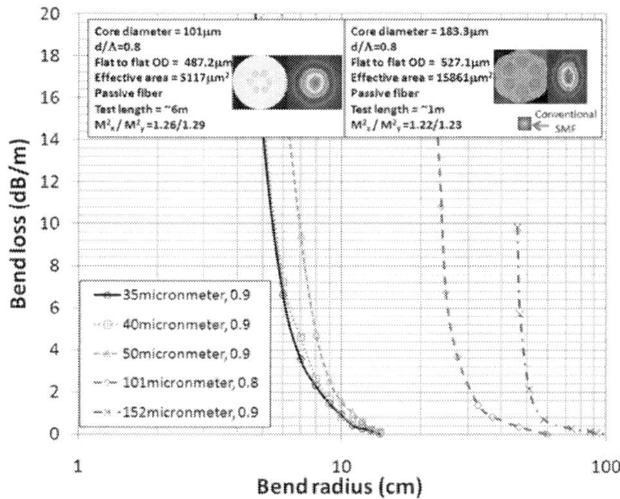

Figure 13. Cross section, measured mode and fiber details are given for the LCF with 101µm core, left inset, and the LCF with 183.3µm core, right inset. Measured bend loss for LCFs with various core diameters [36].

The LCF, shown on the top left inset in Figure 13, had 2ϱ=101µm, and d/Λ=0.9. The effective mode area of the LCF was calculated to be 5117µm² (MFD=80.7µm). A length of the LCF ~6m long was loosely coiled in a 1m coil and the measured M^2 was M^2_x=1.26 and M^2_y=1.29. The LCF, shown on the top right inset, had 2ϱ=183.3µm, and d/Λ=0.8. A conventional single mode optical fiber of the same scale is also shown for comparison. The effective mode area of this LCF was calculated to be 15861µm² (MFD=142.1µm), a record effective mode area for single-mode operation. The Measured M^2 of a 1m straight fiber is M^2_x=1.22 and M^2_y=1.23. Measured mode patterns at the output of the fibers are also shown in Figure 13.

3.3. Polarization maintaining all-glass leakage channel fibers

A PM all-glass LCF was first reported in [37] (see Figure 14). The passive all-glass PM LCF had a core diameter of 50µm. A high d/Λ=0.9 was used for a smaller critical bend radius. The LCF had a refractive index difference between the background and the low index feature of Δn=1.2×10⁻³. The low index features were made of slightly fluorine-doped silica. Two stress elements with a refractive index of ~13 × 10⁻³ below that of the background silica glass were used instead of the regular features on either sides of the core to provide birefringence. The fiber had an outer diameter of ~885µm and was coated with standard acrylic coating. The near field image measured with a single lens is shown in Figure 14(c) for the 1.8m long sample and in Figure 14(d) for the 30m sample. Due to the much higher d/Λ=0.9 used for this fiber, some bending was necessary for fundamental mode operation in a short length of this fiber. The output was robustly single mode in a 30m long sample of this passive LCF coiled in 40cm diameter coils (see Figure 14(d)). The critical bend radius for 1dB/m loss, expected to be ~11cm by FEM simulation, matched very well to the 10.5cm measured. *Polarization extinction ratio* (PER) was characterized at the output of a 1.8m long sample to be >15dB over 1010-1100nm.

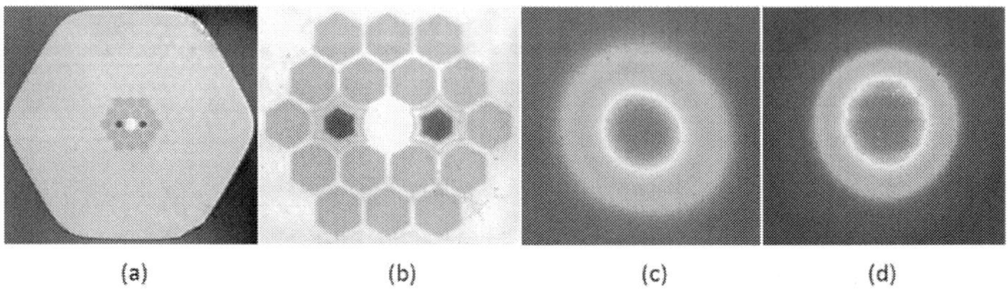

Figure 14. (a) Cross section of the passive PM LCF, (b) magnified cross section, (c) near field from the 1.8m long fiber in a 40cm-diameter coil, (d) near field from the 30m long fiber in a 40cm-diameter coil [37].

3.4. Characterization of mode losses in all-glass leaky channel fibers

Recently, fundamental and higher order mode losses have been characterized and compared to simulations based on the assumption of an infinite cladding [38]. A LCF with a ~50μm core diameter and a hexagonal cladding boundary (see Figure 15(a)) was spliced to a tunable source to ensure launch stability during the measurements. The passive LCF was coated with a low-index polymer to simulate a double-clad fiber with a pump NA of ~0.45. Power in various modes at the output was measured using the S^2 method [39]. The fiber was cut back several times to determine the propagation losses of various modes. The measurement was repeated at various coil diameters. The results are summarized in Figure 15(b) and show remarkable agreement between the measured and simulated losses of fundamental and higher-order modes. LP11 mode loss as high as ~20dB/m was measured, demonstrating the validity of the design. A similar LCF with a circular cladding boundary was also measured, showing significant less higher-order mode losses than those predicted by simulations. It is speculated that the coherent reflection at the circular cladding boundary played a significant role in enhancing the guidance of leaky higher-order modes in this case. It is, therefore, critical to have a non-circular cladding boundary to achieve the maximum possible higher-order mode losses.

3.5. Ytterbium-doped all-glass leakage channel fibers

An ytterbium-doped all-glass LCF with one layer of cladding features was also reported in [37]. The LCF was coated with a low index polymer, providing a pump NA of 0.45. This LCF had pump absorption of 11dB/m at ~976nm. The LCF also had 2ρ=52.7μm and d/Λ=0.8. This gives a simulated effective area of 1548μm² at 1.05μm. The LCF has a rounded hexagonal shape and a flat-to-flat dimension of 254.2μm. The fiber was used to demonstrate amplification of 600ps pulses (with 600μJ pulse energy) to 1MW peak power.

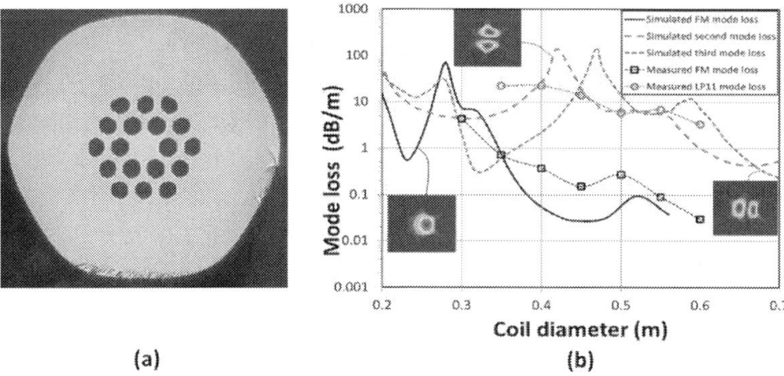

(a)   (b)

Figure 15. (a) The hexagonal LCF used in the measurement and (b) simulated and measured mode loss in the hexagonal Re-LCF [38].

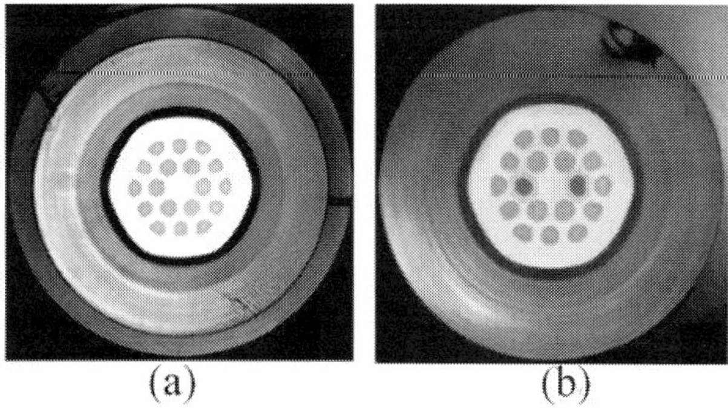

Figure 16. An all glass active LCF (left) and an all glass PM active LCF with highly fluorine-doped pump cladding (right) [37].

Ytterbium double-clad all-glass LCFs with highly fluorine-doped silica as pump cladding were also demonstrated (see Figure 16) [37]. All-glass LCFs have no polymer in the pump path and have independent control of the fiber outer diameters and pump cladding dimensions. This, therefore, enables designs with smaller pump guides for higher pump absorption and, at the same time, with larger fiber diameters to minimize micro and macro bending effects, a much desired feature for large core fibers where intermodal coupling could be an issue due much increased mode density. Stress rods can also be added for PM LCFs (see Figure 16).

The LCFs had a refractive index difference between the background and the low index feature of $\Delta n=1.2\times10^{-3}$. The non-PM LCF (see the left figure in Figure 16) had an inner layer $d/\Lambda{-}0.8$ and outer layer $d/\Lambda=0.7$. It had a 47μm core diameter, a rounded hexagon pump guide with a dimension of 238μm by 256μm, a pump NA of 0.28, pump absorption of ~12dB/m, and an outer diameter of ~538μm coated with standard high index coating, shown as the outermost layer in Figure 16(a). A 3.5m fiber coiled in 53cm diameter was used to demonstrate a slope efficiency of 75% in an amplifier [37]. A single stage gain of 33dB was demonstrated using this fiber. It was also used to directly amplify 15ps pulses to a peak power of ~1MW.

The PM LCF (see Figure 16(b)) had a core diameter of 80μm and had a fluorine-doped pump cladding, providing a pump NA of ~0.28. Low index features with an inner layer d/Λ of 0.8 and an outer layer d/Λ of 0.7 were used. This active PM LCF had a pump guide diameter of ~400μm (flat-to-flat), a fiber outer diameter of ~835μm, and was coated with standard acrylic coating. Pump absorption was estimated to be ~12dB/m. The mode field diameter was measured to be ~62μm. The fiber was used as an amplifier in a single coil 76cm in diameter with a length of straight section at each end, demonstrating a slope efficiency of ~74% and a maximum single-path gain in excess of 30dB [37]. It demonstrated direct amplification of 14.2ps pulses to 190kW peak power with pulse energy of 2.74μJ and negligible SPM spectral broadening. M^2 was measured to be below 1.35 for the entire output power range.

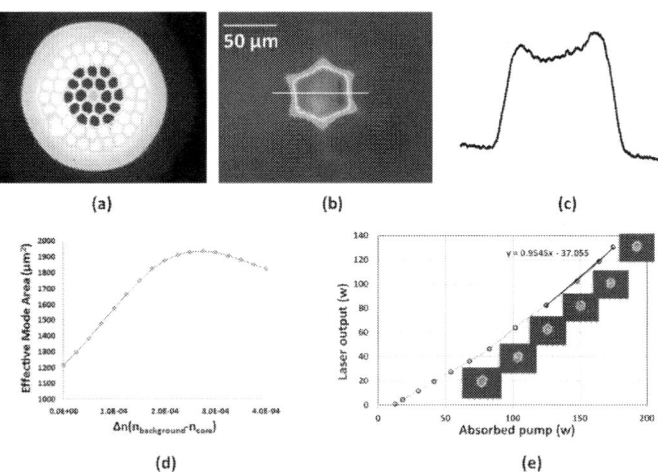

Figure 17. (a) The ytterbium-doped LCF with an index depression in the core center, (b) measured near field intensity of the guided mode, (c) measured mode intensity distribution, (d) simulated effective mode area versus the index depression, and (d) measured laser output and near field patters at various powers [40].

Recently, a flat-top mode has been demonstrated in an ytterbium-doped LCF with a ~50μm core by introducing an area ~30μm in diameter in the core center with a refractive index of ~2×10⁻⁴ lower than that of the background glass (Figure 17(a)) [40]. The flat-top mode (see Figure 17(b) and (c)) increased the effective mode area of the LCF from ~1200μm² to ~1900μm², a ~50% increase (see Figure 17(d)). The LCF also demonstrated near quantum-limited efficiency (see Figure 17(e)). Lasing wavelength was 1026nm and the pump was at 976nm.

4. Higher-order-mode fibers

The concept and demonstration of robust propagation of *higher-order-modes* (HOM) with large effective mode areas were first reported in 2006 [11]. It was argued that the stability of mode propagation in multimode fibers is critically dependent on the effective mode index difference between the propagating mode and its nearest neighbor anti-symmetric mode. This effective mode index difference is actually larger for the higher-order LP_{04}-LP_{07} modes than for the fundamental mode (see Figure 18). In addition, a *long period grating* (LPG), a fairly matured technology, can be used for broad band mode conversion to and from those higher-order LP0n modes.

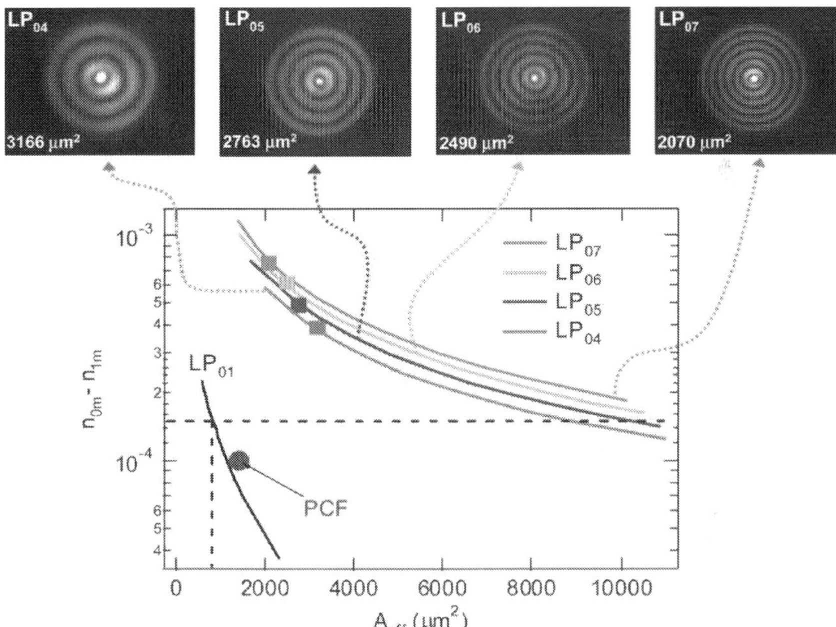

Figure 18. Effective index difference between nearest neighbor anti-symmetric mode versus A_{eff}. Top, near-field images of LPG-excited HOMs after >2m propagation with 7cm bend and with A_{eff} ranging from 2100 to 3200µm² [11].

Mode coupling first requires phase-matching. Perturbations on the fiber need to provide a vector which equals the difference in the propagation constants of the two modes involved. The larger the mode index difference between the two modes, the larger is the wave vector required for phase matching. A large wave vector implies high spatial frequency and small spatial period. For a mode index difference of 1.5×10^{-4}, the required spatial period of the perturbation at 1.55µm is ~10mm. Due to geometric constraints, the spatial frequency distribution of perturbations on fibers usually cuts off at a certain upper frequency limit. This would minimize mode coupling between modes with larger effective mode index differences. The second requirement for mode coupling is that the overlap integral among the two modes and the perturbation needs to be non-zero. This requires the perturbation to break up the mode orthogonality. Although it was not explicitly spelled out in [11], it was assumed that the perturbations on fibers are mostly anti-symmetric. In this case, mode coupling dominates between LP0n and its anti-symmetric counterpart LP1n modes in optical fibers.

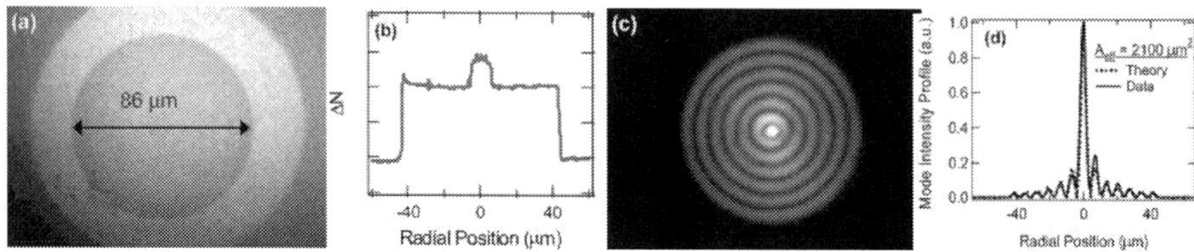

Figure 19. Characteristics of an HOM fiber. The horizontal and vertical scales of the images are identical. (a) Near-field image of a fiber facet, showing the 86μm inner cladding. (b) Refractive-index profile of the HOM fiber, with a core similar to an SMF with an 86μm inner cladding and a down-doped outer trench. (c) Near-field image of the LP_{07} mode at 1600 nm after 12m propagation with a 4.5cm radius bend. (d) Intensity line scan of (c) and theoretical profile: A_{eff}=~2100μm² [11].

The effective mode index difference for LP_{01}, LP_{04}, LP_{05}, LP_{06} and LP_{07} modes with their nearest neighbor anti-symmetric modes are shown versus effective mode area in Figure 18 [11]. The near-field images of the measured HOM modes are shown at the top of in Figure 18 and in Figure 19. It is clear that the effective mode index difference is significantly larger for the LP_{04}-LP_{07} modes than for the LP_{01} mode with the same effective mode area. The effective mode index difference also increases slightly for higher order LP_{0n} modes with the same effective mode area. For the same fiber, the effective mode area is actually larger for lower-mode-order LP_{0n} modes. The schematic of the proposed system is shown in Figure 20(a), where two identical LPGs are required for mode conversion from the LP_{01} mode to the LP_{07} mode and back to the LP_{01} mode. Robust propagation of the LP_{07} mode with an effective mode area of 2070μm² was demonstrated at 1600nm with the arrangement shown in Figure 19(c). The performance of the mode converter with high efficiency over broad bandwidth is shown in Figure 20(b). It was found that the LP_{07} mode suffered negligible bend loss at coil diameters down to 12cm. It was also found that modal stability increases with mode order. In a second paper [12], the effective mode areas of LP_{07} mode was simulated at various bend radii, showing stronger bend resistance than both LP_{03} and LP_{01} modes.

Recently, an erbium-doped HOM amplifier was demonstrated [41]. A small inner core was designed to have a LP_{01} mode MFD of 9μm, allowing for effective excitation of the LP_{01} mode when the HOM fiber was spliced to a single-mode fiber. Higher-order modes expanded to occupy the outer core. The LP_{010} mode had an effective mode area of 2700μm². Both the inner and the outer core were doped with erbium with absorption of ~30dB/m at 1530nm. The amplifier was both seeded at 1564nm and pumped at 1480nm in the LP_{010} mode. The pump was a Raman fiber laser.

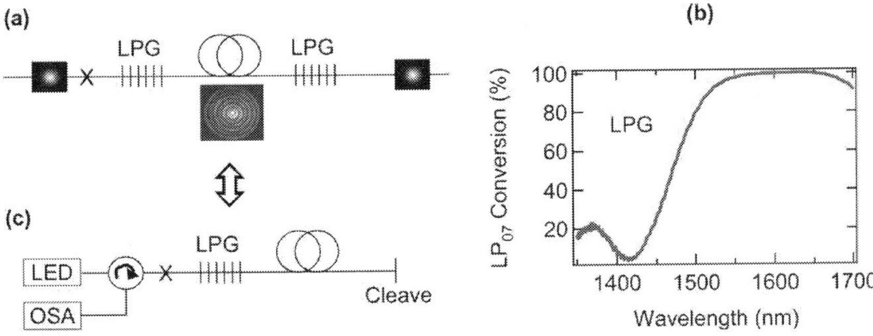

Figure 20. (a) Device schematic: light is coupled into and out of HOM with LPGs whose conversion efficiency being shown in (b). LPG with broadband coupling with efficiency >99% over 94 nm and with peak coupling efficiency as high as 99.93%. (c) Alternative schematic for characterizing HOM fibers: the cleave serves to fold the device propagation path so that the single LPG acts as both the input and the output LPG. X, splice; OSA, optical spectrum analyzer [11].

The difference in effective index between nearest neighbor anti-symmetric modes at $\lambda=1564$ nm as a function of their effective area is plotted as points in Figure 21(a) for the LP_{01} through LP_{010} modes in the fiber. The LP_{02} and LP_{03} modes have a large A_{eff}, but small mode spacing. As the mode order increases, mode spacing increases too, while A_{eff} decreases. Figure 21(b) shows the calculated intensity profiles at $\lambda=1564$nm for the LP_{01} and LP_{010} mode.

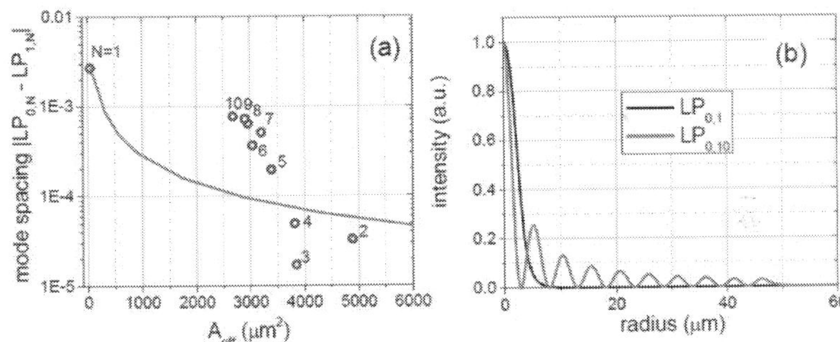

Figure 21. (a) Mode spacing as a function of effective area for the LP_{0n} modes in the HOM fiber (points) compared to a conventional LP_{01} step-index fiber with V=5 (solid curve). (b) Intensity profiles of the LP_{01} and LP_{010} modes. These calculations were done at a wavelength of 1564 nm.

A narrow line width, external cavity laser was amplified to 50mW and combined with the high-power, single mode Raman fiber laser at 1480 nm in a single mode pump/signal combiner. The output of the pump/signal combiner was fusion-spliced to the HOM fiber. The length of the amplifier fiber after the LPG was 2.68 m. The measured slope efficiency at 1564nm was 43.2%. Over 20dB of gain was demonstrated by the amplifier.

5. Chrially-coupled core fibers

A *Chirally-coupled-core* (CCC) fiber was first reported in 2007 [9]. The fiber had a large central core and a smaller side core wound around the central core in a helical fashion (Figure 22). The preform had two parallel cores and the fiber was spun during the draw to form the helical side core. In this first report, the central core had a diameter of 35μm and a NA of 0.07. The side core had a diameter of 12μm and NA of 0.09. Edge-to-edge core separation was 2μm. The helical pitch was 6.2mm. The fiber was measured to be single-mode at 1550nm over a short length of 25cm. It was multimode below 1500nm. The simulation predicted LP_{01} mode loss to be 0.3dB/m and all HOM loss to be >130dB/m for λ>1550nm. The fiber was also confirmed to be polarization-maintaining.

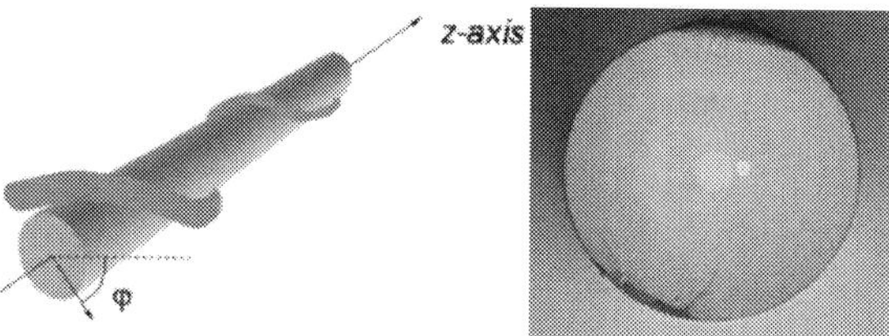

Figure 22. Structure of Chirally-coupled-core fiber [9].

The propagation of modes in the central core was affected by the coupling of modes between the central and side cores. The fiber was designed to operate where there was no fundamental mode coupling with modes in the side core. Higher-order modes in the central core were, however, coupled with the side core modes at the operating wavelength. The modes in the side core had high loss due to the tight bend from the helical arrangement. This led to high loss for the higher-order modes in the central core which were coupled to modes in the side core.

An ytterbium-doped CCC fiber was demonstrated in a subsequent paper [10]. The ytterbium-doped central core had a 33μm diameter and 0.06 NA. The undoped side core had a 16μm diameter and 0.1 NA. The helical pitch was 7.4mm and the edge-to-edge core separation was 4μm. The low index coating provided a pump NA of 0.47. The pump guide had a 250μm diameter. The measured pump absorption was 2dB/m at 915nm. The fiber demonstrated 75% slope efficiency at 1066nm in a laser configuration.

In a more recent paper [42], a more detailed theoretical analysis of quasi-phase-matching (QPM) assisted by spin and orbital angular momentum was given. For two LP modes LP_{l1m1} and LP_{l2m2}, QPM is achieved when

$$\beta_{l1m1} - \beta_{l2m2}\sqrt{1 + K^2 R^2} - \Delta mK = 0 \qquad (2)$$

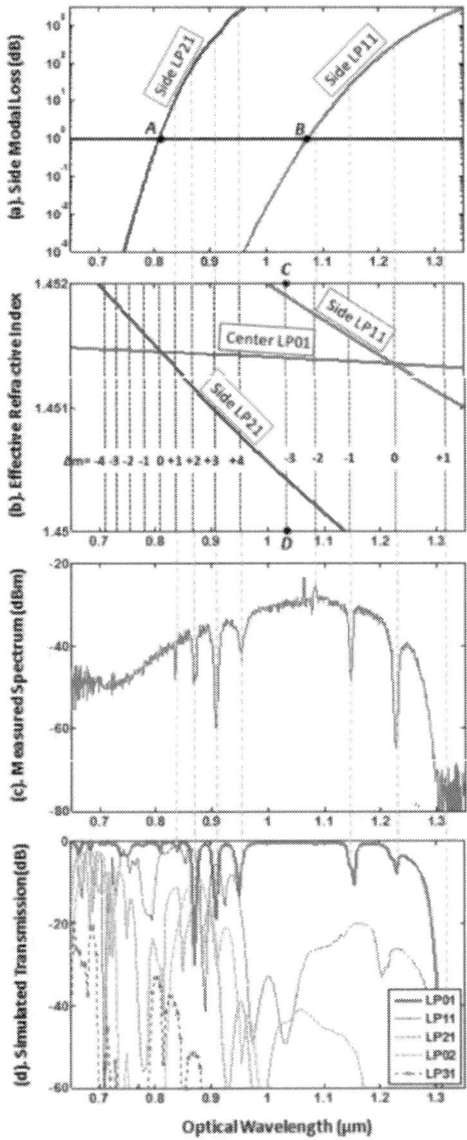

Figure 23. Calculation and measurement of quasi-phase-matching (QPM) for 1.5m-long CCC fiber sample. (a) Calculated loss for side-core LP_{11} and LP_{21} modes as a function of wavelength (top). (b) Calculated refractive indices of inter-
acting modes and calculated QPM resonance positions. (c) Transmission of central core. (d) Simulated transmission of
central core.

where β_{l1m1} and β_{l1m2} are the respective propagation constants of the two modes; $K=2\pi/\Lambda$; and Λ is helical pitch. $\Delta m=\Delta l+\Delta s$; $\Delta l=\pm l1\pm l2$; $\Delta s=-2,-1, 0, 1, 2$. The loss versus wavelength for the LP_{11} and LP_{21} modes in the side core is plotted in Figure 23(a) showing high loss for these modes in certain wavelength regimes. The mode indices of the LP_{11} and LP_{21} modes of the side core and LP_{01} mode of the central core are plotted in Figure 23(b). It can be seen clearly that the LP_{11} mode and LP_{21} modes of the side core naturally phase-match to the LP_{01} mode of the central core at ~1.22μm and ~0.81μm respectively. The QPM by angular momentum from the helical side core extends the phase matching to a number of other wavelengths determined by equation 2. These phase matching wavelengths are plotted in Figure 23(b) as red dotted vertical lines for the LP_{11} mode of the side core and the LP_{01} mode of central core coupling and blue dotted vertical lines for the LP_{21} mode of side core and LP_{01} mode of central core coupling. The measured transmission of the central core is plotted in Figure 23(c). The analysis of the CCC fiber can be simplified significantly in a helical coordination. The Maxwell equations keep the same form in the new coordination system, but the tenor form of permittivity and permeability needs to be transformed [42]. The resulting tenor does not have any z-dependence, which significantly simplifies the analysis. The simulated loss of various modes is shown in Figure 23(d), demonstrating that the narrow peaks in the transmission arise from coupling between the LP_{01} mode in the central core and the LP_{11} and LP_{21} modes in the side core.

The high loss of the LP_{01} mode in the central core at $\lambda>1.3$μm was not explained in the paper. It may be due to the angular momentum assisted coupling between LP_{01} mode in the central core and LP_{01} mode in the side core. This long wavelength cut-off has been used for the suppression of stimulated Raman scattering in fibers [43].

The higher loss for the higher-order-modes in the central core between 1-1.1μm in Figure 23(d) was not clearly explained either. For the LP_{11} mode in the central core, one possible reason for its high loss is coupling to LP_{21} mode of the side core through angular momentum-assisted QPM. Recently, a 60μm core CCC fiber was also demonstrated [44].

6. Photonic bandgap fibers

The guidance of light in a defect in a photonic bandgap fiber is based on fundamentally different principles compared to conventional optical fibers. In a conventional optical fiber, the core has a higher refractive index than that of the cladding. Light is guided through total internal reflection at the core-and-cladding boundary. In a photonic bandgap fiber, the photonic crystal cladding is designed to have photonic bandgaps, where there is an absence of modes propagating in the direction parallel to the fiber axis. Once a defect core is created within such a photonic crystal cladding, light is trapped to propagate only in modes of the defect core as it is forbidden to propagate in the cladding. This leads to significantly lower waveguide loss for the modes in the defect core within the cladding photonic bandgaps. This waveguide loss becomes zero when the photonic crystal cladding is infinite, but there is always a finite loss for practical fibers with finite cladding. This waveguide loss can, however, be significantly lowered by increasing the number of cladding layers.

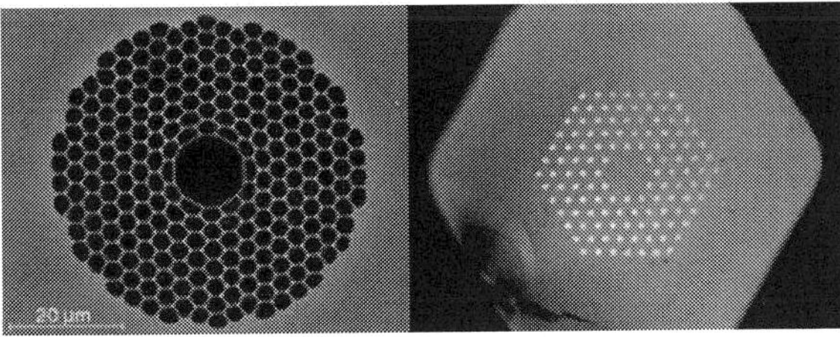

Figure 24. Images of a hollow-core photonic bandgap fiber (left) and an all-solid photonic bandgap fiber (right).

One unique feature of photonic bandgap fibers is that the defect core always has a lower refractive index than that of the higher refractive index material in the cladding. This leads to the possibility of hollow-core photonic bandgap fibers (see Figure 24, left), where light is guided mostly in the air in the hollow-core of the fiber. These fibers have extremely low nonlinearities and are well suited for high power laser delivery. The second type of photonic bandgap fibers are made entirely of glass (see Figure 24, right). These all-solid photonic bandgap fibers have a cladding which consists of a background glass and nodes of slightly higher refractive index. The much lower refractive index contrast (typically just a few percent) in the photonic crystal cladding of the all-solid photonic bandgap fibers still allows photonic bandgaps for paraxial propagation. Another important feature of photonic bandgap fibers is that their transmission is highly wavelength-dependent, i.e. low loss is possible only within the photonic bandgaps of the photonic crystal cladding. This distributive spectral filtering along a fiber can be very useful for range of potential applications including use in fiber lasers at low gain regimes to minimize gain competition, the suppression of stimulated Raman scattering, etc.

Birks et al. clearly explained the origin of the modes in the photonic crystal cladding of an all-solid photonic bandgap fiber in [45]. The density of states (DOS) of modes in the photonic crystal cladding is represented by the shades of grey in Figure 25 [45]. The refractive index of the background glass is represented by the black horizontal line. It is the cut-off line for the modes guided in the cladding nodes. The bandgap regime is represented by the red area. The fundamental mode of the defect core is illustrated by the yellow lines. The modes in the photonic crystal cladding clearly originate from the guided modes of the nodes, which are also labeled at the top of the figure. Above the background index, the modes form relatively narrow bands. Below the background index, the bands of modes significantly broaden, due to strong coupling among nodes as a result of light becoming more spread into the background glass below cut-off. The guidance property can be easily understood once it is recognized that there is a simple analogue to conventional optical fibers [16]. The core index is simply the index of the background glass and the equivalent cladding index is the upper boundary of the photonic bandgap. The effective index of the modes in the defect core (see yellow lines in Figure 25) falls between the core and cladding indices as in conventional optical fibers.

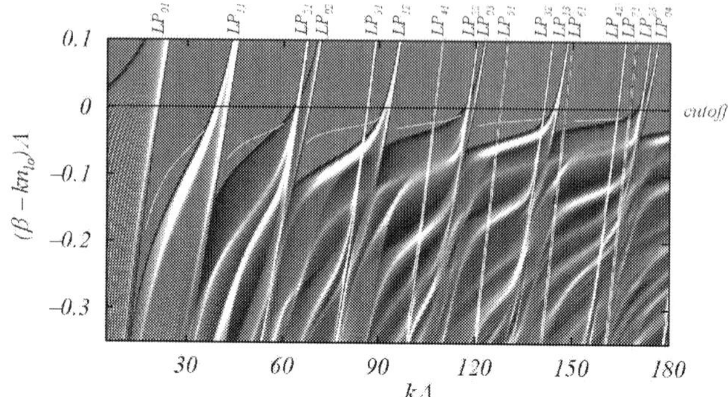

Figure 25. Plots of band structure for an example bandgap fiber. The bandgaps are shown in red. The node modes from which the bands arise are labeled along the top. Density of states (DOS) calculated using the plane-wave method, with light grey corresponding to high DOS. The yellow curve is the "fundamental" core-guided mode [45].

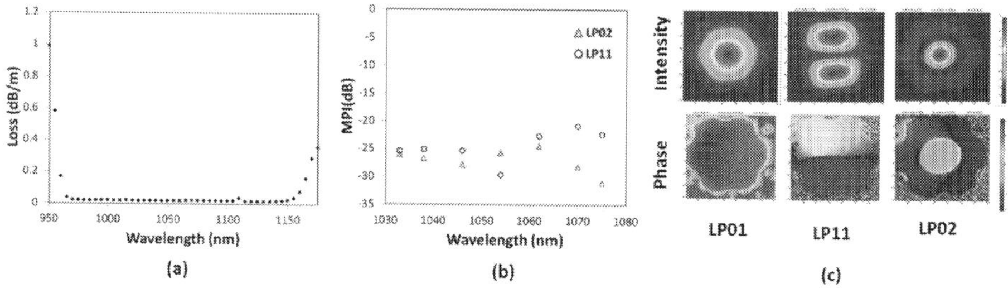

Figure 26. (a) Transmission of the 55μm core all-solid photonic bandgap fiber, (b) measured HOM contents with a S^2 method in a 5m fiber coiled at 70cm diameter, and (c) measured mode images [51].

One interesting application of ytterbium-doped all-solid photonic bandgap fibers is for the lasing of ytterbium at the long wavelength regime of 1150-1200nm [47]. It is normally difficult to lase in ytterbium-doped fibers at these extremely long wavelengths due to strong gain competition from the short wavelength regime of 1030-1070nm. The distributive spectral filtering in a photonic bandgap fiber can be used to minimize gain at short wavelengths, leading to efficient high power lasers in the extremely long wavelength regime.

In all-solid photonic bandgap fibers, a mode is guided only when it falls within the photonic bandgap of the cladding lattice. Guidance can therefore be highly mode-dependent. This provides great potential for creating designs that support only the fundamental mode, i.e. selective mode guidance. Mode area scaling to 20μm mode field diameter using all-solid photonic bandgap fibers was reported in [48]. Recently, all-solid photonic bandgap fibers with up to ~700μm² effective mode area have been demonstrated operating in the first bandgap [49, 50]. More recently, a fiber with a core diameter of ~55μm and an effective mode area of

920μm² at coil diameter of 50cm was demonstrated (see right figure in Figure 24). The fiber operates in the third bandgap with transmission shown in Figure 24(6). At 70cm coil diameter, a 5m fiber showed HOM contents below 25dB (see Figure 26(b) and (c)). At the design coil diameter of 50cm, HOM content was below-30dB [51].

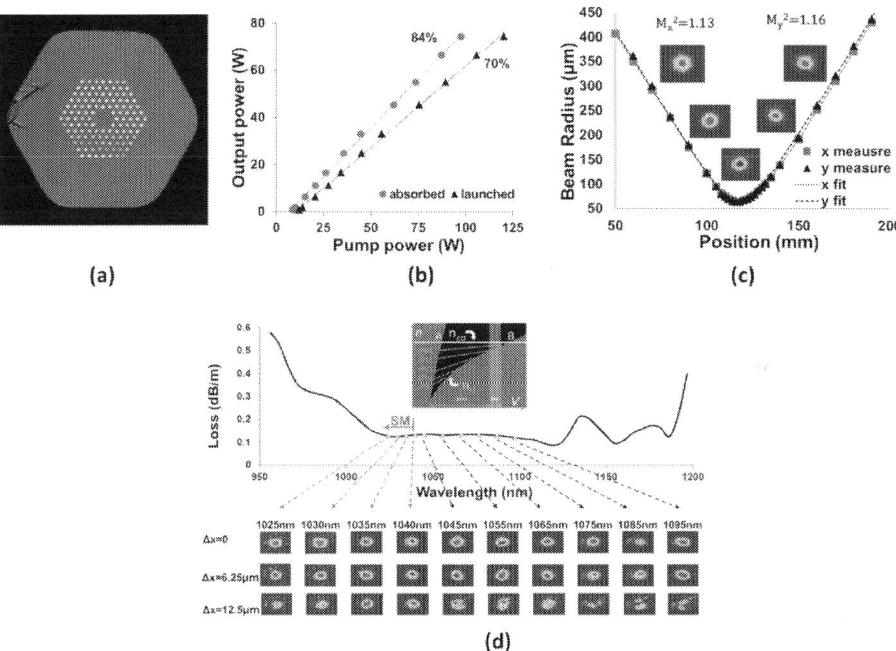

Figure 27. (a) Ytterbium-doped ~50μm core all-solid photonic bandgap fiber, (b) Laser efficiency, (c) M² measurement and (d) mode at various wavelengths across the bandgap [52].

Recently, an ytterbium-doped all-solid photonic bandgap fiber with ~50μm core diameter has also been demonstrated (Figure 27(a)) [52]. The fiber demonstrated high efficiency and excellent mode quality (see Figure 27(b) and (c)). The fiber also demonstrated robust single-mode behavior near the short wavelength edge of the bandgap by monitoring the output when moving away from the optimum launch condition (see Figure 27(d)). This is exactly expected from the dispersive nature of the photonic bandgap cladding.

7. Conclusions

In this chapter, we have briefly introduced a number of emerging fiber technologies for mode area scaling for fiber lasers. These new technologies are critical for the further power scaling of fiber lasers. The basic principles of these technologies were introduced and the latest developments were discussed. This is still a very active area of research. With further development, there is great potential to significantly improve throughput and expand the capabilities of fiber lasers for use in manufacturing, as well as to meet future defense needs.

Author details

Liang Dong[*]

Department of Electrical and Computer Engineering, University of Clemson, Clemson, SC, USA

References

[1] D. Taverner, D.J. Richardson, L. Dong, J.E. Caplen, K. Williams, R.V. Penty, "158µJ pulses from a single transverse mode, large mode-area EDFA", Optics Letters, 22, 378-380 (1997).

[2] H.L. Offerhaus, N.G. Broderick, D.J. Richardson, R. Sammut, J. Caplen and L. Dong, "High-energy single-transverse-mode Q-switched fiber laser based on a multimode large-mode-area erbium-doped fiber", Optics Letters, 23, 1683-1685 (1998).

[3] C. Knight, T. A. Birks, P. St. J. Russell, and D. M. Atkin, "All-silica single-mode optical fiber with photonic crystal cladding," Opt. Lett. 21, 1547-1549 (1996).

[4] T.A. Birks, J.C. Knight and P.St.J. Russell, "Endlessly single-mode photonic crystal fiber," Opt Lett. 22 961-963 (1997).

[5] J.C. Knight, T.A. Birks, R.F. Cregan, P.St.J. Russell and J.P. de Sandro, "Large mode area photonic crystal fibre," Elect. Lett. 34, 1347-1348 (1998).

[6] C.D. Brooks and F. Teodoro, "Multi-MW peak-power, single-transverse-mode operation of a 100µm core diameter, Yb-doped rodlike photonic crystal fiber amplifier," Appl. Phys. 89, 111119 (2006).

[7] W.S. Wong, X. Peng, J.M. McLaughlin and L. Dong, "Breaking the limit of maximum effective area for robust single-mode propagation in optical fibers", Optics Letters, 30, 2855-2857 (2005).

[8] L. Dong, X. Peng and J. Li, "Leakage channel optical fibers with large effective area", Journal of Optical Society of America B, 24, 1689-1697 (2007).

[9] C.H. Liu, G. Chang, N. Litchinitser, A. Galvanauskas, D. Guertin, N. Jacobson and K. Tankala, "Effective single-mode chirally-coupled core fiber," Advanced Solid State Photonics, paper ME2, 2007.

[10] A. Galvanauskas, M.C. Swan, and C.H. Liu, "Effectively-Single-Mode Large Core Passive and Active Fibers with Chirally-Coupled-Core structures," Conference on Lasers and Electro-Optics, paper CMB1, 2008.

[11] S. Ramachandran, J. W. Nicholson, S. Ghalmi, M. F. Yan, P. Wisk, E. Monberg, and F. V. Dimarcello, "Light propagation with ultralarge modal areas in optical fibers," Opt. Lett. 31, 1797-1799 (2006).

[12] J.M. Fini and S. Ramachandran, "Natural bend-distortion immunity of higher-order-mode large-mode-area fibers" Opt. Lett. 32, 748-750 (2007).

[13] M. Koshiba and K. Saitoh, "Applicability of classical optical fiber theories to holey fibers," Opt. Lett. 29, 1739-1740 (2004).

[14] K. Saitoh, Y. Tsuchida, and M. Koshiba, "Endlessly single-mode holey fibers: the influence of core design," Opt. Express 13, 10833-10839 (2005).

[15] B. T. Kuhlmey, R. C. Mcphedran, and C. M. de sterke, "Modal cutoff in microstructured optical fiber," Opt. Lett. 27, 1684-1686 (2002).

[16] J. Zhou, K. Tajima, K. Nakajima, K. Kurokawa, C. Fukai, T. Matsui, and I. Sankawa, "Progress on low loss photonic crystal fibers," Optical Fiber Technology 11 101–110 (2005).

[17] T. P. White, B. T. Kulmey, R. C. McPhedran, D. Maystre, G. Renversez, C. Martijn de Sterke, and L. C. Botten, "Multipole method for microstructured optical fibers. I. Formulation," J. Opt. Soc. Am. B 19, 2322-2330 (2002).

[18] B.T. Kuhlmey, T.P. White, G. Renversez, D. Maystre, L.C. Botten, C. M. de Sterke, and R. C. McPhedran, "Multipole method for microstructured optical fibers. II. Implementation and results," J. Opt. Soc. Am. B 19, 2331-2340 (2002).

[19] W.J. Wadsworth, J.C. Knight and P.St.J. Russell, "Large mode area photonic crystal fiber laser," Conference on Lasers and Electro-Optics, CWC1, 2001.

[20] W.J. Wadsworth, R.M. Percival, G. Bouwmans, J.C. Knight and P.St.J. Russell, "High power air-clad area photonic crystal fiber laser," Opt. Express 11, 48-53 (2003).

[21] D.J. DiGiovanni and R.S. Windeler: "Article comprising an airclad optical fiber," US patent 5907652 (1999).

[22] V. A. Kozlov, J. Hernández-Cordero, R. L. Shubochkin, A. L. G. Carter, and T. F. Morse, "Silica–Air Double-Clad Optical Fiber," IEEE Photonics Technology Letters 12, 1007-1009 (2000).

[23] J.K. Sahu, C.C. Renaud, K. Furusawa, R. Selvas, J.A. Alvarez-Chavez, D.J. Richardson and J. Nilsson, "Jacketed air-clad cladding pumped ytterbium-doped fibre laser with wide turning range," Elect. Lett. 37, 1116-1117 (2001).

[24] K. Furusawa, A. Malinowski, J. H. V. Price, T. M. Monro, J. K. Sahu, J. Nilsson and D. J. Richardson, "Cladding pumped Ytterbium-doped fiber laser with holey inner and outer cladding," Opt. Express 9, 714-720 (2001).

[25] W.J. Wadsworth, R.M. Percival, G. Bouwmans, J.C. Knight, T.A. Birks, T.D. Hedley, and P.St.J. Russell, "Very high numerical aperture fibers," IEEE Photonic Technology Letters 16, 843-845 (2004).

[26] J. Limpert, T. Schreiber, S. Nolte, H. Zellmer, A. Tünnermann, R. Iliew, F. Lederer, J. Broeng, G. Vienne, A. Petersson, and C. Jakobsen, "High-power air-clad large-mode-area photonic crystal fiber laser," Opt. Express 11, 818-823 (2003).

[27] J. Limpert, T. Schreiber, A. Liem, S. Nolte, H. Zellmer, T. Peschel, V. Guyenot, and A. Tünnermann, "Thermo-optical properties of air-clad photonic crystal fiber lasers in high power operation," Opt. Express 11, 2982-2990 (2003).

[28] J. Limpert, A. Liem, M. Reich, T. Schreiber, S. Nolte, H. Zellmer, A. Tünnermann, J. Broeng, A. Petersson, and C. Jakobsen, "Low-nonlinearity single-transverse-mode ytterbium-doped photonic crystal fiber amplifier," Opt. Express 12, 1313-1319 (2004).

[29] J. Limpert, N. Deguil-Robin, I. Manek-Hönninger, F. Salin, F. Röser, A. Liem, T. Schreiber, S. Nolte, H. Zellmer, A. Tünnermann, J. Broeng, A. Petersson, and C. Jakobsen, "High-power rod-type photonic crystal fiber laser," Opt. Express 13, 1055-1058 (2005).

[30] J. Limpert, O. Schmidt, J. Rothhardt, F. Röser, T. Schreiber, and A. Tünnermann, S. Ermeneux, P. Yvernault, and F. Salin, "Extended single-mode photonic crystal fiber lasers," Opt. Express 14, 2715-2720 (2006).

[31] T. Schreiber, F. Röser, O. Schmidt, J. Limpert, R. Iliew, F. Lederer, A. Petersson, C. Jacobsen, K. P. Hansen, J. Broeng, and A. Tünnermann, "Stress-induced single-polarization single-transverse mode photonic crystal fiber with low nonlinearity," Opt. Express 13, 7622-7630 (2005).

[32] O. Schmidt, J. Rothhardt, T. Eidam, F. Röser, J. Limpert, A. Tünnermann, K.P. Hansen, C. Jakobsen, and J. Broeng, "Single-polarization ultra-large-mode-area Yb-doped photonic crystal fiber," Opt. Express 16, 3918-3923 (2008).

[33] L. Dong, J. Li and X. Peng, "Bend resistant fundamental mode operation in ytterbium-doped leakage channel fibers with effective areas up to $3160\mu m^2$", Optics Express, 14, 11512-11519 (2006).

[34] X. Peng and L. Dong, "Fundamental mode operation in polarization-maintaining ytterbium-doped fiber with an effective area of $1400\mu m^2$", Optics Letters, 32, 358-360 (2006).

[35] L. Dong, T.W. Wu, H.A. McKay, L. Fu, J. Li and H.G. Winful, " All-glass large core leakage channel fibers, " IEEE Journal of Selected Topics in Quantum Electronics, 15, invited paper, 47-53(2009).

[36] L. Dong, H.A. Mckay, A. Marcinkevicius, L. Fu, J. Li, B.K. Thomas, and M.E. Fermann, " Extending Effective Area of Fundamental Mode in Optical Fibers, " IEEE Journal of Lightwave Technology, 27, invited paper, 1565-1570(2009).

[37] L. Dong, H.A. Mckay, L. Fu, M. Ohta, A. Marcinkevicius, S. Suzuki, and M.E. Fermann, "Ytterbium-doped all glass leakage channel fibers with highly fluorine-doped silica pump cladding, " Optical Express, 17, 8962-8969(2009).

[38] G. Gu, F. Kong, T.W. Hawkins, P. Foy, K. Wei, B. Samson, and L. Dong, "Impact of fiber outer boundaries on leaky mode losses in leakage channel fibers," Optics Express 21, 24039-24048 (2013).

[39] J. W. Nicholson, A. D. Yablon, S. Ramachandran, and S. Ghalmi, "Spatially and spectrally resolved imaging of modal contents in large-mode-area fibers," Opt. Express 16, 7233-7243(2008).

[40] F. Kong, G. Gu, T.W. Hawkins, J. Parsons, M. Jones, C. Dunn, M.T. Kalichevsky-Dong, K. Wei, B. Samson, and L. Dong, "Flat-top mode from a 50 μm-core Yb-doped leakage channel fiber," Optics Express 21, 32371-32376 (2013).

[41] J.W. Nicholson, J.M. Fini, A.M. DeSantolo, E. Monberg, F. DiMarcello, J. Fleming, C. Headley, D.J. DiGiovanni, S. Ghalmi, and S. Ramachandran, "A higher-order-mode Erbium-doped-fiber amplifier," Opt. Express 18, 17651-17657 (2010).

[42] X. Ma, C.H. Liu, G. Chang, and A. Galvanauskas, "Angular-momentum coupled optical waves in chirally-coupled-core fibers," Opt. Express 19, 26515-26528 (2011).

[43] X. Ma, I,N Hu, and A. Galvanauskas, "Propagation-length independent SRS threshold in chirally-coupled-core fibers," Opt. Express 19, 22575-22581 (2011).

[44] X. Ma, A. Kaplan, and A. Galvanauskas, "Experimental characterization of robust single-mode operation of 50μm and 60μm core chirally coupled core optical fibers," PhotonicsWest, paper 8237-59, 2012.

[45] T. A. Birks, G. J. Pearce, D. M. Bird, "Approximate band structure calculation for photonic bandgap fibers," Opt. Express 14, 9483-9490 (2006).

[46] M.J.F. Digonnet, H.K. Kim, G.S. Kino, and S. Fan, "Understanding Air-Core Photonic-Bandgap Fibers: Analogy to Conventional Fibers," Journal of Lightwave Technology 23, 4146-4177 (2005).

[47] A. Shirakawa, H. Maruyama, K. Ueda, C. B. Olausson, J. K. Lyngsø, and J. Broeng, "High power Yb doped photonic bandgap fiber amplifier at 1150 1200 nm," Opt. Express 17(2), 447–454 (2009).

[48] O. N. Egorova, S. L. Semjonov, A. F. Kosolapov, A. N. Denisov, A. D. Pryamikov, D. A. Gaponov, A. S. Biriukov, E. M. Dianov, M. Y. Salganskii, V. F. Khopin, M. V.

Yashkov, A. N. Gurianov, and D. V. Kuksenkov, "Single-mode all-silica photonic bandgap fiber with 20-micron mode-field diameter," Opt. Express 16, 11735–11740 (2008).

[49] M. Kashiwagi, K. Saitoh, K. Takenaga, S. Tanigawa, S. Matsuo, and M. Fujimaki, "Low bending loss and effectively single-mode all-solid photonic bandgap fiber with an effective area of 650µm².," Opt. Lett. 37, 1292–1294 (2012).

[50] M. Kashiwagi, K. Saitoh, K. Takenaga, S. Tanigawa, S. Matsuo, and M. Fujimaki, "Effectively single-mode allsolid photonic bandgap fiber with large effective area and low bending loss for compact high-power all-fiber lasers," Opt. Express 20, 15061–15070 (2012).

[51] F. Kong, K. Saitoh, D. Mcclane,T. Hawkins, P. Foy, G.C. Gu, and L. Dong, "Mode Area Scaling with All-solid Photonic Bandgap Fibers,", Optics Express 20, 26363-26372 (2012).

[52] G. Gu, F. Kong, T.W. Hawkins, J. Parsons, M. Jones, C. Dunn, M.T. Kalichevsky-Dong, K. Saitoh, and L. Dong, "Ytterbium-doped large-mode-area all-solid photonic bandgap fiber lasers," Optics Express 22, 13962-13968(2014).

Chapter 8

The Influence of Nonlinear Effects Upon Oscillation Regimes of Erbium-Doped Fiber Lasers

Y.O. Barmenkov and A.V. Kir'yanov

Additional information is available at the end of the chapter

http://dx.doi.org/10.5772/59146

1. Introduction

Erbium-doped fiber lasers (EDFLs) are contemporary sources of coherent radiation, attractive for numerous applications requiring both continuous-wave (CW) and pulsed operations, among which telecommunications in a wide wavelength range covering C and L bands ought to be emphasized. Pulsed operation, presenting big interest for practice, is attained in EDFLs by means of standard active and passive Q-switching and mode-locking techniques, capable of enforcing a laser to generate short pulses with durations ranged from hundreds fs to hundreds ns [1]. In the meantime, transients to CW lasing and a laser's relaxation frequency are also of close attention, e.g. when targeting the sensor applications [2, 3]. The detailed knowledge of the processes involved in Erbium-doped fibers (EDFs) to be used, when pumped, as a laser or amplifying medium in each of the referred regimes cannot be overestimated. The present work is a review of some of the most featuring nonlinear-optical effects that affect the oscillation regimes of EDFLs.

In spite of certain advantages (availability at the market, low cost, and easiness of handling), EDFLs demonstrate considerably lower efficiency as compared to the lasers based on Ytterbium-doped fibers. The basic cause is the multi-level energy scheme of Er^{3+} ions, which makes unavoidable absorption of photons at both the pump and laser wavelengths by the ions being at upper Er^{3+} levels, i.e. the "excited-state absorption" (ESA) [4, 5], including the state where 1.5-μm laser transition stems from. In other words, ESA presents a kind of "up-conversion" (UC) loss inherent to EDF; see section 2.

Another, nonlinear in nature, source of UC loss in EDFs and EDFLs on their base relates to so-called "collective" (concentration-related) effects arising in Er^{3+} ion pairs and in more complicate clusters (percentage of which grows with Er^{3+} concentration) [6]; see section 3. Appearance of the latter phenomenon is also associated with the multi-level energy scheme of Er^{3+} ions.

Eventually, the cases of low-and heavily-doped EDFs are to be carefully segregated and properly addressed if one looks for optimization of an EDFL.

One more kind of optical nonlinearity that arises in actively Q-switched (AQS) EDFLs is stimulated Brillouin scattering (SBS). Depending of the EDF length and Q-cell's modulation frequency EDFLs may operate in one of the two QS regimes: either in the "conventional" QS (CQS) one in which QS pulses are composed of several sub-pulses separated by a photon's round-trip time (with negligible pulse jitter) [7, 8], or in the essentially stochastic SBS-induced QS (SBS-QS) one where pulse amplitude is bigger by an order of value as compared with CQS but where pulses suffer severe jitter [9]. The results of an experimental study of the basins the CQS and SBS-QS regimes belong to and the basic spectral features of these regimes are discussed in section 4.

In section 5, the review's conclusions are formulated.

2. ESA in EDF at the pump and laser wavelengths

In this section, we report a study aiming at determination of the ESA's spectral dependence in EDF, covering the most important for applications spectral range, 1.48...1.59 μm, and at ~978 nm, the wavelength that usually is used for pumping EDFs by laser diodes (LDs). In the experiments discussed hereafter a low-doped silica EDF (*Thorlabs M5-970-125*, ~300 ppm of Er^{3+} concentration, Al-Ge-silicate host composition, *NA*=0.24, cut-off wavelength – 0.94 μm) was chosen to avoid possible effect of UC (observed in heavily-doped EDFs; see section 3). [This fiber belongs to M-series (fabricated through the modified chemical vapor deposition (MCVD) process), to be under scope in section 3 where Er^{3+} concentration effects are treated; "M5" signifies that small-signal absorption peaked at 978 nm is 5 dB/m.] The measurements were completed by modeling, a necessary chain in interpretation of the experimental results. The developed modeling also provides an estimate for the so-called ESA parameter, obtained at both pump and laser (signal) wavelengths, and thereafter allows determination of the EDF's net-gain coefficient, "deliberated" from the ESA interfere.

Figure 1 shows the fife level Er^{3+} ion energy scheme upon excitation at the pump ($\lambda_p \approx 980$ nm) and signal ($\lambda_s \approx 1.5$ μm) wavelengths, useful for the discussion presented in this section.

The equations that describe functioning of the Er^{3+} system, in accord to the scheme shown in Figure 1, at excitation at sole or at both excitation wavelengths (λ_s and/or λ_p) in the steady-state are written as follows:

$$\frac{\sigma_{12} I_s}{h\nu_s} N_1 - \frac{\sigma_{21} I_s}{h\nu_s} N_2 - \frac{\sigma_{24} I_s}{h\nu_s} N_2 - \frac{N_2}{\tau_{21}} + \frac{N_3}{\tau_{32}} = 0 \qquad (1)$$

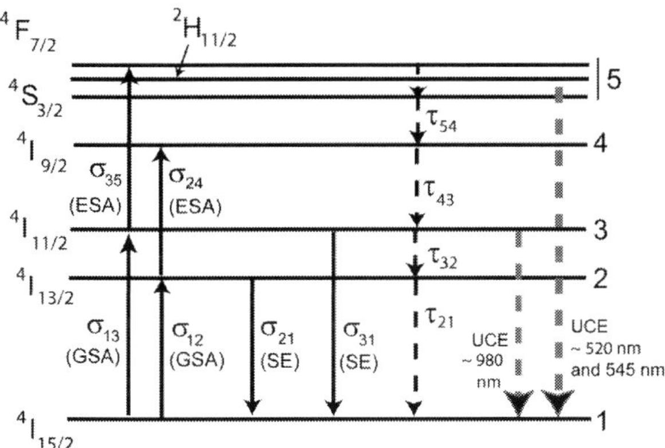

Figure 1. Fife-level scheme of Er³⁺ ion used in modeling. GSA and ESA indicate the ground-state and excited-state absorptions, UCE marks the up-conversion emission transitions (weak but detectable), σ_{ij} and τ_{ij} are, respectively, the cross-sections and decay times for the transitions between the levels i and j. Three closely-spaced energy levels $^4S_{3/2}$, $^2H_{11/2}$, and $^4F_{7/2}$ are regarded as an effective level "5".

$$\frac{\sigma_{13}I_p}{h\nu_p}N_1 - \frac{\sigma_{31}I_p}{h\nu_p}N_3 - \frac{N_3}{\tau_{32}} - \frac{\sigma_{35}I_p}{h\nu_p}N_3 + \frac{N_4}{\tau_{43}} = 0 \tag{2}$$

$$\frac{\sigma_{24}I_s}{h\nu_s}N_2 - \frac{N_4}{\tau_{43}} + \frac{N_5}{\tau_{54}} = 0 \tag{3}$$

$$\frac{\sigma_{35}I_p}{h\nu_p}N_3 - \frac{N_5}{\tau_{54}} = 0 \tag{4}$$

$$N_1 + N_2 + N_3 + N_4 + N_5 = N_0 \tag{5}$$

where N_0 is Er³⁺ concentration in the EDF core, N_i are the populations of the correspondent Er³⁺ levels (i=1 to 5), h is Plank constant, ν_p and ν_s are the frequencies of the pump and signal waves, and I_s and I_p are the pump and signal waves' intensities. The Er³⁺ parameters' values used in modeling are listed in Table 1.

Parameter	Value	Units
Low-signal absorption at 1531 nm (experimental data)	$\alpha_{s0} = 0.016$	cm^{-1}
Low-signal absorption at 977 nm (experimental data)	$\alpha_{p0} = 0.012$	cm^{-1}
Relaxation time for $^4I_{13/2} \rightarrow {}^4I_{15/2}$ transition [10]	$\tau_{21} = 10$	ms
Relaxation time for $^4I_{11/2} \rightarrow {}^4I_{13/2}$ transition[11]	$\tau_{32} = 5.2$	µs
Relaxation time for $^4I_{9/2} \rightarrow {}^4I_{11/2}$ transition [12]	$\tau_{43} = 5$	ns
Relaxation time for $(^4F_{7/2}/^2H_{11/2}/^4S_{3/2}) \rightarrow {}^4I_{9/2}$ transition [13]	$\tau_{54} = 1$	µs
Cross-section of $^4I_{15/2} \rightarrow {}^4I_{13/2}$ transition @ 1531 nm [14]	$\sigma_{12} = 5.1 \times 10^{-21}$	cm^2
Cross-section of $^4I_{15/2} \rightarrow {}^4I_{11/2}$ transition @ 977 nm [14]	$\sigma_{13} = 1.7 \times 10^{-21}$	cm^2

Table 1. EDF "M5" / Er^{3+} parameters used in modeling

2.1. Spectral features of ESA in low-doped EDF within the 1.48 to 1.59-µm range

The first experiment serves to reveal the existence of the ESA process in the EDF at pumping through $^4I_{15/2} \rightarrow {}^4I_{13/2}$ transition, refer to Figure 1. The UC emission (UCE) spectra were recorded in the wavelengths range near 980 nm ($^4I_{11/2} \rightarrow {}^4I_{15/2}$ transition). The pump source was a 12-mW narrow-line LD (*Anritsu Tunics Plus SC*), tunable through the spectral interval, λ_s=1.48...1.59 µm. Experimentally, light from the LD was launched into the EDF (length, 1 m) through a standard 980-nm / 1550-nm wavelength division multiplexer (WDM) while de-multiplexed backward emission from the EDF was registered by an optical spectrum analyzer (OSA, *Ando AQ6315A*). The recorded UCE spectra, at 10-mW in-fiber power and for four different excitation wavelengths λ_s, are shown in Figure 2(a). It is seen that UCE power depends on λ_s and that the maximal UCE signal spectrally matches the peak wavelength of the Er^{3+} ground-state absorption (GSA) contour. The appearance of the UCE spectra at λ_p ~980 nm is very similar to that of $^4I_{11/2} \rightarrow {}^4I_{15/2}$ emission, thus testifying for effective populating of Er^{3+} $^4I_{11/2}$ state at pumping the EDF at λ_s=1.48...1.59 µm and so for the presence of the ESA process ($^4I_{13/2} \rightarrow {}^4I_{9/2}$), followed by fast non-radiative ($^4I_{9/2} \rightarrow {}^4I_{11/2}$) relaxation. Note that no UCE spectral components in the range below 960 nm were detected.

The second experiment, allowing us to reveal the spectral dependence of ESA, was arranged through measurement of the integral lateral emission power collected from the EDF's lateral surface, using a Si photo-detector directly placed above the fiber. Given Si is sensitive from the visible to near-IR (Si band-gap wavelength is ~1.1 µm), the Si photo-detector (PD) does register the UCE signal (centered at ~980 nm) and does not the spontaneous emission (SE) from level $^4I_{13/2}$ (~1.5 µm); a scattered pump-light component was found to be extremely weak. The results obtained at various pump powers are shown in Figure 2(b) (see curves 1–3). For comparison, we also plot in this figure (by curve 4) the integrated backward UCE power measured using OSA (from a 10-cm EDF sample). It is seen that the spectral dependence of the UCE signal is similar to the one of ESA power on the excitation wavelength. In fact, the shape of the presented dependencies is established by a convolution of the known GSA and yet unknown ESA spectra of Er^{3+} (notice that the latter depends on Er^{3+} ions population inversion).

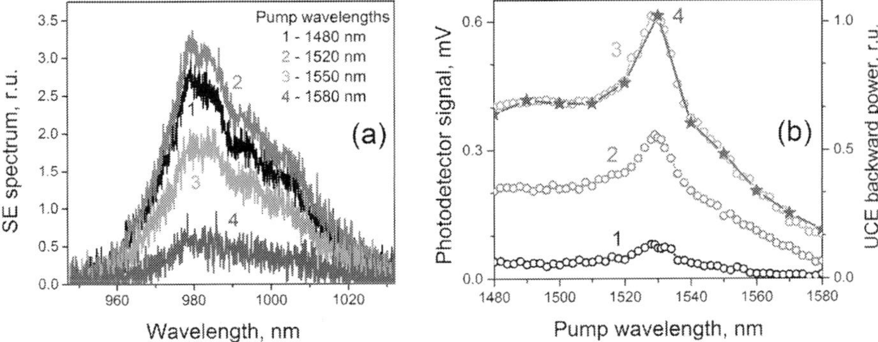

Figure 2. (a) UCE spectra at λ_s ~1.5-µm excitation. (b) Dependencies of PD signal (circles, left scale, curves 1 to 3) and normalized frontal UCE power (stars, right scale, curve 4) on excitation wavelength. Curves 1, 2, and 3 correspond to pump powers of 1, 5, and 10 mW; curve 4 is for 10-mW pump.

Considering the simplified Er^{3+} ion's model and the processes involved at excitation at λ_s ~1.5 µm (see Figure 1 and formulas (1-5)), we obtain simple formula for the normalized population of $^4I_{11/2}$ state, $n_3 = N_3/N_0$ [11]:

$$n_3 = \frac{\varepsilon_s \gamma_s s_s^2}{1 + \xi_s s_s + \varepsilon_s \gamma_s s_s^2},\tag{6}$$

where $\varepsilon_s = \sigma_{24}/\sigma_{12}$ is the ESA parameter at signal wavelength, $\gamma_s = \tau_{42}/\tau_{21} \approx \tau_{32}/\tau_{21}$ ($\tau_{43} \approx 5$ ns is neglected in simulations given its smallness respectively the other decay times), $\xi_s = 1 + \sigma_{21}/\sigma_{12}$ is the spontaneous emission (SE) parameter at the signal wavelength, and $s_s = I_s/I_s^{sat} = P_s/P_s^{sat}$ is the saturation parameter for signal wave (with I_s being the signal wave intensity, $I_s^{sat} = h\nu_s/\sigma_{12}\tau_{21}$ the saturation intensity, $P_s = A_s I_s$, $P_s^{sat} = A_s I_s^{sat}$, P_s and P_s^{sat} the signal wave power and saturation power, respectively, and A_s the area of the Gaussian distribution of the signal wave in the EDF core). Notice that since $\varepsilon_s \gamma_s s_s \ll \xi_s$ ($\gamma_s \approx 7 \times 10^{-4}$ and $\varepsilon_s \leq 0.6$, see below), the term with a second power on s in the denominator of (6) is omitted in further calculus. From formula (6), we find the normalized population n_3 averaged over the fiber core area if the pump wave intensity is taken to obey the Gaussian law, $s_s(r) = s_{s0}\exp[-2(r/w_s)^2]$:

$$\bar{n}_3 = \frac{1}{\pi a^2}\int_0^a n_3(r) \times 2\pi r\, dr = \frac{\varepsilon_s \gamma_s s_{s0}}{2\xi_s}\left(\frac{w_s}{a}\right)^2 \ln\left[\frac{1 + \xi_s s_{s0}}{1 + \xi_s s_{s0}(1 - \Gamma_s)}\right],\tag{7}$$

where s_{s0} is the on-axis saturation parameter that depends on the pump power, P_s, as $P_s = I_s^{sat} s_{s0}(\pi w_s^2/2)$, a is the EDF core radius, w_s is the modal radius of the Gaussian wave, and Γ_s is the mode's to fiber core's radii overlap factor at λ_s. The parameter ξ_s comprises the ESA parameter ε_s, the measured EDF low-signal absorption coefficient $\alpha_{s0} = \Gamma_s \sigma_{12} N_0$, and the full-saturated fiber gain $g_s = \Gamma_s \sigma_{21} N_0 - \Gamma_s \sigma_{24} N_0 = g_{s0} - \varepsilon_s \alpha_{s0}$ (g_{s0} is the EDF net gain at λ_s):

$$\xi_s = 1 + \frac{\sigma_{21}}{\sigma_{12}} = 1 + \frac{g_{s0}}{\alpha_{s0}} = 1 + \frac{g_s}{\alpha_{s0}} + \varepsilon_s. \tag{8}$$

Apparently, UCE power depends on n_3 that in turns depends on ε_s. It is worth wising that the latter can be easily found from the UCE lateral signal, registered using PD (see Figure 2(b)). The results of calculation of the ESA parameters are plotted in Figure 3(a) by filled symbols (left scale), where the ε_s-spectrum is normalized on the GSA peak value (at λ_s=1.531 µm). It is seen that the ESA parameter weakly depends on λ_s within the spectral interval 1.48 µm to 1.55 µm whilst it sharply grows as λ_s approaches ~1.6 µm; this behavior reminiscences the trend reported in [15].

The method to find the absolute values of the ESA parameter ε_s is based on the measurements of the EDF's transmission coefficient $T(\lambda_s, P_s)$, i.e. in function of excitation power, and subsequent comparison of the experimental dependencies with the simulated ones, obtained after integrating, through the fiber length, of the saturation parameter $s_{s0}(z)$, with the latter being obtained from the equation for the excitation wave propagating in the fiber (z is the direction of light propagation) (see [11]):

$$\frac{ds_{s0}}{dz} = -\alpha_{s0}\frac{\varepsilon_s}{\xi_s}\Gamma_s s_{s0} - \alpha_{s0}\frac{\xi_s - \varepsilon_s}{\xi_s^2}\ln\frac{1 + \xi_s s_{s0}}{1 + \xi_s s_{s0}(1 - \Gamma_s)}. \tag{9}$$

Note that the contribution of amplified SE (ASE) to the output spectra was as low as 0.1–1.9% respectively to the overall transmitted power, provided a short (50 cm) EDF piece has been used.

The results of fitting of the experimental EDF transmission curves within the interval 1.48...1.59 µm are shown in Figure 3(a) by asterisks (see right scale). It is evident that these results, as the ones regarding the UCE experiments, almost coincide. The solid curve shown in this figure is the best fit of the experimental data by the polynomial regression.

The knowledge of the ε_s-spectrum for the 1.48...1.59 µm band allows one to find the spectral dependence of the net-gain coefficient g_{s0} in the EDF, which is the real gain in the fiber, in contrast to the gain coefficient g_s, the quantity commonly but uncritically dealt with in experiments and simulations with EDF and EDFL[13]: In fact g_{s0} is diminished by the ESA parameter's value. Figure 3(b) demonstrates the spectral dependences of the three coefficients (g_{s0}, g_s, and α_{s0}), where the spectra for g_s and α_{s0} were measured using standard methods (see

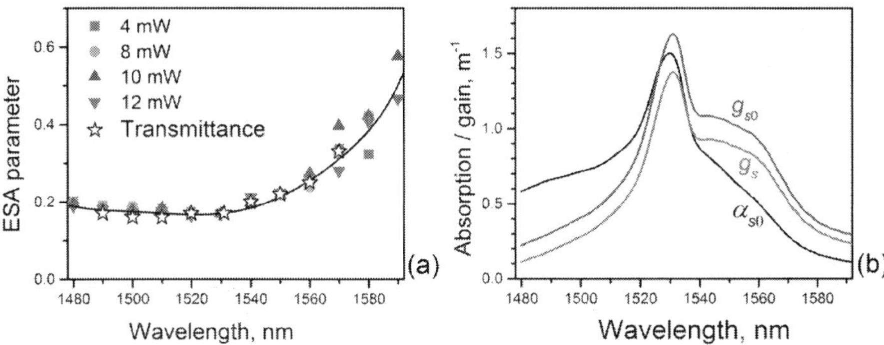

Figure 3. (a) ESA parameter ε_s calculated from the measurements of PD signal (filled symbols) and the ones of the EDF transmission coefficient (stars); solid line is the best polynomial fit. (b) Spectral dependencies of the EDF's small-signal absorption coefficient α_{s0}, full-saturated gain coefficient g_s, and net-gain coefficient g_{s0}.

e.g. Ref. 7) and the spectrum for g_{s0} was calculated using the g_s-and α_{s0}-spectra and the fitted ε_s-spectrum, taken from Figure 3(b). Worth noticing, the data obtained above are universal for any silica-based EDF weakly doped with Er^{3+}, where the "concentration" effects (to be discussed in section 3) are negligible.

2.2. ESA in low-doped EDF at 977 nm

We report here the results of a study of the other ESA process, $^4I_{11/2} \rightarrow {}^4F_{7/2}$ (see Figure 1), which takes the place when EDF is excited simultaneously at the two GSA wavelengths: $\lambda_p \sim 977$ nm (through $^4I_{15/2} \rightarrow {}^4I_{11/2}$ transition) and $\lambda_s \sim 1531$ nm (through $^4I_{15/2} \rightarrow {}^4I_{13/2}$ transition), a situation normally encountered in a diode-pumped (at $\lambda_p \sim 980$ nm) EDFL or EDF-based amplifier (EDFA). Upon simultaneous excitation of EDF at these two wavelengths, population of $^4I_{11/2}$ state increases, resulting in significant growth of the pump-induced ESA loss (when $^4I_{11/2} \rightarrow {}^4F_{7/2}$ transition gets "switched on", see Figure 1). On the contrary, without the presence of a signal wave in the fiber, almost all ions will be at the upper laser level $^4I_{13/2}$, which leads to a lower absorption at λ_p.

The first experiment was focused on the demonstration of the presence of "pump-ESA", when EDF is pumped simultaneously at λ_p=977 nm and at λ_s=1531 nm. It was aimed to reveal of whether UCE is observed (through transitions $^2H_{11/2} \rightarrow {}^4I_{15/2}$ and $^4S_{3/2} \rightarrow {}^4I_{15/2}$ "5"→"3", see Figure 1), as following the ESA process at λ_p. Experimentally, an EDF piece was pumped, using 980 nm / 1550 nm WDM supporting up to 1 W of optical power, by the pump and signal beams from the same EDF's side. These beams were delivered from a standard fibered LD (λ_p=977 nm) and from another semiconductor laser with wavelength λ_s=1531 nm, followed by an EDFA. Figure 4 presents two photographs of lateral emission from the EDF sample when the fiber was pumped, respectively, (*i*) solely at wavelength λ_p (Figure 4(a)) and (*ii*) simultaneously at wavelengths λ_p and λ_s (Figure 4(b)); the incidence pump and signal powers were P_p=P_s=260 mW. One can readily see bright UCE (in the green spectral range, ~520–560 nm) when the EDF is pumped at both the wavelengths, and very weak UCE when it is pumped at λ_p only.

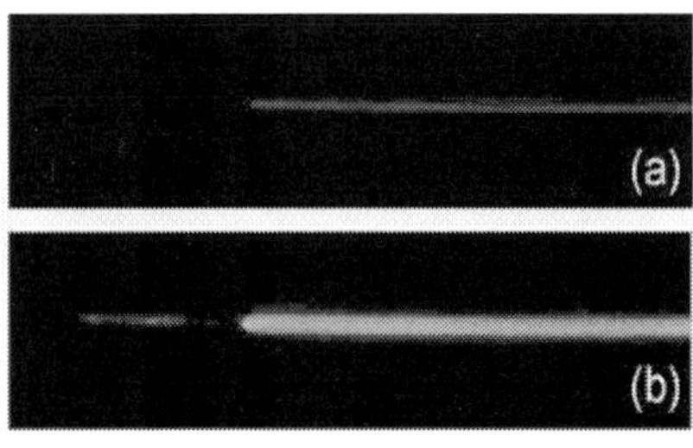
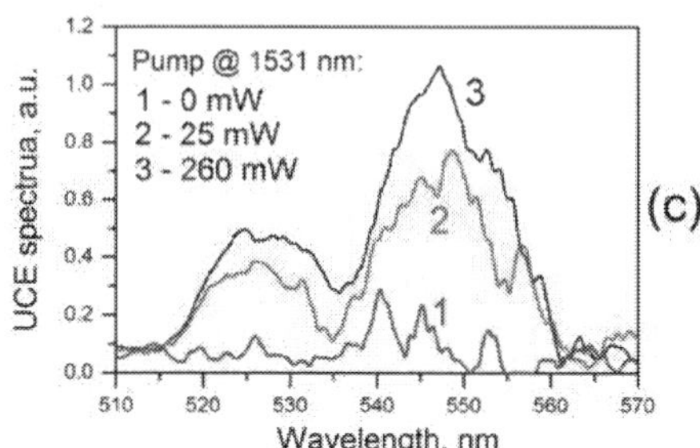

Figure 4. (a) and (b) Photographs of ~520/~545 nm UCE obtained from the EDF's surface at in-core excitation (a) only at λ_p=977 nm and (b) simultaneously at λ_p=977 nm and λ_s=1531 nm. (c) Frontal UCE spectra recorded from a 5-cm EDF sample at in-core excitation at pump wavelength (P_p=260 mW) and variable signal powers (P_s=0,25, and 260 mW).

The UCE spectra, obtained using a 5-cm EDF sample, are shown in Figure 4(c). The spectra were recorded for three signal powers (0, 25, and 260 mW) while fixed pump power (260 mW). One can see from the figure that even low-power (25 mW) signal radiation tremendously enhances UCE (compare curves 1 and 2) and that at further increasing signal power, up to P_s=260 mW, growth of the UCE power becomes slower, demonstrating a saturating behavior (compare curves 2 and 3). Thus, the signal at 1531 nm "ignites" the ESA process at λ_p and thereby increases the number of Er^{3+} ions in the ground state, resulting in an increase of the pump-light absorption. Consequently, there will be a non-negligible population of Er^{3+} ions on $^4I_{11/2}$ level and, as the fact of matter, a more effective ESA process at the pump wavelength, seen as growth of pump-induced loss.

Figure 5 shows the dependence of "green" UCE signal (transition "5"→"3", see Figure 1), detected from the EDF, on powers of pump and signal radiations launched into the fiber (in this case, again, the pump power was kept fixed, P_p=260 mW, and the signal power was varied, P_s=0...290 mW). Experimentally, lateral UCE power $P_{UCE}(P_s)$ was measured using a photomultiplier with a cesium cathode from a short section (~1 mm) of the EDF near its splice with WDM. It is seen that the UCE's behavior demonstrates the already noticed aspects: At very low signal power UCE is extremely weak, whilst upon its increase UCE first strongly enhances and then gets saturated.

Since UCE power is proportional to population of the 5th level of Er^{3+} ions, the experimental data can be fitted well by a simulated curve of normalized population inversion $n_5=N_5/N_0$ (N_5 is the population of the 5th Er^{3+} ion's level) averaged over the fiber core area, which confirms the theory we built. Considering that both the excitation waves (at 977 nm and at 1531 nm) have Gaussian spatial distributions, one can obtain from the steady-state rate equations for the Er^{3+} ion (formulas 1-5) the following expression for the averaged population $\bar{n}_5$:

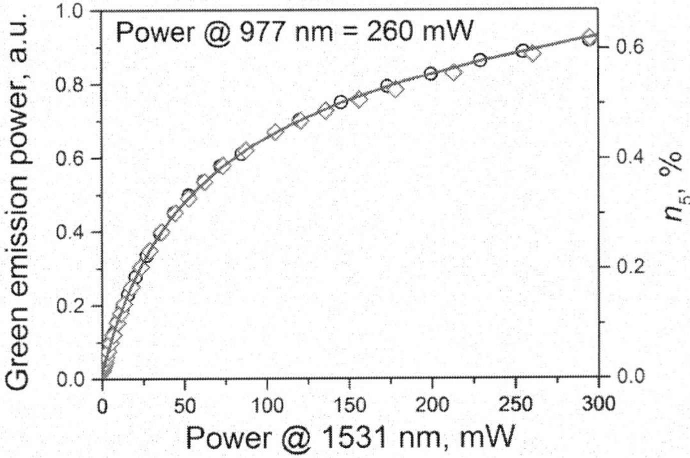

Figure 5. Dependence of UCE power on signal power (left scale): Circles and rhombs correspond to two different experimental series; plain curve is a theoretical fit of normalized inversion of the 5th Er³⁺ level (right scale).

$$\bar{n}_5 = \frac{\varepsilon_p \gamma_2}{a^2} \int_0^a \frac{1 - n_1(r) - n_2(r)}{1 + \varepsilon_p \gamma_2 s_p(r)} s_p(r) 2r dr \tag{10}$$

where the normalized populations $n_1(r)$ and $n_2(r)$ are taken from [16] (found from the same set of rate equations for the Er³⁺ ion); $\varepsilon_p = \sigma_{35}/\sigma_{13}$ is the ESA parameter at λ_p, $\gamma_p = \tau_{53}/\tau_{31}$ (with $\tau_{53} \approx \tau_{54}$ ≈ 1 μs and $\tau_{31} \approx \tau_{21} \approx 10$ ms, see Table 1), and $s_p = I_p / I_p^{sat}$ is the saturation parameter at the pump wavelength (I_p is the pump intensity, $I_p^{sat} = h\nu_p/\sigma_{13}\tau_{21}$, and $h\nu_p$ is the quanta energy at λ_p). As seen from Figure 5, the simulated curve for population of the 5th level ($\bar{n}_5$) fits well the experimental data, thus confirming correctness of the theory.

We found that the best way to find the ESA parameter at the pump wavelength, $\lambda_p = 977$ nm, is to measure the EDF's nonlinear transmission coefficient at the pump wavelength in function of signal power, $T_p(P_s)$, and then compare this data with the simulated ones, obtained from the steady-state rate equations (formulas 1-5) for the fife level Er³⁺ energy diagram (see Figure 1). Considering the Gaussian radial intensity distributions for the pump and signal waves, the equations describing $T_p(P_s)$ at fixed pump power ($P_p = 260$ mW in our case) take the form [16]:

$$\frac{dP_p(z)}{dz} = -\frac{4\alpha_{p0}}{\Gamma_p w_p^2}\left[\int_0^a [n_1(r, z) - (\xi_p - \varepsilon_p)n_3(r, z)]exp\left[-2\left(\frac{r}{w_p}\right)^2\right]rdr + \alpha_{BC}\right]P_p(z) \tag{11}$$

$$\frac{dP_s(z)}{dz} = \frac{4\alpha_{s0}}{\Gamma_s w_s^2}\left[\int_0^a [(\xi_s - \varepsilon_s)n_2(r, z) - n_1(r, z)]exp\left[-2\left(\frac{r}{w_s}\right)^2\right]rdr - \alpha_{BG}\right]P_s(z) \tag{12}$$

where α_{p0} is the small-signal EDF absorption and Γ_p is the pump wave to fiber core overlap factor, both at the pump wavelength, α_{BG} is the small background EDF loss (~3 dB/km for the EDF used), w_p is the radius of the Gaussian distribution of pump wave, and $\xi_p=1+\sigma_{31}/\sigma_{13}$ is the SE parameter at the pump wavelength. The normalized populations of Er^{3+} levels n_1, n_2, and n_3 are found from the rate equations for the Er^{3+} ion (the routine is not discussed here because of its completeness; refer for details to [16]).

Actually, the coefficient adjacent to P_p from the left side (Equation (11)) is the EDF absorption coefficient at the pump wavelength, whereas the one adjacent to P_s from the left side (Equation (12)) is the EDF gain at the signal wavelength; both the coefficients depend on the pump and the signal powers. It is worth of mentioning that the set of equations (11) and (12), added by the corresponding boundary conditions, can be also used for modeling a CW EDFL.

Figure 6 shows the EDF transmissions measured for three different EDF lengths along with the best fits obtained using equations (7) at varying ε_p and ξ_p (and as the result their most relevant values found to be: $\varepsilon_s=0.17$ and $\xi_s=1.08$, both at 1531 nm). As seen from the figure, the best fitting of the experimental EDF transmission at the pump wavelength is provided at $\varepsilon_p=0.95$ and $\xi_p=1.08$. Thus, the ESA cross-section at the pump wavelength is strong enough (being less than the GSA one only slightly), which ought to affect (decrease) an EDFL's efficiency via the pump-induced loss distributed along the active fiber.

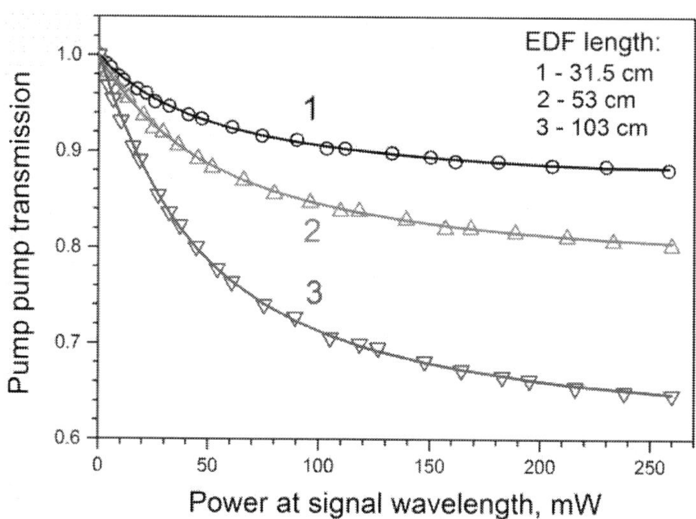

Figure 6. Experimental (symbols) and theoretical (curves) dependencies of pump (λ_p=977 nm) wave's transmission coefficient on signal (λ_s=1531 nm) wave's power; the data are obtained for three different EDF lengths (indicated in the inset). Theoretical curves are the best fits to the experimental data obtained at ε_p=0.95 and ξ_p=1.08.

2.3. Effect of ESA in EDF upon efficiency of CW EDFL.

We discuss hereafter the results of a theoretical study of influence of ESA inherent in Er^{3+} ions upon efficiency of an EDFL assembled in the linear (Fabry-Perot) configuration. The laser setup we shall deal with at modeling is shown in Figure 7. The low-doped EDF discussed above was considered to be an active medium and two fiber Bragg Gratings (FBGs) – to form selective couplers of the cavity, both centered at λ_s=1550 nm, the laser (signal) wavelength. The EDF length was chosen to be 4 m whereas the EDFL cavity length L_c, including the EDF piece and FBGs' tails, to be 6 m. Pump power on the EDF input was fixed in modeling at 200 mW.

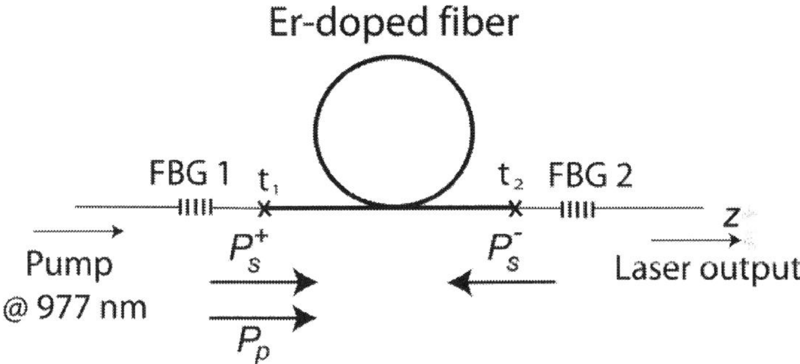

Figure 7. Sketch of the EDFL's geometry: P_s^+ and P_s^- are the laser (signal) waves' powers propagating in the positive (+) and negative (-) z direction, respectively; P_p is the pump power propagating in the positive z direction; t_1 and t_2 are the transmissions on the fibers' splices.

We imply (see Figure 7) that the pump wave at 977 nm (absorption peak of $^4I_{15/2} \rightarrow {}^4I_{11/2}$ transition, see Figure 1) propagates rightward while the laser waves – rightward (marked by "+" super-script) and leftward (marked by "–" superscript), respectively. In the scheme, FBG1 plays the role of a rear 100% reflector (reflection R_1=1), transparent for the pump wave, whereas FBG2 forms an output coupler with reflection R_2 varied for optimizing the laser efficiency. To decrease loss on the fiber splices, both FBGs were considered to be written in a photosensitive fiber with waveguide parameters similar to the EDF's ones. At the laser wavelength the low-signal EDF absorption α_{s0} was taken to be 0.0069 cm⁻¹, according to [17].

The laser is simulated by applying the two contra-propagating laser waves' model discussed in details in [17] with taking into account Gaussian distributions of the laser and pump waves. In this model, the pump wave is described by equation (11) and the signal waves – by equation (12) with a small modification being that powers of the two contra-propagating signal waves are assumed to be governed by the equation.

$$\frac{dP_s^{\pm}(z)}{dz} = \pm \frac{4\alpha_{s0}}{\Gamma_s w_s^2}\left[\int_0^a [(\xi_s - \varepsilon_s)n_2(r,\ z) - n_1(r,\ z)]exp\left[-2\left(\frac{r}{w_s}\right)^2\right]rdr \mp \alpha_{BG}\right]P_s^{\pm}(z) \tag{13}$$

In equation (13) the superscripts "s" stand for the laser wavelength (1550 nm). The model includes also the two contra-propagating waves of SE (not shown in Figure 7) spectrally centered at λ_{se}=1531 nm (this wavelength corresponds to the GSA peak of $^4I_{15/2} \rightarrow {}^4I_{13/2}$ transition), powers of which obey the equation:

$$\frac{d P_{se}^{\pm}(z)}{dz} = \pm\left[g_{se}(z)P_{se}^{\pm}(z) \mp \frac{\Omega}{4\pi} \frac{\alpha_{se0}P_{se}^{sat}}{\Gamma_{se}} \int_0^a n_2(r,\ z)2r dr \right] \tag{14}$$

where the second term on the right side is the SE power generated by a short fiber section dz, α_{se0}=0.016 cm^{-1} is the low-signal absorption at λ_{se}, P_{se}^{sat} is the saturation power at 1531 nm, Γ_{se} is the overlap factor for the SE waves, $\Omega = \pi NA/n^2$ is the fraction of SE photons guided by the EDF core in each direction, and n is the modal refractive index at λ_{se}. The EDF gain at λ_{se} is written as

$$g_{se}\left(z\right) = \frac{4\alpha_{se0}}{\Gamma_{se}w_{se}^2}\left[\int_0^a \left[(\xi_{se} - \varepsilon_{se})n_2(r,\ z) - n_1(r,\ z)\right]exp\left[-2\left(\frac{r}{w_{se}}\right)^2\right]r dr - \alpha_{BG}\right] \tag{15}$$

The boundary conditions are written as:

$$P_p(z=0,\ t)=P_{p0} \tag{16}$$

$$P_s^+(z=0,\ t)=P_s^-(z=0,\ t)R_1 t_1^2 \tag{17}$$

$$P_s^-(z=L_c,\ t)=P_s^+(z=L_c,\ t)R_2 t_2^2 \tag{18}$$

$$P_{se}^+(z=0,\ t)=P_{se}^-(z=L_c,\ t)=0 \tag{19}$$

$$P_s^{out}(t)=P_s^+(z=L_c,\ t)(1-R_2)t_2=0 \tag{20}$$

where P_{p0} is the pump power at the EDF input and P_s^{out} is the EDFL output power. To simplify calculations, we considered that $t_1=t_2=0$ (i.e. no loss on the fiber splices).

The EDFL's efficiency as a function of FBG2's reflectivity R_2, simulated using the laser model described above, is depicted in Figure 8(a). The EDFL was modeled for four different "versions" of the energy level system: Without considering all the ESA transitions (curve 1); with considering the pump ESA only (curve 2) and the signal ESA only (curve 3); with considering all the ESA transitions (curve 4).

The first important observation is that the optimal reflection of output FBG2, at which EDFL demonstrates the maximal efficiency, is drastically decreased when the ESA transitions are accounted for. For instance, the optimal reflection of FBG2 is ≈66% when only the EDF background loss (3.1 dB/km, see Table 1) is present (curve 1), whereas it is ≈11% when all kinds

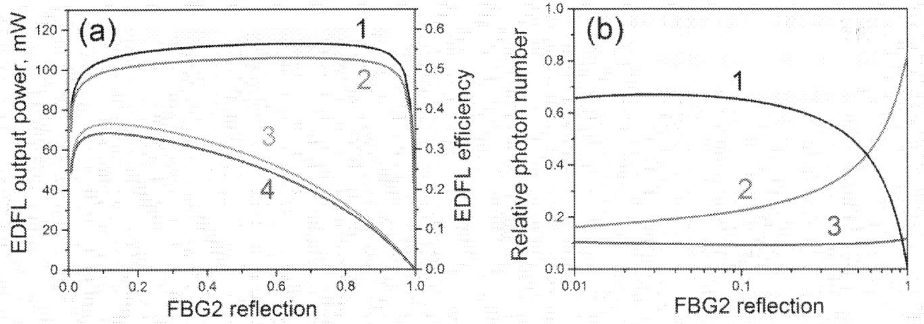

Figure 8. (a) EDFL output power and (b) fractions of the absorbed pump photons spent on the laser output (quantum efficiency) as functions of output (FBG2) reflectivity, R_2. The designations of curves 1 to 4 are given in the text. The small fraction of pump photons spent at the background loss and the ASE contribution are not shown.

of the ESA loss are accounted for (curve 4). The other important fact is that the range of FBG2's reflections, in which the output laser power varies within 10% (with respect to its maximum value if all the ESA transitions are considered), is relatively broad: ~2.5% to ~34%. At the optimal FBG2's reflectivity (R_2=11%) the laser efficiency reaches ~34% (implying all fiber splices are made lossless).

The fractions of pump and signal photons spent on the ESA transitions with respect to the absorbed by EDF pump photons are shown in Figure 8(b). It is seen that the contribution of the signal ESA loss is bigger when FBG2's reflection is bigger (curve 2). Furthermore, if reflection of the output coupler approaches 100%, the absorbed pump power is spent entirely on the ESA transitions (see curves 2 and 3) and no photons at the laser output are produced (see curve 1). At optimal FBG2's reflectivity ($\approx$11%), about 23% of absorbed pump photons are spent on ESA at the laser wavelength and about 9.5% on ESA at the pump wavelength. Note that the sum of the relative photon numbers, as can be revealed from curves 1, 2 and 3, is approximately equal to 1 in the whole range of FBG2's reflections.

The reader is advised here to refer to [18] for comparison of the developed theory with some of the experimental data on laser efficiency of EDFLs based on the EDF of M-type with relatively high Er^{3+} concentrations.

Finally, we conclude that the ESA processes at the laser and pump wavelengths strongly affect an EDFL's efficiency and output coupler's optimal reflectivity, at which the laser output power is maximal.

3. Er^{3+} concentration effects in EDF

In this part, we shall discuss the Er^{3+} concentration effects in EDFs resulting in a reduced efficiency of EDF based lasers and amplifiers, which is associated with the phenomenon of

Er^{3+} ions' clustering that leads, in turn, to non-saturable absorption (NSA) through inhomogeneous up-conversion (IUC).

For our experiments we selected the most representative commercial EDFs fabricated through the MCVD and direct nanoparticle deposition (DND) processes; all the fibers under scope in this section are similar in the sense of Er^{3+} doped core's chemical composition being the most common alumino-silicate glass (in the case of MCVD-EDFs with addition of germanium). The first series of the EDFs (MCVD-based, "M"-series) includes two fibers: M5-125-980 and M12-125-980 (*Fibercore*), hereafter M5 and M12, and the second series (DND-based, "L"-series) – three fibers: L20-4/125, L40-4/125, and L110-4/125 (*Leikki / nLight*), hereafter L20, L40 and L110. These fibers have very similar waveguide parameters and differ mainly in Er^{3+} doping level. [Notice that the EDFs employed in the whole of above experiments, see Section 2, belong to M-series.]

3.1. Absorption and fluorescence spectra

The EDFs' absorption spectra are shown in Figure 9(a) where Er^{3+} transitions $^4I_{15/2} \rightarrow {}^4I_{11/2}$ (within a 940...1020 nm range with a peak at 978 nm) and $^4I_{15/2} \rightarrow {}^4I_{13/2}$ (within a 1400...1600 nm range with a peak at 1.53 µm) are featured. The spectra were obtained using a white light source with fiber output and OSA with 1-nm resolution. It is seen that the absorption spectra of the EDFs of both series have a very similar shape (given by similarity of core glass chemical compositions), differing only in intensity. The ratio of the peaks' magnitudes at 1.53 µm and at 978 nm was measured to be equal to ~1.6, for the M and L EDFs.

Figure 9(b) demonstrates the normalized fluorescence spectra for L fibers (to simplify the picture the spectra for M fibers are not shown), measured at the maximal pump power at 978 nm, P_p ~400 mW, within the 450...1650 spectral range (the area nearby the pump wavelength is cut out). In this experiment the lateral geometry, when fluorescence is captured by a multimode fiber patch cord from the lateral surface of the short fiber samples, was arranged. In spite of Er^{3+} concentration increases in the row of fibers L20 → L40 → L110, the Er^{3+} fluorescence band, centered at ~1.53 µm, is indistinguishable in shape. Although the 1.53-µm band dominates in the EDFs' fluorescence spectra, there also exist the spectral lines at its anti-Stokes side (~450...1100 nm), which evidences the presence of UC. Note that, in contrast to the ~1.53-µm band's stability against Er^{3+} concentration, the higher Er^{3+} content, the more intense the UCE (compare curves 1 – 3 in Figure 9(b)).

To understand the origin of UCE and the dependence of UC intensity upon Er^{3+} concentration in the EDFs, let's refer to Figure 10 in which the scheme of Er^{3+} energy levels and a sketch of the processes involved at the excitation at λ_p=978 nm are presented. UCE (shown in the figure by grey arrows) seems to be mostly associated to Er^{3+} ion clusters being in states $^4I_{11/2}$ and $^4I_{13/2}$, because the ESA process, equally acting for single and clustered Er^{3+} ions, is ineffective at 978-nm excitation.

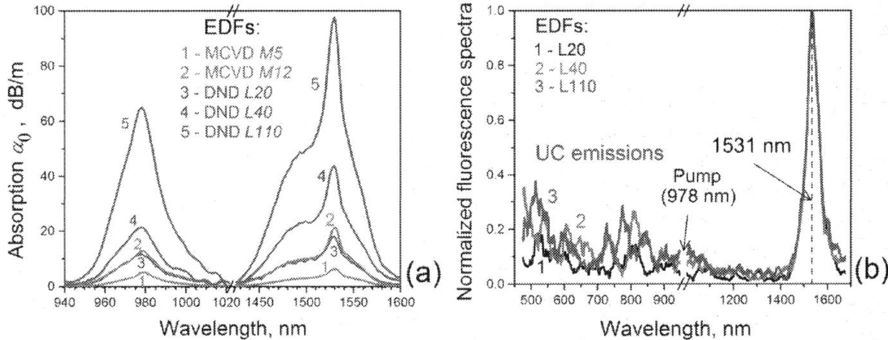

Figure 9. (a) Absorption spectra of the EDFs of L-(blue curves) and M-(red curves) series in the near-IR. (b) Fluorescence spectra of the EDFs of L-series in the VIS...near-IR spectral range at 978-nm pumping.

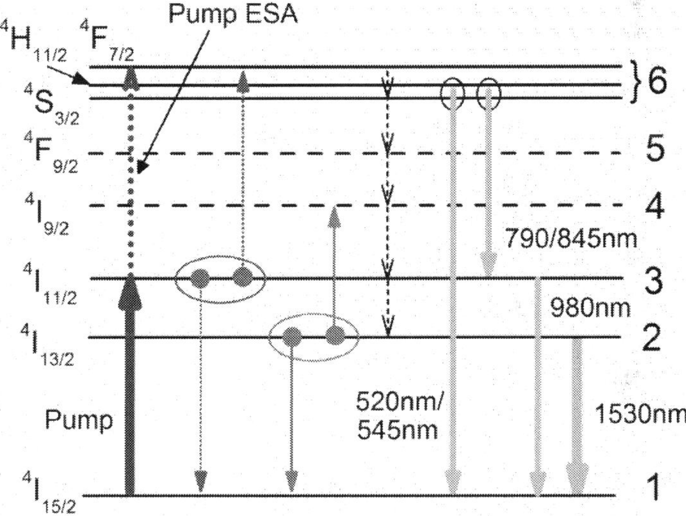

Figure 10. Scheme of Er^{3+} energy levels, applicable for the EDFs with high Erbium content. Functioning of Er^{3+} clusters (shown for simplicity as ion pairs) is sketched by the blue and red arrows for long-living manifolds $^4I_{11/2}$ and $^4I_{13/2}$; the black dotted arrows depict non-radiative relaxations; the grey arrows show the UC and "fundamental" (1.53-μm) emissions; the short-living levels are shown by dashed lines.

3.2. Fluorescence decay kinetics

Like at the fluorescence spectra' measurements discussed above, the kinetics of near-IR fluorescence at ~1.53 μm was detected using the lateral geometry. However, the pump light at 978 nm was in this case switched on / off by applying a rectangular modulation of LD current at Hz-repetition rate. The launched into the EDF samples pump power was varied between zero and ~400 mW. The fluorescence signal was detected either using an InGaAs PD with a Si filter placed between the fiber and a multimode patch cord delivering fluorescence to PD (being so the measurements above ~1.1 μm where the use of Si filter allows cutting off the pump light's spectral component), or using a fast Si-PD with no spectral filtering applied (being in fact the measurements below ~1.1 μm), placed directly above a slit segregating a portion of fluorescence from the EDF's surface. To diminish ASE and re-absorption on the results, we used short (~0.5 cm) EDF pieces.

Typical kinetics of the fluorescence signal, recorded after switching pump light at 978 nm off, are presented in Figure 11 for the heavier doped EDFs M12 (a) and L110 (b); the data were acquired using InGaAs-PD with Si filtering (transmission band above 1.1 μm). We don't present here the results for other, lower doped, EDFs as these showed similar but less featured trends in the decay kinetics.

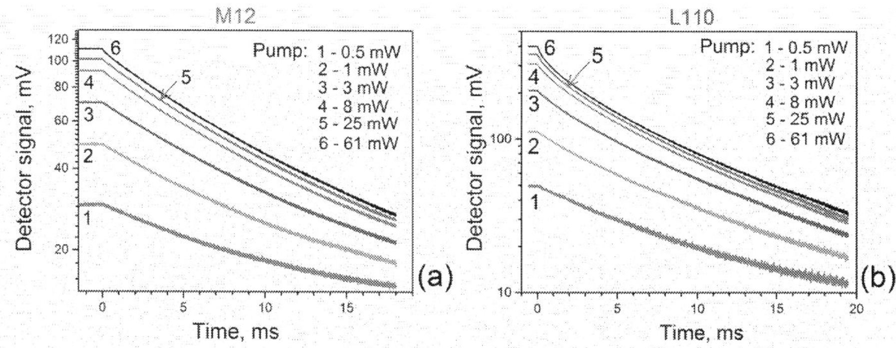

Figure 11. Fluorescence decay kinetics obtained for the EDFs M12 (a) and L110 (b). Curves 1 to 6 are captured for different pump levels (see the insets in the right upper corners). Zero-time corresponds to the moment when the pump light is switched off.

It is seen from Figure 11 that for these two EDFs fluorescence power, corresponding to 1.53-μm spectral band, is saturated (as is saturated GSA of Er^{3+} ions) yet at a few mW of pump power. However the key feature is deviation from the exponential law in the fluorescence kinetics in EDF L110 (see Figure 11(b)). A similar trend occurs but is less expressed in the rest of L and M fibers with lower Er^{3+} concentration; see e.g. Figure 11(a). Another fact deserving attention is the presence of a sharp drop in the fluorescence signal in fiber L110 at high pump powers, which happens after switching pump off (refer to curves 4–6 in Figure 11(b)). Such a feature is present but in a smaller degree also in fibers L40, L20, and M12 (having substantially lower Er^{3+} contents) and almost vanishes in fiber M5 (having the lowest Er^{3+} content). Note that similar fluorescence kinetics were observed in some of the earlier reports, see e.g. [4, 10].

Overview of the fluorescence decays for all the EDFs under scope is provided in Figure 12 (points). These data were obtained at maximal pump power, P_p at 978 nm (400 mW; the high pump power was found to be the right choice for minimizing spatial diffusion of excitation; see e.g. [10]). The dependences in the figure demonstrate the fluorescence decay "tails" obtained after cutting off the short initial segments just after switching pump off (~30 μs), which permits elimination of the influence of non-instant LD power decay (~8 μs).

As seen from Figure 12, 1.53-μm fluorescence decays get more and more deviated from the single exponential law when Er^{3+} concentration increases: The fibers with smaller contents of Er^{3+} ions (M5, M12, and L20) demonstrate decays nearly a single-exponent law whereas fibers L40 and L110 – the decays, apparently different from this law. These features, associated to the Er^{3+} concentration effect, can be addressed in terms of the IUC process – see Section 3.3 where the results of modeling of Er^{3+} fluorescence kinetics are presented. The modeling of the fluorescence kinetics allowed us to get, for each EDF, lifetime τ_0 and constant C_{HUP^*} (characterizing the homogeneous UC process, HUC; see modeling below) and thereafter to build their dependences upon small-signal absorption value α_0 at 1.53 μm (and hereby upon Er^{3+} ions concentration, proportional to α_0).

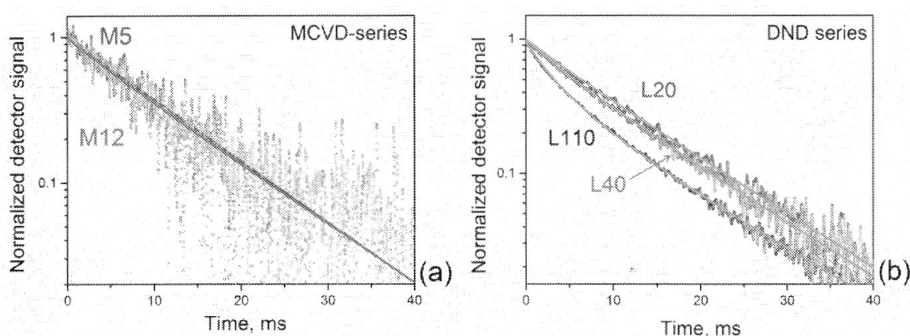

Figure 12. Normalized fluorescence decay kinetics obtained for the EDFs of L-(a) and M-(b) series; points – experimental data (using InGaAs-PD with Si filtering); plain curves are the theoretical fits made using formula (22).

Furthermore, Figure 13 demonstrates the results of the fluorescence kinetics' measurements in the optical band below 1.1 μm within a short (tens of μs) interval just after switching pump light off, for EDFs M12 (a), L40 (b), and L110 (c). The measurements were fulfilled using Si-PD without optical filtering at P_p=400 mW. As is seen from the figure, there are two "fast" components in the PD signal's decay: The shortest ($\approx$8 μs) one, being in fact the setup's technical resolution and originating from the scattered pump light, and the longer one, measured by 21 μs to 26 μs for fibers M12, L40, and L110 (for fibers M5 and L20 this component was not resolved). A similar component was also detected in the 1.53-μm fluorescence kinetics; see Figure 11, which evidences its non-radiative nature. Note that there are known the processes in Er^{3+} -doped materials attributed by similar times [19, 20].

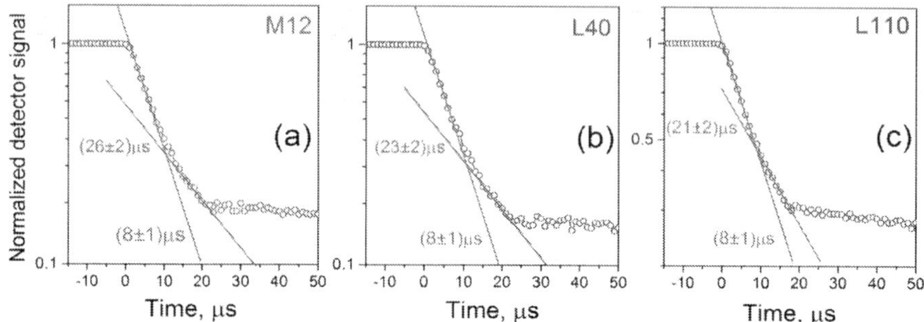

Figure 13. Fluorescence decay kinetics in the EDFs measured using Si-PD: M12 (a), L40 (b), and L110 (c); the short-time components of the fluorescence signals are specified in each plot.

We suggest that the found feature stems from a partial excitation relaxation in Er^{3+} clusters since it is present in the heavier doped EDFs but almost vanishes in the lower doped ones. The magnitude of the short-living component is a function of Er^{3+} concentration (and therefore of α_0), as seen when comparing the plots (a), (b), and (c) in Figure13: The higher Er^{3+} concentration the larger is relative (to the technical, i.e. originated from pump-light scattering) magnitude of this component.

3.3. Nonlinear absorption coefficient

The nonlinear absorption coefficient of a rare-earth doped fiber (e.g. EDF) as a function of pump power $\alpha(P_p)$ contains the useful information about GSA saturation and, consequently, about the fiber's potential as a laser medium. On the other hand, such effects deteriorating laser 'quality' of the fiber as ESA and concentration-related HUC / IUC (lifetime quenching and non-saturable absorption) ought to affect the behavior of $\alpha(P_p)$, too [18].

In the study to be reported hereafter, pump light was delivered to an EDF sample from the same LD operated at 978 nm (used in the measurements of fluorescence spectra and lifetimes); pump power was varied from ≈0.5 to 400 mW. We measured first the nonlinear transmission coefficient of the EDF sample with length L_0, which is defined as $T_{978}=P_P^{out}/P_P^{in}$ where P_P^{in} and P_P^{out} are the pump powers at the EDF's input and output. Then we made a formal re-calculation of the experimental transmission coefficient $T_{978}(P_P^{in})$ in the absorption coefficient, applying formula: $\alpha(P_P^{in})=-ln(T_{978})/L_0$. The EDFs' lengths were chosen such that overall trends in the dependences $\alpha(P_p^{in})$ can be viewed within the whole range of pump powers. The ratio of the EDFs' lengths was such that the optical density (the product $\alpha_0 L_0$) is almost the same for all the samples, which is worth for estimation of the nonlinearity (saturation) of absorption at increasing Er^{3+} concentration in the fibers. Using the OSA, we checked the ratio of pump to ASE powers at the EDFs' outputs; it was found that the ASE contribution is negligible in all samples at $P_p^{in}> 0.5$ mW.

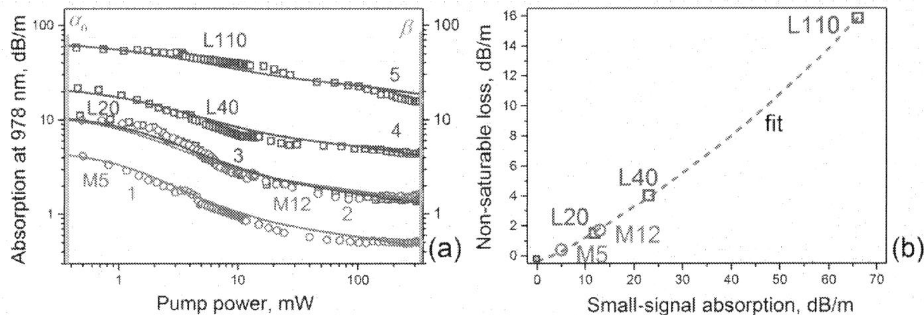

Figure 14. (a) Nonlinear absorption coefficients of the EDFs of L-and M-types vs. pump power at 978 nm. Symbols: experimental points; plain curves: theoretical fits obtained using Eqs. (25-27). Fiber lengths used in experiments and at modeling were, correspondingly: 188.6 (M5, curve 1), 59.4 (M12, curve 2), 43.5 (L20, curve 3), 22.4 (L40, curve 4), and 9.5 (L110, curve 5) cm. (b) Non-saturable absorption β vs. small signal absorption α_0, measured for the entire EDFs' set; the fitting curve is for guiding the eye.

The results obtained by applying the drawn procedure are shown in Figure 14(a) by symbols. Coefficients α_0 and β (saturated absorption at pump wavelength) marked in the upper left and right corners of the figure correspond to the limits of small-signal and saturated pump absorptions. First notice that absorption is "bleached" (in other words, transmission is "saturated") by a more or less similar manner for either fiber. However, as is also seen from the figure, the "residual" absorption (β) rises drastically with increasing Er^{3+} concentration in the fibers' sequences M5→M12 and L20→L40→L110. This trend points out that the ratio between the residual (β) and small-signal (α_0) absorptions is much bigger for the heavier doped fibers (M12, L40 and L110). In fact, pump-induced (looking as residual) absorption β is the measure of nonlinear absorption loss in EDF, as it stems from the modeling results (see Section 3.4 below). Furthermore, the dependence $\beta(\alpha_0)$ plotted in Figure 14(b) allows one to reveal that this pump-induced excessive loss in the EDFs appear as one of the most important Er^{3+} concentration effects.

3.4. Modeling

Firstly, the kinetics of near-IR (~ 1.53 μm) fluorescence decays obtained for the entire set of the EDF samples were modeled, which allows us to find Er^{3+} fluorescence lifetimes τ_0 and HUC coefficient C_{HUP}. The model implicitly implies, in accord to the definition of the HUC process, the only interactions between rather distant single Er^{3+} ions, not forming "chemical", tightly coupled, clusters, which are in turn supposed to interact via the IUC mechanism, discussed further. At this step of modeling we disregard the short-time features in the fluorescence decays, which are assumed to originate from Er^{3+} ions gathered into quenched (weakly fluorescing) clusters. That fact that in the experiments on fluorescence kinetics we used comparatively high pump powers (400 mW) allows us to neglect the intensity dependent HUC contribution and excitation migration effects.

For the normalized population density n_2^s of single (index "s") Er^{3+} ions being in the first excited (laser) state $^4I_{13/2}$, the following rate equation holds [21]:

$$\frac{dn_2^s}{dt} = -\frac{n_2^s}{\tau_0} - C_{HUP}\left(n_2^s\right)^2 \tag{21}$$

where $n_2^s = N_2^s / N_0^s$; N_0^s is the population density of single Er^{3+} ions in the excited state 2 ($^4I_{13/2}$), N_0^s is their concentration, and C_{HUP} [s^{-1}] is the UC parameter, being a product of the "volumetric" HUC constant C_{HUP}^* [$s^{-1}cm^3$] and concentration N_0^s: $C_{HUP} = N_0^s C_{HUP}^*$.

Assuming that pump power at 978 nm is high enough to achieve maximal populating of the excited state $^4I_{13/2}$, i.e. at "infinite" pump power, the part of Er^{3+} ions being in the excited state is $k=\sigma_{12}/(\sigma_{12}+\sigma_{21})$. Furthermore, since in our experimental circumstances (where near-IR fluorescence is detected at ~ 1.53 μm while excitation is at $\lambda_p=978$ nm) the SE process can be disregarded by means of formal zeroing cross-section σ_{21} in the dominator of this ratio (k=1). Implying that $n_2^s(t=0)=1$ and that pump is switched off at $t=0$, equation (21) is solved analytically, giving:

$$n_2^s(t) = \frac{e^{-\frac{t}{\tau_0}}}{1 + \tau_0 C_{HUP}\left(1 - e^{-\frac{t}{\tau_0}}\right)} \tag{22}$$

Formula (21) is a worthy approximation for fitting of the whole of experimental near-IR fluorescence decay kinetics, reported above for $P_p \sim 400$ mW (providing maximal population of manifold $^4I_{13/2}$). The modeling results obtained by using formula (22) are shown by plain curves in Figure 12 (the fitting procedure has been fulfilled until the residual sum R^2 exceeded 0.99) and are seen to be in good agreement with the experimental decay kinetics (points in the figure). The values of constants τ_0 (lifetime of single Er^{3+} ions, found to be ~ 10.8 ms for all the fibers under study), and C_{HUP} (an attribute of the HUC process, determined as the result of fitting) are plotted in Figure 15 in function of the small signal absorption α_0 at 978 nm.

As seen from Figure 15, the parameter C_{HUP} is proportional to Er^{3+} concentration (GSA α_0 at 978 nm). From this figure we found the value of HUC constant: $C_{HUP}^*=2.7\times10^{-18}$ $s^{-1}cm^3$. This value agrees well with the published data for EDFs of similar types; see e.g. Refs. [22-24]. Note that the quantity attributing the HUC phenomenon (C_{HUP}^* constant) should be proportional to the ESA cross-section (see e.g. [25]). As the latter does not depend on Er^{3+} concentration, C_{HUP}^* should be concentration-independent. Indeed, the dependence C_{HUP} vs. α_0 is seen from Figure 15 to be almost linear.

The next step in modeling is simulation of Er^{3+} clusters' contribution on the base of the experimental dependences of nonlinear absorption vs. pump power (see Figure 14(a)). A method to model nonlinear absorption of an EDF is based on the idea that ensemble of Er^{3+} ions in a fiber consists of two independent subsystems, assumed to be single ("s") and clustered

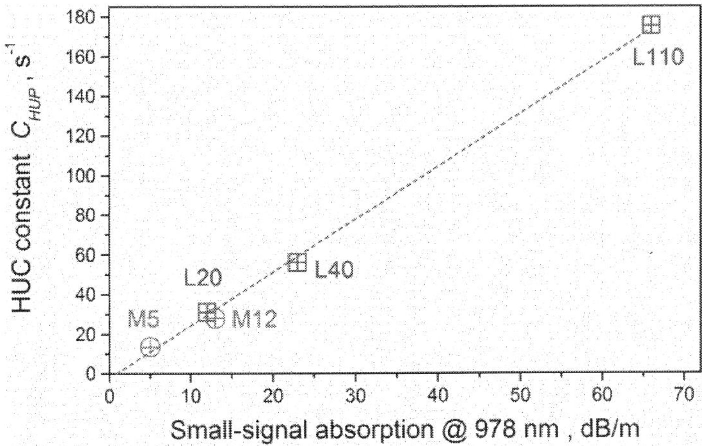

Figure 15. The values of HUC constant C_{HUP} vs. Er^{3+} concentration (in terms of α_0 at 978 nm), obtained for the entire set of EDFs as the result of modeling of the experimental fluorescence kinetic using formula (22); the fitting line in for guiding the eye.

("c") ions. Considering this hypothesis, we generalize the model developed in [21] for propagation of a pump wave through the system of single and paired resonantly absorbing and fluorescing centers (pairs are the simplest case of clusters). The model's generalization signifies here that the clusters' subsystem is meant to comprise an arbitrary number of centers (Er^{3+} ions in our case) whereas the other subsystem – to consist of single species [6, 26].

We assume that a cluster of Er^{3+} ions can occupy only one of the two permitted states – the state <11>, where all ions forming the cluster are in the ground state, and the state <12>, where one excepted ion (an acceptor of energy transferred from the adjacent donor ions within the cluster) is in the excited state. The latter holds because if other cluster's constituents absorb pump photons and move to the excited state, all them except one leave state 2 (down to state 1) immediately, whereas only this excepted one can stay in state 2.

We also take into account the presence of the short time τ_1 (together with much longer τ_0) for Er^{3+} clusters as an essential element of the model, coming from the experimental data.

We consider that the populations of single ($N_{1,2}^s$) and clustered ($N_{1,2}^c$) Er^{3+} ions, satisfy the following relations:

$$N_1^s + N_2^s = N_0^s = (1 - \chi\kappa)N_0 \tag{23}$$

$$N_1^c + N_2^c = N_0^c = \chi\kappa N_0 \tag{24}$$

where κ is the partial weight of clustered ions in ensemble, χ is the effective (averaged) number of Er^{3+} ions in a cluster, and N_0 is the total Er^{3+} ions concentration. The correspondent normalized population densities are defined as $n_{1,2}^{s,c} = N_{1,2}^{s,c}/N_0$, where the lower indices assign, correspondingly, the ground (1 or <11>) and the excited (2 or <12>) states, and $N_0^{s,c}$ are the concentrations of single ions and clusters, respectively.

The balance equations for pump power (P_p) and normalized dimensionless population densities of single and clustered ions being in metastable states 2 and <12>, $n_2^{s,c}$ ($0 \le n_2^{s,c} \le 1$), are as follows [27]:

$$\frac{dP_p(z)}{dz} = -\alpha_{p0}\left[1 - (1 + \xi_s - \varepsilon_s)\left(n_2^s(z) + n_2^c(z)\right)\right]P_p(z) - \alpha_{BG}P_p(z) \tag{25}$$

$$\frac{\alpha_0}{h\nu_p N_0 \Gamma_p A_c}\left[\chi\kappa - (1 + \xi_s)n_2^c(z)\right]P_p(z) - \left(\frac{1}{\tau_0} + \frac{1}{\tau_1}\right)n_2^c(z) = 0 \tag{26}$$

$$\frac{\alpha_{p0}}{h\nu_p N_0 \Gamma_p A_c}\left[1 - \chi\kappa - (1 + \xi_s)n_2^s(z)\right]P_p(z) - \frac{n_2^s}{\tau_0} - C_{HUP}\left(n_2^s\right)^2 = 0 \tag{27}$$

where majority of the quantities have been designated above, parameter $A_c = \pi a^2$ is the EDF core area ($a=1.5$ μm is the core radius), and $\alpha_{BG}=0.03...0.1$ dB/m, depending weakly upon the EDF type). In formulas (25-27) we omit the ASE contribution as negligible in our experiments at pump powers exceeding 0.5 mW (see above). Note that these formulas are written in a general form, applicable not only to the Er^{3+} ion but also to any other resonantly absorbing center, having a three equivalent level system and subjected to the aforementioned concentration effects.

The modeling results are plotted by plain curves 1 to 5 in Figure 16(a). It is seen that they fit well the whole of the experimental data for the EDFs of both (M and L) types. Thus, the IUC process, treated by us as mostly non-radiative relaxation within Er^{3+} ion clusters, is justified as the key mechanism responsible for nonlinear absorption (attributed by coefficient β). The dependence of β upon Er^{3+} concentration in terms of small signal absorption α_{p0} is shown in Figure 16(b) (see curve 1) as the result of modeling; the errors' bars in the curve show uncertainties of fitting the experimental data by the theory.

When making the numerical calculations, we found that, once searching for the best fit of the experiment by the theory, any χ-value ($\chi=2$, 3, and so on) can be used, with κ-value being varied accordingly. Thus, we have concluded that the product $\chi\kappa$ (a relative number of clustered Er^{3+} ions in the system) serves an adjusting parameter at fitting rather than quantities χ and κ separately. Given by the modeling results, useful insight can be made to interrelation between the relative number of clustered Er^{3+} ions $\chi\kappa$ (modeling: Figure 16(a)) and the measured nonlinear (saturated) absorption β (experiment: Figure 14(a)). Placing on the same plot (at double logarithmic scaling) the dependences of these two quantities vs. small-signal absorption α_{p0}, we found that they have the slopes related as ($\sim1.44/\sim0.63$) ≈ 2.3; see Figure 16(b). The found slope's value signifies the average number of Er^{3+} ions in clusters, whose

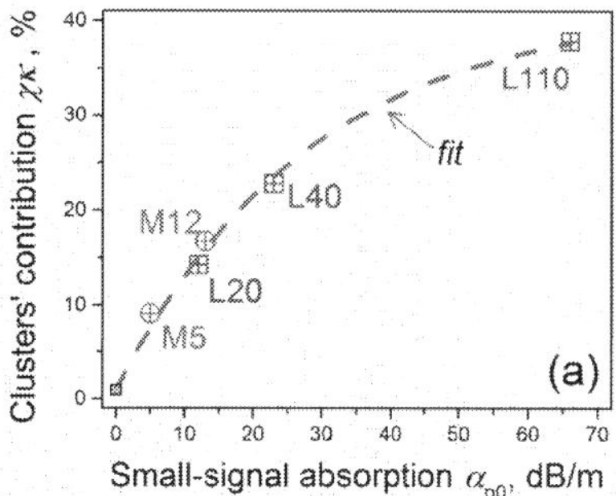
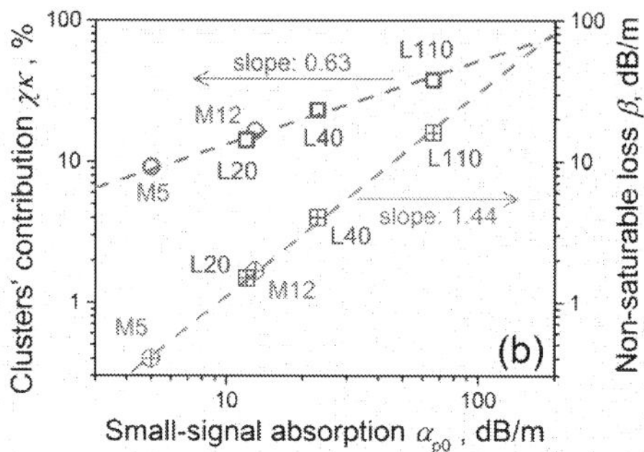

Figure 16. (a) Er^{3+} clusters' contribution $\chi\kappa$ vs. Er^{3+} concentration (in terms of α_{p0}) obtained for the entire set of EDFs of L-and M-types as the result of modeling the experimental dependences of nonlinear absorption coefficient using formulas (25-27); the fitting curve is for guiding the eye. (b) Er^{3+} clusters' contribution $\chi\kappa$ (left scale) and non-saturable absorption loss β (right scale) vs. Er^{3+} concentration (in terms of α_{p0}); the ratio of the slopes attributing the dependences reveals an effective number of Er^{3+} ions in clusters.

presence, according to the model, is responsible for NSA. This result deserves attention since it shows that NSA in the EDFs of both types mostly originates from paired Er^{3+} ions rather than from more complicate aggregates. Thus, in contrast to [28] where the role of "heavier" clusters in the IUC phenomenon is discussed, our results evidence for negligible contribution of Er^{3+} clusters, "heavier" than simple ion pairs.

Notice that the presence in EDFs of NSA at increasing Er^{3+} concentration affects net gain in heavily-doped Er^{3+} fibers, which becomes more and more limited (saturated) as it stems from the presence of single Er^{3+} ions being in the excited state, whereas the clustered ions, in their big part, i.e. $(\chi\kappa-1)$ ions in each cluster, are always in the ground state. As a consequence, efficiency of an EDFL or EDFA is expected to drop down with increasing concentrations of Er^{3+} ions in the active fiber. Some recent studies confirm a severe character of the problem [29-30]. It seems that another problem could be encountered at the use of heavily-doped EDFs for pulsed operation where extensive heating via excitation relaxation within Er^{3+} clusters would affect the dispersive properties of the fibers and deteriorate the regime.

4. Effect of SBS upon operation of actively Q-switched EDFL

Actively Q-switched (AQS) EDFLs based on acousto-optics modulator (AOM), implemented in the Fabry-Perot geometry, usually produce Q-switch (QS) pulses with duration from a few to hundreds ns [1]. The QS pulses are normally composed of a few sub-pulses separated by round-trip time of photon inside the cavity [31, 32]; this AQS regime will be called further "conventional" (CQS). In the meantime, it is known that in certain conditions FLs demonstrate stochastic QS pulsing, which stems, as it will be clearly demonstrated below, from intra-cavity stimulated Brillouin scattering (SBS) [33]. Such pulses, referred further to as SBS-QS ones, are characterized by dramatic increasing of power as compared to CQS pulses but, at the same time, by perceptible jitters [34]. In this section, we show that in certain circumstances SBS-QS pulsing is inherent in AOM-based AQS EDFLs. We also demonstrate that the areas (basins) where CQS and SBS-QS regimes exist are defined by definite values of EDF length and AOM's repetition rate and that the most important condition for turning of the laser to one or another pulsing regime is absence or presence of spurious narrow-line continuous wave (CW) lasing in the intervals where the laser cavity is blocked (AOM is switched OFF).

4.1. Experimental setup

An experimental setup of the QS-EDFL is sketched in Figure 17. The laser cavity consists of a piece of a standard low-doped "M" EDF (*Thorlabs*, M5-980-125), two FBGs (1 and 2) centered at ~1549.4 nm (laser wavelength), which form Fabry-Perot cavity, and a standard down-frequency shifting AOM with fiber outputs (*Gooch & Housego*, operation frequency – 111 MHz), placed nearby FBG2. The full AOM's rise time was measured to be 50 ns, AOM gate was fixed at 2 μs in experiments. FBGs' reflection coefficients were ~ 30% (FBG1) and ~ 100% (FBG2). A long period grating (LPG) tuned to ~1533 nm was used as in-line stop-band filter for neutralizing fiber gain at Er^{3+} SE peak and thus avoiding spurious CW lasing at this wavelength, which might otherwise discharge EDF and thereby reduce QS pulse energy and, via interfering with targeted pulsed lasing at the wavelength selected by FBGs, produce instability of pulsing. The EDF was pumped by a fiber-coupled 976-nm LD through a 980/1550-nm WDM. To decrease the cavity loss, FBG1 and LPG were written in the EDF core after preliminary hydrogenation. The laser signal was registered by OSA with a 50-pm resolution or by 1.2-GHz PD used in-line with a 2.5-GHz oscilloscope. In experiments, pump power was fixed at 500 mW and AOM's repetition rate (f_{AOM}) was varied within a 0...30-kHz range.

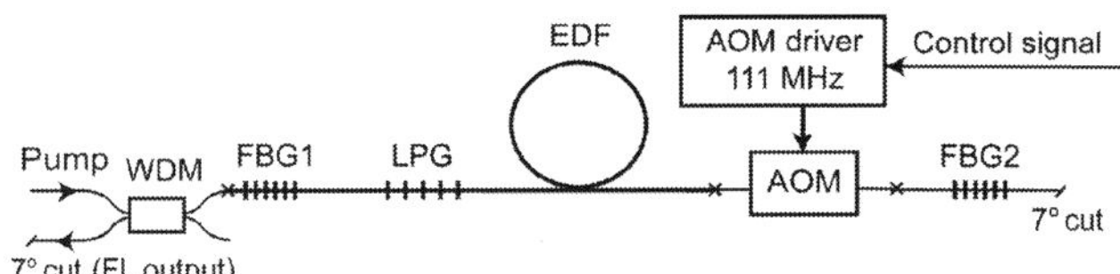

Figure 17. Experimental setup of the QS-EDFL (crosses indicate fiber splices).

4.2. Properties of CQS and SBS-QS regimes

As well-known, AQS FLs operated in CQS regime usually generate pulses consisting of train of sub-pulses (ripples), separated by a time interval equal to a photon's round-trip in the cavity. Such kind operation is fully described by the model of two contra-propagated waves in Fabry-Perot cavity, once considering the laser as a multi-pass amplifier of SE reflected several times by selective mirrors (FBGs) [32]. CQS is observed at any f_{AOM} when EDF length (L_{EDF}) is shorter than some specific value and at larger EDF when f_{AOM} is high. The common features of CQS pulsing observed experimentally and also modeled are as follows: Delay of a QS pulse with respect to the moment of AOM opening increases while its energy and power decrease with increasing AOM's repetition rate. Usually, the first detectable sub-pulse arises in a few round-trips of ASE after AOM got opened. For example, when L_{EDF}=8.8 m and f_{AOM}=8 kHz, the first visible sub-pulse appears at ~250 ns (~2.5 photon round trips; see Figure 18(a)). The RF (FFT) spectrum of pulse train measured at f_{AOM}=16 kHz (see Figure 18(b)) has three peaks centered at 0 MHz, ~10 MHz and ~20 MHz. Width of peak 0 relates to QS pulses width, whereas peaks 1 and 2 correspond to the first and the second harmonics of the round-trip frequency (an inverted interval between sub-peaks, or round-trip time), respectively.

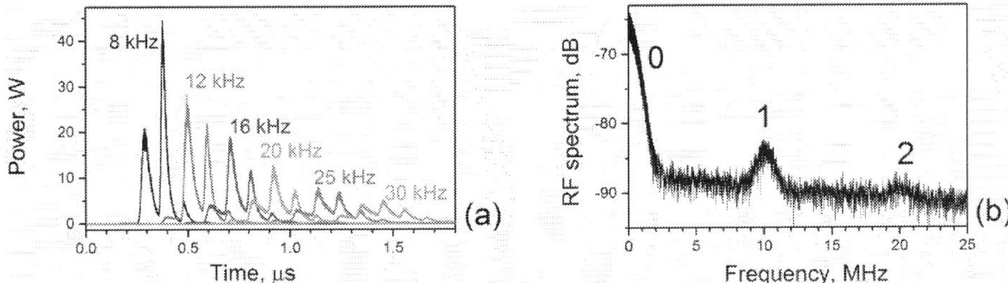

Figure 18. (a) CQS pulses registered on the EDFL output at L_{EDF}=8.8 m and at various f_{AOM}-values. Zero-time in both snapshots corresponds to the moment when AOM gets opened. (b) Averaged RF spectrum of CQS pulsing at f_{AOM}=16 kHz.

If the active fiber is long enough and AOM's repetition rate is not too high the QS EDFL turns into the regime of SBS-induced pulsing. This kind of pulsing is quite different as compared with CQS. Typical SBS-QS pulses are shown in Figure 19(a) for L_{EDF}=8.8 m and f_{AOM}=1 kHz. These pulses, as compared with CQS ones, arise earlier, approximately in ~180...280 ns after the moment of AOM's switching on; they are much narrower (~2.5...10 ns at 3-dB level); the pulses amplitude is more than by 10 dB higher as compared with the one at CQS while their envelop is apparently irregular. Emphasize that no SBS-QS pulses arisen within the intervals between the adjacent AOM's windows, in contrast to SBS-QS pulsing in an ytterbium-doped FL [34].

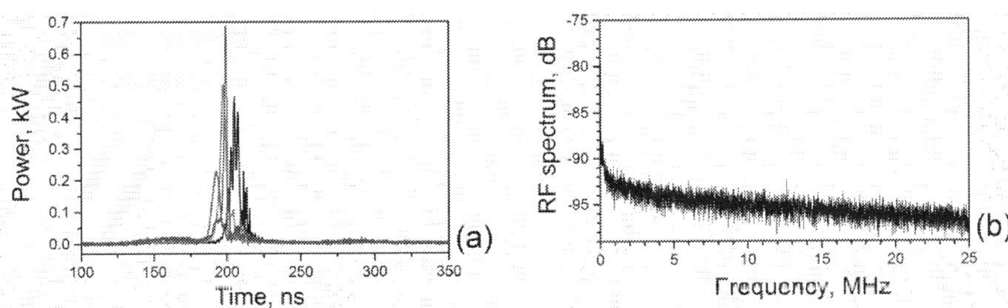

Figure 19. (a) CQS pulses registered on the EDFL output at L_{EDF}=8.8 m and f_{AOM}=1 kHz. (b) Averaged RF spectrum of CQS pulsing at f_{AOM}=1 kHz.

One more detail of SBS-QS is that pulses released in this regime suffer strong amplitude and timing jitters. Apparently, the presence of jittering is an indication of the stochastic nature of the SBS process involved. Furthermore, since the SBS-QS pulses are not composed of sub-pulses spaced by photon round-trip time, their RF (FFT) spectrum does not have peaks at the round-trip frequency (~10 MHz) and its harmonics (see Figure 19(b)).

4.3. Basins of CQS and SBS-QS regimes

To find basins of CQS and SBS-QS regimes existence, we measured the value of f_{AOM} at which the laser transits from one to another QS regime, for different L_{EDF}. The experimental points in the space (f_{AOM}, L_{EDF}) segregating CQS and SBS-QS operations were easily fixed since QS pulses captured at the laser output in these two regimes differ drastically in pulse amplitude, duration and shape (see snapshots in Figures 18(a) and 19(a)). We found that if L_{EDF} is less than or equal to a certain value (5.4 m in our arrangement) the laser operates in CQS regime, at any f_{AOM}. But if EDF is longer than 5.4 m, an operation regime depends on AOM's repetition rate: At low f_{AOM} the laser generates SBS-QS pulses while at high f_{AOM} it turns to CQS operation. The basins of CQS and SBS-QS regimes are illustrated in Figure 20. As seen from the figure, the laser operates in CQS and SBS-QS above and below the border line, respectively (this line schematically marks a transition between the regimes).

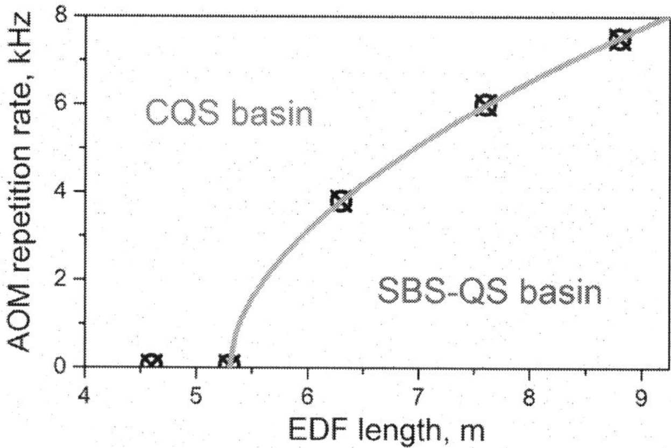

Figure 20. Basins of CQS and SBS-QS regimes; symbols label the experimental points corresponding to the border (solid line) between the basins.

The reason of the AQS-EDFL's switching to CQS or SBS-QS regime is the existence or the absence of spurious narrow-line CW lasing when AOM is blocked. In the last case the cavity is formed by the output reflector FBG1 and by a small reflection from closed AOM (~ –40 dB) ("bad" cavity). The overall loss of this cavity is estimated to be ~45-47 dB, revealing its very low Q-factor. At long EDF and low AOM repetition rates (the area below the border line in Figure 20) the EDF charge is sufficient to provide fiber gain capable of overcoming the cavity loss, which results in arising CW lasing. After switching AOM on, the CW wave starts to propagate along the main laser cavity (formed by FBG1 and FBG2) with simultaneous amplification of its power by the EDF until the latter reaches SBS-threshold and produces a "giant" SBS-QS pulse. Thus, CW spurious lasing arising in the "bad" cavity is a startup mechanism for SBS-QS pulsing. In the area above the border line in Figure 20 the spurious CW lasing is not established since the EDF cannot accumulate gain sufficient to overcome the "bad" cavity's loss.

To confirm the hypothesis that the mechanism "igniting" SBS-QS pulsing relates to arising of the narrow-line CW lasing in "bad" cavity (when AOM is closed), we fulfilled the experiments on measuring the laser's optical spectra.

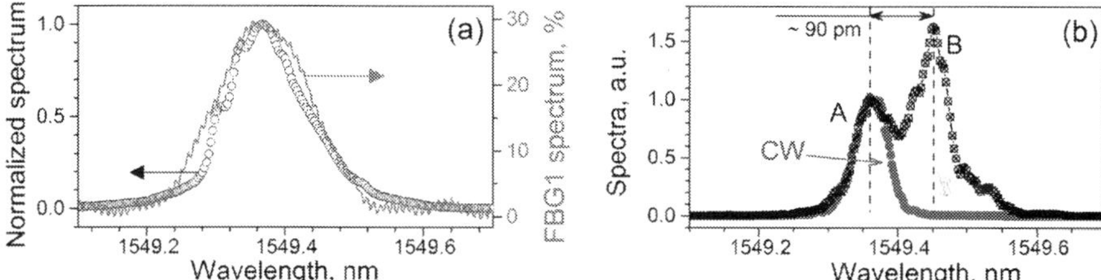

Figure 21. (a) Normalized AQS EDFL spectrum measured at f_{AOM}=8 kHz (open circles, left scale) and FBG1 spectrum (solid line, right scale). (b) Normalized AQS EDFL spectra measured at f_{AOM}=4 kHz (black crossed circles) and at AOM being always blocked (CW lasing, blue circles). In both cases L_{EDF}=7.6 m. OSA resolution is 50 pm.

Firstly, we compared the optical spectra of the laser operated in CQS and SBS-QS regimes; see Figure 21. Comparing the lasing spectra at f_{AOM}=8 kHz (CQS) and f_{AOM}=4 kHz (SBS-QS), one sees that in the former case (Figure 21(a)) the spectrum virtually repeats the reflection spectrum of FBG1 (the one of FBG2 is broader), whereas in the latter case (Figure 21(b)) it consists of two spectral lines A and B, spaced by ~90 pm (~11 GHz, a Brillouin shift at 1550 nm). Both laser lines A and B are narrower (~60 pm) than the lasing (in fact ASE) spectrum at CQS, ~160 pm. Supposedly, line A, centered at the FBG1's peak, corresponds to CW lasing arisen when AOM is in OFF state (between the adjacent AOM's gates) whereas line B – to SBS-QS pulsing. To provide more arguments in favor of this hypothesis, we plot in the same figure (Figure 21(b)) the spectrum of CW lasing, when the main cavity is always blocked (i.e. AOM is continuously in OFF state). It is seen that line A vastly reproduces the CW lasing spectrum. This allows us to reveal that the narrow line A is the signature of CW lasing and that it "ignites" the SBS-process; accordingly, the narrow line B, shifted by ~90 pm to the Stokes side, is the signature of SBS-lasing.

To shade more light on the scenario drawn above, we also measured the spectral width of spurious CW lasing arising when AOM is closed by employing another technique that utilizes a modified delayed self-heterodyne interferometer (DSHI, see Figure 22), described in details in [35-37]. To carry out this, the laser output signal was split into two beams, one of which being passed through an optical frequency shifter (AOM) and then through a long recirculating fiber delay line made using SMF-28 fiber (L_d=20 km in our case). As the result, optical frequency was shifted by f_{AOM}=111 MHz at each pass along the delay line. After that, the signals (un-delayed and delayed) were combined and registered by a fast PD, connected to a RF spectrum analyzer (RFSA). To increase the method's sensitivity, an EDFA was included into the DSHI scheme at the fiber delay line's exit. We should note that the multi-pass self-heterodyne scheme used for estimation the EDFL's line width was chosen because, for correct measurements, the path difference should be much higher that the light source's coherence length (~ 20 km, see below).

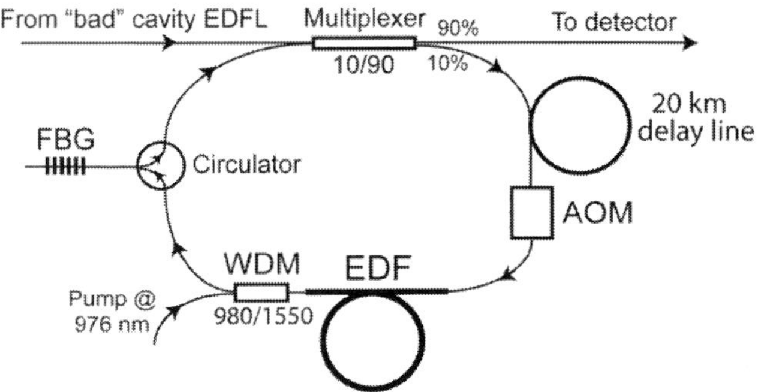

Figure 22. DSHI setup. AOM serves as a 111 MHz frequency shifter (always is open); 100%-FBG filters ASE produced by EDFA (reflects light only at the wavelength of the EDFL's "bad" cavity).

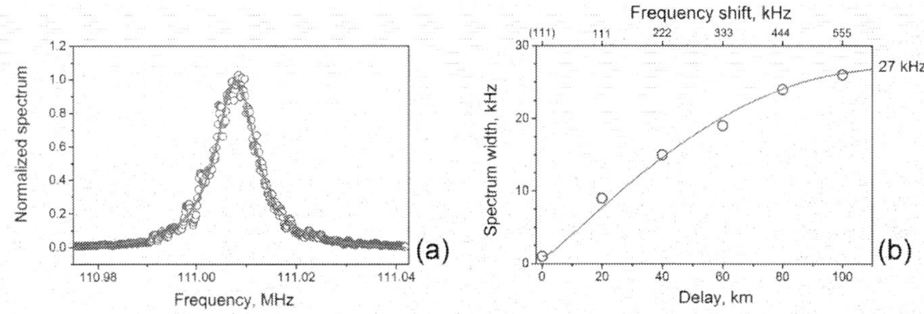

Figure 23. (a) An example of the CW EDFL RF-spectrum obtained from DSHI at 111 MHz; experimental data are shown by circles, solid line is a Lorentzian fit that gives 9.8-kHz width of the spectrum. (b) Spectrum width of CW EDFL measured using the DSHI technique at RF frequencies multiplied by AOM's frequency shift (111 MHz) when the cavity is blocked (circles). Solid line is a fit obtained using the theory presented in [37] (equation 16). The point at zero-delay was obtained with a 111-MHz frequency shift in the absence of delay line, which gives the RFSA's resolution (1 kHz).

Figure 23 shows the spectral width of the signals at the frequencies multiplied by the AOM's frequency shift, measured after fitting the DSHI signal by the Lorentzian law, in function of the delay-line length $L_d \times N$ (N is the number of passes through the line). As shown in [35-37], at high N a signal's width approaches the real value of an EDFL's optical width; in our case, this value is ~30 kHz, which corresponds to coherence length of ~20 km. The earlier reports [38, 39] also revealed a very narrow EDFL's line.

Consequently, a narrow-line CW laser wave developing in "bad" cavity and being a prerequisite of SBS-QS pulsing is highly coherent. This explains why SBS is unavoidably boosted by spurious CW lasing in the "blocked" cavity after the moment of AOM's opening, when the EDF is strongly inverted and thus strongly amplifying. Indeed, "unlimited" in length and therefore "uniform" Brillouin dynamic grating [40], induced in intra-cavity EDF (the fiber is always shorter than the estimated coherence length, ~20 km), is the main cause of SBS-QS pulsing.

5. Conclusions

In this Chapter, we reported some of the important nonlinear-optic features of EDFs, which, on one hand, impact efficiency of CW EDFLs on their base and, on the other hand, underlie the operation regimes established in EDFLs Q-switched using AOMs.

In particular, we showed that strong ESA transitions inherent in the Er^{3+} system at both the pump (~978 nm) and the laser (~1550 nm) wavelengths and ions' clustering inherent in the EDFs heavily doped with Er^{3+} cause unavoidable nonlinear losses that, in turn, strongly reduce efficiency of EDFLs, as compared to efficiency of Ytterbium-doped fiber lasers. We demonstrated as well that for making a correct numerical modeling of an EDFL one needs to consider all kinds of the nonlinear losses intrinsic in EDFs.

We also discussed in details the peculiarities of EDFLs operated in actively Q-switched regime using AOM. We demonstrated that the operation regimes of these lasers strongly depend on EDF length and AOM's repetition frequency. Specifically, at short EDF length or at high AOM's repetition frequency the laser operates in "conventional" Q-switching regime (being in fact multi-pass amplification of Er^{3+} SE) where pulses with relatively moderate power and relatively long in duration are composed of several, stable in time, sub-pulses, separated by a photon's round-trip time in the cavity. Furthermore, if EDF length is long enough and AOM's repetition frequency is not too high, the laser turns to the completely different pulsing regime, characterized by much shorter and much powerful pulses; however, pulses of this type are subjected to noticeable timing and amplitude jitters, originated from the stochastic in nature SBS process, ignited by spurious narrow-line CW lasing in "bad" (at closed AOM) cavity.

Acknowledgements

Authors acknowledge support from the CONACyT, Mexico (Project No. 167945).

Author details

Y.O. Barmenkov* and A.V. Kir'yanov

*Address all correspondence to: yuri@cio.mx

Centro de Investigaciones en Optica, Leon, Guanajuato, Mexico

References

[1] Richardson D. J., Nilsson J., Clarkson W. A. High power fiber lasers: Current status and future perspectives [Invited]. Journal of the Optical Society of America B 2010;27(11): B63–92.

[2] Barmenkov Yu. O., Ortigosa-Blanch A., Diez A., Cruz J. L., Andres M. V. Time-domain fiber laser hydrogen sensor. Optics Letters 2004;29(21): 2461-3.

[3] Arellano-Sotelo H., Barmenkov Yu. O., Kir'yanov A. V. The use of erbium fiber laser relaxation frequency for sensing refractive index and solute concentration of aqueous solutions. Laser Physics Letters 2008;5(11): 825-9.

[4] Laming R. I., Poole S. B., Tarbox E. J. Pump excited-state absorption in erbium-doped fibers. Optics Letters 1988;13(12): 1084-6.

[5] Miniscalco W. J. Erbium-Doped Glasses for Fiber Amplifiers at 1500 nm. Journal of Lightwave Technology 1991;9(2): 234-50.

[6] Blixt P., Nilsson J., Carlinas T., Jaskorzynska B. Concentration-dependent upconversion in Er^{3+} -doped fiber amplifiers: Experiments and modeling. IEEE Transactions Photonics Technology Letters 1991;3(11): 996-8.

[7] Myslinski P., Chrostowski J., Koningstein J. A. K., Simpson J. R. Self-mode locking in a Q-switched erbium-doped fiber laser. Applied Optics 1993;32(3): 286-90.

[8] Escalante-Zarate L., Barmenkov Y. O., Cruz J. L., Andres M. V. Q-switch modulator as a pulse shaper in Q-switched fiber lasers. IEEE Photonics Technology Letters 2012;24(4): 312-4.

[9] Barmenkov Y. O., Kir'yanov A. V., Cruz J. L., Andres M. V. Dual-kind Q-switching of erbium fiber laser. Applied Physics Letters 2014;104(9): 091124.

[10] Nykolak G., Becker P. C., Shmulovich J., Wong Y. H., DiGiovanni D. J., Bruce A. J. Concentration-dependent $^4I_{13/2}$ lifetimes in Er^{3+} -doped fibers and Er^{3+} -doped planar waveguides. IEEE Photonics Technology Letters 1993;5(9): 1014-6.

[11] Guzman-Chavez A. D., Barmenkov Y. O., Kir'yanov A. V. Spectral dependence of the excited-state absorption of erbium in silica fiber within the 1.48–1.59 μm range, Applied Physics Letters 2008;92(19): 191111.

[12] Bellemare A. Continuous-wave silica-based erbium-doped fibre lasers. Progress in Quantum Electronics 2003;27(4): 211–66.

[13] Krug P. A., Sceats M. G., Atkins G. R., Guy S. C., Poole S. B. Intermediate excited-state absorption in erbium-doped fiber strongly pumped at 980 nm. Optics Letters 1991;16(24): 1976-8.

[14] Barnes, W. L. Laming, R. I. Tarbox, E. J.; Morkel, P. R. Absorption and emission cross-section of Er^{3+} doped silica fibers. IEEE Journal of Quantum Electronics 1991;27(4): 1004-10.

[15] Bolshtyansky M., Mandelbaum I., Pan F. Signal Excited-State Absorption in the L-Band EDFA: Simulation and Measurements. Journal of Lightwave Technology 2005;23(9): 2796-9.

[16] Barmenkov Y. O., Kir'yanov A. V., Guzman-Chavez A. D., Cruz J. L., Andres M. V. Excited-state absorption in erbium-doped silica fiber with simultaneous excitation at 977 and 1531 nm. Journal of Applied Physics 2009;106(8): 083108.

[17] del Valle-Hernandez J., Barmenkov Y. O., Kolpakov S. A., Cruz J. L., Andres M. V. A distributed model for continuous-wave erbium-doped fiber laser. Optics Communications 2011;284(22): 5342–7.

[18] Kir'yanov A. V., Barmenkov Y. O., Il'ichev N. N. Excited-state absorption and ion pairs as sources of nonlinear losses in heavily doped Erbium silica fiber and Erbium fiber laser. Optics Express 2005;13(21): 8498-507.

[19] Kik P. G., Polman A., Exciton–Erbium interactions in Si nanocrystal-doped SiO_2. Journal of Applied Physics 2000;88(4): 1992–8.

[20] Kik P. G., Brongersma M. L., Polman A. Strong exciton-Erbium coupling in Si nanocrystal-doped SiO_2, Applied Physics Letters 2000;76(17): 2325–7.

[21] Kir'yanov A. V., Dvoyrin V. V., Mashinsky V. M., Barmenkov Y. O., Dianov E. M. Nonsaturable absorption in alumino-silicate bismuth-doped fibers. Journal of Applied Physics 2011;109(2): 023113.

[22] Berkdemir C., Ozsoy S. On the temperature-dependent gain and noise figure of C-band high-concentration EDFAs with the effect of cooperative upconversion. Journal of Lightwave Technology 2009;27(9): 1122–7.

[23] Jung M., Chang Y. M., Jhon Y. M., Lee J. H., Combined effect of pump excited state absorption and pair-induced quenching on the gain and noise figure in bismuth oxide-based Er^{3+} -doped fiber amplifiers, Journal of the Optical Society of America B 2011;28(11): 2667–73.

[24] Valles J. A., Bordejo V., Rebolledo M. A., Diez A., Sanchez-Marin J. A., Andres M. V. Dynamic characterization of upconversion in highly Er-doped silica photonic crystal fibers. IEEE Journal of Quantum Electronics 2012;48(8), 1015–21.

[25] Hehlen M. P., Cockroft N. J., Gosnell T. R., Bruce A. L., Nikolak G., Shmulovich J. Uniform upconversion in high concentration Er^{3+}-doped soda lime silicate and aluminosilicate glasses. Optics Letters 1997;22(11): 772–4.

[26] Myslinski P., Nguyen D., Chrostowski L., Effects of concentration on the performance of Erbium-doped fiber amplifiers. Journal of Lightwave Technology 1997;15(1): 112–9.

[27] Kir'yanov A. V., Barmenkov Y. O., Sandoval-Romero G. E., Escalante-Zarate L. Er^{3+} Concentration effects in commercial erbium-doped silica fibers fabricated through the MCVD and DND technologies. IEEE Journal of Quantum Electronics 2013;49(6): 511-21.

[28] An H. L., Pun E. Y. B., Liu H. D., Lin X. Z. Effects of ion clusters on the performance of heavily doped Erbium-doped fiber laser. Optics Letters 1998;23(15): 1197–9.

[29] Lim E.-L., Alam S. U., Richardson D. J., High energy in-band pumped Erbium doped fibre amplifiers. Optics Express 2012;20(17): 18803–18.

[30] Lim E.-L., Alam S. U., Richardson D. J. Optimizing the pumping configuration for the power scaling of in-band pumped Erbium doped fiber amplifiers. Optics Express 2012;20(13), 13886–95.

[31] Lim E. L., Alam S. U., Richardson D. J. The multipeak phenomena and nonlinear effects in Q-switched fiber laser, " IEEE Photonics Technology Letters 2011;23(23): 1763–5.

[32] Kolpakov S. A., Barmenkov Y. O., Guzman-Chavez A. D., A. V. Kir'yanov, Cruz J. L., A. Diez, Andres M. V. Distributed model for actively Q-switched erbium-doped fiber lasers. IEEE Journal of Quantum Electronics 2011;47(7): 928–34.

[33] Fotiadi A. A., Megret P., Self-Q-switched Er-Brillouin fiber source with extra-cavity generation of a Raman supercontinuum in a dispersion shifted fiber. Optics Letters 2006;31(11), 1621–3.

[34] Barmenkov Y. O., Kir'yanov A. V., Andres M. V. Experimental study of the nonlinear dynamics of an actively Q-switched ytterbium doped fiber laser. IEEE Journal of Quantum Electronics 2012;48(11): 1484–93.

[35] Tsuchida H. Simple technique for improving the resolution of the delayed self-heterodyne method. Optics Letters 1990;15(11): 640–2.

[36] Dawson J. W., Park N., Vahala K. J., An improved delayed self-heterodyne interferometer for linewidth measurements. IEEE Photonics Technology Letters 1992;4(9): 1063–5.

[37] Horak P., Loh W. H. On the delayed self-heterodyne interferometric technique for determining the linewidth of fiber lasers. Optics Express 2006;14(9): 3923–8.

[38] Chernikov S. V., Taylor J. R., Kashyap R., Coupled-cavity erbium doped lasers incorporating fiber grating reflectors. Optics Letters 1993;18(23): 2023–5.

[39] Choi K. N., Taylor H. F. Spectrally stable Er-fiber laser for applications in phase-sensitive optical time-domain reflectometry. IEEE Photonics Technology Letters 2003;15(3): 386–8.

[40] Song K. Y. Operation of Brillouin dynamic grating in single-mode optical fibers. Optics Letters 2011;36(23): 4686–8.

Chapter 9

Writing of Long Period Fiber Gratings Using CO_2 Laser Radiation

João M.P. Coelho, Catarina Silva, Marta Nespereira,
Manuel Abreu and José Rebordão

Additional information is available at the end of the chapter

http://dx.doi.org/10.5772/59153

1. Introduction

The development of optical fiber gratings (OFGs) had made significant advances both in terms of research and development of optical communications and sensors. OFGs are intrinsic devices that allow modulate the properties of light propagation within the fiber. Grating structures are comparatively simple and in its most basic form, consist on a periodic modulation of the properties of an optical fiber (usually the refraction index of the core). Its application as a sensing element is advantageous because of the intrinsic characteristics of the fiber sensors, such as remote sensing, electromagnetic immunity, weight and compactness, and capability for real time sensing and low cost [1].

Among OFGs, long period fiber gratings (LPFGs) are one of the most important fiber-based sensors. They were first presented by Vengsarkar and co-workers in 1996 [1] as band-rejection filters. Since then, LPFG technology has been receiving continuously growing attention from scientific community. Due to their spectral characteristics, LPFGs have found many applications in both optical communications and sensing systems. In the optical communications field, they have been demonstrated as gain equalizers [1], dispersion compensators [2], optical switches [3], components in wavelength division multiplexing (WDM) systems [4], band rejection filters and mode converters [5]. For optical sensing applications, LPFGs have been implemented as a temperature [6], strain [7] and refractive index sensor [8-10]. As element sensor a LPFG exhibits high sensitivity to the refractive index (RI) of the material surrounding the cladding surface. Other strengths are their low insertion losses, low back-reflection, polarization independence, relatively simple fabrication, and remote sensing easily multiplexed.

These gratings devices operate in transmission mode and are manufactured with periods typically in the range from 100 μm to 1000 μm [6]. Their large modulation period promotes the coupling of the light from the fundamental core mode to co-propagating cladding modes in a single-mode fiber [1]. The light coupled to the cladding decays quickly due to the absorption and scattering by the coating over the cladding.

A commonly-used optical fiber typically consists of a core and a cladding. In Figure 1 it is schematized the coupling mode that occurs in a LPFGs inscribed in single-mode fiber (SMF). As a result, the transmission spectrum of a LPFG has a series of resonant loss peaks (attenuation bands) centered at discrete wavelengths. A typical (theoretical) example is shown in Figure 2.

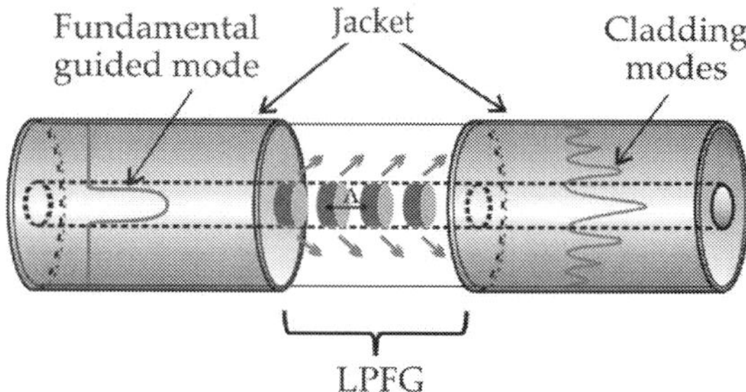

Figure 1. Schematic diagram of a mode coupling in a long period grating [11].

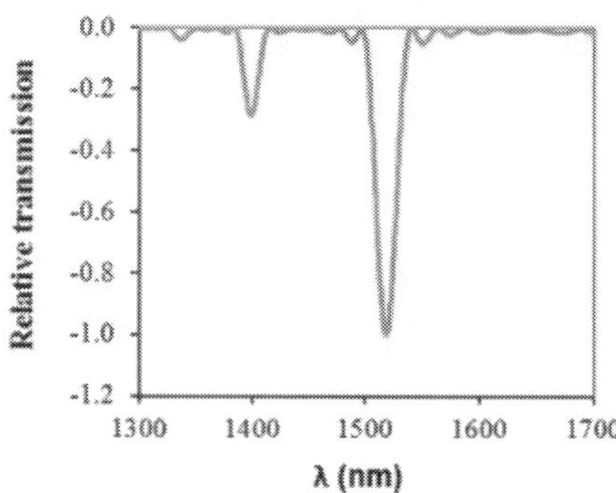

Figure 2. Theoretical example of the spectrum of a 500 μm-LPFG inscribed in a Corning SMF-28 fiber.

The resonant wavelength at which coupling takes place satisfies the phase matching condition

$$\lambda_{res}^{m} = \left(n_{eff,co} - n_{eff,cl}^{m} \right) \Lambda \tag{1}$$

in which $n_{eff,co}$ and $n_{eff,cl}$ are the effective refractive indices of the fundamental core mode and the m_{th} cladding mode, respectively; λ_{res} is the center wavelength of the transmission resonance; and Λ is the period of refractive index modulation [12].

Both resonant wavelength and attenuation amplitude of LPFGs are sensitive to a several physical parameters: temperature, strain, external refractive index, fiber dimensions, grating pitch, etc. These physical parameters affect the coupling between the core and cladding modes, which could lead to both amplitude and wavelength shift of the attenuation bands in the transmission spectrum [13]. The measurement of these spectral parameters in response to the environment surrounding the grating region is the basis of sensing with these devices.

In particular, LPFGs exhibits high sensitivity to changes in the RI of the medium surrounding the fiber due to the dependence of the phase matching condition upon the effective refractive index of the cladding modes. This characteristic makes these structures very attractive for sensing applications and several configurations, as well as applications, of LPFG devices for the measurement of physical and chemical quantities have been studied [6-13].

As mentioned before, LPFGs are created by inducing a periodic refractive index modulation (typically 10^{-4} [14-19]) in the core of an optical fiber with period lengths of several hundred micrometers. This can be made by permanent modification of the refractive index of the optical fiber's core or by physical deformation of the fiber.

Since the first demonstration of these devices by writing the grating with ultraviolet (UV) laser light through an amplitude mask in 1996 [1], several methods have been developed to create and improve the quality of the LPFGs. The conventional UV writing method is based on the exposure of photosensitive optical fibers to UV light through an amplitude mask, phase mask or by interferometry [8-9]. In germanium-doped (Ge-doped) silica fibers, UV light changes the refractive index of the core fiber, being this effect related with the formation of Ge-associated defects [20]. However, this method has some inherent limitations. It requires complex and time-consuming processes, including annealing and hydrogen loading for making the fibers photosensitive, and different amplitude masks when different dimensions are required. Also, the masks need replacement after prolonged usage and the required laser equipment is expensive.

There are, however, many non-photochemical methods available for gratings inscription. These include ion beam implantation [21], applying mechanical pressure [22], electric-arc discharge [23, 24] and irradiation by femtosecond laser pulses [25] or CO_2 laser beam [26, 27]. Among these methods, the latter is particularly flexible, as it can be applied to different types of fibers and the writing process can be computer-controlled to fabricate complicated gratings profiles without using amplitude masks. Furthermore, the use of infrared radiation as showed that the resulting interaction mechanisms are more efficient and allow creating devices with particular characteristics.

Taking this in consideration, this chapter addresses the application of CO_2 laser radiation in writing LPFGs and the physical principles involved in the process. A special emphasis will be given to the modulation of the refractive index resultant from the interaction between the mid-infrared (MIR) radiation (emitted by these lasers) and a conventional Ge-doped SMF.

In section 2 it will be described the fabrication process for applying the MIR radiation, starting with a review on the use CO_2 lasers in the creation of LPFGs. Experimental work is presented as well as the physical principles that may be responsible to induce the periodic refractive index modulation in the fiber's core.

Section 3 will address the subject of simulating the thermal mechanical processes involved in the process. Analytical and numerical models will be analysed and compared. In particular, a 3D finite element method (FEM) model will be presented, including the temperature dependence of the fiber's main parameters.

In section 4, it is presented a practical example of writing LPFGs on a Ge-doped fiber using a CO_2 laser. A comparison between calculated data and experimental data is made, and future work towards a full understanding of the physical processes is foreseen.

2. CO_2 laser induced LPFG

The use of CO_2 lasers to write LPFGs has emerged as an important alternative. Compared with other LPFG fabrication methods, this irradiation technique provides many advantages, including high thermal stability, more flexibility, lower insertion loss, lower cost. This section addresses the application of the CO_2 laser irradiation in creation of LPFG and the physical principles involved in this process.

The first results of the application of the 10.6 µm wavelength radiation emitted by CO_2 lasers for the fabrication of LPFG in a conventional fiber were published in 1998 [26-28]. Since then, different experimental methodologies have been described. The most common is the point-to-point technique using a static asymmetrical irradiation with a CO_2 laser emitting in a specific mode (continuous wave, CW, or pulsed) and a lens focusing the beam on the fiber. Akiyama [28] used continuous wave emission, while Davis [26] applied laser pulses with powers of about 0.5W.

In Figure 3 it is presented a schematic representation of a typical LPFG fabrication system based on the point-to-point technique employing a CO_2 laser. The optical fiber with its buffer stripped is placed on a motorized translation stage. In order to keep the fiber straight during the writing process, a small weight is applied at the end of the fiber and the laser beam is focused on the fiber. In this technique the periodicity of the LPFG writing is accomplished by moving the fiber along its axial direction via a computer controlled translation stage, which also controls the CO_2 laser beam emission. A broadband source and an optical spectrum analyzer (OSA) are employed to monitor the evolution of the spectrum during the laser irradiation. This method has the advantage of requiring a simpler setup, although the irradiation occurs on just one of the sides of the fiber. Also, the translation stage movement can generate vibrations that may be transmitted to the fiber, affecting the quality and repeatability of the LPFG. This problem can be solved using a system where the beam delivery system is moving instead of the fiber [29] or if the fiber is maintained static and a two-dimensional galvanometric mirrors system scans the beam over its surface [29-30]. Some authors reported hybrid methods, combining point-by-point and scanning [31,32]. In this chapter we will consider the method presented by Alves *et al.* [32] that combines a translation stage to move the fiber synchronized with a one-dimensional scan over a cylindrical lens.

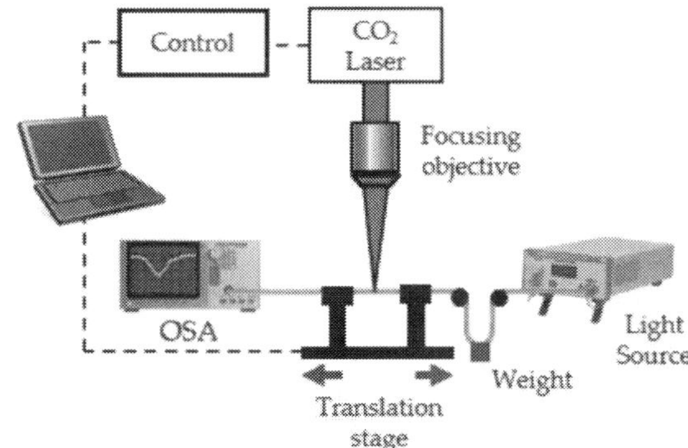

Figure 3. Schematic diagram of a typical LPFG fabrication system based on the point-to-point technique using a CO$_2$ laser.

Since the silica glass has strong absorption around the wavelength of the CO$_2$ laser radiation, the beam intensity is gradually attenuated along the incident direction, resulting in asymmetric RI modulation. Such distribution could cause coupling of the core mode to both the symmetric cladding modes and the asymmetric cladding modes [33]. As a result, high fiber grating birefringence and high polarization-dependent loss can become inevitable [30].

In order to solve the birefringence problem, different methods have been proposed. Single–side and symmetric exposures to the laser radiation were compared by Oh *et al.* [34], demonstrating that the polarization-dependent loss of the single-side exposure (1.85 dB at 1534 nm) could be significantly reduced to 0.21 dB by applying the second method. Nevertheless, due to its simplicity, the single-size exposure is the most common applied methodology. Grubsky [35] proposed a simple fabrication method for obtaining high-quality LPFGs with a CO$_2$ laser using a reflector to make the fiber's exposure axially uniform. Zhu *et al.* [36] used a high frequency CO$_2$ laser system to write LPFGs with uniform RI modulation by introducing twist strains to the fiber and then exposing the two sides of the fiber to the laser radiation. Experimental results showed that twisted LPFGs (T-LPFGs) exhibited clear spectra, low insertion losses, and low polarization-dependent losses when compared with those created by exposure to a single-side CO$_2$ laser beam [37].

Gratings inscription was also achieved through the use physical deformations (or geometrical deformation). Wang *et al.* [38] proposed a new method for writing an asymmetric LPFG by means of carving periodic grooves on one side of a fiber with a focused CO$_2$ laser beam. The periodic grooves do not cause a large insertion loss because these grooves are totally confined within the cladding and have no influence on the light transmission in the fiber's core. The grooves enhance the efficiency of grating fabrication and introduce unique optical properties (extremely high strain sensitivity). In 2006, Su *et al.* [39] demonstrated the possibility of producing long-period grating in multimode fibers by deforming the geometry of the fiber periodically with a focused CO_2 laser beam and applied them to strain measurements.

Besides conventional single-mode fibers (SMFs), CO_2 laser have been used to write LPFGs in other types of fiber, such as boron doped SMF [40,41], and photonic crystal fibers (PCFs) [42,44].

There have been many studies focused on the understanding of the physical mechanisms involved in the CO_2 laser writing process for different kinds of fibers. Most of the existing works consider that the main mechanisms responsible for creating a refractive-index change in the CO_2 laser irradiation-induced LPFGs are residual stress relaxation processes [26,29,35,40-44]. In these processes, heat created by the absorption of laser energy in the material play an important role and, as will be explained next, modelling the writing process requires considering both thermal and mechanical processes.

3. Thermo-mechanical modelling

As mentioned before, the main effect behind LPFG fabrication using CO_2 lasers is heating, where the refraction index change is achieved by irradiating a fiber submitted to a tensile stress. The high absorption of the glass material to the MIR radiation emitted by these lasers leads to an excess of energy due to the excitation of the lattice which is transformed into heat, increasing the material's temperature from its surface to its bulk by heat conduction. This effect depends on the irradiation time and on the thermal diffusivity of the material, it is localized and periodically induced in the fiber's length, being responsible for the creation of the gratings.

Considering the temperature, T, changing with time, t, (transient regime) due to the action of a heat source $Q(r,t)$, the resulting energy balance leads to the heat conduction equation:

$$\left[\frac{\partial \rho}{\partial t} + \rho \nabla \cdot \vec{u}\right] \int C_p \, dT + \rho C_p \left(\frac{\partial T}{\partial t} + \vec{u} \cdot \nabla T\right) - \nabla \cdot K \nabla T = q(T) + Q(r,t) \tag{2}$$

where r represents the geometric coordinates (depending on the geometry) and being ρ the density, $\vec{u}$ the velocity vector, C_p the specific heat and K the thermal conductivity. The convective and radiative heat flux is represented by [45]:

$$q(T) = h(T_{ext} - T) + \sigma_B \varepsilon \left(T_{amb}^4 - T^4\right) \tag{3}$$

being T_{ext} the external temperature, T_{amb} the ambient temperature, h the heat transfer coefficient, σ_B the Stefan-Boltzmann constant and ε the surface emissivity.

If enough energy is applied, differences in the thermal expansion coefficients and viscosity of core and cladding lead to residual thermal stresses and draw-induced residual stresses, and refractive index change results from frozen-in viscoelasticity [46]. The analysis of these effects is complex and highly dependent on the physical characteristics of the different materials composing the optical fiber. For simplifying the subject, we will consider silica-based single mode fibers since they are at the base of most LPFGs manufactured using CO$_2$ lasers. Also, from the different irradiation methodologies explained in the previous section, we will consider the coordinate referential illustrated in Figure 4. The main interfaces between regions of interest in the fiber, illustrated in Figure 4(b), are represented by a point in the upper surface (relative to laser incidence), two points in the cladding/core interface and one point in the bottom interface. In order to simplify the plots regarding calculated data in section 4, and since early work demonstrated low variation in temperature between the cladding/core interfaces [47-49], we use the core's middle point instead.

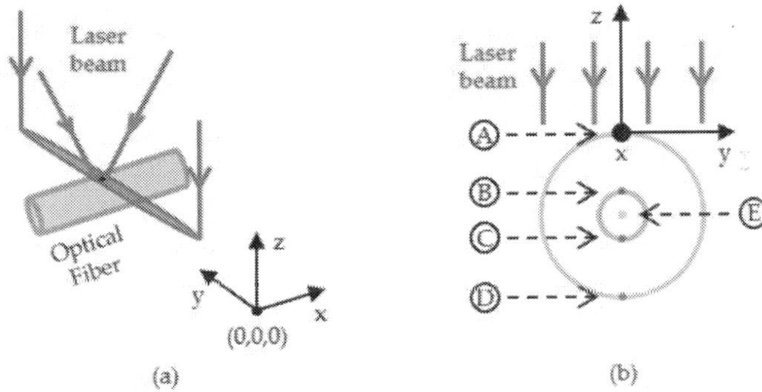

Figure 4. (a) Schematic of coordinates used in this work and (b) optical fiber cross-section indicating the considered referential, the interfaces between the different regions, and points of interest: A – irradiated surface, B – core/cladding interface (upper), C – core/cladding interface (lower), D – bottom surface and E – middle point. The origin of the reference system is in the middle of the laser line.

3.1. Laser heating modelling

When considering an homogeneous isotropic material, the condition of mass conservation, and introducing the thermal diffusivity k, given by $K/(\rho C_p)$, Equation (2) can be simplified to:

$$\left(\frac{\partial T}{\partial t} + \vec{u} \cdot \nabla T\right) - k\nabla \cdot \nabla T - q(T) = \frac{Q(r,t)}{\rho C_p} \tag{4}$$

For a laser beam incident on a surface and propagating in the z-direction:

$$Q(r,t) = a_T(1-R)I(r)\varphi(t) \tag{5}$$

where a_T is the attenuation coefficient of the material, R its reflectance and $I(r)$ the irradiance. For continuous wave emission with a duration t_{on}:

$$\varphi(t) = \begin{cases} 0, & \text{if } t \leq 0 \vee t > t_{on} \\ 1, & \text{if } 0 < t \leq t_{on} \end{cases} \tag{6}$$

Accordingly with the irradiation geometry schematized in Figure 4, and considering that the laser beam has an elliptical Gaussian distribution at the surface being irradiated, then [50]:

$$I(r) = I(x,y,z) = \frac{2a_T P}{\pi r_x r_y} \exp\left[-2\left(\frac{x^2}{r_x^2} + \frac{y^2}{r_y^2}\right) - a_T|z|\right] \tag{7}$$

where r_x and r_y are the values of the ellipse's semi-minor and semi-major axis, respectively.

If one neglects radiative and convective losses and considers temperature dependent material properties, then it is possible to obtain analytical expressions to the temperature. Typically [47,50,51], the heat equation is solved numerically using the Green's function method and the temperature can be obtained [49]:

$$T(x,y,z,t) = \frac{(1-R)P}{4\pi K r_x r_y} \int_0^{\sqrt{4kt}} \Psi(x,y,s) \cdot \left[\exp\left(a_T|z|\right) erfc\left(\frac{a_T s}{2} + \frac{|z|}{s}\right) + \right.$$
$$\left. + \exp\left(-a_T|z|\right) \cdot \sqrt{a^2 + b^2} \cdot erfc\left(\frac{a_T s}{2} - \frac{|z|}{s}\right)\right] ds \tag{8}$$

with

$$\Psi(x,y,s) = \frac{a_T s}{\dfrac{s^2}{(r_x r_y)} + 1} \cdot \exp\left[\frac{x^2}{r_x^2 + s^2} - \frac{y^2}{r_y^2 + s^2} + \frac{(a_T s)^2}{4}\right] \tag{9}$$

From (8) an on-axis approximation can be used [51] and the temperature rise can be approximated through simple analytical expressions:

$$\Delta T(0,0,\infty) = \frac{(1-R)P}{2\sqrt{\pi r_x r_y} K} \tag{10}$$

for the steady state condition ($t \gg r_x^2 r_y^2 / 4k$) and

$$\Delta T(0,0,t) = \frac{(1-R)P}{\pi^{3/2}K\sqrt{r_x\,r_y}}\tan^{-1}\left(\sqrt{\frac{4kt}{r_x\,r_y}}\right) \tag{11}$$

under transient conditions. Although Yang *et al.* [51] proved that these equations can be used to study the thermal transport in CO$_2$ laser irradiated fused silica, when modelling the effects in writing LPFGs, the full 3D temperature distribution is required and a numerical approach is mandatory [48]. Using a dedicated Finite Element Method (FEM) program, although computationally demanding, allows not only considering all the physical phenomena, but also including the temperature dependence of the materials and even the simulation of consecutive periods. If two laser shots are considered, one centred in (0,0,0) and at *t*=0 s, and the other at *x*=Δx and $t=t_2=t_{on}+\Delta t$. In the latter case ($t=t_2$), equations (6) and (7) are altered accordingly to

$$\varphi(t') = \begin{cases} 0, & \text{if } t' \leq 0 \vee t' > t_{on} \\ 1, & \text{if } 0 < t' \leq t_{on} \end{cases} \text{, with } t' = t - \left(t_{on} + \Delta t\right) \tag{12}$$

$$I(r) = I(x,y,z) = \frac{2a_T P}{\pi r_x r_y}\exp\left[-2\left(\frac{(x-\Delta x)^2}{r_x^2}+\frac{y^2}{r_y^2}\right)-a_T|z|\right] \tag{13}$$

3.2. Residual elastic stresses modelling

Residual elastic stresses are considered those that are frozen into the fiber [52] and have an important impact on the production of LPFG since they affect the refractive index of an optical fiber. When using the considered MIR radiation, two categories of residual elastic stresses must be considered: thermal and draw-induced stresses.

3.2.1. Residual thermal stresses

Residual thermal stresses appear as the optical fiber is cooled down from high temperatures and regions with different thermal expansion coefficients contract differently in time. Dopants introduced increase the differences in viscosity and thermal expansion coefficients.

A solution for the resulting residual thermal stresses of an initially unstressed axisymmetric cylinder heated at a given temperature, T, can be obtained using the radial coordinate r [52,53]

$$\sigma_x = \frac{E}{1-\nu}\left[\frac{2\nu}{r_c^2}\int_{r=0}^{r_c}\alpha T r\,dr - \alpha T\right] \tag{14}$$

$$\sigma_r = \frac{E}{1-\nu}\left[\frac{1}{r_c^2}\int_{r=0}^{r_c}\alpha Tr\,dr - \frac{1}{r^2}\int_{r=0}^{r}\alpha Tr\,dr\right] \tag{15}$$

$$\sigma_\theta = \frac{E}{1-\nu}\left[\frac{1}{r_c^2}\int_{r=0}^{r_c}\alpha Tr\,dr + \frac{1}{r^2}\int_{r=0}^{r}\alpha Tr\,dr - \alpha T\right] \tag{16}$$

being r_c the radius (cladding or core), E the Young's modulus and ν the Poisson's ratio. This solution is valid when the elastic properties can be considered constant, which doesn't applies for optical fibers. Thus, and similarly to the approximated solution for heating, given by equations (8) and (9), not taking in consideration the temperature dependence of the different parameters can lead to non-accurate results. Again, the use of numerical methods, with particular focus on FEM, are appropriated and give the opportunity to combine both heating and thermal stresses models. In this case, by solving equation (4), the thermally-induced residual stresses, σ_T, can be obtained considering the constitutive equations for a linear isotropic thermoelastic material and the stress tensor obtained.

3.2.2. Draw-induced stresses

Residual stress effects on the refractive indices of fibers were reported for the first time by Hibido *et al.* in 1987 [54,55] regarding undoped silica-core single mode fibers. Considering a fiber with core and cladding, having different viscosities due to different dopants concentrations, during the draw process the higher viscosity glass will solidify first and support the draw tension. The low-viscosity glass solidifies conforming with the elastically stretched high-viscosity glass. Then, when the draw force is released at room temperature, the high-viscosity glass cannot contract due to the already solidified low-viscosity glass. The resulting residual axial elastic stresses can be expresses as

$$\sigma_{x,1} = F\left(\frac{E_1}{A_1 E_1 + A_2 E_2}\right) \tag{17}$$

and

$$\sigma_{x,2} = \frac{F}{A_2}\left(\frac{A_1 E_1}{A_1 E_1 + A_2 E_2}\right) \tag{18}$$

being F the draw tension, A the cross-section area and E the Young's module for the considered regions. The indexes 1 and 2 in these equations represent the regions of low-viscosity and high-viscosity glasses, respectively.

Similarly to these stresses, frozen-in inelastic strains were also found if a fiber is rapidly cooled to room temperature while under tension (as in the considered case) [46,52]. Using the equivalent elastooptic relations [52], the isotropic perturbation on the refractive, Δn, index is

$$\Delta n = -\frac{n^3}{6}\chi\left(T_F\right)p\,\sigma \tag{19}$$

In the later expression, n is the nominal refractive index, $\chi(T_F)$ is the relaxation compressibility at the fictive temperature, T_F, p is the orientation average elastooptic coefficient, and σ the overall residual stresses (in MPa) in the fiber's axial direction. Accordingly with Yablon [52], stresses in the other directions can be neglected.

Besides stress-related refractive index change, localized heating can induce micro-deformation of the fiber and changes in the glass structure. The latter is likely to occur in the core for which the fictive temperature (the glass structure doesn't change below the fictive temperature) is lower [56,57]. As reported for a Ge-doped core (e.g. the fictive temperature ranges from 1150 K to 1500 K.

4. A practical example

To illustrate the application of the theory and also correlate it with experimental data, we will consider a common example of LPFG writing using NIR radiation: a standard single-mode fiber, SMF-28 [58], consisting of a core of 3.5 mol% Ge-doped SiO$_2$, is irradiated by a CW 10.6 μm wavelength CO$_2$ laser. For fused silica, $n \approx 1.45$ (in the near-infrared), $\chi(T_F) \approx 0.0568$ GPa^{-1} and $p \approx 0.22$, which allows simplifying equation (19):

$$\Delta n \approx -6.35 \times 10^{-6}\,\sigma \tag{20}$$

Also, in this case, equations (17) and (18) represent, respectively, the residual axial elastic stresses at the core (low-viscosity glass) and the cladding (high-viscosity glass) [46].

The particular conditions used in both simulation and experimental works, as well as the obtained results, will be detailed next.

4.1. Simulation

The simulation of the writing process was made implementing a 3D FEM model using the COMSOL Multiphysics program. Whenever relevant, the material's dependence with temperature was considered and the proper geometry and FEM parameters defined.

Besides COMSOL, two other programs were used: Matcad and a simulation tool developed in MatLab by Baptista [60] based on the three layer model developed by Erdogran [14,60]. The latter was used to apply the refractive index data obtained in the FEM model to simulate the transmission spectrum of a LPFG. Matcad was used to solve equations (8) and (11) and compare the temperature data with that obtained through the FEM model.

4.1.1. Fiber's characteristics

When light interacts with matter, one of the main parameters is the absorption coefficient, a_T. As it can be deduced from the formulae of section 3, it plays a major role in the process of heating the fiber. Besides its dependence with the wavelength of the light, it varies with temperature. This variation is important and for the 10.59 μm CO_2 laser wavelength (λ_1), within 298 K–2,073 K temperature range can be obtained by [61]:

$$a_T(T) = \frac{4\pi}{\lambda_1}\left[1.82 \times 10^{-2} - 10.1 \times 10^{-5}\left(T - 273.15\right)\right] \tag{21}$$

For the thermal conductivity, heat capacity, density and emissivity, the temperature dependence was modelled using native COMSOL functions for a Corning fused silica glass (7940) [48]. The doping effect on most of the parameters was disregarded mainly because the Ge concentration in the fiber's core is very low [62]. However, for the Young's modulus and Poisson's ratio (Figure 5), the function behaviour was extrapolated [63]. Also, both the heat transfer coefficient and reflectivity were considered constant and equals to 418.68 W m^{-2} K^{-1} [45] and 0.15 [51], respectively.

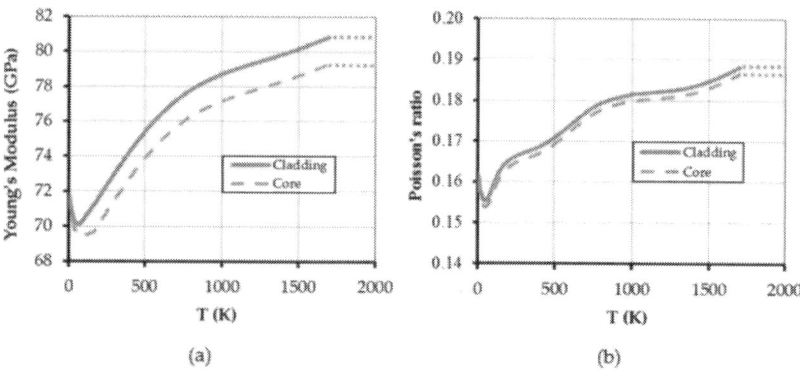

Figure 5. Variation of (a) Young's module and (b) Poisson's ratio with temperature for both fused silica (from COMSOL materials library) cladding and Ge-doped fused silica (extrapolated) core glasses.

4.1.2. Implementation

The physical problem was mathematically solved using the FEM model implemented using the COMSOL Multiphysics 3.5 program to create the transient heat conduction and (mechanical) stress-strain models under the conditions of this study. In order to introduce some of the complexity of stress-related issues regarding the processing of the optical fibers, the residual axial elastic stresses were implemented considering Equations (9) and (10) and the total resulting stress was obtained adding the thermally-induced residual stresses obtained with the program.

As illustrated in Figure 4, the implemented geometry consists of a set of (concentric) cylinders with radius of curvatures accordingly with the characteristics of the core and cladding of the fiber previously described. To avoid the influence of the external boundaries on the irradiated and analysed zones, the overall length for the geometry was set as 13 mm. However, to reduce the computational load and loosen the mesh dimensions in the volumes not affected heat source, the cylinders were implemented as three separate sets; the central one, where the laser incidence will be simulated, has a 1.7mm length. The outer set of cylinders are asymmetric since the second laser shot will be simulated just in the positive x-direction.

Table 1 presents the 3D geometry data and the mesh statistics and Figure 6 shows the implemented mesh, with particular focus on the central irradiation zone. Both outer boundary surfaces are defined as thermally isolated, being one of them fixed. The ambient temperature was considered to be 295 K and equal to the external temperature, T_{ext} in Equation (3).

	Central geometry		External geometries			
	Cladding	Core	Cladding		Core	
Geometry			Left	Right	Left	Right
Length (mm)	1.7		5.15	6.15	5.15	6.15
Radius (μm)	62.5	4.1	62.5		4.1	
Mesh (tetrahedral)						
# elements	26,122	1,795	26,262	26,825	1,710	1,795
min. quality	0.0495	0.1659	0.2287	0.2284	0.2004	0.1956
volume ratio	8.47×10^{-4}	0.1010	0.0028	0.0036	0.4906	0.3593

Table 1. Geometry data and mesh statistics.

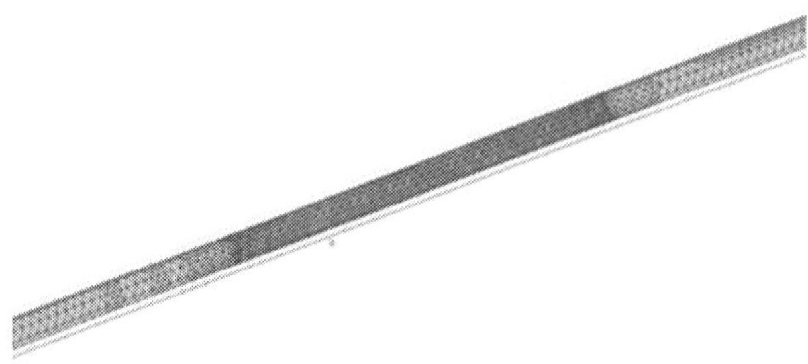

Figure 6. Image of the mesh implemented at the central zone of the FEM geometry. The coloured region corresponds to the central geometry with finer mesh (Table 1).

4.2. Experimental methodologies

The implemented irradiation methodology combines a translation stage to move the fiber synchronized with a one-dimensional scan over a cylindrical lens [32]. Figure 7 shows a schematic of the setup and Figure 8 a picture of its implementation. The light source is a Synrad 48-2 CO_2 laser emitting a 3.5 mm diameter CW laser beam with a wavelength of 10.6 μm. Two mirrors direct the beam towards the focusing lens, a ULO Optics ZnSe cylindrical lens (50 mm focal length) which produces a 0.15 mm x 1.75 mm (measured using the knife-edge method [64]) elliptical spot on the fiber with its longer axis perpendicular to the fiber's axis. One of the mirrors is a galvanometric mirror (Cambridge Technology 6860) which allows a scan over the lens (and, consequently, over the surface of the fiber).

A linear translation stage moves axially the fiber so the periodic refractive index change is accomplished. During the process, one of its ends is fixed and a weight (typically ~50 g ± 0.5 g) hangs on the other, thus creating a tension. Thus, this system acts like a XY writing system. The advantage of the translation stage is its long range with micrometric precision (10 cm with a repeatability of 1 μm). However, it has the disadvantage of having relative low speed when compared with a galvanometric system. Combining the two systems, we have the benefit of having a long X-axis range with a fast Y-axis range and, by moving the laser spot very fast over the fiber (Y-axis), lower interaction times can be achieved. Also, since the laser is not always over the fiber, it's possible to have the laser emitting continuously, which prevents the transients during the laser start-up and allows easier control of its power. Uncertainties regarding the irradiation data are: power, ±0.5 W; duration, ±1 ms.

To monitor the LPFG fabrication process, a broad band light source (Thorlabs S5FC1005S) and an OSA (Agilent 86140B) were used. The irradiated zones were analysed using an optical microscope (Zeiss AxioScope A1) with a maximum amplification of 1,000×. The LPFG period is limited by the laser spot size and by the translation stage minimum movement. In the considered setup, since the translation stage can have 1 μm steps, the laser spot size (150 μm) constitutes the major limitation.

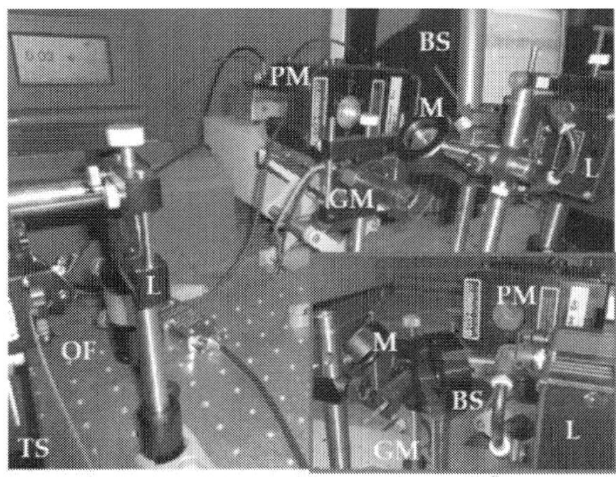

Figure 7. (a) Top and (b) lateral schematic views of the optical setup. BS – beam splitter; CL-cylindrical lens; GM – galvanometric mirror; M – mirror; OF – optical fiber; PM – power meter; TS – translation stage.

Figure 8. Picture of the setup implemented. The inset shows a detail view of the laser output area. BS – beam splitter; CL – cylindrical lens; GM – galvanometric mirror; L – laser; M – mirror; OF – optical fiber; PM – power meter; TS – translation stage.

4.3. Results and analysis

An example of the temperature distribution resulting from considering 6 W laser power, irradiation duration of 0.6 s and 47 g weight (F=0.461 N) is shown in Figure 9(a). The result of simulating a second irradiation (equivalent to have a LPFG period), 1 s after the first, at a distance of 500 μm, is shown in Figure 9(b).

Figure 10 shows the equivalent plots of temperature with time for the irradiated front, core (middle) and the back surfaces (Figure 4), at x=y=0 m and x=500 μm, y=0 m. These plots show that the (spatial) proximity between shots raises the temperature even when they are not under direct irradiation. As the distance reduces (shorter LPFG periods), this secondary heating increases (Figure 11). This is particularly important when the second shot is applied because

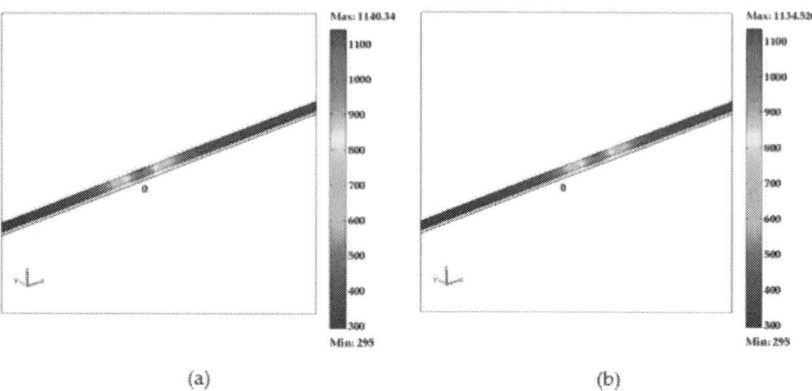

(a)   (b)

Figure 9. Temperature distribution in the implemented 3D geometry for the laser irradiation of an optical fiber (P=6 W; t_{on}=0.6 s; F=0.461 N; Δt=1 s; Δx=500 μm), at (a) t=0.6 s and (b) t=2.2 s. Colour bar values are in K.

of the possibility of annealing, which could (totally or partially) relieve the previously induced internal stresses.

Similarly, as we analyse the temperature distribution at the fiber's axial direction plotted in Figure 12(a) (for the same conditions of Figure 10), the superposition of thermally affected areas in both shots underlines one critical aspect when writing LPFGs: the influence of the grating's period on the pitch width, and consequent "softening" of the spatial refractive index gradient. As it can be seen in Figure 12(b), as the period decreases, the interception's x-coordinate values decreases and the temperature at that point increases.

The practical consequence of this behaviour it's not clear at this time and further research is necessary, but these phenomena can be responsible by the reduction of the success rate in producing LPFGs with shorter periods found in other studies [65].

The importance of using the FEM simulation instead of the analytical equation (11) or the approximated integral equation (8) is observed in Figure 13 for the temperature at the core (middle). In fact, disregarding radiative and convective losses, adding to the important variation on the parameters values with temperature, deviates the solution by a significant amount (about 200 K at its maximum values, between each solution).

As expected, as the laser power increases, the temperature also increases. From the data obtained using the FEM model and plotted in Figure 14, a 3rd order polynomial can be used to relate these two parameters. Considering

$$T(P) = a_3 P^3 + a_2 P^2 + a_1 P + a_0 \tag{22}$$

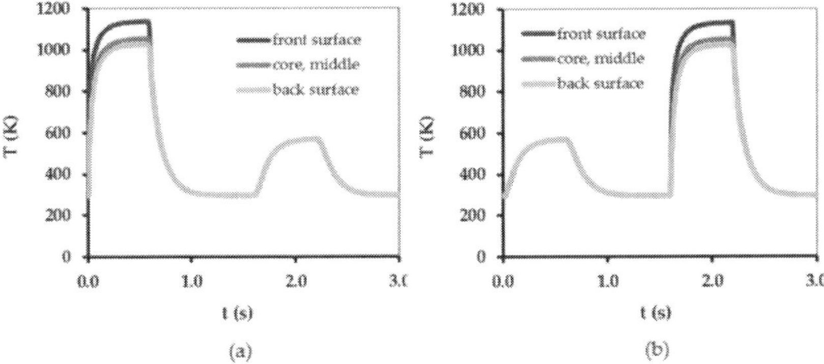

Figure 10. Plots of the temperature evolution during laser irradiation and cooling at (a) x=0 m and (b) x=Δx. (P=6 W; t_{on}=0.6 s; F=0.461 N; Δt=1 s; Δx=500 μm).

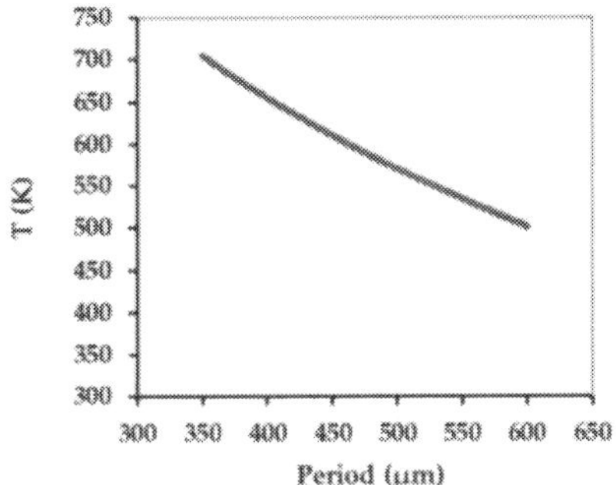

Figure 11. Maximum temperature at x=0 m when the laser is irradiating at x=Δx for different values of Δx (LPFG period). (P=6 W; t_{on}=0.6 s; Δt=1 s; F=0.461 N).

(with the laser power, P, in W, and the temperature, T, in K), Table 2 gives the equivalent parameters for the different zones of the fiber.

For the considered example, at t=0.6 s, axial residual thermal stresses values along z-axis were determined as having a maximum of about −0.8 MPa. Axial elastic stresses act in the opposite direction and were calculated as being $\sigma_{x,1}$=35 MPa and $\sigma_{x,2}$=0.153 MPa. This results in refractive index changes (difference between final and initial values) for the core and cladding of the order of -2×10^{-4} and 4×10^{-6}, respectively. Figure 15 shows the calculated (maximum) changes for different laser power. Under the considered conditions, it is clearly observed that for the cladding, a well-defined step occurs between 4.5 W and 5 W. Regarding the core, although its refractive index shows minor variations, it is possible to observe the beginning of the contribution of thermal stresses around 5 W.

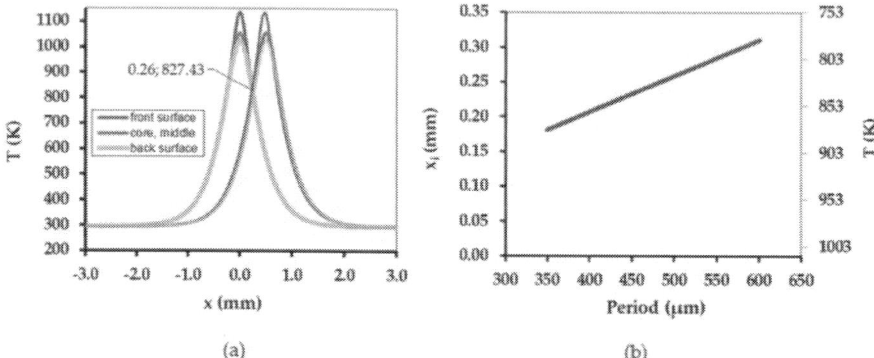

Figure 12. (a) Plots of the temperature distribution at the fiber's axial direction simulated for t=0.6 s and t=2.1 s, with Δx=500 μm (the label shows the position, in mm, and the temperature, in K, of the interception between plots), and (b) the variation on the interception point for different values of Δx (LPFG period). (P=6 W; t_{on}=0.6 s; Δt=1 s; F=0.461 N).

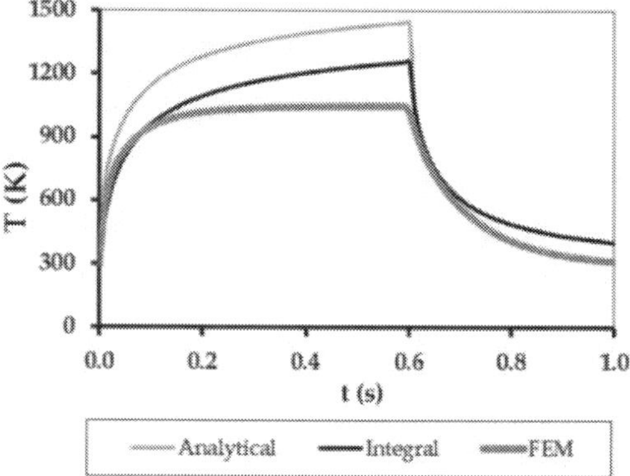

Figure 13. Plots of the temperature evolution during laser irradiation and cooling obtained using equations (11), analytical, and (8), integral, and the FEM model. (P=6 W; t_{on}=0.6 s; F=0.461 N).

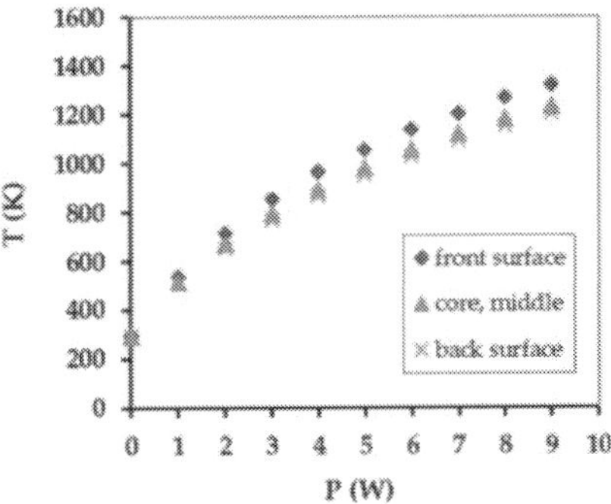

Figure 14. Maximum temperature calculated for different laser powers, at x=0 m. (P=6 W; t_{on}=0.6 s; F=0.461 N).

	a_3	a_2	a_1	a_0
Front surface	1.21	-26.64	256.4	295
Core, middle	1.06	-22.91	224.7	302
Back surface	0.97	-21.18	212.8	302

Table 2. Coefficients for equation (22), depending on the analysis point being considered (Figure 4).

Another parameter to take in consideration is the draw force due to the weight applied on the fiber. Figure 16 shows the behaviour of the refractive index change with the applied force (and corresponding weight), considering P=6 W and t_{on}=0.6 s. This parameter affects mainly the core and a linear relation (with F in N) can be obtained:

$$\Delta n_{core} = \left(-4.9388\,F + 0.0166\right) \times 10^{-4} \tag{23}$$

A microscope photograph of an example of an irradiated fiber under the experimental conditions considered in this work is shown in Figure 17. The imaged zone comprises one 500 μm period of a 25 mm grating irradiated with 6 W, for a duration of 0.6 s (each pulse) and subjected to a weight of 47 g. It is also visible a (small) micrometric deformation of the fiber in each irradiated region. Figure 18 shows the spectral transmission of the written LPFG, comparing the experimental data with the simulated spectrum, obtained using the refractive index changes calculated by the FEM model and using the simulation tool developed by Baptista [59].

In spite the relative spectral transmission data agreement, the experimental work demonstrated that different operational parameters can influence the resulting LPFG. In particular, the way the weight is positioned and laser power fluctuations can easily change the final result.

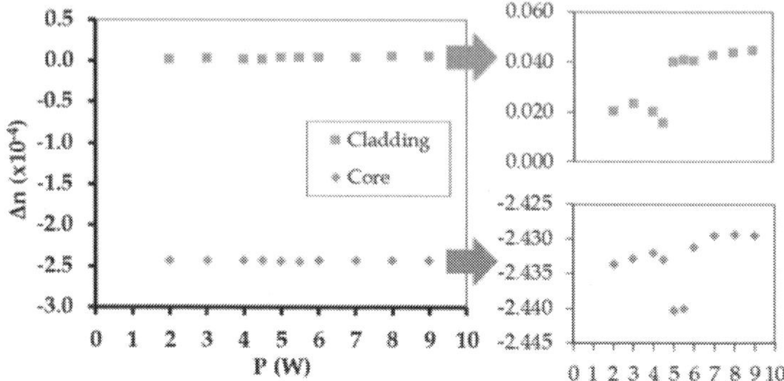

Figure 15. Calculated (maximum) refractive index change at the core and cladding for different applied laser powers. (t_{on}=0.6 s; F=0.461 N).

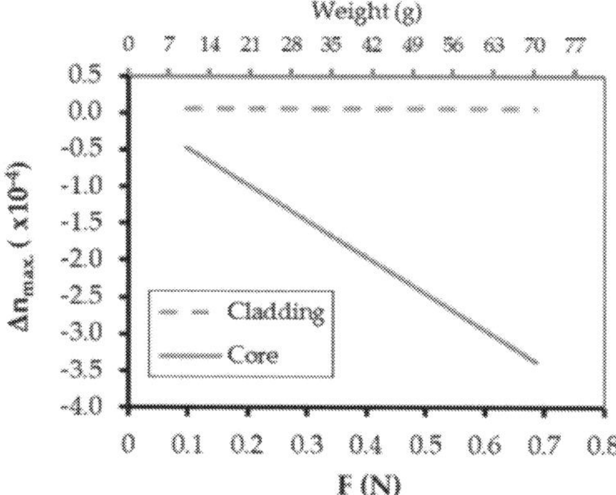

Figure 16. Calculated refractive index change (maximum change for core and cladding) for different applied draw tensions. (P=6 W; t_{on}=0.6 s; F=0.461 N).

In general, the process involves monitoring the transmission spectrum and iterative action on the length of the LPFG so a well-defined resonance is obtained. In our setup, the feedback on the emitted laser power reduces the problem, but not completely.

Although theoretically we could simulate considering different laser powers and weights, experimentally it was observed that using lower laser powers (<5 W) no LPFGs were obtained. Also, using higher laser powers (>8 W) or higher weights (typically >60 g, F > 0.6 N) tapering occurs, a phenomena not included in our model.

The values obtained by the model are in agreement with those estimated by other authors for the refractive index modulations necessary for achieving a fiber optic grating. Temperatures

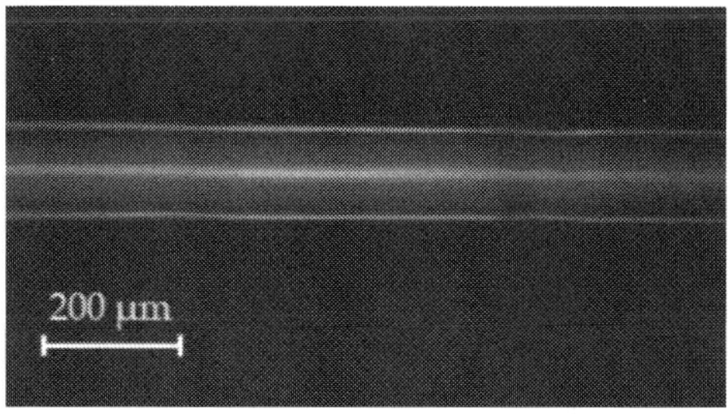

Figure 17. Picture showing two irradiated zones from a 25 mm LPFG with 500 μm period written on a SMF-28 optical fiber. ($P \approx 6$ W; t_{on}=0.6 s; Δt=1 s; $F \approx 0.5$ N)

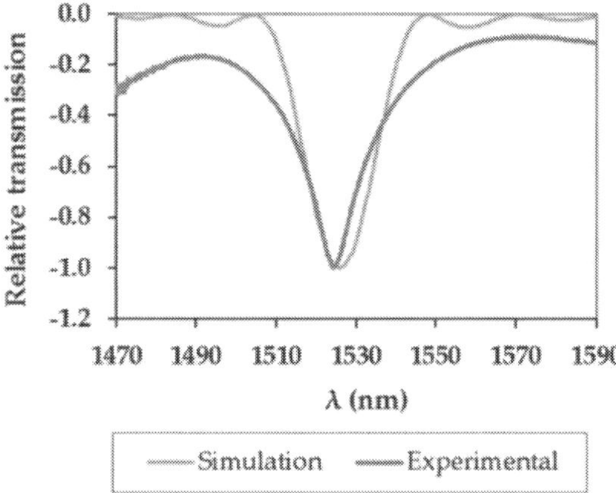

Figure 18. Experimentally obtained and simulated relative normalized spectral transmission under the same conditions considered in Figure 17.

calculated are similar to those obtained by other authors for arc-induced LPFG (e.g. in the range 1,100 K–1,400 K according to [56]) and the refractive index changes are within the overall range mentioned in other works [46,52,56]. Also, the behaviour of the refractive index change as the applied drawing force increases complies with recent experimental indications that the refractive index of the core decreases while the opposite occurs in the cladding, and that this change occurs primarily in the core [66,67].

Nevertheless, a complete model of the complex physical phenomena involved, in particular regarding the refractive index change dependence on stress, requires further research. It's expected that future work will focus on experimental measurements of temperature, stresses and refractive index changes induced by the MIR laser radiation. Also, although published works can contribute in assessing the validity of the results, the influence of specific characteristics of the fibers is a well-recognized issue. In particular, the effect of pre-existing stresses (typically from the fiber manufacture or preparation), differences in the materials, or other unaccounted phenomena can influence the performance of the FEM model when compared with real data. Similarly, the impact of the several approximations considered (e.g., transverse stresses are neglected), unaccounted phenomena like eventual changes on the glass polarizability and using standard material data must be analysed in detail, as well as the influence of the experimental data uncertainties on the model.

5. Conclusions

In Summary, the CO_2 laser irradiation technique is a highly efficient, low cost and versatile technique to write high-quality LPFGs in different types of optical fibers, such as conventional single mode fibers, polarization-maintaining fibers, and photonic crystal fibers. This technique offers a number of advantages over other fabrication techniques. It eliminates the need of using a mask as well as the need for pre-hydrogenation of the fiber and consequent post-thermal annealing to stabilize the gratings. The LPFGs induced by CO_2 laser exhibits unique grating properties, such as high thermal stability.

Although simplifications can lead to analytical equations, FEM modelling allows more realistic simulations of the physical processes involved in the writing of LPFGs using MIR radiation. The 3D model presented simulates the writing of one period and allows the analysis of both thermal and stress data. All the main practical parameters are considered as inputs and thermal dependence of the material's data is included.

The model performance was evaluated by considering a practical example of writing LPFGs on a Ge-doped fiber. Different analysis were presented and it was demonstrated that refractive index changes predicted by the FEM model led to transmission spectra with resonance peaks similar to those obtained experimentally. So, although additional work should be performed to further validate the analysis done (mainly regarding the characterization of stresses acting in the optical fiber and experimentally measuring refractive index changes), the FEM results are in accordance with literature and experimental data.

Acknowledgements

This work was partially supported by FEDER funding through the Programa Operacional Factores de Competitividade – COMPETE and by national funding by the FCT – Portuguese Fundação para a Ciência e Tecnologia through the project PTDC/FIS/119027/2010. The authors gratefully acknowledge José Luis Santos, Orlando Frazão, Pedro Jorge and Paulo Caldas from INESC-Porto for their advices and crucial contributions. A special thanks to David Castro Alves, Fernando Monteiro and António Oliveira for their support to the experimental activities described in this chapter.

Author details

João M.P. Coelho[1,2], Catarina Silva[1], Marta Nespereira[1], Manuel Abreu[1] and José Rebordão[1]

1 Laboratório de Óptica, Lasers e Sistemas, Faculdade de Ciências, Universidade de Lisboa, Pólo do Lumiar, Estrada do Paço do Lumiar, Lisboa, Portugal

2 Instituto de Biofísica e Engenharia Biomédica, Faculdade de Ciências, Universidade de Lisboa, Campo Grande, Lisboa, Portugal

References

[1] Vengsarkar A, Lemaire P, Judkins J, Bhatia V, Sipe J. Long-Period Fiber-Grating-Based Gain Equalizers. Optics Letters 1996; 21(5):335-338.

[2] Das M, Thyagarajan K. Dispersion compensation in transmission using uniform long period fiber gratings. Optics Communications 2001; 190:159–163.

[3] Eggleton BJ, Slushe RE, Judkins JB, Stark JB, Vengsarkar AM. All-optical switching in long-period fiber gratings. Optical Letters 1997; 22(12): 883-885.

[4] Zhu C, Lu Y, Lacquet BM, Swart PL, Spammer SJ. Wavelength-tunable add/drop multiplexer for DWDM using long period gratings and fiber stretchers. Optics Communications 2002; 208(9): 337-344.

[5] Vengsarkar AM, Lemaire PJ, Judkins JB, Bhatia B, Erdogan T, Sipe JE. Long-period fiber gratings as band-rejection filters. In Optical Fiber Communication Conference, OFC95, San Diego, CA, Feb. 1995, PD4-2, 1995.

[6] Khaliq S, James S, Tatam R. Enhanced sensitivity fibre optic long period grating temperature sensor. Measurement Science and Technology 2002; 13(5):792–795.

[7] Patrick H, Chang C, Vohra S. Long Period Fiber Gratings for Structural Bend Sensing. Electronics Letters 1998;34(18):1773–1775.

[8] James S, Tatam R. Optical fibre long-period grating sensors: Characteristics and application. Measurement Science and Technology 2003;14:R49-R61.

[9] Bhatia V, Vengsarkar A. Optical fiber long-period grating sensors. Optics Letters 1996,21(9). 692-694.

[10] Patrick H, Kersey A, Bucholtz F. Analysis of the response of long-period fiber gratings to external index of refraction. Journal of Lightwave Technology 1998;16(9): 1606–1612.

[11] Silva C, Coelho J, Caldas P, Jorge P. Fibre Sensing System Based on Long-Period Gratings for Monitoring Aqueous Environments. In Yasin, M., Harun S, Arof H (Eds.) Fiber Optic Sensors. Ridjeka: InTech 2012. p317–342.

[12] Erdogan T. Fiber grating spectra. Journal of Lightwave Technology 1997;15(8): 1277-1294.

[13] Bhatia V. Applications of long-period gratings to single and multi-parameter sensing. Optics Express 1999;4(11): 457-466.

[14] Erdogan T. Cladding-mode Resonances in Short-and Long-period Fiber Grating Filters. Journal of the Optical Society of America A 1997;14(8): 1760-1773.

[15] Costa RZV, Kamikawachi RC, Muller M, Fabris JL. Thermal characteristics of long-period gratings 266 nm UV-point-by-point induced. Optics Communications 2009;282(5): 816-823.

[16] Zhang Y, Chen X, Wang Y, Cooper K, Wang A. Microgap Multicavity Fabry-Pérot Biosensor. Journal of Lightwave Technology 2007;25(7): 1797-1804.

[17] Lee CE, Gibler WN, Atkins RA, Taylor HF. In-Line Fiber Fabry-Pérot-Interferometer with High-Reflectance Internal Mirrors. Journal of Lightwave Technology 1992;10(10): 1376-1379.

[18] Nespereira M, Silva C, Coelho J, Rebordão J. Nanosecond Laser Micropatterning of Optical Fibers. In: Costa M. (ed.) International Conference on Applications of Optics and Photonics: proceedings of SPIE on CD-ROM: 8001, 3-7 May 2011, Braga, Portugal. Bellingham: SPIE 2011.

[19] Kawasaki BS, Hill KO, Johnson DC, Fujii Y. Narrow-band Bragg reflectors in optical fibers. Optical Letters 1978;3(2): 66–68.

[20] Douay M, Xie WX, Taunay T, Bernage P, Niay P, Cordier P, Poumellec B, Dong L, JF Bayron, Poignant H, Delevaque E. Densification involved in the UV-based photosensitivity of silica glasses and optical fibers. Journal of Lightwave Technology 1997;15(8):1329-1342.

[21] Fujimaki M, Nishihara Y, Ohki Y, Brebner JL, Roorda S. Ion-implantation-induced densification in silica-based glass for fabrication of optical fiber gratings. Journal of Applied Physics 2000;88(10): 5534-5537.

[22] Cardenas-Sevilla GA, Monzon-Hernandez D, Torres-Gomez I, Martinez-Rios A. Mechanically induced long-period fiber gratings on tapered fibers. Optics Communications 2009;282(14):2823-2826.

[23] Rego G, Okhotnikov O, Dianov E, Sulimov V. High-temperature stability of long period fiber gratings produced by using an electric-arc. Journal of Lightwave Technology 2001;19(10): 1574-1579.

[24] Humbert G, Malki A. High performance bandpass filters based on electric arc-induced π-shifted long-period fibre gratings. Electronics Letters 2003;39(21): 1506-1505.

[25] Nikogosyan D N. Long-period Gratings in a standard telecom fibre fabricated by high-intensity fentosecond UV and near-UV laser pulses. Measurement Science and Technology 2006;17(5): 960-967.

[26] Davis DD, Gaylord TK, Glytis EN, Kosinski SG, Mettler SC, Vengsarkar AM. Long Period Fibre Grating Fabrication with Focused CO_2 Laser Pulses. Electronics Letters 1998;34(3): 302-303.

[27] Kim BH, Ahn T, Kim DY, Lee BH,Chung Y, Un-Chul Paek, Won-Taek Han. Effect of CO_2 laser irradiation on the refractive-index change in optical fibers. Applied Optics 2002;41(19): 3809-3815.

[28] Akiyama M, Nishide K, Shima K, Wada A, Yamauchi R. A Novel Long-period Fiber Grating Using Periodically Releases Residual Stress of Pure-silica Core Fiber. In: Optical Fiber Communication Conference (OFC), San José, CA, pp. 276-277, Feb. 1998. [Techn. Dig. Conf. Opt. Fiber Commun., 276, 1998].

[29] Wang Y. Review of Long Period Fiber Gratings Written by CO_2 Laser. Journal of Applied Physics 2010;108(8): 081101.

[30] Rao YJ, Zhu T, Ran ZL, Wang YP, Jiang J, Hu AZ. Novel Long-period Fiber Gratings Written by High-frequency CO_2 Laser Pulses and Applications in Optical Fiber Communication. Optics Communications 2004;229(1-6): 209–221.

[31] Chan H, Perez E, Alhassen F, Tomov I, Lee H. Ultra-Compact Long-Period Fiber Grating and Grating Pair Fabrication using a Modulation-Scanned CO_2 Laser. In: Conference on Optical Fiber Communication and the National Fiber Optic Engineers Conference (OFC/NFOEC), Anaheim, CA, pp. 1-3, Mar. 2007. [Proc. of Nation Fiber Optic Engineers Conference, Optical Society of America, paper JWA5, 2007].

[32] Alves DC, Coelho J, Nespereira M, Monteiro F, Abreu M, Rebordão JM. Automation methodology for the development of LPFG using CO_2 laser radiation. In: Costa M. (ed.) 8th Iberoamerican Optics Meeting and 11th Latin American Meeting on Optics, Lasers, and Applications: proceedings of SPIE 8785, 18 November 2013: 6 pages.

[33] Ryu HS, Park YW, Oh ST, Chung YJ, Kim DY. Effect of asymmetric stress relaxation on the polarization-dependent transmission characteristics of a CO_2 laser-written long-period fiber grating. Optical Letters 2003;28(3): 155–157.

[34] Oh ST, Han WT, Paek UC, Chung Y. Azimuthally Symmetric Long-period Fiber Gratings Fabricated with CO_2 Laser. Microwave and Optical Technology Letters 2004;41(3): 188-190.

[35] Grubsky V, Feinberg J. Fabrication of axially symmetric long-period gratings with a carbon dioxide laser. IEEE Photonics Technology Letters 2006;18(21): 2296–2298.

[36] Zhu T, Chiang K, Rao Y, Shi C, Song Y, Liu M. Characterization of long period fiber gratings written by CO_2 laser in twisted single mode fibers. Journal of Lightwave Technology 2009. 27(21): 4863–4869.

[37] Shang RB. Fabrication of twisted long period fiber gratings with high frequency CO_2 laser pulses and its bend sensing. Journal of Optics 2013;15(7): 075402.

[38] Wang YP, Wang DN, W. jin, Rao YJ, Peng GD. Asymmetric long period fiber gratings fabricated by use of CO_2 laser to carve periodic grooves on the optical fiber. Applied Physics Letters 2006;89: 151105.

[39] Su L, KS Chiang, Lu C. CO_2-laser-induced long-period gratings in graded-index multimode fibers for sensor applications. *IEEE* Photonics Technology Letters 2006;18(1): 190–192.

[40] Kim CS, Han Y, Lee BH, Han WT, Paek UC, Chung Y. Induction of the refractive index changes in B-doped optical fibers through relaxation of the mechanical stress. Optics Communications 2000;185(4-6): 337-342.

[41] Kim BH, Park Y, Ahn TJ, Lee BH, Chung Y, Paek UC, Han WT. Residual stress relaxation in the core of optical fiber by CO_2 laser radiation. Optical Letters 2001;26(21): 1657-1659.

[42] Kakarantzas G, Birks TA, Russell PS. Structural long-period gratings in photonic crystal fibers. Optical Letters 2002;27(12): 1013–1015.

[43] Lee HW, Liu Y, ChiangKS. Writing of long-period gratings in conventional and photonic-crystal polarization-maintaining fibers by CO_2-laser pulses. IEEE Photonics Technology Letters 2008;20(2):132–134.

[44] Liu Y, Chiang KS. Recent development on CO_2-laser written long-period fiber gratings. In: V, Ming-Jun Li, Ping Shum, Ian H. White, Xingkun Wu, (eds.) Passive Components and Fiber-based Devices, proceedings of SPIE 7134: pp. 713-437,. Hangzhou, China October 26, 2008.

[45] Grellier AJC, Zayer NK, Pannell CN. Heat transfer modelling in CO_2 laser processing of optical fibers. Optics Communications 1998;152(4-6): 324-328.

[46] Yablon AD, Yan MF, Wisk P, DiMarcello FV, Fleming JW, Reed WA, Monberg EM, DiGiovanni DJ, Jasapara J. Refractive Index Perturbations in Optical Fibers Resulting from Frozen-in Viscoelasticity. Applied Physics Letters 2004;84(1):19-21.

[47] Coelho J, Nespereira M, Silva C, Rebordão J. Modeling refractive index change in writing long-period fiber gratings using mid-infrared laser radiation. Photonic Sensors 2013;3(1): 67-73, 2013.

[48] Coelho J, Nespereira M, Silva C, Rebordão J. 3D Finite Element Model for Writing Long-Period Fiber Gratings by CO_2 Laser Radiation. Sensors 2013;13(8): 10333-10347.

[49] Coelho J, Nespereira M, Silva C, Pereira D, Rebordão J, Advances in optical fiber laser micromachining for sensors development, In: Sulaiman Wadi Harun and Ham-

zah Arof (Eds.) Current Developments in Optical Fiber Technology. Ridjeka: Intech; 2013. p375-401.

[50] Coelho J, Abreu MA, Carvalho-Rodrigues F. Modelling the spot shape influence on high-speed transmission lap welding of thermoplastics films. Journal of Optics and Lasers in Engineering 2008;46(1): 55-61.

[51] Yang S, Matthews M, Elhadj S, Draggoo, V, Bisson S. Thermal transport in CO_2 laser irradiated fused silica: in situ measurements and analysis. Journal of Applied Physics 2009;106: 103-106.

[52] Yablon AD. Optical and Mechanical Effects of Frozen-in Stresses and Strains in Optical Fibers. IEEE Journal of Selected Topics in Quantum Electronics 2004;10(2): 2004.

[53] Timoshenko SP, Goodier JN. Theory of elasticity, 2nd Ed.; New York: McGraw-Hill; 1970. p409-410.

[54] Hibino Y, Hanawa F., Abe T, Shibata S. Residual stress effects of refractive indices in undoped silica-core single-mode fibers. Applied Physics Letters 1987;50(22): p1565-1566.

[55] Hibino Y, Edahiro T, Horiguchi T, Azuma Y, Shibata N. Evaluation of residual stress and viscosity in SiO_2-core/F-SiO_2 clad single mode fibers from Briouin gain spectra. Journal of Applied Physics 1989;66(9): 4049-4052.

[56] Rego, GM. Arc-induced long-period fibre gratings. Fabrication and their application in communications and sensing. Ph.D. dissertation, Dept. Elect. Comp. Eng., Univ. of Porto, Porto, Portugal, 2006.

[57] Lancry, M, Réginier, E, Poumellec, B. Fictive temperature in silica-based glasses and its application to optical fiber manufacturing. Progress in Material Sciences 2012;57: 63-94.

[58] Corning® SMF-28 optical fiber product information. New York: Corning Inc.: 2002.

[59] Baptista, FDV Simulação do Comportamento Espectral de Redes de Período Longo em Fibra Óptica. Ms. dissertation, Centro de Ciências Exactas e da Engenharia, Univ. of Madeira, Funchal, Portugal, 2009.

[60] Erdogan, T. Cladding-mode resonances in short-and long-period fiber grating filters: errata. Journal of the Optical Society of America A 2000;17: p2113.

[61] McLachlan A, Meyer F. Temperature dependence of the extinction coefficient of fused silica for CO_2 laser wavelengths. Applied Optics 1987;26(9):p1728–1731.

[62] André P, Rocha A, Domingues F, Facão M. Thermal effects in optical fibers. In Marco Aurélio dos Santos Bernardes (Ed.) Developments in Heat Transfer; Ridjeka: Intech; 2011. p1–20.

[63] Clowes J, Syngellakis S, Zervas M. Pressure sensitivity of side-hole optical fiber sensors. IEEE Photonics Technology Letters 2009;10(6): 857–859.

[64] Siegman AE, Sasnett MW, Johnston TF. Choice of clip level for beam width measurements using knife-edge techniques. IEEE Journal of Quantum Electronics 1991;27(4): 1098–1104.

[65] Nespereira M, Alves, DC, Monteiro, F, Abreu, M, Coelho J, Rebordão JM. Repeatability analysis on LPFGs written by a CO_2 laser. In: Costa M. (ed.) II International Conference on Applications of Optics and Photonics: proceedings of SPIE 9286, 26-30 May 2014, Aveiro, Portugal. Bellingham: SPIE 2014.

[66] Li Y, Wei T, Montoya JA, Saini SV, Lan X, Tang X, Dong J, Xiao H. Measurement of CO_2-laser-irradiation-induced refractive index modulation in single-mode fiber toward long-period fiber grating design and fabrication. Applied Optics 2008;47(29): 5296-5304.

[67] Hutsel MR, Gaylord TK. Residual-stress relaxation and densification in CO_2-laser-induced long-period fiber gratings. Applied Optics 2012;51(25): 6179-6187.

Chapter 10

SiGe Based Visible-NIR Photodetector Technology for Optoelectronic Applications

Ashok K. Sood, John W. Zeller, Robert A. Richwine,
Yash R. Puri, Harry Efstathiadis, Pradeep Haldar,
Nibir K. Dhar and Dennis L. Polla

Additional information is available at the end of the chapter

http://dx.doi.org/10.5772/59065

1. Introduction

This chapter covers recent advances in SiGe based detector technology, including device operation, fabrication processes, and various optoelectronic applications. Optical sensing technology is critical for defense and commercial applications including telecommunications, which requires near-infrared (NIR) detection in the 1300-1550 nm wavelength range. [Here we consider the NIR wavelength band to span approximately 750-2000 nm; the upper portion of this band, e.g., 1400 nm and longer wavelengths, is sometimes elsewhere designated short-wave infrared (SWIR).] Although silicon (Si) photodetectors have been widely used to detect in the visible to short NIR wavelength regime, the relatively large Si band gap of 1.12 eV, corresponding to an absorption cutoff wavelength of ~1100 nm, hinders the application of Si photodetectors for longer wavelengths vital for medium-and long-haul optical fiber communications.

Group III-V compound semiconductors possess the advantages of high absorption efficiency, high carrier drift velocity, excellent noise characteristics, and mature design and fabrication technology for optical devices, and are commonly used in IR detection related devices [1]. InGaAs based IR photodetectors have been developed for NIR (up to ~1700 nm) applications, InSb for 3-5 μm applications, and HgCdTe for 1-3, 3-5 and 8-14 μm applications [2]; the spectral responses of these and various other IR detector material systems are shown in Figure 1. While it is possible to integrate III-V semiconductor materials on Si by wafer bonding or epitaxy [3], III-V based detectors normally require cooling (typically down to 77 K), and incorporating III-V materials into the prevalent silicon process is at the expense of high cost and increased

complexity. In addition, there is the potential of introducing doping contaminants into the silicon, since III-V semiconductors act as dopants for Group IV materials [4].

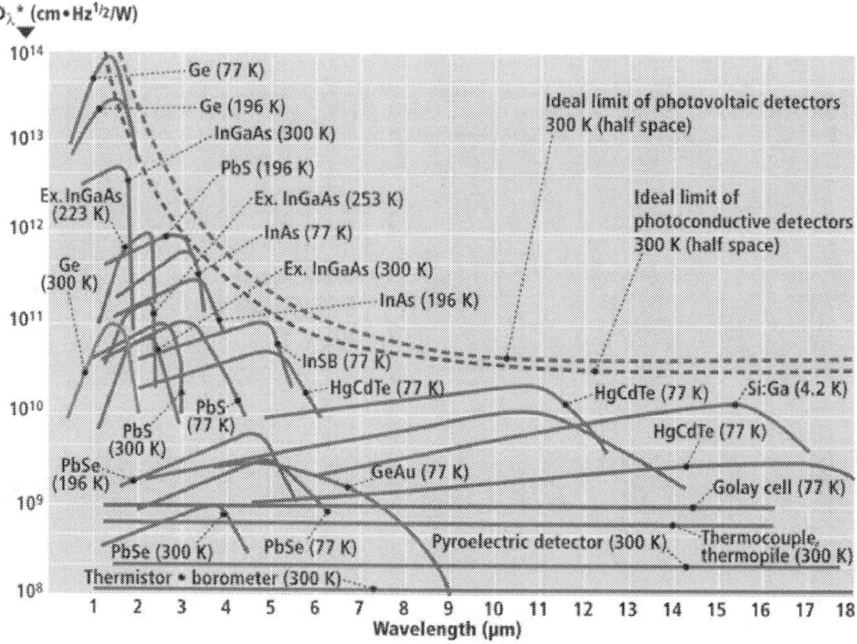

Figure 1. Spectral response characteristics of various IR detectors of different materials/technologies. Detectivity (vertical axis) is a measure of signal-to-noise ratio (SNR) of an imager normalized for its pixel area and noise bandwidth [5].

Germanium (Ge) is a Group IV material as is silicon, and thus avoids the cross contamination issue [6]. Ge, which can now be produced with extremely high purities, has an absorption spectrum similar to that of InGaAs, and can be alloyed with Si to improve the mobility and/or velocity of mobile carriers [7]. Ge forms a covalent bond with Si, and a number of SiGe based alloys involving the addition of hydrogen or oxygen are known. Amorphous alloys such as a-Si$_x$Ge$_y$H$_z$ are characterized by the absence of long-range order, but often possess a considerable degree of short-range order which is referred to as chemical ordering. A useful table listing Si and Ge bond and defect energies is found in Ref. [8]. One property of Ge that is of particular interest is the nature of its band gap. Though Ge like Si is predominately an indirect band gap (E_g=0.66 eV) material, it has a direct band gap of 0.80 eV that is only 140 meV above its indirect band gap [9]. This enables Ge to transform from an indirect gap material towards a direct gap material through the incorporation of tensile strain, as will be discussed in more detail in Section 5.2.

Consequently, strained absorption layers composed of Ge/SiGe can provide much higher optical absorption and enhanced transport properties over the ~1300-1600 nm wavelength range than layers of pure Si, enabling SiGe based photodetectors with extended NIR capabilities. (Although potential drawbacks of Si-Ge integration exist including lattice mismatch between the materials and a relatively low thermal budget for Ge, the growth processes can be adjusted to compensate in each case.) While detectors based on Ge crystals have been used for NIR detection for many years, these have required cooling down to 77 K, making them expensive and limiting their use [10]. Detectors incorporating epitaxially grown Ge/SiGe on Si substrates can operate at room temperature (RT), thus offering substantially reduced cost and size, weight, and power (SWaP). Furthermore, SiGe photodetectors can be designed to exhibit low dark currents (nA range) and dark current densities comparable to those of large area Group III-V detectors, with accordingly high signal-to-noise ratios (SNRs) [11]. Consequently, SiGe based devices have become promising and practical candidates for many applications requiring detection of radiation at visible to NIR wavelengths.

Perhaps the most important advantage of SiGe based devices is that SiGe epitaxial growth processes are compatible with both front-and back-end silicon complementary metal-oxide-semiconductor (CMOS) fabrication technologies. Consequently, SiGe detector devices can be heterogeneously combined with CMOS circuitry using widely installed manufacturing infrastructure used for production of CMOS integrated circuits (ICs). In addition, SiGe photodetectors and Si CMOS receiver circuits can be simultaneously fabricated and then monolithically integrated [12]. Fabricated SiGe detectors can be incorporated directly with low noise Si readout integrated circuits (ROICs) to yield low-cost and highly uniform IR focal plane arrays (FPAs) to maximize the fill factor, as will be discussed in Section 7. This allows SiGe detector based imaging devices to be produced much less expensively and with less difficulty than those based on III-V detectors. An attractive feature of CMOS-compatible SiGe IR detectors/imagers is that they can be fabricated on large diameter (up to 450 mm) Si wafers [13], further decreasing costs and maximizing production output.

2. Applications of SiGe detector technology

2.1. Telecommunications

The relatively recently realized capability of growing Ge epitaxially on Si has enabled the incorporation of Ge in an expanded variety of detector applications. A primary application for SiGe NIR detectors involves optical telecommunications networks. Due to fundamental physical advantages over copper as well as improved bandwidth, power dissipation, cost, and noise immunity, fiber optic based communications have been utilized to enhance available bandwidths for services such as internet, cable television, and telephone, e.g., using fiber-to-the-premises (FTTP) network architectures [14]. By replacing electrical wires with optical fibers, data rates can be enhanced from 10 Mb/s up to the order of 10 Gb/s with much lower power budgets [15].

Monolithic integration of optics with Si electronics is a primary means to realize low-cost and high performance interconnections, and Ge is a promising material to bridge low-cost electronics with the advantages of optics [12]. Unlike their Si based counterparts, photodetectors with tensile strained Ge/SiGe layers can provide high optical absorption over the entire C band (1530-1565 nm) and most of the L band (1565-1625 nm). The L band is commonly utilized by dense wavelength division multiplexing (DWDM) systems, and it has been determined that expanding the detection limit from 1605 nm to 1620 nm can enable 30 additional channels for long-haul optical telecommunications [16]. The performance of SiGe based photodetectors operating at extended NIR wavelengths is now comparable to or in some cases exceeds the performance of InGaAs based devices that have traditionally been used in telecommunications networks [10].

2.2. Optical interconnects

Conventional copper interconnects become bandwidth limited above 10 GHz due to frequency-dependent losses such as skin effects and dielectric losses from printed circuit board substrate materials [18]. In addition, RC delay and heat dissipation issues originating from metal interconnects on Si ICs have become increasingly problematic as feature sizes continue to shrink in accordance with Moore's Law [15]. Consequently, recent years have seen a rapid advancement in the adaption of Si based optical interconnects from rack-to-rack and board-to board to chip-to-chip as well as to on-chip applications. The latter two applications require a large number of high-speed, low-cost photodetectors densely integrated with Si electronics [12].

Figure 2. The compatibility of SiGe technology with standard CMOS processing makes new types of optoelectronic ICs possible. Shown here is a new IC technology from IBM designated *CMOS Integrated Silicon Nanophotonics* [17].

While compound semiconductor devices offer high performance due to their excellent light emission and absorption properties, the process of integrating them in optical interconnects is generally very complicated, as well as costly due to the overhead associated with manufacturing in a separate facility combined with the costs associated with packaging and assembling [19]. On the other hand, SiGe based photodetectors have been demonstrated that provide nearly all of the characteristics desirable for integrated optoelectronic receivers [20]. SiGe detectors offer high speeds (10 Gb/s and greater), high sensitivity, a broad detection spectrum, and the potential for monolithic integration with IC CMOS fabrication technology as will be discussed in Section 7.2. Thus, SiGe technology holds much promise for optical interconnects in next generation ICs (Figure 2) to overcome bottlenecks inherent in conventional microelectronic devices.

2.3. Further commercial and military applications

The detection of visible-NIR radiation offered by SiGe based sensors and imaging devices operating at RT make them useful for a variety of additional industrial, scientific, and medical applications. Applications requiring low-cost NIR capable sensors include medical thermography for cancer and tumor detection during diagnosis and surgery, machine vision for industrial process monitoring, sorting of agricultural products, biological imaging techniques such as spectral-domain optical coherence tomography, and imaging for border surveillance and law enforcement [21]. SiGe based NIR sensors/imagers also provide a low-cost solution for a wide range of military applications. These military applications include, but are not limited to, day-night vision, soldier robotics, plume chemical spectra analysis, biochemical threat detection, and night vision for occupied and autonomous vehicles [13].

An additional military application of particular significance is hostile mortar fire detection and muzzle flash (Figure 3). Muzzle flashes, which approximate a blackbody spectrum from 800 K to 1200 K [22], consist of an intermediate flash and, unless suppressed, a brighter secondary flash [23]. Such incendiary events produce large amounts of energy in the NIR spectral region. The ability to image flashes from hostile fire events combined with target detection capability [e.g., by using spectral tags (chemical additives) for identification of friendly fire] provides a vital function in the battlefield that can be key to saving the lives of soldiers as well as making good strategic decisions such as knowing when and where to attack [24]. The realization of small and low-cost SiGe devices that can detect hostile fire sources therefore has the potential to greatly benefit our armed forces.

Another commercial application in view involves very small form factor SiGe based visible-NIR cameras. Since imaging has become a core feature to most mobile phone users and manufacturers, the industry puts much effort into related performance improvements and optimization of camera manufacturing methods. Wafer-level packaging of CMOS image sensors and wafer-level optics provide a cost-effective means of potentially equipping future generations of camera smartphones with visible-NIR imaging capability with smaller form factors [26]. Developing such miniature cameras based on SiGe integrated CMOS technology will require demonstrating small pixel and format NIR detector arrays that enable wide field-of-views. Producing a practical NIR imager will likewise involve further refining the thermal,

Figure 3. SiGe technology is associated with a number of military applications involving NIR sensitivity, including muzzle flash detection [25].

mechanical, and optical analyses of encapsulation and optical materials to enable compatibility with NIR FPA manufacturing.

3. SiGe sensor performance modeling

3.1. Performance model overview

This section deals with modeling of SiGe NIR FPA imaging performance over a wide range of light levels that can occur for day-night operation [13]. The model predicts detector dark currents, photocurrents, and readout and background noise associated with a novel small pixel, low-cost SiGe visible-NIR prototype camera. This type of imager, based on the ability to grow NIR-sensitive SiGe layers on silicon to form pixels utilizing existing high quality and low-cost semiconductor and electronic architectures, is intended to provide NIR night vision capability in addition to visible operation.

A fairly large matrix of variables, which include NIR background, pixel size, focal length, f-number, integration time, spectral bandpass, dark current level, and readout noise level, require a significantly complex model to perform the necessary design trade studies. This model predicts values for sensor noise equivalent intensity (NEI) and SNR, and also generates simulated 30 and 60 Hz NIR image sequences. The model was designed to assist in the development of miniature NIR or visible-NIR cameras and FPA designs and predict NEI and SNR performance of image or video quality (resolution and noise), so as to aid in the design of SiGe detector based camera optics, FPA formats, readout electronics, and pixel size.

3.2. Variable NIR background

The NIR background radiance between overcast dark night and full daylight varies by about eight orders of magnitude, spanning approximately 0.1 mlux to 25,000 lux. For daytime operation, spectral filtering, aperture reduction, and/or integration time reduction are required to prevent saturation of an FPA. The night radiance over the visible-NIR wavelength range spanning 400-1750 nm can also be quite varied: $\sim 1.0 \times 10^{-9}$ W/cm^2 for overcast rural settings, $\sim 1.5 \times 10^{-9}$ W/cm^2 for overcast urban conditions, $\sim 1.2 \times 10^{-8}$ W/cm^2 for clear night sky rural conditions, and up to $\sim 3.1 \times 10^{-8}$ W/cm^2 for clear moonlit night skies.

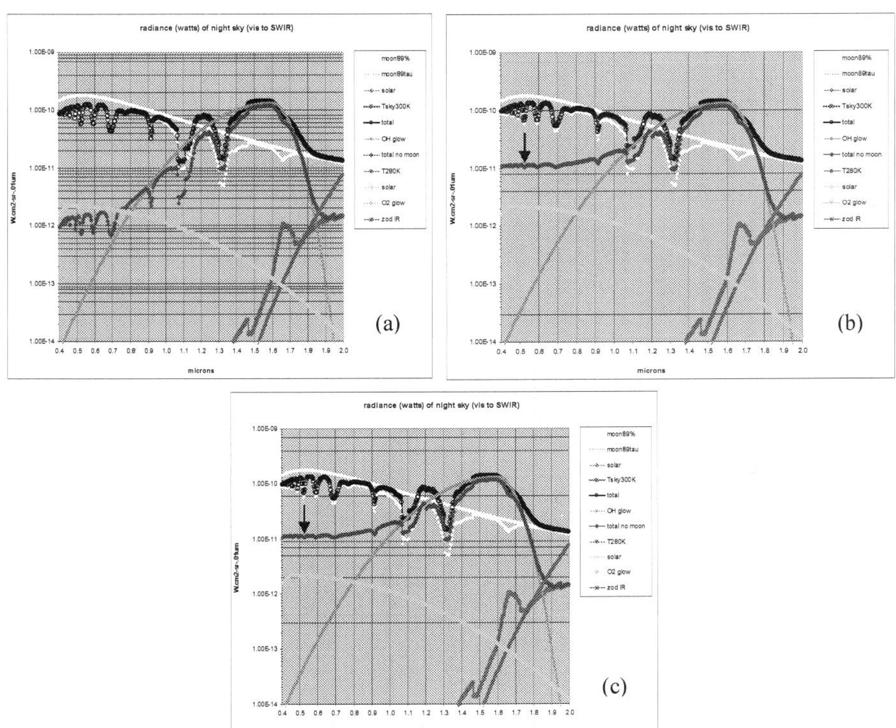

Figure 4. Visible to NIR spectral radiance of night sky based on astronomical data [27] along with data from M. Vatsia [28] plotted in (a), and data from R. Littleton [29] plotted in (b); in (c), radiance includes light pollution (Toronto). In these plots, airglow is shown in orange, moonlight in white, zodiacal IR in beige, lower atmosphere radiance in light blue, total radiance transmitted to ground in black, and total radiance to ground from airglow alone in dark blue.

Since the primary source of illumination in the NIR regime is upper atmosphere airglow, the imaging performance NIR cameras typically degrades when used in dark night overcast conditions or under a thick canopy. Moonlight and light pollution that exist in more urban settings can also help to illuminate terrain, but such illumination occurs mostly at shorter wavelengths. This situation is shown in Figure 4 in which the lunar radiance is mainly significant in the 400-1200 nm region, and is particularly evident in Figure 4(c) which shows the effects of city light pollution (where the radiance level is derived from a Toronto, Canada based spectral measurement). These spectral radiances have been modeled in order to determine the electron noise level in NIR FPAs for a given FPA pixel size, spectral band, integration time, and set of optics. This provides the background limited performance (BLIP) conditions to which dark currents and the readout noise must be added.

Basic atmospheric transmittance and path radiance capabilities have been included in the model. The percent cloud cover, which attenuates the airglow and celestial sources as well as the specified solar scattering level, can be taken into account along with aerosol visibility (5 km or 23 km). The transmittance from scene to sensor assumes a horizontal path at the earth's surface. The attenuation effects of the atmosphere on the images were computed, and subsequently a path radiance based on the ambient NIR background was reinserted into the images. This effectively compensated for the loss of scene brightness and contrast with attenuation and the overall increase in brightness due to path radiance. The images in Figure 5 illustrate the loss of contrast with increasing range.

Figure 5. Atmospheric effects on clear night sky. NIR images at distances of 1, 10, and 15 km (left to right) for 5 km visibility conditions.

3.3. Image quality metrics

Figure 6 displays images from a simulated camera to illustrate the effects of resolution and SNR on image quality and potential image identification. For 30 or 60 Hz image sequences, the eye can integrate some of the frames which allows for slightly better identification than is seen in these single images. Motion of an object over the field-of-view also aids in identification, since the eye can compensate for pixilation when viewing a moving object.

Figure 6. Effects of SNR and resolution on imagery; middle column is typical of an $f/1.5$, 60 Hz, 15 μm pixel broadband NIR imager. The rows (top to bottom) show reduction in resolution (1X, 2X, and 4X). The columns (left to right) show no noise, SNR=3.5 or NEI=2.2×10^{10} photons/s-cm^2, and SNR=1 or NEI=7.7×10^{10} photons/s-cm^2.

3.4. Maximizing the signal

The available signal is determined based on the FPA integration time, quantum efficiency (QE), optics f-number (defined as the ratio of lens focal length to diameter of the entrance pupil), and visible-NIR background in the chosen spectral band. The general SNR is calculated as [13]:

$$\text{SNR} = \frac{e_{sig}}{\tilde{e}_{noise}} = \frac{e_{bk,NIR}}{\sqrt{e_{n,bk}^2 + e_{n,read}^2 + e_{n,dark}^2}} \tag{1}$$

where the noise level consists of background, read, and dark current noise. The signal as measured by collected electrons e_{sig} is given by

$$e_{sig} = t_i G \tau_o f \eta \frac{A_{det}}{4F_\#^2} \int_{\lambda_1}^{\lambda_2} \Phi_{bk} d\lambda \tag{2}$$

where t_i is the integration time, G is the gain, τ_o is the optics transmittance, f is the fill factor, η is the QE, A_{det} is the detector area, $F_\#$ is the f-number, and Φ_{bk} is the visible-NIR background in photons/sec-cm^2. The background consists of airglow, moonlight and light pollution sources:

$$\int_{\lambda_1}^{\lambda_2} \Phi_{bk} d\lambda = \int_{\lambda_1}^{\lambda_2} \left[t_a \left(\Phi_{airglow} + \Phi_{moon} + \Phi_{solar} \right) + \Phi_{lightpol} \right] d\lambda \tag{3}$$

where the transmittance τ_a occurs from ground to space.

3.5. Minimizing noise

The NIR camera noise is a combination of the background noise, readout noise (typically consisting of a few electrons to tens of electrons), and dark current noise. In designing the optics and setting the parameters of the FPA, the dark current and readout noise must be maintained below the background noise. The f-number is the metric for signal and background level noise; however, the focal length required for identifying the object of interest at the desired range must first be specified before the f-number can be determined. It is necessary to know the focal length along with the f-number in order to obtain an adequately accurate SNR for the specific night sky background, which in turn is used to determine the required optical aperture.

The readout noise $e_{n,read}$ is normally in the range of 10-50 electrons. The background noise is simply the square root of the background electrons collected:

$$e_{n,bk} = \sqrt{t_i G \tau_o f \eta \frac{A_{det}}{4F_\#^2} \int_{\lambda_1}^{\lambda_2} \Phi_{bk} d\lambda} \tag{4}$$

Likewise, the dark current noise may be expressed as

$$e_{n,dark} = \sqrt{\frac{t_i GI_{dark}}{q}} = \sqrt{\frac{t_i GJ_{dark} A_{det}}{q}} \tag{5}$$

If the dark current, which is a function of the bias voltage, is further reduced by decreasing the negative bias, uniformity and responsivity may be degraded.

In addition to these basic temporal noises, NIR cameras exhibit spatial noise. Although calibrations usually reduce the spatial noise to levels below that of the temporal noise, spatial noise varieties such as random pattern noise, fixed row and column noise, temporal row and column noise, and frame fluctuation noise all can be observed in the images. The model incorporates all the noise types described in a three-dimensional (3D) noise model, the concept of which is illustrated in Figure 7.

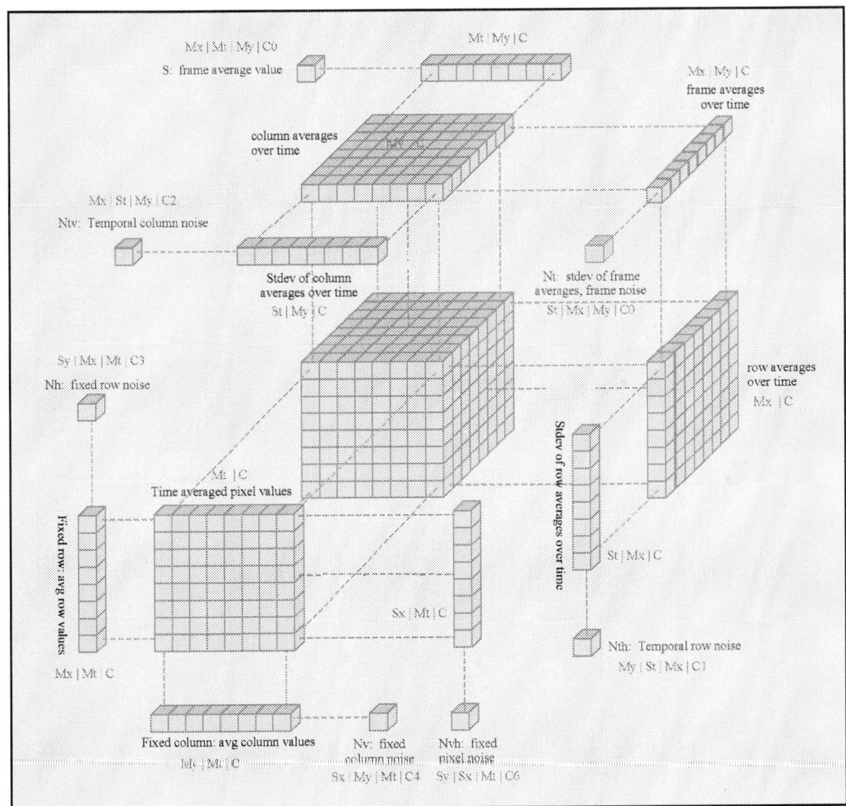

Figure 7. Method of deriving 3D noise from the model illustrated.

The background SNR formula for specific dependencies is expressed as

$$SNR_{bk} = \sqrt{t_i G\tau_o f\eta \frac{A_{det}}{4F_\#^2} \int_{\lambda_1}^{\lambda_2} \Phi_{bk} d\lambda} \qquad (6)$$

while the dark current SNR is given by the formula

$$SNR_{dark} = \frac{\sqrt{qt_i GA_{det} \tau_o f\eta}}{4F_\#^2 \sqrt{J_{dark}}} \int_{\lambda_1}^{\lambda_2} \Phi_{bk} d\lambda \qquad (7)$$

The sensor NEI condition is

$$\int_{\lambda_1}^{\lambda_2} \Phi_{bk} d\lambda = NEI = \frac{\sqrt{e_{n,read}^2 + e_{n,bk}^2 + e_{n,dark}^2}}{t_i G\tau_o f\eta \dfrac{A_{det}}{4F_\#^2}} \qquad (8)$$

Typical NEIs are in the 8×10^8 to 5×10^9 photon/s-cm^2 range and vary with integration time and f-number, where nominal values are in the ranges of 16 to 33 ms and $f/1.0$ to $f/1.5$, respectively. The SNRs in view are for single frames. As was noted, ability of the human eye/brain to integrate some of the frames when perceiving 30 to 60 Hz imagery improves detection and identification performance compared to single static frame viewing. The improvement for random temporal noise dominated images generally varies with the square root of the number of frames the eye can integrate. Eye integration is complicated and varies with light level, resolution, and other factors. This improvement is not only limited by the temporal duration of the eye integration, but also by the underlying spatial noise of the sensor image. Since the level of spatial noise including row and column noise is often marginally below the level of the temporal noise, the eye integration improvement ceases when temporal noise abatement becomes equal to the spatial noise.

The SNR can also be improved by spatially binning pixels, but this is at the expense of sacrificing spatial resolution. This SNR improvement is generally proportional to the square root of the number of binned pixels. Thus, implementing 2×2 binning improves the SNR by a factor of 2. Adding these two phenomena to the previously derived SNR gives:

$$SNR = \frac{e_{sig}}{\tilde{e}_{noise}} = \frac{e_{bk,swir} N_{fs} N_{ps}}{\sqrt{N_{fs} N_{ps} [e_{n,bk}^2 + e_{n,read}^2 + e_{n,dark}^2]_{temporal} + N_{fs} N_{ps} e_{n,spatial}}} \qquad (9)$$

This expression shows the improvement in SNR with the square root of the product of summed pixels and summed frames for the temporal noise part, and demonstrates the ineffectiveness of summing toward improving the SNR of spatial noise dominated imagery.

3.6. Predicting NEI and SNR

NEI vs. operating temperature for pixel sizes of 30, 20, 10 and 5 μm is shown plotted in Figure 8 using a diffusion expression derived from low dark current density data. For these simulations, the following parameters were employed: integration time of 33 ms, gain of unity, read noise of 10 electrons rms, dark current residual nonuniformity calculated for a temperature delta of 0.1 K, optics with $f/1.25$, QE of 80%, and wavelength range spanning 1000-1750 nm. It can be seen that the NEI due to dark current (blue squares) increases as the pixel size is decreased, since the dark current noise is a function of linear pixel size while light collection is a function of pixel area. The NEI improves with pixel size because the number of photons collected increases with detector area while the BLIP noise increases in proportion to the square root of the detector area. For the sensor modeled, the BLIP NEI for 30 μm pixels is 1.5×10^9 photons/s-cm^2, significantly lower than the value of 9×10^9 photons/s-cm^2 determined for 5 μm pixels.

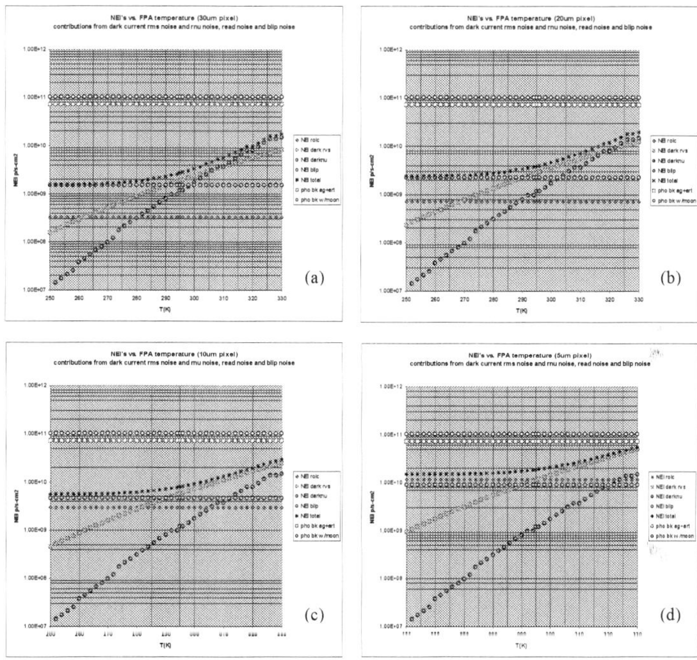

Figure 8. NEI vs. operating temperature for visible-NIR sensors with various pixel sizes: (a) 30 μm, (b) 20 μm, (c) 10 μm, and (d) 5 μm.

Minimizing dark current in SiGe detectors, especially for those with smaller pixels, is a driving requirement. Figure 9 shows the SNR vs. dark current density for 7.5 and 12 µm pixels as a function of dark current density. The level lines are the readout and background SNRs and the slanted lines are the dark current SNRs. The background SNR is the best attainable SNR. The intersection of the lines thus signifies the dark current level where the dark current SNR is equal to the readout or background SNR. In Figure 9(a), the yellow lines signify the background or BLIP SNRs for clear skies with no moonlight for the two pixel sizes, with the upper yellow line showing the SNR for 12 µm pixels and 0.89 moonlight conditions. In Figure 9(b), the readout noise SNR (red squares) has been added for both large and small pixels (based on 10 noise electrons per integration).

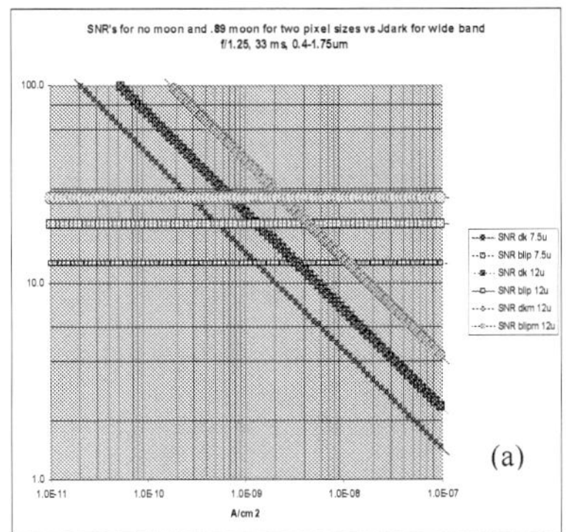
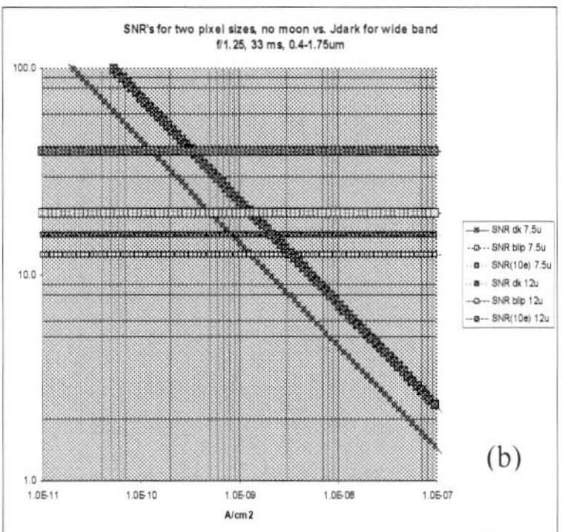

Figure 9. (a) SNR vs. dark current density for 7.5 and 12 µm pixels for dark sky, 0.89 moonlight conditions, *f*/1.25, 33 ms integration time, and 400-1750 nm bandpass; (b) shows the addition of read noise based SNR (10 electrons).

Dark current appears to be the performance limiting factor for small pixel NIR FPAs operating at RT. The performance can be improved by compensating for the dark currents using lookup tables, though nonuniformity due to uncompensated variance in dark current over the FPA must also be characterized. While utilizing lookup tables should smooth out most of the nonuniformity, there will be residual nonuniformity as a result of the FPA pixels' dark current to temperature difference ratios at RT in combination with the temperature increment used in the lookup tables. The dark current residual nonuniformity must be kept below the average dark current noise level to preserve the performance, as illustrated in Figure 9.

3.7. SiGe imager performance based on modeling results

Miniature SiGe detector based FPAs that can be incorporated into handheld cameras or inserted into smartphones require *f*/2 to *f*/3 optics and pixels approximately 5-7 µm in size. These detectors will consequently have reduced light collection with relatively high dark currents at RT. Such higher *f*-numbers, which are about twice those characteristic of an ideal NIR camera, effectively reduce the signal and SNR by about a factor of four. While small optics of *f*/1 to *f*/1.5 are possible, these may require an increase in optics diameter. Another method to improve the SNR in dark current limited NIR sensors is to incorporate a microlens array (e.g., having 20 μm lens centers) to focus light from the scene onto 5-7 μm detector pixels. This maintains the signal strength while reducing detector dark current, enabling small pixel sizes within a larger cell that allows for extra on-chip signal processing electronics.

Predictions based on the modeling that has been detailed in this section are summarized as follows: Imaging under rural night sky conditions becomes challenging for small pixel, small optics designs, and dark currents can significantly impact performance in an uncooled NIR camera. A small NIR camera will respond well to minimal amounts of illumination from a direct NIR source, such as one imaged in indoor or shorter-range outdoor environments. In addition, the performance limitations of small uncooled NIR cameras are not found to be problematic for live fire detection and identification applications. Overall, these findings indicate that low-cost, small pixel, uncooled detectors based on growth of SiGe on Si are potentially advantageous for imaging in indoor or low light level outdoor environments.

4. Operation and performance of SiGe photodetectors

A photodetector may be basically defined as a device that converts an optical signal (photons) into an electrical one (electrons). There exist three primary classes of semiconductor based photodetectors: avalanche photodiodes (APDs), metal-semiconductor-metal (MSM) detectors, and p-i-n (*pin*) detectors. [An additional type of detector device, known as a metal-insulator-semiconductor (MIS) photodetector featuring an insulator layer inserted between metal and semiconductor layers [30], has been developed as well but is not that common.] Detector devices of each of these three main classifications based on SiGe technology have been demonstrated. For the visible-NIR detection applications in view, SiGe based MSM and *pin* devices are the best suited, with each having associated advantages and disadvantages that will be discussed in some detail throughout the remainder of this section.

4.1. Avalanche Photodiodes (APDs)

APDs, which are commonly employed in high bitrate optical communication systems, achieve high built-in gain through avalanche multiplication, and require high bias voltages (~20 V/μm) to achieve desired ionization rates and provide detection of low power signals with high sensitivity [12]. Under sufficiently high external bias, the electrical field in an APD's depletion region causes photogenerated electrons from the absorption layer to undergo a series of impact ionization processes. This enables a single absorbed incoming photon to generate a large number of electron/hole pairs (EHPs), which effectively amplifies the photocurrent and improves the sensitivity, providing a QE potentially greater than unity. SiGe APDs typically have separate absorption-charge-multiplication (SACM) structures (see Figure 10), in which light is absorbed in an intrinsic Ge film and electrons are multiplied in an intrinsic Si film; such structures allow optimization of both QE and multiplication gain [10].

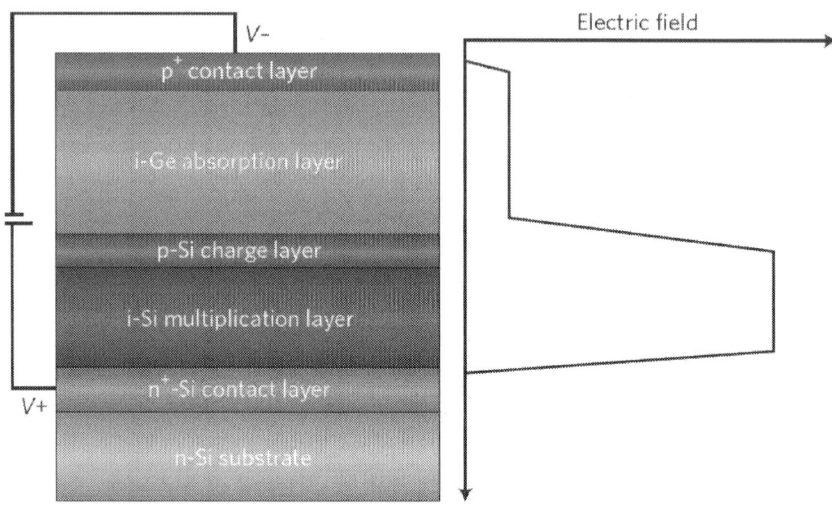

Figure 10. Schematic cross-section of a SiGe based APD device with a separate absorption-charge-multiplication (SACM) structure and its internal electric field distribution [10].

The most important performance metrics for APDs are ionization ratio (which should be minimized), internal electric field distribution, excess noise factor, gain-bandwidth product, and sensitivity [4]. The device structure of a basic APD is similar to that of a *pin* photodetector. Compared with their *pin* counterparts, APDs offer 5-10 dB better sensitivity and higher SNR due to their internal multiplication gain, as well as high bandwidth-efficiency products [4,31]. However, the comparatively low operation bandwidths and the requirement of very high bias voltages limit the integration potential of SiGe based APDs into practical CMOS-based devices.

4.2. Metal-Semiconductor-Metal (MSM) photodetectors

MSM photodetectors comprise two back-to-back Schottky contacts and feature a closely spaced interdigitated metal electrode configuration on top of an active light absorption semiconductor layer [32]. The material, physical, and electrical properties of MSM devices are depicted in Figure 11(a), (b), and (c), respectively. MSM detectors are photoconductive devices not functional under zero bias, and require sufficient external bias for the semiconductor layer to become fully depleted. The Schottky junctions present in MSM detectors exhibit rectified current-voltage (I-V) characteristics as do *pn* junctions, but occur at the metal-semiconductor rather than semiconductor-semiconductor interfaces. Also, while *pn* junctions allow both electrons and holes to flow under forward bias, Schottky junctions allow only majority carriers to flow.

Advantages of MSM detectors include low capacitance and consequent low RC delay, which enables high-speed operation. Detection bandwidths for SiGe based MSM devices are comparatively high, making them suitable for fast optical fiber communications. In addition, since MSM detectors are inherently planar and require only a single photolithography step, they are relatively easy to fabricate, boosting their potential for practical integration. However, the external QE and effective responsivity in MSM devices are generally lower than those in *pin* detectors due to shadowing of the metal electrodes, which typically occupy 25-50% of the surface area [18].

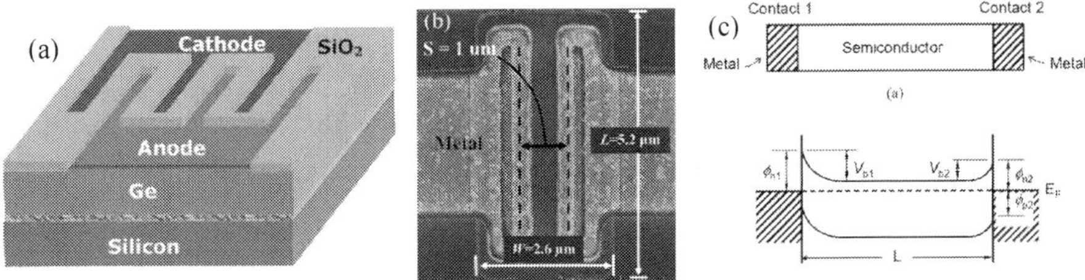

Figure 11. (a) Cross-section of MSM photodetector fabricated on Ge layer grown on Si substrate [33]. (b) Scanning electron microscope (SEM) image of an evanescent waveguide-coupled Ge-on-SOI MSM photodetector [34]. (c) Schematic diagram of MSM structure and corresponding energy band diagram at thermal equilibrium [12].

In addition, high dark current associated with SiGe based MSM devices, primarily as a result of hole injection over the Schottky barrier [35], is a significant problem that raises the noise floor and increases standby power consumption [10]. This dark current may include current associated with thermally generated electron-hole pairs and carrier injection over the Schottky barriers, since SiGe MSM detectors typically have poor Schottky contacts with Ge [12]. While techniques to suppress dark current in MSM devices, such as dopant segregation and utilizing an intermediate layer of amorphous Ge and SiC, have suppressed dark current in detection devices significantly [36], MSM detectors generally still exhibit higher levels of dark current than comparable SiGe based *pin* devices [37] often resulting in an inferior level of performance [4].

4.3. *Pin* photodetectors

As their name may suggest, *pin* photodetectors consist of an intrinsic (*i*) region sandwiched between heavily doped p^+ and n^+ semiconductor layers. A typical *pin* photodetector design cross-section is depicted in Figure 12(a). The p^+/n^+ regions may be formed by implantation, *in situ* doping, or consist of a highly doped monocrystalline Si substrate [15,38]. The depletion layer in which all absorption occurs is almost entirely defined by the thicker highly resistive intrinsic region. This is in contrast to a common *pn* photodiode, in which the width of the depletion region (usually thinner than that in *pin* devices) is governed by the applied external electric field.

SiGe based *pin* photodetector structures can exhibit significant built-in electric fields of several kV cm⁻¹ inside the *i*-Ge/SiGe layers, which overcome recombination processes at lattice defects, improving the device quality and enabling smaller devices [10]. The uniform electric field F in the intrinsic region in a *pin* photodetector is given approximately by [41]:

$$F = \left(V_{bi} - V\right) / w_D \tag{10}$$

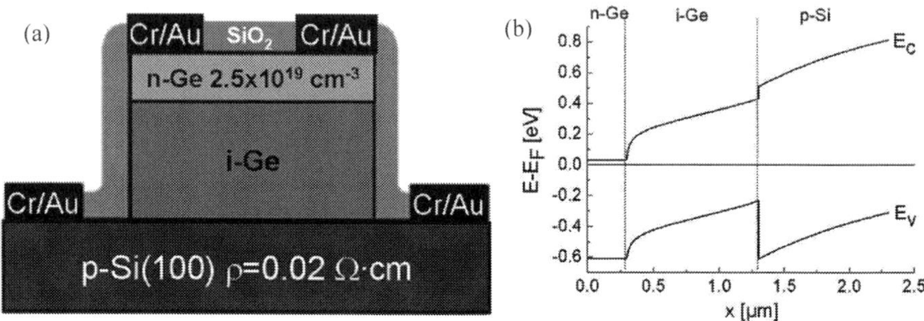

Figure 12. (a) Cross-sectional schematic view of SiGe based *pin* photodetector structure [39]. (b) Band diagram of *pin* Ge/Si heterojunction [40].

where V is the applied voltage, V_{bi} is the built-in voltage, and w_D is the thickness of the depletion or intrinsic region. Upon absorption of a photon in the intrinsic region of energy $hv > E_g$, EHPs are created and immediately separated by the external electric field resulting from reverse biasing the device, leading to generation of photocurrent. Absorption outside the intrinsic region will also result in photocurrent if the minority carriers manage to diffuse to the intrinsic region. Since there are few charge carriers in the intrinsic region, the space charge region reaches completely from the *p*-type to the *n*-type region. Figure 12(b) shows the energy band diagram of a *p*-Si/*i*-Ge/*n*-Ge heterojunction.

The intrinsic/depletion region thickness, which is normally made substantially larger than that of the p^+ and n^+ regions, can be tailored to optimize detector performance [4]. Having a relatively thick depletion region with a strong internal electric field causes most of the generated EHPs to be transferred to the p^+/n^+ regions and collected as a result of carrier drift rather than diffusion. (Since the valence band offset is much larger than the conduction band offset, the higher barrier in the valence band limits the movement of holes, so the conductivity in *pin* heterojunctions is due to electrons in the conduction band [40].) This in turn results in less carrier recombination at dislocations or point defects, leading to higher collection efficiency [10]. Because of this as well as other factors, a thicker depletion region in a *pin* photodetector is associated with higher responsivity and QE. On the other hand, having a thinner intrinsic region in a *pin* detector reduces the transit time, thereby enhancing the response bandwidth [4]. However, a thinner intrinsic layer also effects a larger capacitance, which produces greater RC delay that can have a limiting effect on the detector speed of operation [12].

There are two main classifications of SiGe based *pin* photodetectors structures: normal incidence (NI) and lateral. As might be expected, in the former type of device light is incident vertically or normal to its top or bottom surface, while in the latter type the photons are incident horizontally or laterally. In addition, a substantial portion of SiGe *pin* detectors now have waveguide-coupled (WC) designs, which circumvent alignment issues by featuring either evanescent coupling or butt-coupling between integrated SiGe detectors and Si optical waveguides. Practically all lateral (and some NI) *pin* photodetectors are WC. The advent of practical WC SiGe *pin* photodetectors was relatively recent— virtually all such detector devices have been reported within the past decade. While lengths of early WC *pin* photodetectors were typically on the order of 100 μm to ensure full absorption of light around 1550 nm [4], more recent devices are have been designed smaller (~10X) to reduce the RC delay and maximize potential bandwidths [42].

4.4. Reported performance of SiGe *pin* photodetectors

Due to their comparative ease of fabrication, performance advantages, and prevalence, the focus throughout the remainder of this chapter centers primarily on SiGe based *pin* photodetectors and associated technology. Performance specifications reported for NI and WC *pin* SiGe photodetectors are given in Table 1 and Table 2, respectively. In Table 3 and Table 4, typical ranges of performance results for MSM detectors and typical APD specifications, respectively, are presented for comparison.

WL (μm)	Resp. (A/W)	DC Dens. (mA/cm^2)	DC (μA)	BW V$_{Bias}$ (V)	BW (GHz)	V$_{Bias}$ (V)	Pub. Year	1st Author	Ref.
1.3	0.13	0.2	0.2	-1	2.3	-3	1998	S. Samavedam	[43]
1.55	0.33	30	12	-1	~0.4	-4	2000	L. Colace	[20]
1.55	0.75	15	0.14	-1	2.5	-1	2002	S. Fama	[44]
1.55	0.035	100	0.31	-1	38.9	-2	2005	M. Jutzi	[37]
1.55	0.56	10	0.79	-1	8.5	-1	2005	J. Liu	[15]
1.55	—	375	0.075	—	39	-2	2006	M. Oehme	[45]
1.3	0.45	6.4	0.20	-1	8.8	-2	2006	M. Morse	[46]
1.55	0.28	180	0.57	-1	17	-10	2006	Z. Huang	[47]
1.55	0.20	~200	~10	-1	10	-1	2006	L. Colace	[48]
1.55	0.037	27	0.035	-1	15	-1	2007	T. Loh	[49]
1.55	1.0	130	~0.1	-1	49	-2	2009	S. Klinger	[50]
1.55	0.8	—	0.042	-1	36	-3	2009	D. Suh	[51]

Table 1. Reported performance specifications of NI *pin* SiGe photodetectors.

Compared to NI detectors, WC SiGe *pin* photodetectors generally have similar bandwidths (up to 47 GHz), higher responsivity and QE, and dark currents and dark current densities that are comparable in magnitude. Among *pin* WC detectors, lateral structures have demonstrated higher QE values compared to NI devices due to less light consumption in the highly doped region, while achieving similar response speeds [42].

Resp. (A/W) @ 1.55 μm	DC Dens. (mA/cm²)	DC (μA)	BW V$_{Bias}$ (V)	BW (GHz)	V$_{Bias}$ (V)	Pub. Year	1st Author	Ref.
0.87	1.3×10³	0.9	-1	7.5	-1	2007	D. Ahn	[54]
0.89	51	0.17	-2	31.3	-2	2007	T. Yin	[55]
1.0	0.7	0.0002	-1	4.5	-3	2008	M. Beals	[56]
0.65	—	0.06	-1	18	-1	2008	J. Wang	[42]
1.0	60	~1	-1	42	-4	2009	L. Vivien	[57]
1.1	1.6×10⁴	1.3	-1	32	-1	2009	D. Feng	[52]
0.8	—	0.072	-1	47	-3	2009	D. Suh	[58]
1.1	28	1.3	-1	36	-3	2010	D. Feng	[59]
0.78	40	0.003	-1	45	-1	2011	C. DeRose	[60]
0.8	71	0.025	-1	45	0	2013	L. Virot	[61]

Table 2. Performance specifications of WC *pin* SiGe photodetectors.

In comparison to SiGe based MSM devices, SiGe *pin* detectors offer high bandwidths, low noise, and high responsivities. Responsivities reported for some *pin* devices are as high as ~1 A/W (and even greater for certain WC detectors), and typically are substantially better than those of MSM devices [50,52]. SiGe *pin* detectors generally also offer higher responsivities and lower dark currents than SiGe APDs. SiGe *pin* detector devices have demonstrated responsivities at 1310 and 1550 nm that are similar to those of commercially available InGaAs photodetectors [53].

Parameter	Best	Typical	Worst
Responsivity @ 1.55 μm	0.8-1.2 A/W	0.53-0.75 A/W	0.10-0.14 A/W
Dark Current Density	85-100 mA/cm²	0.6-2.0 A/cm²	650-1000 A/cm²
Dark Current	0.011-0.020 μA	4-10 μA	90-4000 μA
Bandwidth	36.5-40.0 GHz	10-25 GHz	1.0-4.3 GHz

Table 3. General performance specification ranges of MSM type SiGe photodetectors.

Bandwidths of Ge/SiGe *pin* photodetectors have improved from several gigahertz to close to 50 GHz in recent years, and are presently comparable to those of MSM devices. Currently the highest *pin* detector bandwidth is 49 GHz at-2 V reverse bias as reported by Klinger *et al.*, and three reported WC detectors can operate above 45 GHz, fast enough to accommodate future 40 Gb/s telecommunications applications [50]. Techniques considered to enhance bandwidths further include limiting the thickness of the Ge/SiGe intrinsic layers in *pin* photodiodes to reduce carrier transition times, and altering the device structure to limit undesirable parasitic effects.

Parameter	Best	Typical	Worst
Responsivity @ 1.55 μm	0.80 A/W	0.40 A/W	0.17 A/W
Dark Current	~10 μA	~50 μA	~100 μA
Bandwidth	~35 GHz	~12 GHz	~5 GHz
Gain-bandwidth Product	350 GHz	105 GHz	50 GHz

Table 4. General performance specifications of SiGe APDs.

Reported dark currents and dark current densities in SiGe *pin* detectors are both on average approximately two orders of magnitude lower compared to those of MSM devices, with dark current densities of certain *pin* devices as low as in the $\mu A/cm^2$ range [42]. The dark current and dark current density results presented were almost entirely measured for devices biased at 1 V. To minimize dark current and operating power further, there has recently been increasing research interest in the development of lower bias or even zero-bias *pin* photodiodes. Zero-bias SiGe *pin* photodetectors have demonstrated responsivities at 0 V bias nearly equivalent to the saturated value at 2 V bias [15], as well as 3 dB bandwidths as high as 25 GHz [38].

5. Design objectives of SiGe pin photodetectors

5.1. $Si_{1-x}Ge_x$ photodetector design parameters

While some early attempts to develop SiGe IR detectors concentrated on potential LWIR applications [62,63], in this chapter we focus solely the development of devices for applications involving detection in the NIR band (up to ~1700 nm). The most straightforward method by which to adjust the cutoff wavelength of a SiGe photodetector in order to tune its range of response is to modify the $Si_{1-x}Ge_x$ alloy composition. Si and Ge have the same type of crystallographic structure and the materials can thus be alloyed with varying Ge concentrations. For $Si_{1-x}Ge_x$ alloys, the lattice constant does not exactly follow Vegard's law. The relative change of the lattice constant is given by [63]:

$$a_{Si_{1-x}Ge_x} = 0.5431 + 0.1992x + 0.0002733x^2 \, (nm)$$

(11)

The concentration of Ge in a layer of $Si_{1-x}Ge_x$ may be accurately measured using characterization techniques such as X-ray diffraction (XRD). As the Ge concentration is increased, the band gap of the material is reduced, and therefore the cutoff wavelength of a detector will increase (extending its operational wavelength range) assuming all other factors remain constant. However, from a practical device fabrication standpoint, depositing pure Ge or SiGe with very high Ge concentration entails certain technical challenges; for instance, as predicted by Equation (11), a higher Ge concentration of $Si_{1-x}Ge_x$ grown on Si results in a larger lattice mismatch between the materials. This can lead to Stranski-Krastanov growth in which islands form to relieve the misfit strain, which in turn leads to rougher surfaces [12]. However, as will be discussed in the following section, the incorporation of even small amounts of tensile strain can be utilized to extend the operating range of a SiGe photodetector having an absorption layer of a given Ge concentration further into the NIR regime. In addition, modification of parameters such as the doping concentration and growth temperature can be undertaken to further fine-tune the spectral response of a device. Thus, there are multiple factors that more or less influence the operational wavelength range of a SiGe based detector. These must be properly balanced in the process of designing and developing a detector device that exhibits required and optimal performance characteristics for a given application(s).

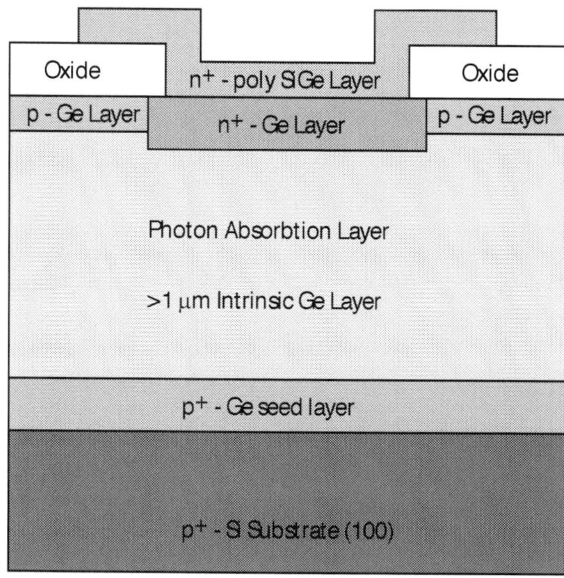

Figure 13. SiGe *pin* photodetector structure used to evaluate impact of various fabrication methodologies [64].

A diagnostic *pin* photodetector device structure designed to evaluate the impact of various fabrication methodologies to reduce leakage currents and produce higher detector perform-ance is shown in Figure 13 [64]. The structure, from the bottom up, consists of a p^+Si substrate, thin p^+Ge seed layer, thicker i-Ge layer, and top n^+polysilicon layer with an underlying n^+doped Ge region. The polysilicon layer covers sections of oxide deposited on the sides of the top surface of the cylindrical detector, under which shallow p-Ge regions may form. This structure can be used to help to assess the following: ability to grow high quality/low defect density Ge on Si; layer thicknesses necessary for minimal topological and defect density requirements; and isolation of defect states at the Ge/oxide interface from the signal carrying layers. Also of relevance is the determination of the optimum doping level and thickness of the lighter doped

p-type Ge regions under the oxide to isolate interface states and lateral leakage current that could result between the highly doped n^+Ge region below the polysilicon and the p^-Ge regions.

5.2. Incorporating strain to improve NIR detection

Strains and consequent stresses normally arise during epitaxial growth of thin films on substrates of different compositions and/or crystal structures. Internal strains and stresses can result from a mismatch in the lattice constants of the individual layers, which is illustrated in Figure 14. If the lattice mismatch between two materials is less than ~9%, the initial layers of film will grow pseudomorphically, i.e., the films strain elastically in order to maintain the same interatomic spacing. As the film grows thicker, the increasing strain will create a series of misfit dislocations separated by regions of relatively good fit.

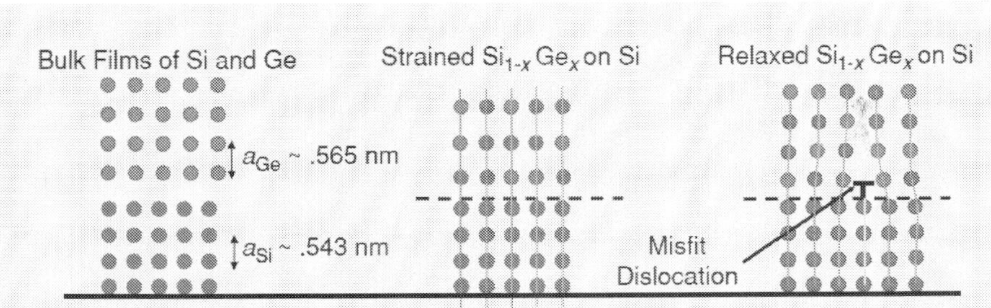

Figure 14. Relationship between lattice mismatch of Si and Ge and misfit dislocations that degrade detector performance [18].

Since the lattice constant of Ge exceeds of that of Si by 4.18%, very thin Si$_{1-x}$Ge$_x$ ($x > 0$) layers grown on a Si substrates are initially compressively strained. Near perfect epitaxial growth of such a strained heteroepitaxial layer can be achieved if its thickness does not exceed a critical thickness for stability. Since the pseudomorphic critical thickness for growth of Ge on Si with strain due to lattice mismatch is less than 1 nm, a Ge layer that is grown with a thickness that is substantially larger than this limit will relax through the formation of misfit dislocations [7,10].

However, the difference in thermal expansion coefficients between the layers can also play a significant role in the development of strain following epitaxial growth. Since Ge has a larger thermal expansion coefficient than Si, when the temperature cools to RT after growth the consequent reduction in the lattice constant of a deposited Ge/SiGe layer will be suppressed by the Si substrate [65]. This results in the generation of residual tensile strain in the Ge/SiGe layer normally within the range of 0.15-0.30% [9,66]. The changes in band gap energy and absorption that occur with the introduction of strain are depicted in Figure 15(a) and (b), respectively.

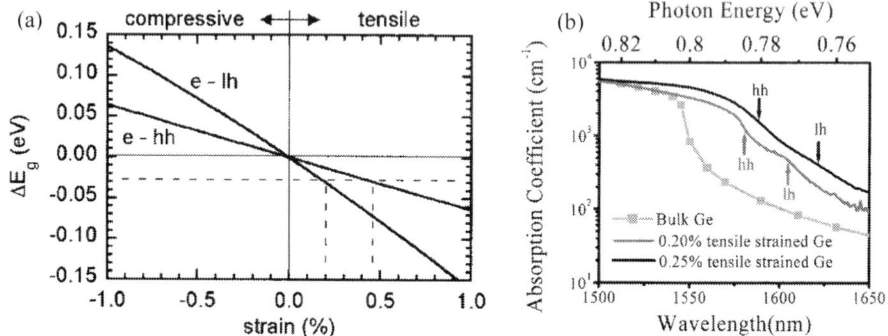

Figure 15. (a) Calculated change in direct band gap energy as a function of strain in Ge [65]. (b) Absorption spectra of bulk Ge, and 0.20% and 0.25% tensile strained Ge [53].

The presence of this biaxial tensile stress in Ge causes the valence subbands to split, where the top of the valence band comprises the light hole band. The light hole band energy increases and consequently both the direct and indirect gaps shrink, with the direct gap shrinking more rapidly. Thus, with the increase of tensile strain, Ge transforms from an indirect gap material towards a direct gap material. This stress-induced shift in valence subbands is depicted in Figure 16(a).

Upon application of tensile strain, e.g., of 0.2%, the direct band gap of Ge reduces from 0.80 eV for unstrained material to ~0.77 eV, which effectively increases the corresponding cutoff wavelength from 1550 to 1610 nm [15,65]. As shown in Figure 16(b), this provides greater sensitivity for sensor operation at NIR wavelengths of 1600 nm and above due to the higher absorption coefficient (~5X) and recombination rates of the strained Ge over this range [9]. This extended operational range is very useful for telecommunications, since strained layer SiGe based sensors can operate over most or all of the L band spanning 1560-1620 nm, as well as for other applications requiring detection of longer wavelengths in the NIR regime.

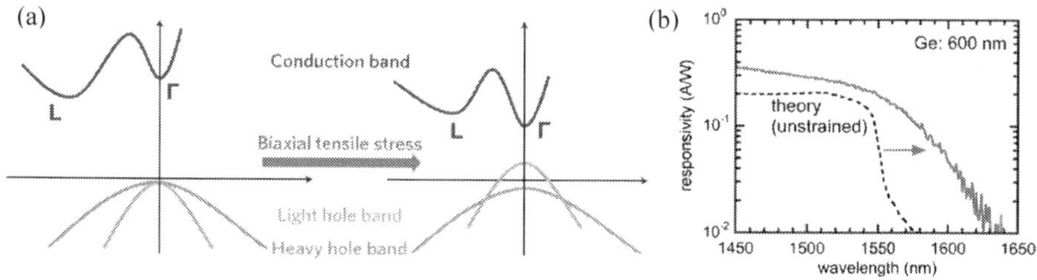

Figure 16. (a) Shift of Ge from indirect gap toward direct gap material with application of tensile strain [10]. (b) Comparison of responsivity spectra for *pin* photodetectors having unstrained and strained Ge layers [67].

5.3. Reducing dark current

The growth of Ge on Si can be characterized as Stranski-Krastanov growth, an example of which is shown in Figure 17(a). For film thicknesses below the critical thickness, a 2D wetting layer is formed, beyond which a transition to 3D islanding growth mode occurs to relieve the built-in strain in the Ge layers [66]. Defects and threading dislocations arising during Stranski-Krastanov growth typically form recombination centers. At RT, dark current in *pin* photodetectors, i.e., the current measured under reverse bias with no illumination, is mainly due to generation current through such traps [68]. Higher levels of dark current result in increased power consumption that reduces detector performance, and shot noise associated with this leakage current can also degrade the SNR [5] and lower sensitivity for NIR systems [1]. Figure 17(b) shows typical I-V characteristics SiGe devices where the dark current at negative bias increases proportionally to device size.

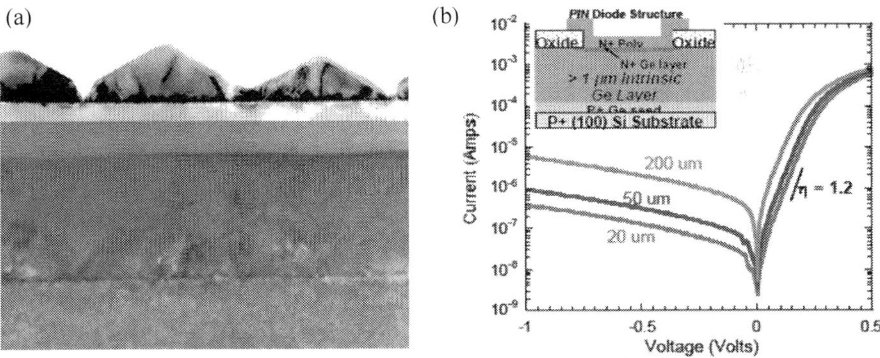

Figure 17. (a) Ge-on-Si Stranski-Krastanov epitaxial growth [10]. (b) Measured RT I-V characteristics for large area diodes with 20, 50 and 200 μm unit cells; the inset shows a schematic of the device cross-section [24].

Since dark current can be particularly high in SiGe based photodetectors, a major research thrust has been to reduce the dark current to the greatest extent possible in order to enhance sensitivity and boost overall device performance. (It is noted that in SiGe *pin* photodetectors dark current increases with applied electric field and does not saturate, and thus the measured dark current is usually specified at a given reverse bias, e.g.,-1 V.) The goal is to limit the dark current to levels acceptable for high-speed operation usually considered to be not more than 1 μA [54] (or dark current densities of 1-10 mA/cm^2), above which the transimpedance amplifier noise will be exceeded [35] and the SNR reduced [1]. However, a precise value of the required dark current is dependent upon the particular speed of operation and the amplifier design. Thermionic emission limits the dark current density in SiGe photodetectors down to ~10^{-2} mA/cm^2 at RT, which is around two orders of magnitude higher than that of standard InGaAs based photodetectors [10].

Various approaches have been proposed to further reduce the dark current in SiGe detectors by several orders of magnitude, including superlattice structures [24], incorporation of quantum dots [63], use of buried junctions [69], and graded compositional layer designs [68]. Dark current generally scales with device area, so reducing the overall size of SiGe detector devices is one means of limiting leakage current for a given photodetector design. For the fabrication of SiGe *pin* detectors, a two-step growth process and high temperature anneal (which will be covered in Section 6) can reduce threading dislocations and thus resultant dark current and dark current density [43,70]. The effects of a buffer layer grown by two-step growth and high temperature annealing on dark current density are shown in Figure 18(a) and (b), respectively. The splitting of the valence bands in Ge due to the presence of tensile strain also lowers the density of states for holes, leading to reduction of intrinsic carrier density that can likewise contribute to reduced reverse dark current in devices [65]. Further methods to effect reductions in dark current include improving surface passivation and/or utilizing smaller selective growth regions during device fabrication [63].

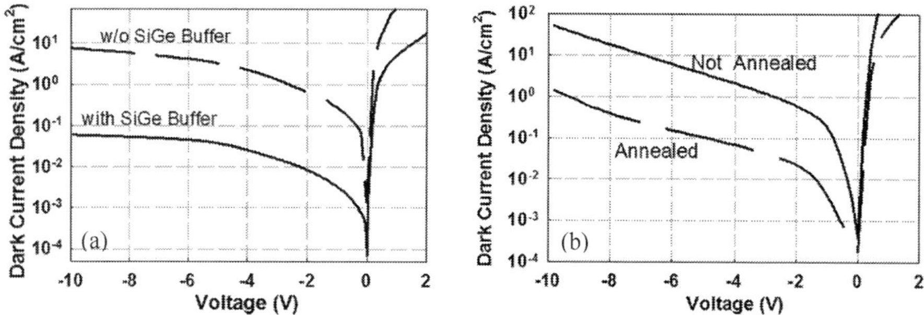

Figure 18. Effect of (a) buffer layer grown by two-step growth, and (b) high temperature annealing, on dark current density characteristics [68].

6. Fabrication of SiGe *pin* photodetectors

6.1. SiGe detector growth methods

Epitaxial growth of Si/SiGe using gas precursors has been utilized for the past three decades [10]. Selective growth of Ge/SiGe epitaxial films, using mask layers such as SiO_2 and Si_3N_4, generally requires the formation of vertical sidewalls [usually by reactive ion etching (RIE)] to minimize faceting and enhance trench filling [9]. An early method for growing Ge on Si, first proposed by Luryi *et al.* in 1984 [71] and later optimized by other groups, involved using graded SiGe buffer layers to reduce the density of threading dislocations arising in the Ge layer. Such graded structures lead to an optimized relaxation of the graded layers, where existing threading dislocations are more effectively utilized to relieve stress [43]. However, this method requires significant time and resources as well as necessitates films at least 6 μm thick that are associated with large residual surface roughness, which are problematic for the fabrication of practical, cost effective-devices [4,72].

In recent years the most prevalent and useful method to deposit Ge/SiGe layers to form functional *pin* detector devices has involved a two-step growth process where the growth temperature is ramped up between the growth steps. This technique was first applied to epitaxial grown Ge on Si by Colace *et al.* in 1998 [73], and it has since been commonly adopted for Ge epitaxial growth. This method most often involves deposition of Ge/SiGe on intrinsic Si, but growth of Ge on silicon-on-insulator (SOI) surfaces has also been demonstrated [20].

6.2. Two-step growth process overview

The two-step growth process commonly used for fabricating NI *pin* detectors consists of initial low temperature (LT) epitaxial growth of Ge/SiGe to form a thin strain-relaxed layer, followed by relatively high temperature (HT) growth to form the thicker absorbing film, and a subsequent HT anneal [16,70,74]. In general, the growth steps are primarily designed to prevent islanding. The first LT growth step is crucial in governing the film crystalline quality and the surface morphology and also the final strain state in the Ge films [66]. Ge/SiGe films grown using this process have been shown to have reduced rms surface roughness of less than 1 nm [15,70]. In addition, the HT anneal reduces threading dislocations arising from lattice mismatch between the Si and Ge to enable a higher quality Ge film with reduced dark current [9]. Figure 19(a) shows a cross-sectional view of a Ge layer grown on Si with a close-up view of the Ge-Si interface, while Figure 19(b) comprises a top view of the fabricated *pin* photodetector.

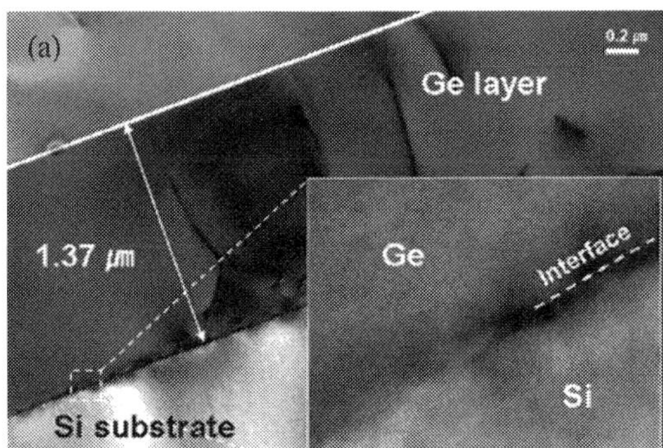

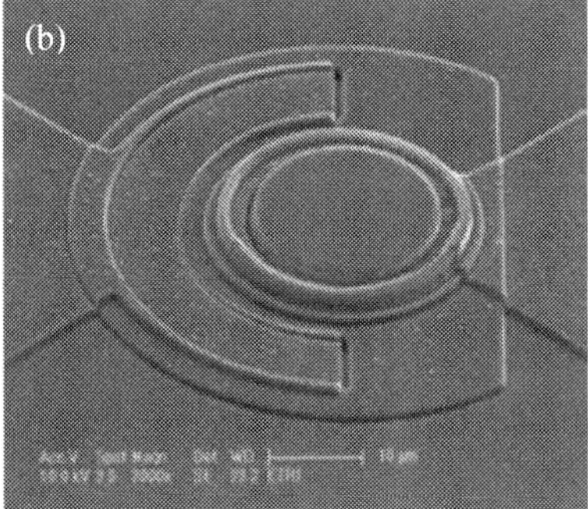

Figure 19. (a) Cross-sectional transmission electron microscope (TEM) image of Ge grown on Si substrate, where the inset shows Ge atoms coherently matched up to the Si substrate on a lattice scale; and (b) top view SEM image of fabricated Ge *pin* photodetector device [51].

The epitaxial growth in this two-step process is usually performed using a variant of the chemical vapor deposition (CVD) method. The most commonly employed variant is ultrahigh vacuum CVD (UHV-CVD) [53,75], in which the operating pressures are high enough to control oxygen background contamination levels. However, SiGe based devices have also been grown using low-pressure CVD (LP-CVD) [74] more broadly utilized by industry [7], low-energy plasma-enhanced CVD (LEPE-CVD) [48], reduced pressure CVD (RP-CVD) [46], and rapid thermal CVD (RT-CVD) [40]. These CVD based methods enable high control of layer and multi-layer thickness and suitability for future large wafer-scale fabrication. The two-step process is likewise compatible with the molecular beam epitaxy (MBE) method, which has been employed in fewer but still a considerable number of instances [37,45,70,76]. Primary advantages of MBE are allowance of lower thermal budgets [66] and tight control over doping profiles [1].

6.3. LT growth

In the first LT (slower growth rate) step of this low/high temperature growth process, fully planar homoepitaxial deposition of a thin Ge/SiGe seed or buffer layer on a Si wafer is performed to ensure smooth surface morphology and to avoid islanding of the film [10]. Si wafers with (100) orientation are associated with lower leakage currents than (001) oriented wafers [1]. The Ge seed layer is deposited on the surface of the substrate, which is often highly doped to facilitate the future requirement of low resistance ohmic contacts. This seed layer is designed to prevent strain release from undesirable 3D island growth, reduce surface roughness, and enhance the migration of threading dislocations (Figure 20) to decrease their proliferation. Buffer/seed layer thicknesses in the range of 30-75 nm are most optimal to withstand the temperature ramp and homoepitaxially grow Ge films with smooth surface morphologies [77] with reduced threading dislocation densities [68]; for layers less than 30 nm thick, islanded surfaces have a tendency to form [74]. The first ~0.7 nm (i.e., below the critical thickness) of the buffer layer will be strained due to the 4.18% lattice mismatch between it and the underlying Si substrate, after which a progressive strain relaxation takes place, and a fully strain-relaxed Ge epilayer is produced for growth beyond a few additional nanometers [66]. Therefore, this layer, assuming it is of sufficient thickness, is initially predominately relaxed.

The seed layer growth temperature influences adatom processes on the surface, crystalline growth, surface morphology, abruptness of doping transitions, and relaxation processes [68]. Temperatures employed for seed layer deposition are predominately in the 300-400°C range, and usually from 320-360°C [9]. Depositing seed layers at temperatures below 300°C can lead to crystallographic defect formation, while temperatures above 400°C have been found to produce surface roughening due to increased surface mobility of Ge [7]. At such relatively low growth temperatures, the low surface diffusivity of Ge kinetically suppresses undesired islanding that can otherwise result [10].

In situ doping (e.g., with boron) of this layer can be utilized to enhance the seed growth rate and lower the Ge/Si interfacial oxygen level [7]. Seed layer doping has been found to scale linearly with boron doping levels up to 10^{20} cm^{-3} [74]. It has also been determined to reduce series resistance under forward bias and lower dark current under reverse bias [63].

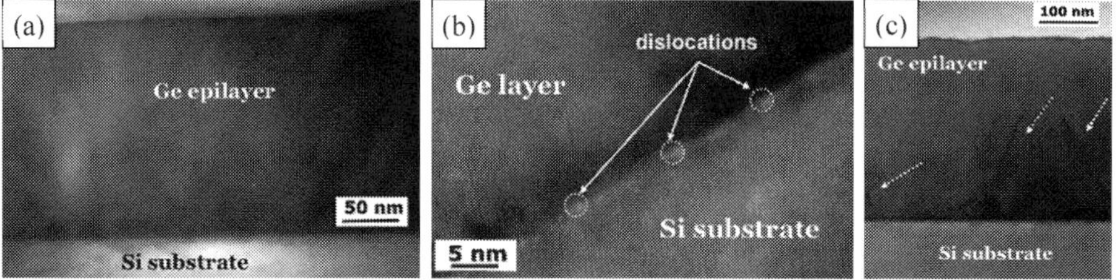

Figure 20. (a) Cross-sectional TEM image of Ge layer grown on Si; (b) enlargement of layer in (a) near the interface region, showing a high density of misfit dislocations; and (c) TEM image of the Ge layer grown following a two-step growth process, where threading dislocations are less evident [66].

6.4. HT growth

In the subsequent HT step of the growth process, a layer of intrinsic Ge or SiGe serving as the *pin* detector absorption region is grown in the temperature range of 500-800°C above the relaxed buffer/seed layer. The growth temperatures utilized are most commonly in the 600-700°C range, which has been found to produce a satisfactory growth rate and sufficient degree of tensile strain while also providing a smooth high crystal quality Ge/SiGe film [10]. Using a layer composed of Ge (rather than SiGe) maximizes the potential cutoff wavelength of device, but high Ge content $Si_{1-x}Ge_x$ ($x \geq 0.8$) can likewise be effective. In general, as this *i*-Ge/SiGe layer is made thicker, the transit time increases which reduces the device bandwidth, while the responsivity rises due to higher absorption and junction capacitance is reduced [15]. However, topological and defect density concerns limit the *i*-layer thickness for practical *pin* devices to 2 μm or less [64]. The doping level of this intrinsic SiGe layer is typically three to four orders of magnitude lower than that of the highly doped $n^+$and p^+layers of a *pin* detector [76]. Upon cooling following the HT growth, this layer becomes strained due to the difference in the thermal expansion coefficients between the Ge/SiGe layer material and the Si substrate. In one instance the level of strain present after cooling was found to rise as the layer thickness increased up to ~150 nm, and then remain essentially constant as its thickness grew further [66].

6.5. HT anneal

Following the LT/HT growth steps, HT *in situ* annealing, often cyclic in nature, is usually performed. The cyclic annealing process enables a reduction in sessile threading dislocation density by transforming sessile threading dislocations to glissile ones, which due to thermal stress glide effectively annihilate dislocations [78]. This annealing process is intended to reduce the threading dislocation density by up to two orders of magnitude (e.g., from 10^9 to 10^7 cm^{-2} [9]) and thereby diminish resultant dark/leakage currents, and also to enhance the strain/stress of the *i*-Ge layer [48]. To have an optimal effect, the high anneal temperature is usually chosen to be marginally less than the Ge melting temperature (939°C), with the low temperature at least 50°C below high anneal temperature and above ambient [78]. Typical cyclic anneal temperatures span the 700-900°C range.

Cyclic annealing for up to 10 cycles compared to a single cycle was found in multiple cases to further reduce the dislocation density by a significant degree [70]. On the other hand, a single anneal cycle can result in lower boron diffusion out from the p$^+$SiGe layer while still maintaining acceptably low dislocation density [47]. High and low cyclic annealing durations are most commonly 10 min; however, a single anneal lasting up to 2 h has been found to be equally effective in certain cases [68]. As the anneal time increases, Ge/Si interdiffusion can become an issue and limit tensile strain [66]. An alternate approach involves a hydrogen ambient, by which annealing at ~800ºC for 30 min can effectively reduce surface roughness and threading dislocation density attributed to enhanced atomic mobility from the annealing [79].

6.6. Subsequent fabrication steps

Following selective two-step LT/HT growth and annealing, additional processing steps are required in the development of a practical Ge/SiGe *pin* photodetector device. The top contact of the detector can comprise a thin (100-200 nm) layer of polysilicon deposited on the intrinsic Ge/SiGe layer, *in situ* doped with phosphorus [53]. This forms the n^+layer that provides a conductive path to the opposite site of the detector. Free-carrier absorption, which can be significant in Si at NIR wavelengths, was modeled using the Drude equation [18] for a polysilicon layer of 200 nm thickness with dopant concentration of 10^{19} cm^{-3}, and was found to have an acceptably minor impact on performance. However, if the doping level is increased significantly the layer thickness will need to be reduced, and vice versa, in order to prevent free-carrier absorption from significantly degrading detector performance.

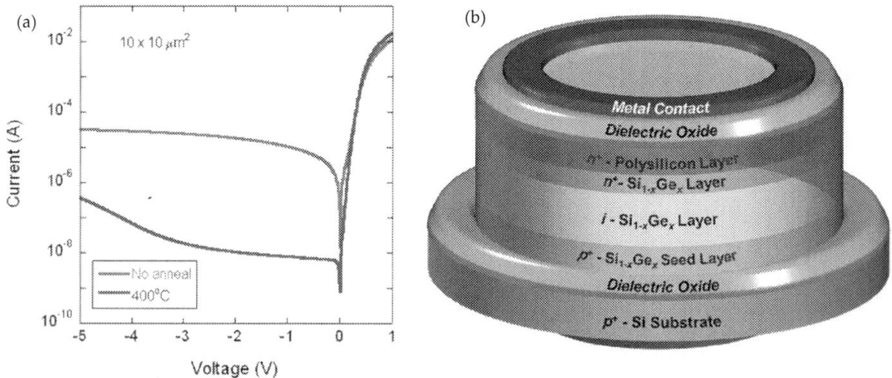

Figure 21. (a) I-V characteristics for 10×10 µm² SiGe *pin* detector devices with and without 400°C post-metallization anneal in N₂; at-1 V, the dark current is reduced by ~1000X with 400°C by the annealing [24]. (b) Schematic showing composition of a prospective SiGe *pin* photodetector device after fabrication.

Following deposition of the polysilicon layer, an activation anneal can be performed, which serves to out-diffuse dopant atoms from the polysilicon layer into the underlying Ge/SiGe to form a vertical *pin* junction [63]. A passivation layer (e.g., of SiO₂) may then be deposited at relatively low temperatures using a CVD based process, which serves to reduce leakage currents and isolate active elements [1]. This oxide layer can be patterned using a photolithographic process to open a window to the underlying $Si_{1-x}Ge_x$ surface. The next prospective step in this process, nearly completing the photodetector design, involves sputter depositing metal (e.g., aluminum or titanium) to form low resistance top and bottom contacts. Silicidation annealing may subsequently be performed at temperatures in the 600-900°C range to ensure highly conductive ohmic contacts enabling higher photocurrent [53]; this has also been observed to marginally increase the tensile strain in the Ge/SiGe layer [16]. Following this metallization process, the samples may be annealed in nitrogen, as shown in Figure 21(a), which has been found reduce dark current by up to three orders of magnitude for small area Ge/SiGe *pin* photodetectors [24]. A potential design layout of a fabricated SiGe *pin* photodetector device having undergone these processing steps is depicted in Figure 21(b).

7. Practical integration SiGe detectors for imaging arrays

7.1. IR FPA and ROIC technology

Because of the compatibility of Ge growth methods with standard silicon based CMOS processes, photodetectors developed through selective epitaxial growth of Ge/SiGe can be heterogeneously integrated with CMOS circuitry using manufacturing infrastructure already widely installed for the production of BiCMOS and CMOS integrated circuits. In addition, unlike charge-coupled device (CCD) based imagers that require specialized and relatively complicated processing techniques, CMOS based imagers can be built on fabrication lines designed for commercial microprocessors. This has enabled the resolution of CMOS imagers to continue to increase rapidly due to the ongoing transition to finer lithographies as predicted by Moore's Law. This in turn has led to higher circuit density and levels of integration, better image quality, lower voltages, and lower overall system costs for CMOS devices in comparison with traditional CCD based solutions [80].

The term *focal plane array* (FPA) refers to a 2D assemblage of individual detector pixels located at the focal plane of an imaging system [21]. FPAs convert optical images into electrical signals that can then be read out and processed and/or stored in digital format. Staring array FPAs, in which the associated optics serve solely to focus the visual image onto the detectors in the array, are scanned electronically usually using circuits integrated with the arrays. The electrical output from the array can be either an analog or digital signal, which in the latter case requires the inclusion of analog-to-digital conversion electronics. CMOS based silicon addressing circuits, the dominant technology for large scale arrays, are mature with respect to fabrication yield and attainment of near-theoretical sensitivity.

Readout integrated circuits (ROICs) enable a FPA to be fully functional by accumulating photocurrent from each pixel to provide parallel signal processing circuitry for readout. ROIC functions include pixel deselecting, antiblooming on each pixel, subframe imaging, and output preamplification [80]. In monolithically integrated ROICs, both detection of light and signal readout (multiplexing) is performed in the detector material in the spacing between the pixels rather than in an external readout circuit [75]. Advantages of this approach include reduced number of processing steps, increased yields, and reduced costs. Another common architecture for IR FPAs uses a hybrid based approach, in which the individual pixels are directly connected with readout electronics providing for multiplexing [21]. Some benefits of this method are the potential for near 100% fill factor, increased signal processing area, and the ability to optimize the detector and multiplexer independently.

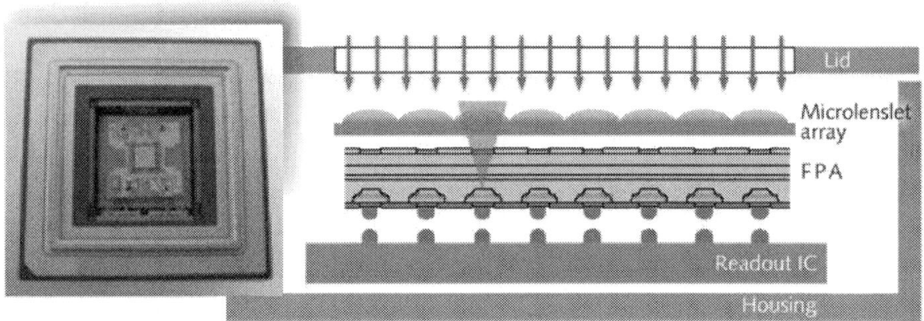

Figure 22. Photograph and schematic of a focal plane array (FPA) consisting of a hybridized chip stack with readout integrated circuit (ROIC), pixel array, and microlens array [81].

ROICs comprise input cells or unit cells, which in the case of hybrid FPAs consist of the areas located directly under each pixel that are connected to the pixels through indium bumps that bond the aligned FPA and ROIC together [82]. This procedure allows multiplexing the signals from thousands of pixels onto a few output lines. FPAs can utilize either frontside illumination, where photons pass through the ROIC, or backside illumination, where photons pass through a transparent detector array substrate; the latter is often the most advantageous since ROICs typically have areas of metallization and other opaque regions that effectively reduce optical area of the structure [21]. ROICs are processed in standard commercial foundries, and can be custom designed to feature any type of circuit that will fit in the unit cells, though this space is often quite limited. Microlenses deposited above each pixel arrays concentrate incoming light into photosensitive regions, and thus offer a means of further improving sensitivity for devices having relatively low fill factors. A typical indium bonded hybrid architecture FPA utilizing a microlens array is depicted in Figure 22.

7.2. Integration of SiGe technology in CMOS processes

The progressing technological development of low dark current SiGe detector arrays has made possible the fabrication of high density large format SiGe NIR FPAs. The frontside illumination process flow shown sequentially in Figure 23 was developed by DRS Technologies and provides various potential steps for the processes for fabrication and integration of FPA pixels with SOI wafers [63]. In this process, the SOI wafers have a thin, high quality Si layer on top, and a buried oxide layer below. The detector p^+base layer and intrinsic (i) layer are deposited by SiGe epitaxy (*Step 1*). Vias into SiGe are then etched by RIE, where the buried oxide layer provides the etch stop (*Step 2*). Next, an oxide layer is deposited to provide dielectric isolation for the via structure. Doped polysilicon deposition completes the top n^+layer of the *pin* photodiode structure and provides a conductive path to the opposite site of the detector (*Step 3*). A Si handle is bonded to the detector wafer (*Step 4*) to provide support during the etching and thinning. The Si wafer is then etched (*Step 5*), where the buried oxide layer again provides an etch stop (selectivity > 1000:1). Next, vias are opened though the oxide to access the top n^+detector layer and the p^+base layer (*Step 6*). This step is followed by via metal fill. The detector wafer will then be ready for direct CMOS bonding to interconnects (*Step 7*); this enabling technology is associated with low temperatures, high density, ultra-small pitch, and pixel scale reductions. Following bonding, the handle wafer is removed (*Step 8*). Finally, a pixel isolation etch completes the detector array (*Step 9*).

1. Grow Si-Ge epi on Si SOI wafers

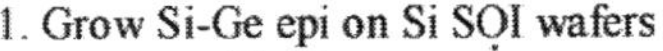

2. Etch via to buried oxide

3. Deposit oxide/doped poly and pattern

4. Bond to silicon handle wafer

5. Etch Si wafer to oxide etch stop

6. Etch vias to p+/n+ and metal fill

7. Detector/CMOS wafer bonding

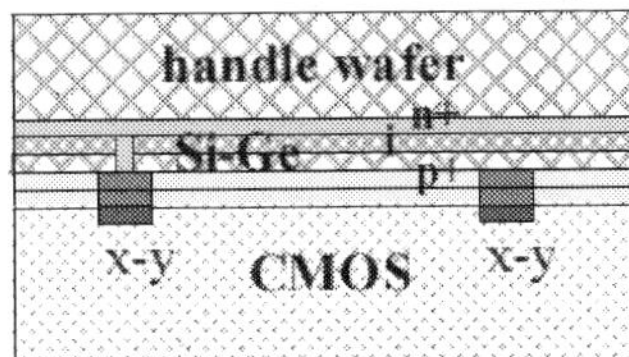

8. Etch handle wafer to oxide etch stop

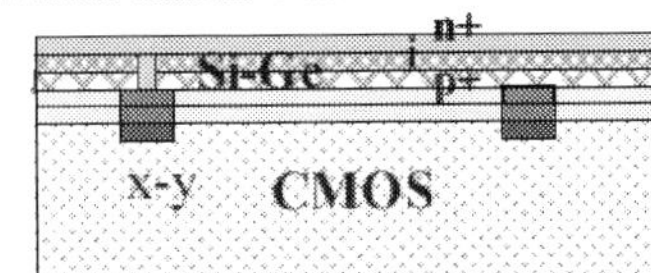

9. Pixel isolation etch

Figure 23. Process flow for fabrication of NIR front side illuminated SiGe FPA and CMOS ROIC for SiGe detection and imaging applications [63].

The first monolithic integration of Ge NIR photodetectors in a CMOS process that produced multichannel, high-speed optical receivers was reported by Masini *et al.* in 2007 [83]. With this approach, the Ge epitaxy step was integrated at the end of the frontend processing and before the contact module. RP-CVD deposition of Ge at a temperature of 350°C without HT annealing was performed to avoid potential damage resulting from high temperature epitaxy. Since high temperatures are necessary for gate oxide growth, the Ge module was inserted after the gate processing. The integrated WC *pin* detectors, which are depicted with the Si CMOS circuit in Figure 24, operated at 1550 nm at a bandwidth of 10 Gb/s with sensitivity greater than-14 dBm.

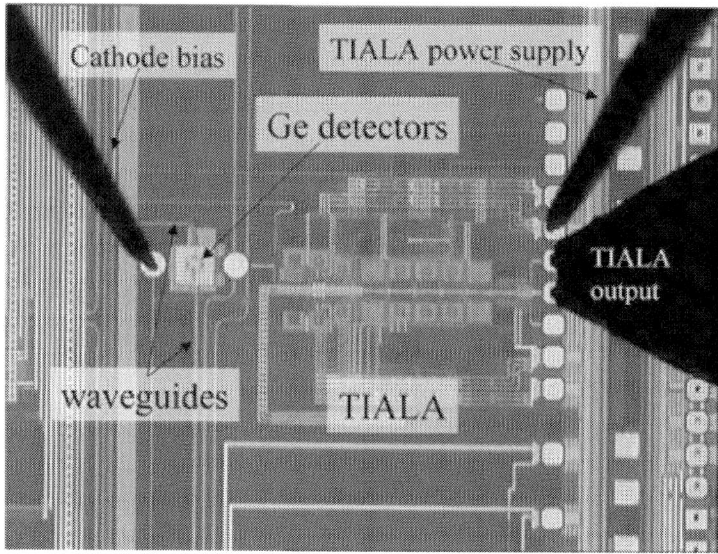

Figure 24. Integration approach for monolithic fabrication of a Ge pin photodetector and Si CMOS circuit on a common SOI platform [83].

In 2010, Ang *et al.* [19] developed and monolithically integrated highly efficient WC *pin* Ge-on-SOI photodetectors with a CMOS circuits. They utilized an "electronic-first, photonic-last" approach to avoid Ge degradation and cross contamination to fabricate both vertical and lateral devices, where the former were found to offer superior performance in relation to bandwidth and dark current density. A closely matched integrated CMOS inverter circuit was demonstrated capable of performing logic functions. A high temperature (800°C) prebake treatment was used before the Ge epitaxy growth, which was found to not have any observable detrimental impact on the operation of the CMOS devices. The vertical *pin* detectors achieved a responsivity of 0.92 A/W at 1550 nm, QE of 73%, bandwidth of 11.3 GHz, and dark current of 0.57 µA. Figure 25(a) and (b) show the design of the evanescent coupled Ge photodetectors in vertical and lateral *pin* configurations, respectively, while Figure 25(c) shows the integration approach for monolithically fabricating the Ge *pin* photodetector and Si CMOS circuit.

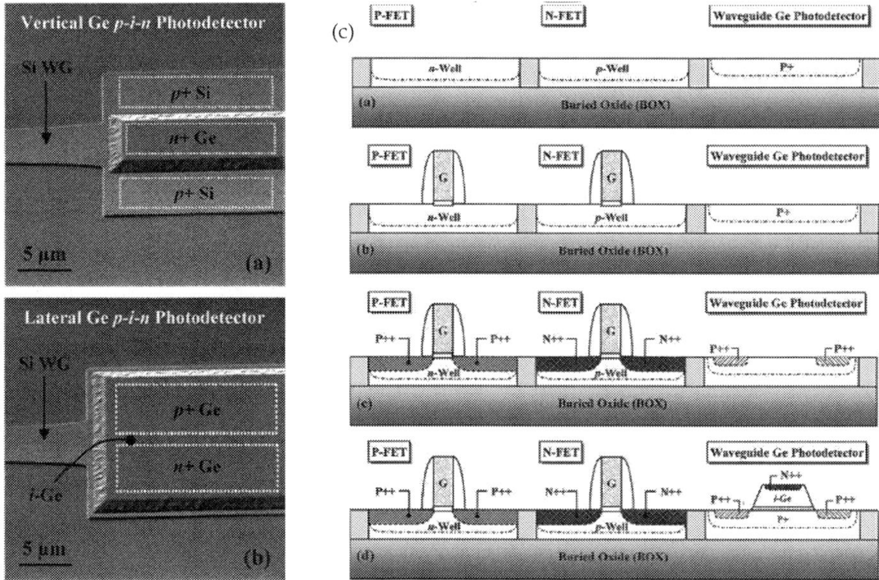

Figure 25. (a) SEM micrograph showing design of evanescent coupled Ge photodetector featuring vertical *pin* configuration; (b) Ge photodetector design with lateral *pin* configuration; and (c) schematic showing "electronic-first and photonic-last" integration approach for monolithically fabricating the WC Ge *pin* photodetector and Si CMOS circuit on a common SOI platform [19].

7.3. Development of SiGe detector arrays for imaging

IR FPAs have traditionally been based on conventional materials utilized for IR detection including HgCdTe, InSb, InGaAs, and VOx [64]. SiGe FPAs for NIR detection are relatively new to the scene. SiGe based FPAs with associated ROICs can leverage low-voltage, deeply scaled, nanometer class IC processes that enable high yield of low-power, high-component density designs with large dynamic, on-chip digital image processing (for SWaP-efficient sensor designs) and high-speed readouts. A common objective is to produce large format NIR FPAs that are very compact.

Colace *et al.* in 2007 [6] demonstrated an optoelectronic chip incorporating a fully functional 2D array of 512 polycrystalline heterojunction Ge pixels integrated on CMOS electronics and operating in the NIR. The ROIC was serial and made use of an 8-bit data bus and 9-bit address bus. In order to compensate for a significant level of dark current, the chip was equipped with offset control and a dark current cancellation circuit. It also featured addressing and signal processing electronics including 64 analog-to-digital converters. This integrated circuit, which operated up to and above 1550 nm, was found to exhibit good photoresponse and could acquire simple images. The chip, its architecture, and its cross-section are illustrated in Figure 26.

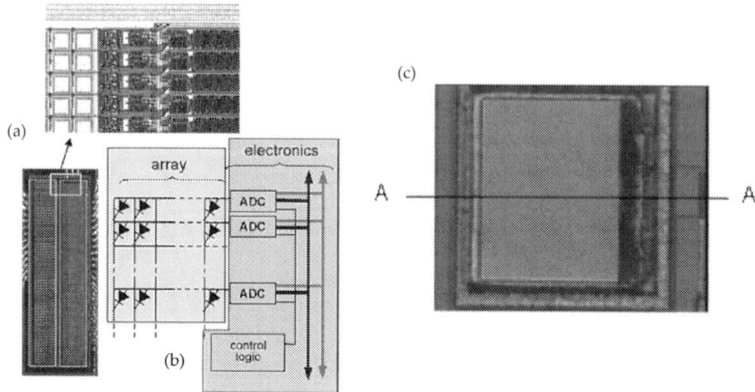

Figure 26. (a) Photograph of optoelectronic chip featuring 2D Ge photodetector array, showing zoomed-in image of a portion of the array; (b) chip architecture, comprising readout electronics; and (c) photograph of individual pixel [6].

In 2010 Vu *et al.* [75] developed arrays of both vertical and lateral *pin* photodetectors that were integrated into electronic-photonic FPAs. Layers of metal were employed to connect the detector electrodes to transimpedance amplifiers and CMOS circuits, and light signals were then coupled from waveguides or inserted directly into the side of the Ge intrinsic layer via optical fibers. A responsivity of 1.11 A/W at 1550 nm, bandwidth of 15 GHz, and dark current density on the order of 100 nA/cm^2 were achieved for the NI photodetectors. The integrated chips were produced in a standard CMOS foundry, where the fabrication processes was optimized for manufacturability.

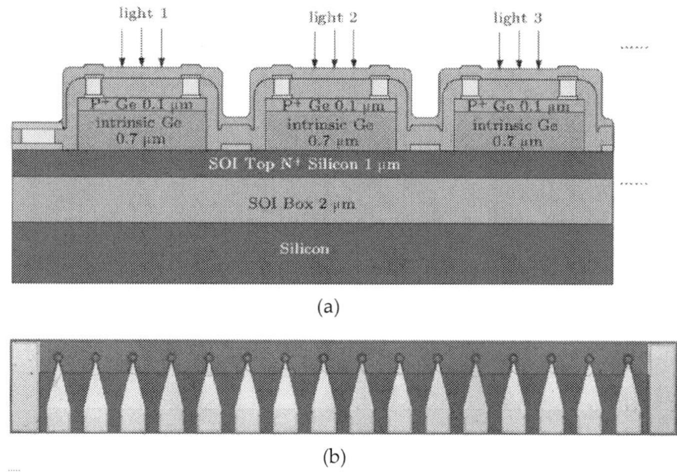

Figure 27. (a) Schematic cross-section, and (b) top view, of a linear photodetector array consisting of 16 detectors [84].

In 2014, Chong *et al.* [84] reported a parallel system of 16 element *pin* photodetector arrays, shown in Figure 27, for parallel optical interconnect applications. The detectors comprised Ge absorption layers epitaxially grown on a SOI substrate by UHV-CVD using the two-step LT/HT growth process, and incorporated a plasma etched double mesa vertical structure to reduce parasitic capacitance. The array featured responsivities of 0.38 and 0.23 A/W at 1310 and 1550 nm, respectively, with a very low dark current density of ~5 mA/cm² with no applied bias and a bandwidth of up to 8 GHz.

8. Optoelectronic properties of Si/Ge based nanostructures

8.1. Quantum confinement and strain in Si/Ge nanostructures

A growing number of optoelectronic devices including photodetectors are being developed that employ low-dimensional nanostructures (NSs), particularly quantum wires (Q-wires) and quantum dots (QDs), to enhance their performance. NSs offer unique optical and electronic properties resulting from the quantum confinement of electrons and holes. Such quantum confinement in NSs, which is directly affected by their dimensions, has a substantial impact on band gap energy. The quantum confinement effect (QCE) causes the band gap of crystalline Si and Ge Q-wires and QDs to increase as their diameters are reduced according to the relation $E_g \sim 1/R^a$, where E_g is the band gap and a falls between 1 and 2 [85]. Figure 28(a) and (b) show a comparison of band gap energy reported by various groups as a function of QD diameter in crystalline Si and Ge QD structures, respectively, where it can be seen that the band gap energies of the Si QDs follow a much more uniform and predictable pattern than those of their Ge counterparts. As a result of larger exciton energy, the QCE effect is stronger in Ge than in Si, and the electronic properties Ge QDs can thus be more easily modulated by the QCE than can those of Si QDs [85,88]. In addition, making the length scale in NSs small enough produces uncertainty in the momentum **k** vectors that consequently allows the **k** selection rules to be broken, causing the band gap to change from indirect toward direct that allows electron-hole recombination to take place without the need of phonons [86].

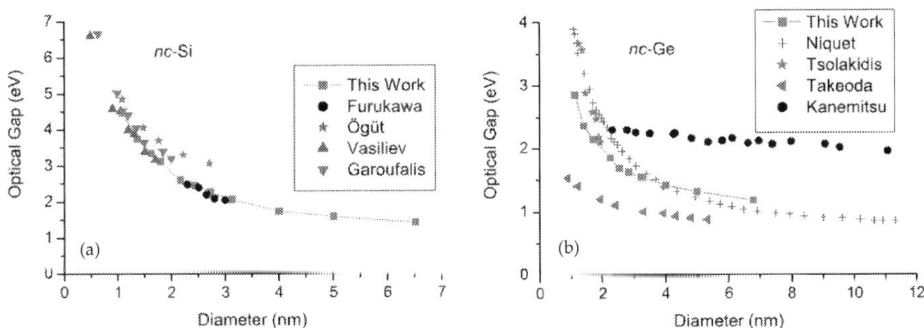

Figure 28. Comparison of band gap energy as a function of QD diameter for (a) Si and (b) Ge [87].

As was shown to be the case with bulk SiGe alloys, strain alters the intrinsic interatomic distances and thus affects the band gap energy, and also impacts the effective masses and mobility [85]. However, their reduced dimensionality and small size allow NSs to tolerate relatively large stress and strain without introducing significant dislocations or other defects that could undermine their electrical properties. Due to the nature of their geometry, NWs — especially the core-shell variety — experience tensile stress due to bending in addition to that caused by lattice mismatch [86]. By applying an external tensile strain of around 2.8% to Si-core/Ge-shell NWs, a transformation from direct band gap to indirect one can likewise be achieved [89].

8.2. Photodetectors based on SiGe nanowires and quantum dots

In addition to the enhancement of performance properties due to the QCE and strain, detectors based on one-dimensional Q-wires [i.e., nanowires (NWs)] offer potentially greater sensitivity primarily due to larger surface-to-volume ratios [90]. There is still progress to be made in this area, however, as Ge NW based photodetectors currently have significantly longer photocurrent rise and decay kinetics and associated time constants than those based on bulk Ge/SiGe. As illustrated in Figure 29(a), the optical absorption of $Si_{1-x}Ge_x$ NWs is largely affected by the material concentration, with the band gap (and thus SiGe NW based photodetector response) shifting to lower energies and longer wavelengths as x increases. As might be expected based on the previous discussion on QDs, a shift to lower energies was observed with increasing NW diameter for both Si and Ge NWs, again evidencing the potential for tuning the optical properties of NS based photodetector devices by varying the constituent NS sizes.

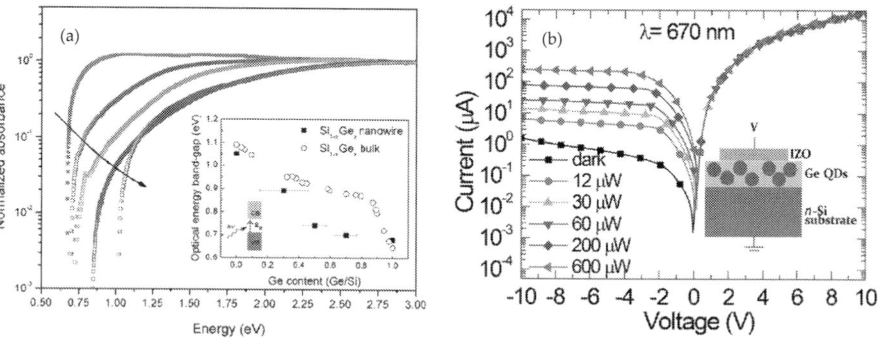

Figure 29. (a) Optical absorption spectra vs. band gap energy of $Si_{1-x}Ge_x$ NWs of five representative compositions (following the arrow, the spectrum corresponds to Ge, $Si_{0.3}Ge_{0.7}$, $Si_{0.5}Ge_{0.5}$, $Si_{0.7}Ge_{0.3}$, and Si NWs, respectively); the inset summarizes variation of optical band edges with the known values from bulk $Si_{1-x}Ge_x$ crystals [91]. (b) I-V characteristics for amorphous Ge QD photodetector at different illumination powers; the inset shows a schematic of the device [92].

In the past few years, a number of detector devices comprising QDs, which exhibit quantum confinement in all three dimensions, have been developed. QD detectors offer the advantages of increased sensitivity to normally incident radiation as a result of breaking of the polarization selection rules, large photoelectric gain associated with a reduced capture probability of photoexcited carriers due to suppression of electron-phonon scattering, and small thermal generation rate resulting from the zero-dimensional character of the electronic spectrum that renders improved SNR [93]. Compared to Si QDs, Ge QDs have higher absorption coefficients due to localized defect states [92]. SiGe QD detectors have been reported that operate up to the LWIR regime; however, the responsivity of these devices is typically much greater at NIR wavelengths, i.e., below 2000 nm [93]. The response at NIR wavelengths of photodetectors comprising Ge QDs grown on SiGe has been attributed to interband transitions between electrons in Ge/SiGe layers and holes in the Ge QDs. Figure 29(b) shows the I-V characteristics of a Ge QD photodetector exposed to different intensities of visible illumination. Ge QD based photodetectors have recently demonstrated peak responsivities as high as 4 A/W at-10 V bias and response times down to ~40 ns [92].

9. Summary

This chapter has covered the operation, design, fabrication, and applications of SiGe based photodetector technology. A model to predict SiGe sensor performance over a wide range of light levels has been presented, which indicates that a low-cost, small pixel, uncooled SiGe based detector will respond well to small amounts of illumination from a direct NIR source. The operation and relative performance characteristics of Ge based avalanche photodiodes (APDs), metal-semiconductor-metal (MSM) detectors, and *pin* detectors have been discussed. SiGe *pin* photodetectors offer performance advantages including high responsivities, high bandwidths, low bias voltage requirements, and low dark current compared to other types of SiGe detectors. The impact of detector dark current and techniques for reducing it in *pin* photodetector devices have been examined. The nature and impact of strain and stress on extending SiGe based detector response to longer NIR wavelengths were also discussed.

Installed infrastructure and heterogeneous integration can be leveraged to fabricate small feature CMOS-compatible SiGe based *pin* detector array devices exhibiting optimal properties for NIR detection. A common fabrication process for SiGe based *pin* photodetectors incorporating two-step low/high temperature epitaxial growth of Ge/SiGe layers on Si substrates followed by a high temperature anneal and additional processing steps has been outlined, which was found to reduce threading dislocation density and thereby improve device quality. In addition, fabricated SiGe detectors can be directly integrated with low noise Si readout integrated circuits to yield low SWaP, low cost, and highly uniform IR focal plane arrays (FPAs) that can function as imaging devices. Various integrated SiGe based FPA imagers have been demonstrated that exhibit enhanced functionality and performance characteristics. Finally, the impact of the quantum confinement effect and strain on band gap in low dimensional nanostructures was analyzed, and the characteristics of photodetectors based on quantum dots and nanowires were discussed.

Author details

Ashok K. Sood[1*], John W. Zeller[1,2], Robert A. Richwine[1], Yash R. Puri[1], Harry Efstathiadis[2], Pradeep Haldar[2], Nibir K. Dhar[3] and Dennis L. Polla[4]

*Address all correspondence to: aksood@magnoliaoptical.com

1 Magnolia Optical Technologies, Woburn, MA, USA

2 State University of New York College of Nanoscale Science and Engineering, Albany, NY, USA

3 Defense Advanced Research Project Agency, Arlington, VA, USA

4 College of Science and Engineering, University of Minnesota, Minneapolis, MN, USA

References

[1] DiLello, Nicole Ann. "Fabrication and characterization of germanium-on-silicon photodiodes." Ph.D. dissertation, Massachusetts Institute of Technology, 2012.

[2] A. K. Sood, R. A. Richwine, Y. R. Puri, N. DiLello, J. L. Hoyt, T. I. Akinwande, N. Dhar, S. Horn, R. S. Balcerak, and T. G. Bramhall, "Development of low dark current SiGe-detector arrays for visible-NIR imaging sensor," *Proc. SPIE* 7660, 76600L (2010).

[3] G. Masini, L. Colace, G. Assanto, H. C. Luan, and L. C. Kimerling, "High-performance p-i-n Ge on Si photodetectors for the near infrared: from model to demonstration," *IEEE Trans. Electron. Dev.* 48(6), 1092-1096 (2001).

[4] J. Wang and S. Lee, "Ge-photodetectors for Si-based optoelectronic integration," *Sensors* 11(1), 696-718 (2011).

[5] Image Courtesy of Hamamatsu. (*Source*: http://www.vision-systems.com)

[6] L. Colace, G. Masini, V. Cencelli, F. DeNotaristefani, and G. Assanto, "A near-infrared digital camera in polycrystalline germanium integrated on silicon," *IEEE J. Sel. Top. Quant. Electron.* 43(4), 311-315 (2007).

[7] Olubuyide, O. Oluwagbemiga. "Low pressure epitaxial growth, fabrication and characterization of Ge-on-Si photodiodes." Ph.D. dissertation, Massachusetts Institute of Technology, 2007.

[8] H. Efstathiadis and F. W. Smith, "Thermodynamic modelling of the local atomic order in amorphous alloys based on silicon and germanium," *Philosophical Mag. B* 70(3), 547-556 (1994).

[9]   J. Liu, R. Camacho-Aguilera, J. T. Bessette, X. Sun, X. Wang, Y. Cai, L. C. Kimerling, and J. F. Michel, "Ge-on-Si optoelectronics," *Thin Sol. Films* 520(8), 3354-3360 (2012).

[10] J. Michel, J. Liu, and J. C. Kimerling, "High-performance Ge-on-Si photodetectors," *Nature Photon.* 4(8), 527-534 (2010).

[11] J. Osmond, G. Isella, D. Christina, R. Kaufmann, M. Acciarri, and H. von Kanel, "Ultralow dark current Ge/Si(100) photodiodes with low thermal budget," *Appl. Phys. Lett.* 94, 201106 (2009).

[12] Okyay, Ali Kemal. "Si-Ge photodetection technologies for integrated optoelectronics." Ph.D. dissertation, Stanford University, 2007.

[13] A. K. Sood, R. A. Richwine, G. Pethuraja, Y. R. Puri, J. U. Lee, P. Haldar, and N. K. Dhar, "Design and development of wafer-level short wave infrared micro-camera," *Proc. SPIE* 8704, 870439 (2013).

[14] L. Colace, G. Masini, G. Assanto, H. C. Luan, K. Wada, and L. C. Kimerling, "Efficient high-speed near-infrared Ge photodetectors integrated on Si substrates," *Appl. Phys. Lett.* 76, 1231-1233 (2000).

[15] J. Liu, J. Michel, W. Giziewicz, D. Pan, K. Wada, D. D. Cannon, S. Jongthammanurak, D. T. Danielson, L. C. Kimerling, J. Chen, F. O. Ilday, F. X. Kartner, and J. Yasaitis, "High-performance, tensile-strained Ge p-i-n photodetectors on a Si platform," *Appl. Phys. Lett.* 87, 103501 (2005).

[16] J. Liu, D. D. Cannon, K. Wada, Y. Ishikawa, S. Jongthammanurak, D. T. Danielson, J. Michel, and L. C. Kimerling, "Silicidation-induced band gap shrinkage in Ge epitaxial films on Si," *Appl. Phys. Lett.* 84, 660-662 (2004).

[17] Image courtesy of IBM Research. (*Source*: http://phys.org)

[18] B. Jalali, M. Paniccia, and G. Reed, "Silicon photonics," *IEEE Microwave Mag.*, 58-68 (2006).

[19] K. W. Ang, T. Y. Liow, M. B. Yu, Q. Fang, J. Song, G. Q. Lo, and D. L. Kwong, "Low thermal budget monolithic integration of evanescent-coupled Ge-on-SOI photodetector on Si CMOS platform," *IEEE J. Sel. Top. Quant. Electron.* 16(1), 106-113 (2010).

[20] S. J. Koester, J. D. Schaub, G. Dehlinger, and J. O. Chu, "Germanium-on-SOI infrared detectors for integrated photonic applications," *IEEE J. Sel. Top. Quant.. Electron.* 12(6), 1489 (2006).

[21] A. Rogalski, "Progress in focal plane array technologies," *Prog. Quant. Electron.* 36, 342-473 (2012).

[22] D. B. Law, E. M. Carapezza, C. J. Csanadi, G. D. Edwards, T. M. Hintz, and R. M. Tong, "Multi-spectral signature analysis measurements of selected sniper rifles and small arms," *Proc. SPIE* 2938 (1997).

[23] C. Callan, J. Goodman, M. Cornwall, N. Fortson, R. Henderson, J. Katz, D. Long, R. Muller, M. Ruderman, and J. Vesecky, "Sensors to support the soldier." No. JSR-04-210. Mitre Corp., McLean, VA (2005).

[24] A. K. Sood, R. A. Richwine, A. W. Sood, Y. R. Puri, DiLello, J. L. Hoyt, T. I. Akinwande, N. K. Dhar, R. S. Balcerak, and T. G. Bramhall, "Characterization of SiGe-detector arrays for visible-NIR imaging sensor applications," *Proc. SPIE* 8012, 801240 (2011).

[25] Courtesy of U.S. Marine Corps; photo by Sgt. Timothy Brumley. (*Source*: http://www.defense.gov)

[26] M. Zoberbier, S. Hansen, M. Hennemeyer, D. Tönnies, R. Zoberbier, M. Brehm, A. Kraft, M. Eisner, and R. Völkel, "Wafer level cameras—novel fabrication and packaging technologies," Int. Image Sens. Workshop (2009).

[27] C. Leinert, "The 1997 Reference of Diffuse Night Sky Brightness," *Astronomy and Astrophysics Supplement Series* 127 (1998).

[28] M. Vatsia, "Atmospheric Optical Environment," Research and Development Technical Report ECOM-7023, Sept. 1972.

[29] R. Littleton, "Spectral irradiance of the night sky for passive low light level imaging applications," *Proc. of MSS*, Passive Sensors (2005).

[30] C. H. Lin and C. W. Liu, "Metal-insulator-semiconductor photodetectors," *Sensors* 10(10), 8797-8826 (2010).

[31] J. C. Campbell, S. Demiguel, F. Ma, A. Beck, X. Guo, S. Wang, X. Zheng, X. Li, J. D. Beck, M.A. Kinch, A. Huntington, L. A. Coldren, J. Decobert, and N. Tscherptner, "Recent advances in avalanche photodiodes," *IEEE J. Sel. Top. Quant. Electron.* 10, 777-787 (2004).

[32] P. R. Berger, "MSM photodiodes," *IEEE Potentials* 15(2), 25-29 (1996).

[33] A. K. Okyay, A. M. Nayfeh, K. C. Saraswat, T. Yonehara, A. Marshall, and P. C. McIntyre, *Opt. Lett.* 31, 2565-2567 (2006).

[34] K. W. Ang, S. Zhu, M. Yu, G. Q. Lo, and D. L. Kwong, *IEEE Photon. Technol. Lett.* 20, 754-756 (2008).

[35] Wang, Jian. "Fabrication and characterization of germanium photodetectors." Ph.D. dissertation, National University of Singapore, 2011.

[36] H. Zang, S. Lee, M. Yu, W. Loh, J. Wang, G. Lo, and D. Kwong, "High-speed metal-germanium-metal configured PIN-like Ge-photodetector under photovoltaic mode and with dopant-segregated Schottky-contact engineering," *IEEE Photon. Technol. Lett.* 20, 1965-1967 (2008).

[37] K. Ang, S. Zhu, J. Wang, K. Chua, M. Yu, G. Lo, and D. Kwong, "Novel silicon-carbon (Si: C) Schottky barrier enhancement layer for dark-current suppression in Ge-on-SOI MSM photodetectors," *IEEE Electron. Dev. Lett.* 29, 704-707 (2008).

[38] M. Jutzi, M. Berroth, G. Wohl, M. Oehme, and E. Kasper, "Ge-on-Si vertical incidence photodiodes with 39-GHz bandwidth," *IEEE Photon. Tech. Lett.* 17, 1510-1512 (2005).

[39] C. Xu, R. T. Beeler, L. Jiang, G. Grzybowski, A. V. G Chizmeshya, J. Menendez, and J. Kouvetakis, "New strategies for Ge-on-Si materials and devices using non-conventional hydride chemistries: the tetragermane case," *Semicond. Sci. Technol.* 28, 105001 (2013).

[40] H. D. Yang, Y. H. Kil, J. H. Yang, S. Kang, T. S. Jeong, C. J. Choi, T. S. Kim, and K. H. Shim, "Fabrication of PIN photo-diode from p-Ge/i-Ge/n-Si hetero junction structure," *Mat. Sci. Semi. Proc.* 17, 74-80 (2014).

[41] E. Kasper, "Prospects and challenges of silicon/germanium on-chip optoelectronics," *Front. Optoelectron. China* 3(2), 143-152 (2010).

[42] J. Wang, W. Loh, K. Chua, H. Zang, Y. Xiong, S. Tan, M. Yu, S. Lee, G. Lo, and D. Kwong, "Low-voltage high-speed (18 GHz/1 V) evanescent-coupled thin-film-Ge lateral PIN photodetectors integrated on Si waveguide," *IEEE Photon. Technol. Lett.* 20(17), 1485-1487 (2008).

[43] S. B. Samavedam, M. T. Currie, T. A. Langdo, and E. A. Fitzgerald, "High-quality germanium photodiodes integrated on silicon substrates using optimized relaxed graded buffers," *Appl. Phys. Lett.* 73(15), 2125-2127 (1998).

[44] S. Fama, L. Colace, G. Masini, G. Assanto, and H. C. Luan, "High performance germanium-on-silicon detectors for optical communications," *Appl. Phys. Lett.* 81, 586-588 (2002).

[45] M. Oehme, J. Werner, E. Kasper, M. Jutzi, and M. Berroth, "High bandwidth Ge p-i-n photodetector integrated on Si," *Appl. Phys. Lett.* 89, 071117 (2006).

[46] M. Morse, O. Dosunmu, G. Sarid, and Y. Chetrit, "Performance of Ge-on-Si p-i-n photodetectors for standard receiver modules," *IEEE Photon. Tech. Lett.* 18, 2442-2444 (2006).

[47] Z. Huang, N. Kong, X. Guo, M. Liu, N. Duan, A. L. Beck, S. K. Banerjee, and J. C. Campbell, "21-GHz-bandwidth germanium-on-silicon photodiode using thin SiGe buffer layers," *IEEE J. Sel. Top. Quant. Electron.* 12, 1450-1454 (2006).

[48] L. Colace, M. Balbi, G. Masini, G. Assanto, H. C. Luan, and L. C. Kimerling, "Ge on Si p-i-n photodiodes operating at 10 Gbit/s," *Appl. Phys. Lett.* 88(10), 101111 (2006).

[49] T. H. Loh, H. S. Nguyen, R. Murthy, M. B. Yu, W. Y. Loh, G. Q. Lo, N. Balasubramanian, D. L. Kwong, J. Wang and S. J. Lee, "Selective epitaxial germanium on silicon-

on-insulator high speed photodetectors using low-temperature ultrathin $Si_{0.8}Ge_{0.2}$ buffer," *Appl. Phys. Lett.* 91, 073503 (2007).

[50] S. Klinger, M. Berroth, M. Kaschel, M. Oehme, and E. Kasper, "Ge-on-Si p-i-n photo-diodes with a 3-dB bandwidth of 49 GHz," *IEEE Photon. Technol. Lett.* 21(13), 920-922 (2009).

[51] D. Suh, S. Kim, J. Joo, and G. Kim, "36-GHz high-responsivity Ge photodetectors grown by RPCVD," *IEEE Photon. Technol. Lett.* 21, 672-674 (2009).

[52] D. Feng, L. Shirong, P. Dong, N. N. Feng, H. Liang, D. Zheng, C. C. Kung, J. Fong, R. Shafiiha, J. Cunningham, A. V. Krishnamoorthy, and M. Asghari, "High-speed Ge photodetector monolithically integrated with large cross-section silicon-on-insulator waveguide," *Appl. Phys. Lett.* 95(26), 261105 (2009).

[53] J. Liu, D. D. Cannon, K. Wada, Y. Ishikawa, S. Jongthammanurak, T. Danielson, J. Michel, and L. C. Kimerling, "Tensile strained Ge p-i-n photodetectors on Si platform for C and L band telecommunications," *Appl. Phys. Lett.* 87(1), 011110 (2005).

[54] D. Ahn, C. Hong, J. Liu, W. Giziewicz, M. Beals, L. Kimerling, J. Michel, J. Chen, and F. X. Kartner, "High performance, waveguide integrated Ge photodetectors," *Opt. Express* 15, 3916-3921 (2007).

[55] T. Yin, R. Cohen, M. M. Morse, G. Sarid, Y. Chetrit, D. Rubin, and M. J. Paniccia, "31GHz Ge n-i-p waveguide photodetectors on silicon-on-insulator substrate," *Opt. Express* 15, 13965-13971 (2007).

[56] M. Beals, J. Michel, J. F. Liu, D. H. Ahn, D. Sparacin, R. Sun, C. Y. Hong, L. C. Kimerling, A. Pomerene, D. Carothers, J. Beattie, A. Kopa, A. Apsel, M. S. Rasras, D. M. Gill, S. S. Patel, K. Y. Tu, Y. K. Chen, and A. E. White, "Process flow innovations for photonic device integration in CMOS," *Proc. SPIE* 6898, 689804 (2008).

[57] L. Vivien, J. Osmond, J. Fédéli, D. Marris-Morini, P. Crozat, J. Damlencourt, E. Cassan, Y. Lecunff, and S. Laval, "42 GHz pin Germanium photodetector integrated in a silicon-on-insulator waveguide," *Opt. Express* 17, 6252-6257 (2009).

[58] D. Suh, J. Joo, S. Kim, and G. Kim, "High-speed RPCVD Ge waveguide photodetector," *Proceedings of the 6th IEEE International Conference on Group IV Photonics (GFP '09)*, San Francisco, CA, USA, pp. 16-18 (2009).

[59] N. Feng, P. Dong, D. Zheng, S. Liao, H. Liang, R. Shafiiha, D. Feng, G. Li, J. Cunningham, and A. Krishnamoorthy, "Vertical pin germanium photodetector with high external responsivity integrated with large core Si waveguides," *Opt. Express* 18, 96-101 (2010).

[60] C. T. DeRose, D. C. Trotter, W. A. Zortman, A. L. Starbuck, M. Fisher, M. R. Watts, and P. S. Davids, "Ultra compact 45 GHz CMOS compatible germanium waveguide photodiode with low dark current," Opt. Express. 19(25), 24897-24904 (2011).

[61] L. Virot, L. Vivien, J. M. Fédéli, Y. Bogumilowicz, J. M. Hartmann, F. Boeuf, P. Crozat, D. Marris-Morini, and E. Cassan, "High-performance waveguide-integrated germanium PIN photodiodes for optical communication applications," *Photon. Res.* 1(3), 140-147 (2013).

[62] B. Y. Tsaur, C. K. Chen and S. A. Marino, "Long-wavelength Ge_xSi_{1-x}/Si heterojunction infrared detectors and focal plane array," *Proc. SPIE* 1540, 580-595 (1991).

[63] A. K. Sood, R. A. Richwine, Y. R. Puri, N. DiLello, J. L. Hoyt, N. Dhar, R. S. Balcerak, and T. G. Bramhall, "Development of SiGe arrays for visible-near IR imaging applications," *Proc. SPIE* 7780,77800F (2010).

[64] N. K. Dhar, R. Dat, and A. K. Sood, "Advances in infrared array detector technology," Chapter in *Optoelectronics: Advanced Materials and Devices*, edited by S. L. Pyshkin and J. M. Ballato.

[65] Y. Ishikawa, K. Wada, D. D. Cannon, J. Liu, H. C. Luan, and L. C. Kimerling, "Strain-induced band gap shrinkage in Ge grown on Si substrate," *Appl. Phys. Lett.* 82(13), 2044-2046 (2003).

[66] T. K. P. Luong, M. T. Dau, M. A. Zrir, M. Stoffel, V. Le Thanh, M. Petit, A. Ghrib, M. El Kurdi, P. Boucaud, H. Rinnert, and J. Murota, "Control of tensile strain and inter-diffusion in Ge/Si(001) epilayers grown by molecular-beam epitaxy," *J. Appl. Phys.* 114(8), 083504 (2013).

[67] Y. Ishikawa and K. Wada, "Germanium for silicon photonics," *Thin Sol. Films* 518, S83-S87 (2010).

[68] Z. Huang, J. Oh, S. K. Banerjee, and J. C. Campbell, "Effectiveness of SiGe buffer layers in reducing dark currents of Ge-on-Si photodetectors," *IEEE J. Quantum. Electron.* 43, 238-242 (2007).

[69] V. T. Bublik, S. S. Gorelik, A. A. Zaitsev, and A. Y. Polyakov, "Calculation on the binding energy of Ge-Si solid solution," *Phys. Status Solidi* 65, K79-84 (1974).

[70] H. C. Luan, D. R. Lim, K. K. Lee, K. M. Chen, J. G. Sandland, K. Wada, and L. C. Kimerling, "High-quality Ge epilayers on Si with low threading-dislocation densities," *Appl. Phys. Lett.* 75(19), 2909-2911 (1999).

[71] S. Luryi, A. Kastalsky, and J. Bean, "New infrared detector on a silicon chip," *IEEE Trans. Electron. Dev.* ED-31, 1135-1139 (1984).

[72] Y. Y. Fang, J. Tolle, J. Tice, A. V. G. Chizmeshya, J. Kouvetakis, V. R. D'Costa, and J. Menéndez, "Epitaxy-driven synthesis of elemental Ge/Si strain-engineered materials and device structures via designer molecular chemistry," *Chem. Mater* 19, 5910-5925 (2007).

[73] L. Colace, G. Masini, F. Galluzzi, G. Assanto, G. Capellini, L. Di Gaspare, E. Palange, and F. Evangelisti, "Metal-semiconductor-metal near-infrared light detector based on epitaxial Ge/Si," *Appl. Phys. Lett.* 72, 3175-3178 (1998).

[74] O. O. Olubuyide, D. T. Danielson, L. C. Kimerling, and J. L. Hoyt, "Impact of seed layer on material quality of epitaxial germanium on silicon deposited by low pressure chemical vapor deposition," *Thin Sol. Films* 508(1), 14-19 (2006).

[75] V. A. Vu, D. E. Ioannou, R. Kamocsai, S. L. Hyland, A. Pomerene, and D. Carothers, "PIN germanium photodetector fabrication issues and manufacturability," *IEEE Trans. Semi. Manufact.* 23, 411-418 (2010).

[76] M. Oehme, J. Werner, O. Kirfel, and E. Kasper, "MBE growth of SiGe with high Ge content for optical applications," *Appl. Surf. Sci.* 254, 6238-6241 (2008).

[77] M. Halbwax, D. Bouchier, V. Yam, D. Débarre, L. H. Nguyen, Y. Zheng, P. Rosner, M. Benamara, H. P. Strunk, and C. Clerc, "Kinetics of Ge growth at low temperature on Si(001) by ultrahigh vacuum chemical vapor deposition," *J. Appl. Phys.* 97, 064907 (2005).

[78] H. C. Luan and L. C. Kimerling. "Cyclic thermal anneal for dislocation reduction," U.S. Patent No. 6,635,110, 21 Oct. 2003.

[79] D. Choi, Y. Ge, J. S. Harris, J. Cagnon, and S. Stemmer, "Low surface roughness and threading dislocation density Ge growth on Si," *J. Crys. Growth*, 310(18), 4273-4279 (2008).

[80] A. Rogalski, "Infrared detectors: status and trends," *Prog. Quant. Electron.* 27, 59-210 (2003).

[81] M. Itzler and M. Entwistle, "Geiger-mode focal plane arrays enable SWIR 3D imaging," *Laser Focus World*, March 2011.

[82] A. Hairston, J. Stobie, and R. Tinkler, "Advanced readout integrated circuit signal processing," *Proc. SPIE* 6206, 62062Z (2006).

[83] G. Masini, G. Capellini, J. Witzens, and C. A. Gunn, "A four-channel, 10 Gbps monolithic optical receiver in 130 nm CMOS with integrated Ge waveguide photodetector," Proc. 4th IEEE Int. Conf. Group IV Photon., Tokyo, Japan, pp. 1-3 (2007).

[84] L. Chong, X. Chun-Lai, L. Zhi, C. Bu-Wen, and W. Qi-Ming, "Sixteen-element Ge-on-SOI PIN photo-detector arrays for parallel optical interconnects," *Chin. Phys. B* 23(3), 038507 (2014).

[85] M. Amato, M. Palummo, R. Rurali, and S. Ossicini, "Silicon-germanium nanowires: chemistry and physics in play," *Chem. Rev.* 114, 1371-1412 (2014).

[86] E. G. Barbagiovanni, D. J. Lockwood, P. J. Simpson, and L. V. Goncharova, "Quantum confinement in Si and Ge nanostructures," *J. Appl. Phys.* 111, 034307 (2012).

[87] C. Bulutay, "Interband, intraband, and excited-state direct photon absorption of silicon and germanium nanocrystals embedded in a wide band-gap lattice," *Phys. Rev. B*
76, 17 (2007).

[88] S. Cosentino, S. Mirabella, M. Miritello, G. Nicotra, R. Lo Savio, F. Simone, C. Spine-
lla, and A. Terrasi, "The role of the surfaces in the photon absorption in Ge nanoclus-
ters embedded in silica," *Nanoscale Res. Lett.* 6, 135 (2011).

[89] S. Huang and Li. Yang, "Strain engineering of band offsets in Si/Ge core-shell nanowires," *Appl. Phys. Lett.* 98, 093114 (2011).

[90] C. Soci, A. Zhang, X. Y. Bao, H. Kim, Y. Lo, and D. Wang, "Nanowire photodetectors," *J. Nanosci. Nanotech.* 10, 1430-1449 (2010).

[91] J. E. Yang, C. B. Jin, C. J. Kim, and M. H. Jo, "Band-gap modulation in single-crystal-
line $Si_{1-x}Ge_x$ nanowires," *Nano Lett.* 6, 2679-2684 (2006).

[92] P. Liu, S. Cosentino, S. T. Le, S. Lee, D. Paine, A. Zaslavsky, D. Pacifici, S. Mirabella,
M. Miritello, I. Crupi, and A. Terrasi, "Transient photoresponse and incident power
dependence of high-efficiency germanium quantum dot photodetectors," *J. Appl.
Phys.* 112, 083103 (2012).

[93] A. Yakimov, V. Kirienko, V. Armbrister, and A. Dvurechenskii, "Broadband Ge/SiGe
quantum dot photodetector on pseudosubstrate," Nanoscale Res. Lett. 8, 217 (2013).

Passive Optical Network Technology

Chapter 11

Self-Coherent Reflective Passive Optical Networks

S. Straullu, S. Abrate and R. Gaudino

Additional information is available at the end of the chapter

http://dx.doi.org/10.5772/58999

1. Introduction

Fiber optic access networks are the necessary infrastructural approach for a real broadband delivery, allowing the fiber to arrive closer to the final customer, eventually up to the premises equipment; such infrastructures, depending on the depth of reach of the fiber, are usually referred to as FTTX (Fiber To The X), where X stands for H (Home), B (Building), C (Curb) or Cab (Cabinet). They are the basis for broadband access networks, enabling high-speed Internet access, digital TV broadcast (IPTV), video on demand (VOD) and other services. Comparing with copper technologies, like for example DSL (digital subscriber line), higher bandwidths (up to several 10 Gbit/s) and higher distances (up to several tens of km) are possible. FTTX infrastructures can address both residential and business access, and support mobile back-hauling.

Access networks can be divided in two main categories: Active Optical Networks (AONs) and Passive Optical Networks (PONs):

- AONs are a point-to-point (P2P) network structure, where all users have their own fiber optic line that is terminated on an optical access node, which can be designed differently depending on specifications;

- PONs feature a point-to-multi-point (P2MP) tree-like architecture (as shown in Fig. 1 below) to provide broadband access, based on a network composed by an Optical Line Terminal (OLT) at the service provider's Central Office (CO) and several Optical Network Units (ONUs) at the users' premises. PONs allow only passive elements along the Optical Distribution Network (ODN) that connects the OLT to the ONUs. The main advantages are that PONs do not require electrical power in the outside plant to power the distribution elements, thereby lowering operational costs and complexity and, moreover, they allow to highly reduce the number of required optical ports in the CO compared to P2P solutions.

The OLT is the interface between backbone and access networks, and it is also responsible for the enforcement of any media access control (MAC) protocol for upstream bandwidth arbitration. The ONU cooperates with the OLT in order to control and monitor all PON transmission and to enforce the MAC protocol for upstream bandwidth arbitration; it also acts as the residential gateway, coupling the ODN with the in-home network.

The ODN consists of the distribution fibers and all the passive optical distribution elements, mainly optical splitters and/or wavelength-selective elements (WDM filters), that are located in sockets or cabinets. The splitting ratio in most cases is between 1:8 and 1:128, and can be performed in lumped or cascaded elements. The ODN power budget is usually taken as a reference for PON reach calculations, and is computed as the difference between the transceiver back-to-back power budget (i.e. OLT transmitter directly coupled into ONU receiver or vice versa) and the passive optical equipment necessary inside the OLT and the ONU to perform all multiplexing of PON signals into a single fiber. Hence, the ODN power budget considers the remaining power budget that can be spent for the distribution fibers and the distribution elements in the remote nodes.

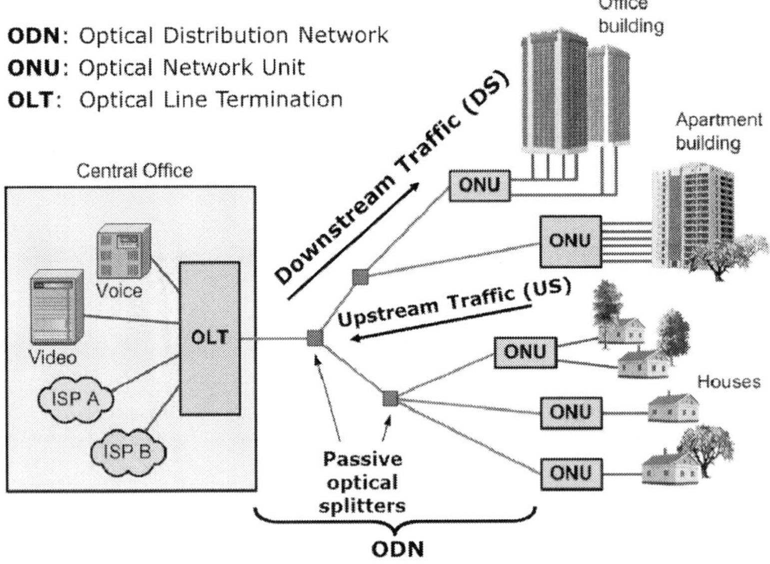

Figure 1. Passive Optical Network architecture

During the last 20 years, the Full Service Access Network (FSAN) and the Ethernet in the First Mile alliance (EFM) working groups, in cooperation with the International Telecommunications Union (ITU) and the Institute of Electrical and Electronics Engineers (IEEE) standardization bodies, defined several PON standards, as summarized in Figure 2.

For the purposes of this Chapter, we will not go through all these standards, since none of these has been approached with self-coherent reflective architectures that will be presented in this Chapter, but we will mention only the most recent one, called NG-PON2 (Next Generation PON 2), in progress of standardization at the time of writing and with still open technological issues. In particular, we will address only physical layer aspects. The definition of such standard started in 2010 by FSAN.

Since about the 70% of the total investments in deploying PONs is due to the optical distribution networks, it is crucial for the NG-PON2 evolution to be compatible with the deployed networks and to reuse the outside plant. Moreover, NG-PON2 technology must outperform existing PON technologies in terms of ODN compatibility, bandwidth, capacity and cost-efficiency. For this reason, the initial "wish list" for NG-PON2 was originally very demanding, requiring not only an increase of bit rates with respect to previous generations, but also in terms of total number of users per PON tree and reach.

The major original general requirements of NG-PON2 can be summarized as follows:

- increase the aggregate rate to 40 Gbit/s in downstream or upstream;

- increase the reach to 40-60 km (or more);

- increase the splitting ratio up to 256 users (or more);

- compatibility with the attenuation classes defined in XG-PON, as reported in Table 1;

- increase the effective bit rate per user (in both directions) to a value closer to 1 Gbit/s per ONU.

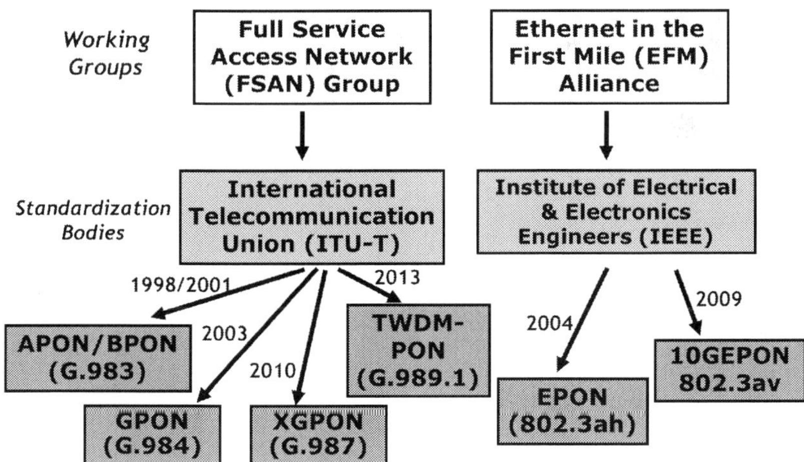

Figure 2. PON standards

	'Nominal1' class (N1 class)	'Nominal2' class (N2 class)	'Extended1' class (E1 class)	'Extended2' class (E2 class)
Minimum loss	14 dB	16 dB	18 dB	20 dB
Maximum loss	29 dB	31 dB	33 dB	35 dB

Table taken from G.987.2

Table 1. Classes for optical XG-PON path loss

Many PON technologies have been proposed to provide broadband optical access beyond 10 Gbit/s, like for example:

- the 40 Gigabit time-division multiplexed PON (XLG-PON) proposal [1] which increases the single carrier serial downstream bit rate of a 10 Gigabit PON (XG-PON) to 40 Gbit/s;

- a set of orthogonal frequency-division multiplexed (OFDM)-based PON proposals which employ quadrature amplitude modulation and fast Fourier transform to generate digital OFDM signals for transmission [2, 3];

- a group of WDM-PON proposals which provide a dedicated wavelength channel at the rate of 1 Gbit/s to each ONU with different WDM transmit or receive technologies [4, 5];

- the time-and wavelength-division multiplexed PON (TWDM-PON) proposal which stacks multiple XG-PONs using WDM [6].

The first proposal, based on a single wavelength (per direction) at very high bit rate, was finally discarded, since it was perceived as too expensive. Indeed, it had to include a duo-binary modulation and a downstream transmission around 1300 nm to avoid dispersion compensation. Moreover, it was not scalable, thus no future upgrades above 40 Gbit/s downstream seemed feasible on a single wavelength.

The proposals based on OFDM are very interesting and technologically advanced solutions, but they do not easily guarantee a complete compatibility with existing deployed networks and devices.

Regarding the WDM-PON architectures, a key point is whether the WDM filters are inside the ODN or in the ONU.

The "pure" WDM-PON approach, based on WDM filters inside the ODN, generates some constraints to Telecom operators:

- all passive splitters installed in the already deployed PON (brown field scenario) need to be substituted with Arrayed Waveguide Grating (AWG) filters;

- limited backward compatibility with already deployed standards (GPON, EPON, XG-PON, etc.);

- a dedicated wavelength per user is a "circuit" per individual user, thus no sharing or statistical multiplexing is possible.

For these reasons, this solution seems to be out of question for NG-PON2 standard. Among all of the aforementioned proposals, TWDM-PON has attracted the majority support from global vendors and was selected by the FSAN community in the April 2012 meeting as a primary solution to NG-PON2. TWDM-PON increases the aggregate PON rate by stacking XG-PONs via multiple pairs of wavelengths; since an XG-PON system offers the access rates of 10 Gbit/s in downstream and 2.5 Gbit/s in upstream, a TWDM-PON system with four pairs of wavelengths will then be characterized by:

- 4 wavelengths (upgradable to 8) at 100 GHz WDM spacing:

- 1595-1600 nm for downstream (up to 1605 nm with 8 wavelengths);

- 1535-1540 nm for upstream (up to 1545 nm with 8 wavelengths);

- 40 Gbit/s in downstream (10 Gbit/s per wavelength) and 10 Gbit/s in upstream (2.5 Gbit/s per wavelength);

- up to 64 ONUs (16 per wavelength);

- up to 40 km;

- up to 35 dB of power budget;

- full backward compatibility.

The basic TWDM-PON architecture is shown in Figure 3. Four XG-PONs are stacked by using four pairs of wavelengths (e.g., wavelength pairs of $\{\lambda 1, \lambda 5\}$, $\{\lambda 2, \lambda 6\}$, $\{\lambda 3, \lambda 7\}$ and $\{\lambda 4, \lambda 8\}$, in Figure 3). ONUs are equipped with tunable transmitters and receivers. The tunable transmitter is tunable to any of the four upstream wavelengths; the receiver is tunable to any of the four downstream wavelengths. In order to achieve power budget higher than that of XG-PON, optical amplifiers (OAs) are used at the OLT side to boost the downstream signals as well as to pre-amplify the upstream signals. The ODN remains passive since OAs are placed at the OLT side, together with WDM Mux/DeMux.

Therefore, the TDWM-PON key features are:

- colorless ONUs;

- splitter-based ODN.

Most of the TWDM-PON components are commercially available in access networks today. Comparing to previous generations of PONs (e.g., GPON, XG-PON), the only, but technically very challenging, new components in TWDM-PON are the tunable receivers and tunable transmitters at the ONU.

For what concerns the tunable receiver, it should tune its wavelength to any of the TWDM-PON downstream wavelengths by following the OLT commands. This function can be implemented by using candidate technologies such as:

- thermally tuned Fabry–Perot (FP) filter [7];

- angle-tuned FP filter, injection-tuned silicon ring resonator [8];

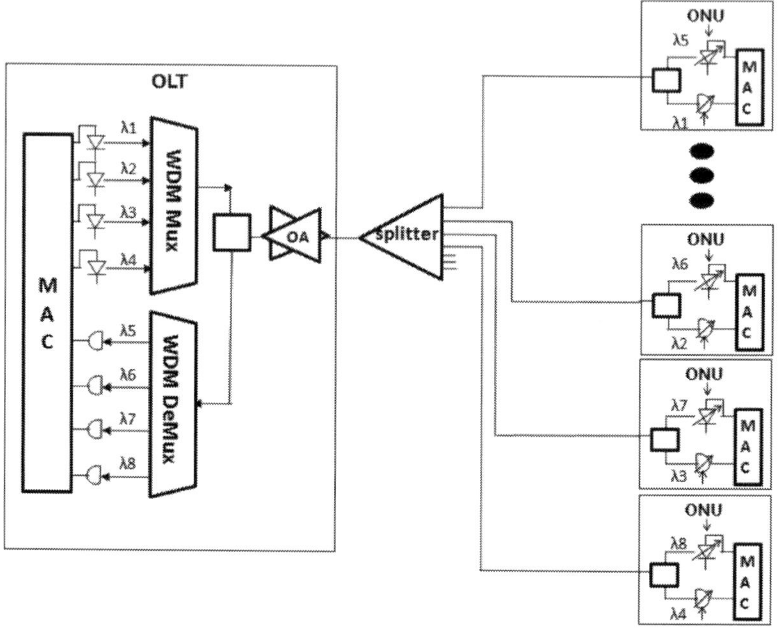

Picture taken from [6]

Figure 3. TWDM-PON system diagram

- liquid crystal tunable filter [9];

- thermally tunable FP detector [10].

The ONU tunable transmitter can tune its wavelength to any of the upstream wavelengths. The related implementation technologies are:

- distributed feedback (DFB) laser with temperature control (TC) [11];

- DFB laser with partial TC [12];

- multi-section distributed Bragg reflector laser (electrical control) without cooling [13];

- external cavity laser (ECL) with mechanical control without cooling [14];

- ECL with thermo/electro/piezo/magneto-optic control without cooling [15, 16].

All these solutions are today under consideration for NG-PON2, but none has yet completely demonstrated to be achievable at the (very low) target prices for ONU. Purpose of this Chapter is to describe a solution for the upstream transmission that avoids the need for a tunable laser at the ONU side: it is based on self-coherent reflective PON architectures as a possible technological approach to the NG-PON2 requirements. In the following we will then first show the concept of the reflective PON, describing the key components it needs, the problems it solves and the limitations it encounters; then, we will propose the self-coherent detection enhancement and will give an overview of the latest research results available in literature.

2. The reflective PON approach

A very basic schematic view of a WDM-PON architecture is depicted in Figure 4. It makes use of integrated multichannel transmit (TX) and receive (RX) arrays in the OLT, and tunable laser (T-TX) and filters (TOF) in the ONUs. From the OLT, all downstream wavelengths are broadcast via the ODN; therefore, a tunable filter is necessary at each ONU in order to select the correct wavelength. Similarly, each ONU has to be provided with a tunable laser for the correct upstream wavelength selection.

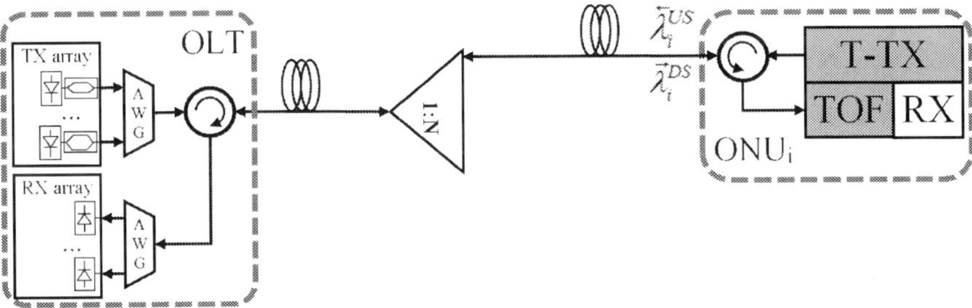

Figure 4. WDM-PON architecture

One of the few physical layer problems over which FSAN is still working today before the final NG-PON2 Recommendation release, is related to these two devices at the ONU. Indeed, they should both have a precision compatible with a 100 GHz wavelength grid, and be able to tune on (at least) four wavelengths, adopting some protection systems during their switch-on time, in order to avoid interferences with other channels (due to the uncontrolled wavelength transmission at the laser power on); moreover, they should operate on a very wide temperature range and they should have a target price compatible with equipment to be installed at the customer premises. Moreover, in the longer term, if more than four wavelengths will be used (for instance for WDM overly), this issue will be particularly critical.

An alternative solution to overcome this problem is represented by reflective PON architectures, whose key idea is to generate the unmodulated upstream wavelengths at the OLT side, and modulate them in reflection at the ONU side, as schematically depicted in Figure 5. With respect to the setup depicted in Figure 4, the OLT is equipped by an additional multichannel transmit array, for the unmodulated seed signals generation, and the tunable laser at the ONU side is replaced by a reflective transmitter (R-TX). This way, the costs and the technical issues (the wavelength control in particular) related to the tunable lasers are moved from the ONU to the OLT, where they can be more easily managed.

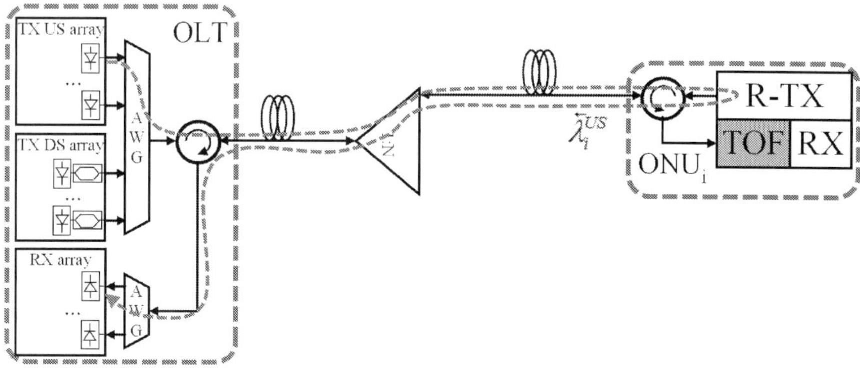

Figure 5. Scheme of a possible solution for Reflective-PON ONUs

2.1. Key components for the reflective ONU

Many variants of WDM reflective PON architectures can be found in literature, all around the common denominator of avoiding expensive tunable lasers at the ONU by means of using a reflective modulator for the upstream transmission that sends back a centrally-generated seed signal. One of the most common approaches adopted in order to obtain this effect at the ONU side is based on the use of Reflective Semiconductor Optical Amplifier (RSOA), which combine the amplification and modulation capabilities in a single device. They are composed by a high reflectivity coating on one facet, along with a curved waveguide and ultra-low reflectivity coating on the other facet, to produce a highly versatile reflective gain medium [17]. Today, few optical devices manufacturers propose commercial RSOA devices (e.g. as CIP, MEL and Kamelian), since they are just used in the reflective PON scenario and at a research level.

Figure 6 schematically depicts a RSOA in a typical low-cost TO-CAN package, while Figure 7 shows the electrical bandwidth measurement result of a typical commercial RSOA (butterfly-packaged solutions may have higher bandwidth, but their cost seems too high for application in the extremely cost-sensitive scenario of ONU). From this graph, it is possible to notice that the 3 dB electrical bandwidth of the device under test is about 500 MHz; anyway, the signal obtained by modulating the RSOA bias current I_b with a 1 Gbit/s OOK modulation (shown in the inset of Figure 7) is received without inter-symbol interference.

Since 2000, the interest of RSOA as upstream transmitters for WDM-PON applications based on reflective ONU has grown-up. The first results have been proposed in [18], where the uplink data stream was reflected and modulated via the RSOA at 1.25 Gbit/s, as then further developed and investigated in several later works (e.g. [19-22]).

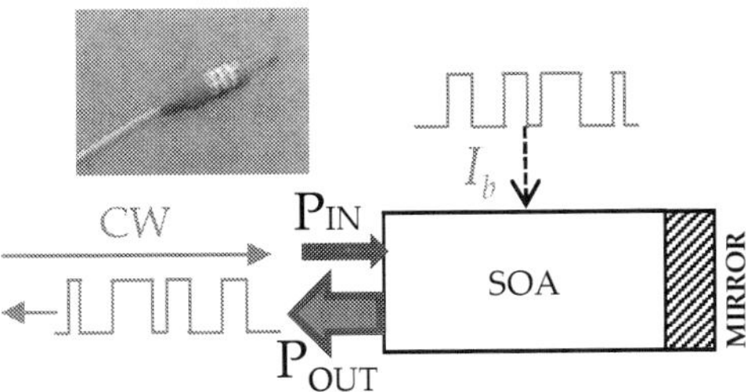

Figure 6. Schematic view of a RSOA

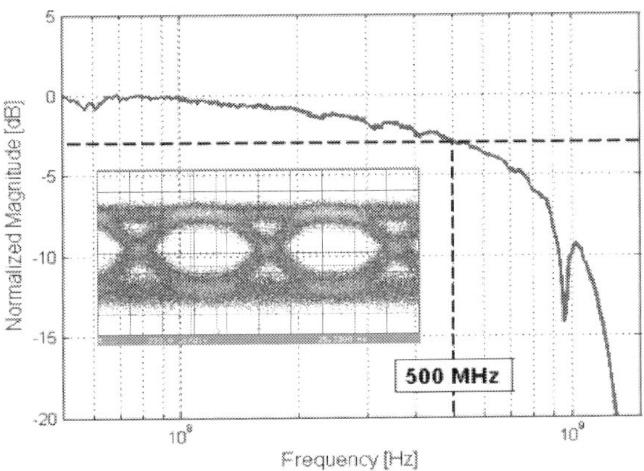

Figure 7. Characterization of a RSOA

In order to increase the upstream data rate up to 10 Gbit/s in such an architecture, using this low-bandwidth devices, different approaches have been proposed by research groups, like the use of optical filtering and electronic equalization [23-25] or the adoption of advanced modulation formats [26].

Another device that can be adopted as reflective transmitter at the ONU side is the Reflective Electro-Absorption Modulator (REAM). EAM are semiconductor devices usually made in the form of a waveguide with electrodes for applying an electric field in a direction perpendicular to the modulated light beam. Their principle of operation is based on a change in the absorption spectrum caused by the applied electric field, which changes the band-gap energy without involving the excitation of carriers by the electric field: the so called Franz-Keldysh effect [27]. Reflective EAM include an EAM and a mirror, as schematically depicted in Figure 8, and they can be used to reflect and modulate the incoming light by means of the applied electrical signal (V_b). Thanks to EAM, a modulation bandwidth close to 10 GHz can be achieved, as shown in

Figure 9, which makes these devices useful for optical fiber communication at 10 Gbit/s and above.

Differently from the RSOA, these devices do not amplify the light but, if coupled in cascade to a Semiconductor Optical Amplifier (SOA), they represent a very interesting solution for the ONU of a reflective PON architecture, since they provide a high-speed modulation capability, combined with the linear amplification of signal provided by the SOA. This solution has been proposed in several works, like for example [28, 29], in order to achieve a 10 Gbit/s upstream transmission in a WDM-PON scenario.

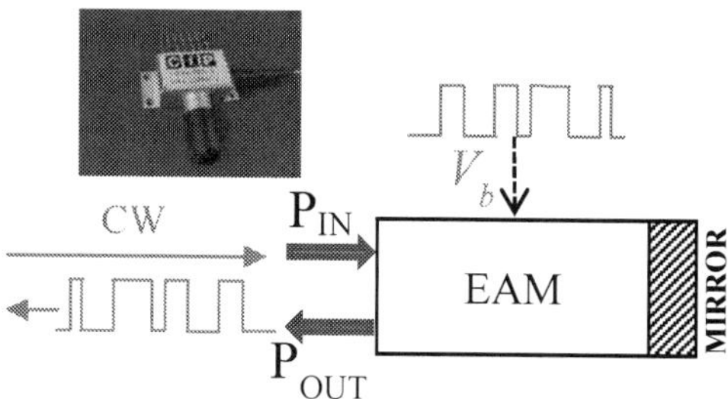

Figure 8. Schematic view of a REAM

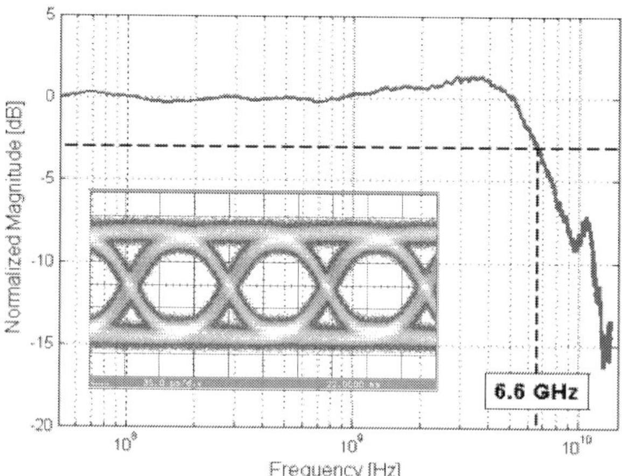

Figure 9. Characterization of a REAM

At the time of writing this Chapter, another solution for the reflective transmitter implementation is under analysis inside the FP7 EU STREP project titled "FABULOUS" [30]. It consists on a Mach-Zehnder based subsystem allowing polarization independent reflective carrier suppressed optical modulation for application in frequency division multiple access (FDMA) PON. The architecture of this subsystem, presented at first in [31], relies on using a Mach-Zehnder modulator (MZM) simultaneously in both directions within a polarization diversity loop made of a polarization beam splitter (PBS) looped on itself through the modulator. The MZM provides RF access to both the electrodes independently, allowing modulation to be efficient in both the forward and the backward directions. On one electrode of the MZM, RF drive power is applied in the forward direction while, on the other electrode, the modulation is applied in the backward direction (the same signal is applied to both sides). The two polarizations of the incoming optical signal are split through the PBS and sent to the MZM in counter-propagating directions (the MZM having polarization maintaining fibers on both the input and the output). After being modulated, they are recombined through the PBS and sent back towards the OLT, as schematically depicted in Figure 10.

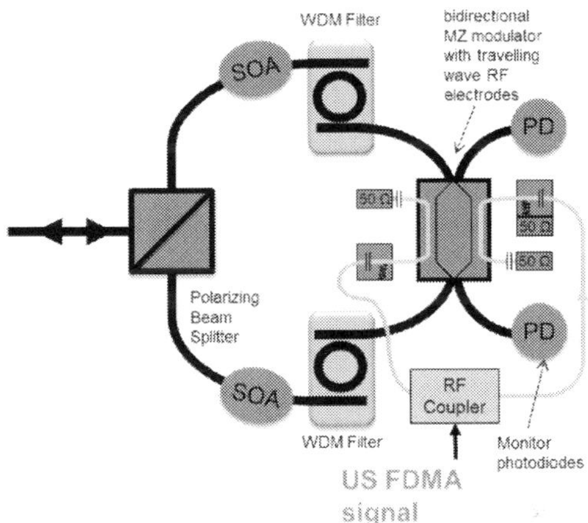

Figure 10. Architecture of R-MZM subsystem

2.2. Transmission issues in reflective PON

The reflective WDM-PON architectures proposed in literature in the last ten years seem to be completely incompatible with TWDM-PON for a set of different reasons such as:

- most of them require the use of AWGs in the ODN, while as previously mentioned most telecom operators believes that backward compatibility in PON is a must, and thus the splitter-based architecture should be maintained also in NG-PON2;

- the achievable ODN loss is limited to typically 20-25 dB, due to several spurious effects such as the Rayleigh back-scattering and the low received optical power;

- most of them use a dedicated wavelength for each ONU, and do not support burst-mode TDMA in the upstream.

In a reflective architecture like the one depicted in Figure 5, the upstream signal undergoes the ODN path loss twice (from the OLT to the ONU as CW seed signal and back to the OLT after the ONU reflection). The optical power at the OLT input is thus determined by the following formula:

$$S = \vec{P}_{CW} - 2L_{ODN} + G_{RONU} \tag{1}$$

where $\vec{P}_{CW}$ is the optical power of the CW signal at the input of the fiber, L_{ODN} is the ODN loss and G_{RONU} is the gain of the optical amplifier installed at the ONU. Since the TWDM-PON standard fixed the maximum value for the launched optical power at the ODN input to +11 dBm per wavelength and the gain of the optical amplifier at the ONU is typically of the order of $G_{RONU} = 20\,dB$, the optical power at the OLT input, for the lowest ODN class N1 ($L_{ODN} = 29\,dB$), is of the order of $S = -27\,dBm$, which is lower than the sensitivity of the standard direct-detection receivers. Moreover, it is well known that, in such an architecture, the spurious back reflections for a SMF fiber are of the order of 30-35 dB below the forward propagating signal, due to the concentrated reflections on components and the Rayleigh backscattering.

Rayleigh backscattering is an unavoidable phenomenon in optical fiber propagation. It is a fundamental loss mechanism arising from random density fluctuations frozen in the fiber during manufacturing. There is a growing interest in understanding Rayleigh backscattering since it can be a limiting factor in various optical systems. It must be taken into account in the performance computation of bidirectional lightwave system, especially in a wavelength-reuse system, like for example the reflective PON architecture depicted in Figure 11.

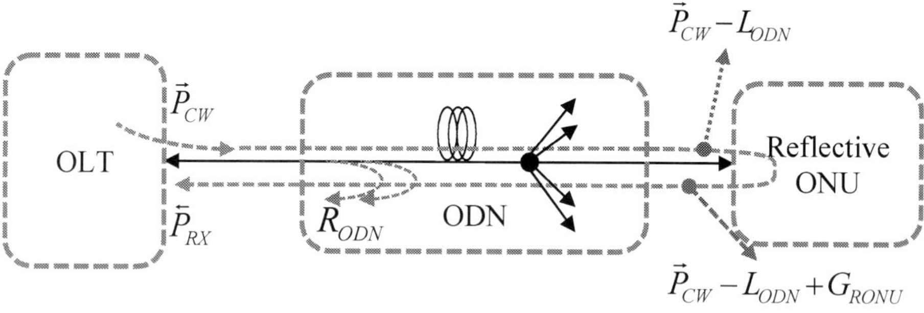

Figure 11. Rayleigh backscattering effect in reflective PON architecture at the OLT

In such a system, the received optical power is composed by the useful signal S, as specified above, and the interference signal I due to reflections, given by:

$$I = \vec{P}_{CW} - R_{ODN} \tag{2}$$

where R_{ODN} is the power of the reflections. Therefore, the signal to interference ratio is given by:

$$\left(\frac{S}{I}\right)_{dB} = -2L_{ODN} + G_{RONU} + R_{ODN} \tag{3}$$

For a standard direct-detection receiver, even if the best solutions proposed in literature to mitigate the Rayleigh backscattering effect are adopted, the signal to interference ratio should be greater than 5-10 dB. This sets an important limit to the maximum achievable ODN loss. Indeed, the maximum reachable ODN loss is lower than 25 dB, since the optical power of the spurious back reflections is about 30-35 dB below the forward propagating signal.

To improve the performance, one could in principle increase the G_{RONU}, but there are techno-logical component issues that limit the maximum reachable gain of optical amplifiers; in addition, another issue can arise from the Rayleigh backscattering that is generated at the ONU side, as depicted in Figure 12 and explained in details in [32]. This means that, in order to satisfy the NG-PON2 requirements with a reflective PON architecture, direct-detection at the OLT side is out of question.

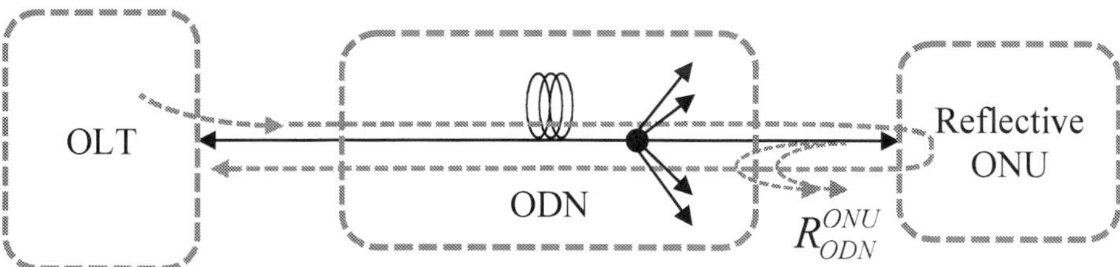

Figure 12. Rayleigh backscattering effect in reflective PON architecture at the ONU

2.3. Self-coherent detection in RPONs

Because of the aforementioned problems related to direct-detection, a coherent detection at the OLT seems to be a must in order to satisfy the NG-PON2 requirements, whether a reflective architecture approach is adopted.

The fundamental concept behind coherent detection is to use the product of the electrical fields of the modulated signal light (centred at f_S) and a CW local oscillator (centred at f_{LO}). This produces a frequency down-conversion of the signal to the frequency $f_{IF} = |f_S - f_{LO}|$, as schematically depicted in Figure 13.

A possible implementation of coherent detection in a reflective PON architecture is depicted in Figure 14: in this case, the CW light source placed at the OLT side is used both as a feed to be sent downstream to the reflective ONU and as a local oscillator for the OLT coherent

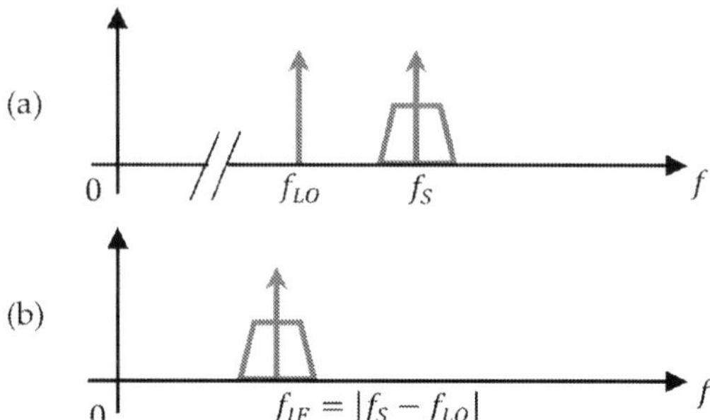

Figure 13. Spectra of (a) the optical signal and (b) the down-converted IF signal

receiver, executing a self-coherent detection. Therefore, after the optical-to-electrical conversion, the signal is down-converted to baseband.

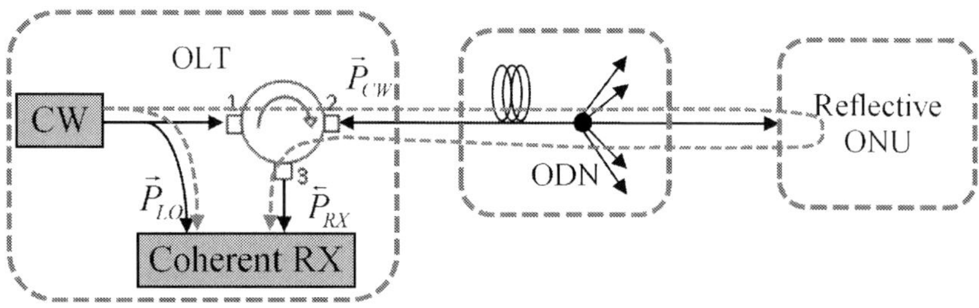

Figure 14. Scheme of a possible solution for self-coherent OLT in a reflective-PON architecture

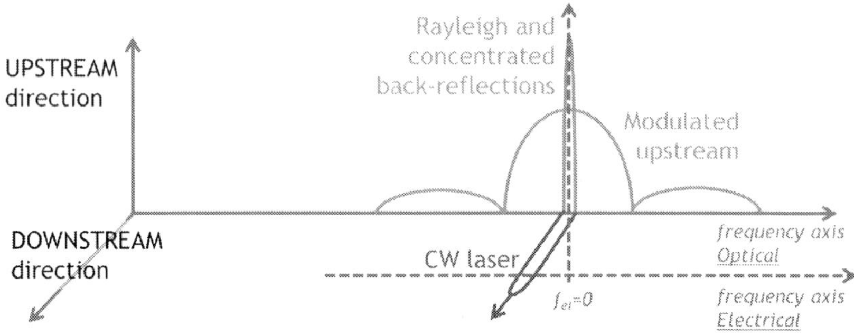

Figure 15. Spectral analysis of signal and reflections in reflective PON

As demonstrated by the results available in literature presented in the following, this setup supports higher ODN loss values with respect to the limit of reflective PONs based on direct-detection, in particular for the following reasons:

- much better sensitivity than direct-detection;

- much larger resilience to spurious back reflections.

Indeed, using a coherent receiver, the Rayleigh backscattering reflections appear as added close to the "electrical" DC, as depicted in Figure 15, thus they can be easily filtered out by electrical high-pass filters.

3. Latest research results

WDM reflective PON "variant", characterized by self-coherent detection at the OLT side, was initially proposed in [33], where a RSOA-based WDM PON architecture, employing a novel self-homodyne receiver and a novel signal processing technique to eliminate the penalty caused by the back-reflection of the seed light, is proposed. In addition, they successfully demonstrated a long-reach fiber-loopback system of over 100 km. These results indicate that the proposed architecture could provide an attractive solution to realize a long-reach WDM PON with high-split ratio.

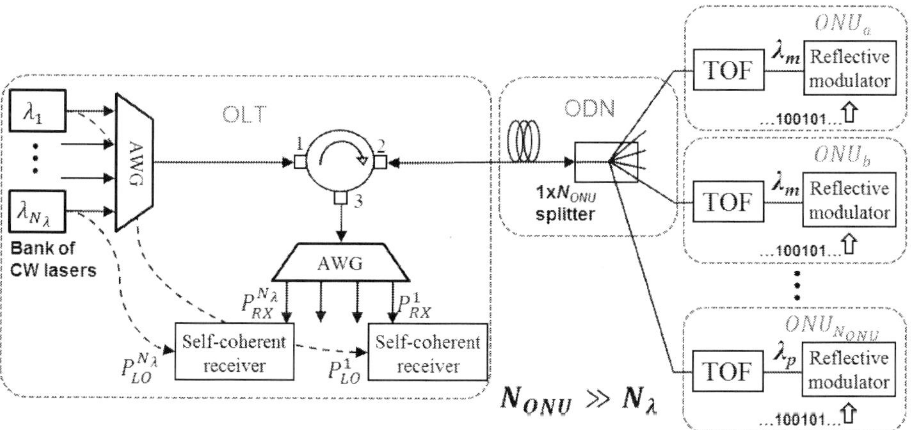

Figure 16. Scheme of self-coherent reflective PON architecture (only upstream transmission)

A possible implementation of the upstream transmission in a self-coherent reflective PON is shown in Figure 16; this architecture may offer the following advantages:

- the upstream wavelengths comb accuracy is completely set by the OLT and not by each individual ONU. In this solution, each ONU needs to tune its optical filter by locking it on one of the N_λ already existing wavelength;

- subsequent upgrades to dense-WDM (DWDM) seem more feasible when using a number of wavelengths N_λ significantly higher than 4, and possibly a narrower frequency spacing (such as 50 GHz), since it is possible to avoid tunable lasers at the ONU;

- for each upstream wavelength, the required CW laser and coherent receiver at the OLT are shared by several ONUs (of the order of N_{ONU}/N_λ) so that their cost may be more reasonably justified.

In addition to WDM, a second level of multiplexing is necessary in order to overcome the limit of a dedicated wavelength per user; this can be obtained sharing time, basing on Time Division Multiplexing (TDM) in the downstream and Time Division Multiple Access (TDMA) in the upstream, or frequency (FDM and FDMA) between users.

3.1. TDM-WDM results

Basing on the architecture depicted in Figure 16, the results presented in [34] experimentally demonstrate the simultaneous transmission of two independent TDMA ONUs with upstream bit rate equal to 2.5 Gbit/s, working in TDMA burst mode and with performance compatible with E2 XG-PON class, thus compatible with US TWDM-PON requirements. At the OLT, the signal demodulation is performed thanks to a self-coherent receiver and a custom burst-mode digital-signal processing technique. In particular, a conventional DSP solution based on Viterbi-Viterbi carrier-phase estimation and LMS adaptive equalization [35, 36] has been modified to work in burst mode operation, focusing on the alignment procedures on the received packets and on the convergence speed of the LMS algorithm [37, 38].

The reflective ONU of such architecture is emulated as depicted in Figure 17. The CW signal reaching the ONU is reflected and modulated by means of a REAM. The SOA placed in front of the REAM amplifies the optical signal twice, first on the feed CW downstream, and then on the reflected and modulated upstream signal. In order to emulate the ONU wavelength selection functionality, a TOF is placed between the SOA and the REAM. This TOF is also useful to partially filter out the ASE noise and, thanks to its approximately 4 dB insertion loss, to avoid excessive SOA saturation in the upstream direction. The SOA+REAM structure, even though not yet commercial, has been integrated in several research projects (such as in [40]). The 2.5 Gbit/s upstream signal is generated at the ONU by driving the REAM by pure NRZ signal, while the optical packets are generated by switching on and off the SOA.

Each packet contains a short dummy-pattern at the beginning and end of each burst, 127 bits of sync-pattern and 1000 bits payload (8B/10B coded), as shown in Figure 18. The dummy-pattern is useful for "absorbing" the rise and fall-time of the SOA acting as a gate, and also the transient effect of the coherent receiver AC-coupling. The sync-pattern is used for identifying

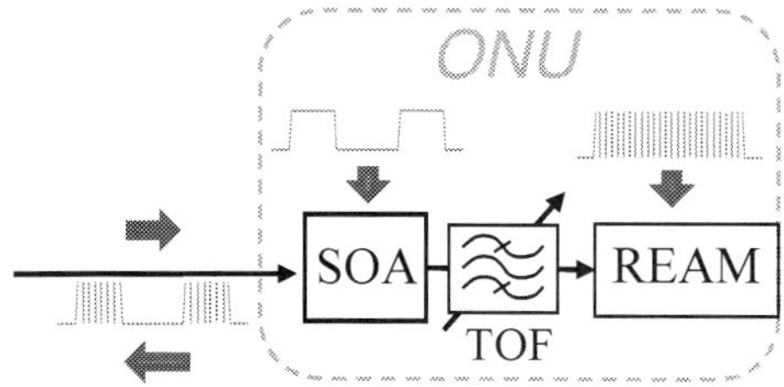

Figure 17. Reflective ONU implementation

the start of each packet and also for training the LMS algorithm, before switching to the LMS "decision-directed" mode, which should carry out the burst payload elaboration. This system emulates a TDMA transmission from two independent ONUs with 25 ns guard-time, where the ONU 1 is the reference ONU and the ONU 2 is the interfering ONU, reproducing the worst-case condition in terms of interference for such a system.

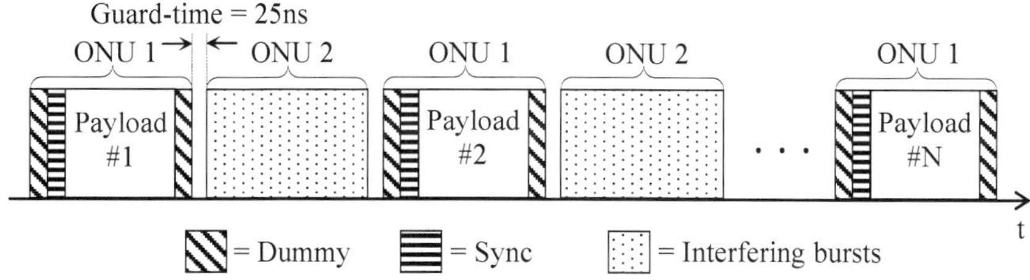

Figure 18. Time relations of the useful and interfering optical US packets for two ONUs

The authors of [34] expressed the performance of the system in terms of Bit Error Rate (BER) as a function of the ODN loss and the acceptable differential optical path loss (DOPL).

The results show that a BER<10^{-3} (a reasonable Forward Error Correcting code threshold for simple FECs, as those used in XG-PON) is obtained up to 35 dB of ODN loss and more than 15 dB of DOPL, as required by XG-PON class E2.

The same setup was also tested increasing the upstream bit rate and using commercial DFB lasers rather than ECL lasers at the OLT side [39]. The bit rate was set to 10 Gbit/s per wavelength (from the previous 2.5 Gbit/s rate), a bit rate currently under consideration in FSAN for the point-to-point WDM (PtP-WDM) for symmetric traffic, and for the TWDM-PON longer term evolution. Replacing the ECL lasers with DFB lasers would provide a significant cost reduction at the OLT. As reported in [39], in these working conditions, the FEC threshold is reached after 31 dB and for a DOPL higher than 15 dB, satisfying the requirement of N2 class.

3.2. FDM-WDM results

One of the main drawbacks of the more traditional TDMA-PON approach deployed nowadays, is that it does not scale well above 10 Gbit/s per wavelength in term of cost/complexity and power efficiency, mostly due to the fact that each ONU should work on the aggregated bit rate.

On the contrary, FDM/FDMA approach allows ONUs to operate at low DSP speeds (and so reduce their cost and power consumption); indeed, the speed at which the ONUs operates is equal to the maximum speed that the customers are allowed to communicate at (e.g. 1 Gbit/s) which is much smaller than the aggregated line rate (e.g. 20 Gbit/s).

The work presented in [41] proposes an innovative reflective PON approach for the upstream transmission using a special configuration based on FDMA, where each ONU is assigned a portion of the available electrical spectrum to perform an high spectral efficiency M-QAM modulation format as depicted in Figure 19.

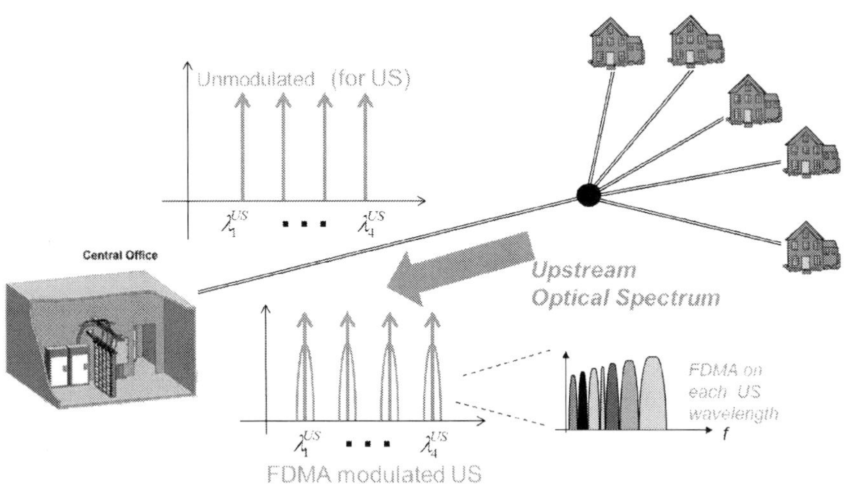

Figure 19. FDMA reflective PON architecture

The results presented in [41] demonstrate that this architecture targets an upstream capacity of 32 Gbit/s per optical carrier, outperforming NG-PON2 with a tenfold increase in the upstream capacity, with power budget and max reach overcoming ITU-T class N2, as defined in the XG-PON standard (up to 40 km and 31 dB ODN loss). An important feature of this approach consists in the fact that, due to the absence of optical sources at the customer premises, the ONU can be realized as a Photonic Integrated Circuit with an unprecedented level of integration. This holds true for self-coherent reflective PONs in general, but the highest level of integration is to date relevant to the architecture shown in [41], where the Mach

Zehnder Modulator, the two tunable filters (for multi-wavelength transmission) and the polarizing beam splitter that characterize the proposed ONU will be realized on silicon platform, the two SOA will be suitable for photonic integration and finally the electrical driver will be flip-chipped on top; for a detailed description see [42, 43].

4. Conclusions

In this Chapter, we have given a general description of the self-coherent reflective PON architecture as a possible technological approach to the NG-PON2 requirements. We have shown that, thanks to reflective ONU and the self-coherent detection at the OLT side, the performance required by NG-PON2 are satisfied even without the need for a tunable laser at the ONU side. If properly integrated, the reflective ONU structure can be less expensive than the TWDM-PON solution in term of CAPEX, and have a much simpler wavelength control. The only significant increase in cost in the proposed solution is due to the presence of one coherent receiver per each upstream wavelength (a cost to be shared among all the users with the same wavelength). While the cost of coherent receivers used today for long-haul is likely still too high for an OLT, it can be observed an enormous effort in the long-haul technical community to make small-form factor and low-power coherent receivers possible, so that it is reasonable to envision a sharp decrease in coherent receiver costs that may allow the use of coherent detection at OLT in the near term.

Author details

S. Straullu[1*], S. Abrate[1] and R. Gaudino[2]

*Address all correspondence to: straullu@ismb.it

1 Istituto Superiore Mario Boella, Torino, Italy

2 Politecnico di Torino, Torino, Italy

References

[1] E. Harstead, et al., "Technologies for NGPON2: Why I think 40G TDM PON (XLG-PON) is the clear winner", in Proc. Opt. Fiber Commun. Conf./Nat. Fiber Opt. Eng. Conf. Workshop, Mar. 2012

[2] X. Hu, et al., "Energy-efficient WDM-OFDM-PON employing shared OFDM modulation modules in optical line terminal", Opt. Exp., vol. 20, no. 7, pp. 8071–8077, Mar. 2012.

[3] W. Wei et al., "Optical orthogonal frequency division multiple access networking for the future internet", IEEE/OSA J. Opt. Commun. Netw., vol. 1, no. 2, pp. 236–246, Jul. 2009.

[4] H. K. Lee et al., "A WDM-PON with an 80 Gb/s capacity based on wavelength-locked Fabry-Perot laser diode", Opt. Exp., vol. 18, no. 17, pp. 18077–18085, Aug. 2010.

[5] K. Grobe et al., "PON in adolescence: from TDMA to WDM-PON", IEEE Commun. Mag., vol. 46, no. 1, pp. 26–34, Jan. 2008.

[6] Y. Luo at al., "Time-and Wavelength-Division Multiplexed Passive Optical Network (TWDM-PON) for Next-Generation PON Stage 2 (NG-PON2)", IEEE J. LIGHTWAVE TECH. 31 no. 4, 587-593 (2013).

[7] M. Zhu et al.,"Upstream multi-wavelength shared PON with wavelength-tunable self-seeding Fabry-Perot laser diode," in Proc. Opt. Fiber Commun. Conf./Nat. Fiber Opt. Eng. Conf., Mar. 2011, pp. 1–3.

[8] D. Liang et al., "Low threshold electrically-pumped hybrid silicon microring lasers," IEEE J. Sel. Topics Quantum Electron., vol. 17, no. 6, pp. 1528–1533, Nov./Dec. 2011.

[9] A. Kato et al., "Tunable optical filter with cascaded waveguide Fabry-Pérot resonators featuring liquid crystal cladding," Photon. Technol. Lett., vol. 24, no. 4, pp. 282–284, Feb. 2012.

[10] R. Murano et al., "Tunable 2.5 Gb/s receiver for wavelength-agile DWDM-PON," presented at the presented at the Opt. Fiber Commun. Conf./Nat. Fiber Opt. Eng. Conf., 2008, Post Deadline Paper.

[11] G. W. Yoffe et al., "Widely-tunable 30 mW laser source with sub-500 kHz linewidth using DFB array," in Proc. IEEE Lasers Electro-Optics Soc., Nov. 2008, pp. 892–893.

[12] D. Ton et al., "2.5-Gb/s modulated widely tunable laser using an electroabsorption modulated DFB array and MEMS selection," IEEE Photon. Technol. Lett., vol. 16, no. 6, pp. 1573–1575, Jun. 2004.

[13] Y. Liu et al., "Uncooled DBR laser directly modulated at 3.125 Gb/s as athermal transmitter for low-cost WDM systems," IEEE Photon. Technol. Lett., vol. 17, no. 10, pp. 2026–2028, Oct. 2005.

[14] P. Ossieur et al.,"Demonstration of a 32 512 Split, 100 km Reach, 2 32 10 Gb/s hybrid DWDM-TDMA PON using tunable external cavity lasers in the ONUs," J. Lightw. Technol., vol. 29, no. 24, pp. 3705–3718, Dec. 2011.

[15] F. Wei et al. "Tunable external cavity diode laser with a PLZT electrooptic ceramic deflector," IEEE Photon. Technol. Lett., vol. 23, no. 5, pp. 296–298, Mar. 2011.

[16] J. Zheng et al., "Optically tunable ring external-cavity laser," in Proc. IEEE Photon. Conf., Oct. 2011, pp. 644–645.

[17] http://www.freepatentsonline.com/7046435.html

[18] P. Healey et. al., "Spectral slicing WDM-PON using wavelength-seeded reflective SOAs", Electronics Letters, 37, 19, 1181 – 1182 (2001).

[19] H.C. Shin et al., "Reflective SOAs optimized for 1.25 Gbit/s WDM-PONs", Lasers and Electro-Optics Society (LEOS), Vol. 1 (2004).

[20] L. Wooram et al., "Bidirectional WDM-PON based on gain-saturated reflective semi-conductor optical amplifiers", Photonics Technology Letters, IEEE, 17, 11, 2460 – 2462 (2005).

[21] C. Arellano et al, "Bidirectional single fiber transmission based on a RSOA ONU for FTTH using FSK-IM modulation formats", Optical Fiber Communication Conference, 2005, OFC/NFOEC, Vol. 3 (2005).

[22] E. Wong et al., "Low-cost WDM passive optical network with directly-modulated self-seeding reflective SOA", Electronics Letters, 42, 5, 299 – 301 (2006).

[23] I. Papagiannakis et al., "Investigation of 10-Gb/s RSOA-Based Upstream Transmission in WDM-PONs Utilizing Optical Filtering and Electronic Equalization", Photonics Technology Letters, IEEE, 20, 24, 2168 – 2170 (2008).

[24] K.Y. Cho et al., "10-Gb/s Operation of RSOA for WDM PON", Photonics Technology Letters, IEEE, 20, 18, 1533 – 1535 (2008).ù

[25] G. Qi et al., "Low-Bandwidth RSOA Using Partial Response Equalization", Photonics Technology Letters, IEEE, 23, 20, 1442 – 1444 (2011).

[26] T. Duong et al., "Experimental demonstration of 10 Gbit/s upstream transmission by remote modulation of 1 GHz RSOA using Adaptively Modulated Optical OFDM for WDM-PON single fiber architecture", ECOC 2008.

[27] http://users.df.uba.ar/bragas/Web roberto/Papers/Ralph Franz Keldish.pdf

[28] E.K. MacHale et al., "Extended-Reach PON Employing 10Gb/s Integrated Reflective EAM-SOA", ECOC 2008.

[29] D. Smith et al., "Colourless 10Gb/s reflective SOA-EAM with low polarization sensitivity for long-reach DWDM-PON networks", ECOC 2009.

[30] http://www.fabulous-project.eu/

[31] B. Charbonnier et al., "Reflective polarisation independent Mach-Zenhder modulator for FDMA/OFDMA PON", Electronics Letters, 46, 25, 1682 – 1683 (2010).

[32] C. Arellano et al., "Reflections and Multiple Rayleigh Backscattering in WDM Single-Fiber Loopback Access Networks", IEEE J. LIGHTWAVE TECH. 27 no. 1, 12-18 (2009).

[33] S.P. Jung et al., "Demonstration of RSOA-based WDM PON employing self-homo-dyne receiver with high reflection tolerance", Optical Fiber Communication Conference, OFC/NFOEC (2009).

[34] S. Straullu et al., "Compatibility between coherent reflective burst-mode PON and TWDM-PON physical layers", OPEX 22, Issue 1, pp. 9-14 (2014), http://dx.doi.org/10.1364/OE.22.000009

[35] K. Y. Cho et al., "10-Gb/s operation of RSOA for WDM PON," IEEE PHOTON. TECHNOL. LETTER, 20, 18, 1533 – 1535 (2008).

[36] S. J. Savory, "Digital filters for coherent optical receivers," OPEX 16, 2, 804-817 (2008).

[37] F. Vacondio et al., "Experimental demonstration of a PDM QPSK real-time burst mode coherent receiver in a packet switched network," ECOC, Tu.3.A.1, (2012).

[38] F. Vacondio, et al. "Flexible TDMA Access Optical Networks Enabled by Burst-Mode Software Defined Coherent Transponders", ECOC, We.1.F.2, London, UK, (2013).

[39] S. Straullu et al., "TWDM-PON-compatible 10 Gbps Burst-mode coherent reflective ONU achieving 31 dB ODN loss using DFB lasers", ECOC, Cannes, France (2014).

[40] A. Naughton et al., "Optimisation of SOA-REAMs for hybrid DWDM-TDMA PON applications", Optics Express, 19, 26 (2011).

[41] S. Straullu et al., "Reflective FDMA-PON with 32 Gbps upstream capacity per wavelength and more than 32 dB ODN loss", ECOC, Cannes, France (2014).

[42] S. Abrate et. al., "FDMA-PON architecture according to the FABULOUS European Project" Proc. SPIE 8645, Broadband Access Communication Technologies VII, 864504 (2013).

[43] S. Menezo et al., "Reflective silicon Mach-Zehnder modulator with Faraday rotator mirror effect for self-coherent transmission", OFC/NFOEC 2013.

Chapter 12

Recent Advances in Wavelength-Division-Multiplexing Plastic Optical Fiber Technologies

David Sánchez Montero, Isabel Pérez Garcilópez,
Carmen Vázquez García, Pedro Contreras Lallana,
Alberto Tapetado Moraleda and
Plinio Jesús Pinzón Castillo

Additional information is available at the end of the chapter

http://dx.doi.org/10.5772/59518

1. Introduction

Growing research interests are focused on the high-speed telecommunications and data communications networks with increasing demand for accessing even from the home, due to the huge successes during the last decade of new multimedia services (high-definition (HD), three-dimensional visual information (3D) or remote "face-to-face communication") which forecast requirements for data transmission speed more than 40Gbps by 2020, which can be achievable only with optical network [1]. Regarding this data transmission capability, Polymer Optical Fiber (POF) technology has emerged as a useful medium for short-reach distances scenarios such as Local Area Networks (LANs), in-home and office networks, automotive and avionic multimedia buses or data center connections among others. However, its potential capacity for communication needs a greater exploitation to meet user requirements for higher-data rates.

The strong increase of bandwidth demand presents an increasing challenge for service operators to delivery their high-quality service to the end user's device. At this moment, commercially available progressive service plans range between the 50-100Mbps while premium services typically range around 100-150Mbps. And it should be reminded that the bandwidth in the local loop is forecasted to grow with an average of 20-50% annually. Recent interests are focused on gigabit-order data transmission, being desirable, at the same time, to introduce optical fiber networks even to the customer's premises for covering more than 10Gbps in the near future, introducing the concept of FTTx (Fiber To The Home/Node/

Building/Curb) deployments. There is a worldwide consensus that the optical fiber solution provides enough bandwidth to attend user's demand at the required transmission distances in the short-reach domain (typically up to 200m).

In this optical fiber network deployment scenario, POF offers several advantages over conventional silica multimode optical fiber over short distances. Such fiber type can provide an effective solution as its great advantage is the even potential lower cost associated with its easiness of installation, splicing and connecting. This is due to the fact that POF have higher dimensions, larger numerical aperture (NA) and larger critical curvature radius in comparison with glass optical fibers [2]. Moreover, it is more flexible and ductile, making it easier to handle. Consequently, POF termination can be realized not only faster but also cheaper than in the case of multimode silica optical fiber [3]. To summarize, POFs have multiple applications in sensor systems at low or competitive cost compared to the well–established conventional technologies [4].

To date, most used POF type is the step index POF (SI-POF) but many variants have been manufactured and tested showing different performances between them [5]. SI-POF is made of polymethyl-methacrylate (PMMA), (also called standard POF) and it has 980 μm core diameter, 10 μm cladding thickness and 0.5 NA. However, SI-POF suffers from high modal dispersion, which reduces the usable bandwidth to typically 50 MHz × 100 m [6]. And it is only used in the visible spectrum range (VIS), where it can provide acceptable attenuation (e.g. 100 dB/Km at 650 nm) [5]. This is because of the large attenuation due to the high harmonic absorption loss by carbon-hydrogen (C-H) vibration (C-H overtone). However, improvements in the bandwidth of POF fiber can be obtained by grading the refractive index, thus introducing the so-called Graded-Index POFs (GIPOFs). Although firstly developed PMMA-GIPOFs were demonstrated to obtain very high transmission bandwidth compared to that of SI counterparts [7], the use of PMMA is not still attractive due to its strong absorption at the near-infrared (near-IR) to infrared (IR) regions. As a result, PMMA-based GIPOFs can only be used at a few wavelengths in the visible portion of the spectrum. Today, unfortunately, almost all gigabit optical sources operate in the near-infrared (typically 850nm or 1300nm), where PMMA and similar polymers are essentially opaque. Reduction of loss has been achieved by using amorphous perfluorinated polymers for the core material. This new type of POF has been named perfluorinated GIPOF (PF GIPOF), and has a relative low loss wavelength region ranging from 650nm to 1300nm (even theoretically in the third transmission window) [8]. Consequently, available off-the-shelf light sources for silica fiber based systems can be used with PF-GIPOF systems. In 1998, the PF-based GIPOF had an attenuation of around 30dB/km at 1310nm. Attenuation around 20dB/km was achieved only three years after and lower and lower values of attenuation are being achieved. The theoretical limit of PF-based GIPOFs is ~0.5 dB/km at 1250-1390nm [9]. Nevertheless, although these losses are coming down steadily due to ongoing improvements in the production processes of this still young technology, the higher than silica attenuation inhibits their use in relative long link applications, being mainly driven for covering in-building optical networks link lengths for in-building/home optical networks (with link lengths less than 1 km), and thus the loss per unit length is of less importance. In addition, PF-GIPOF can provide a bandwidth per length product ~400MHz x km at both 850nm and 1300nm, respectively, and can support bit rates of 40Gbps up to 200m for any launch condition [10]. This fact is due to the PF-GIPOF low material dispersion characteristics (even lower compared to silica multimode optical fibers) [11].

Although POFs reveal a cost effective solution for short-reach optical deployments, their bandwidth characteristics still limit the reach distances and the capacity to attend future end users' transmission requirements. These facts hamper the desired integration of multiple broadband services into a common multimode fiber access or in-building/home network. Overcoming the bandwidth limitation of such fibers requires the development of techniques oriented to extend the capabilities of POF networks to attend the consumer's demand for multimedia services. Different efficient and advanced modulation formats and/or adaptive electrical equalization schemes can alternatively be applied. Considering the industry's extensive experience and the large economies of scale, orthogonal frequency division multiplexing (OFDM) [12], subcarrier multiplexing (SCM) [13] and discrete multitone modulation (DMT) [14] are seen as promising technologies for low-cost, reliable, and robust Gigabit transmission through hundreds of meters of POF. Particularly, DMT modulation has been demonstrated to achieve near-optimum performance and to enable highly spectral efficient transmission at high bit-rates over silica multimode fibers (MMFs) and POFs [15, 16]. Initially, transmission with SI–POF has been realized with only one channel, typically at 650 nm, reaching data rates of 100 Mb/s over links of 275 m [17]; even multi–gigabit transmission over links of 50 m has been reached [18]. Commercial systems with data rates of 1 Gbit/s via up to 50 m of SI–POF with a single channel have also been reported [19]. Moreover, theoretical simulations with data rates of 1.25 Gbit/s, 2.1 Gbit/s [20] and 6.2 Gbit/s [21], via up to 50 m of SI-POF using a single channel with NRZ, CAP-64 and QAM512 modulations, respectively, have been demonstrated.

After exploiting the capabilities of a single channel, the next step to increase the capacity of an individual POF is to use multiple channels over a single fiber what is well known as wave-length division multiplexing (WDM). In the last years, WDM techniques over POFs are being proposed to expand the usable bandwidth of POF-based systems. For instance, Jončić et al. [22] firstly reported a 10 Gbit/s transmission over 25 m of SI–POF using offline-processed NRZ modulation. Beyond that, same authors achieved data rates up to 14.77 Gbit/s, with 4 channels via up to 50 m of a SI–POF link using offline-processed discrete multitone modulation [23]. In the same way, many POF based sensors implement self–referencing schemes by transmitting different wavelengths over a single fiber [4]. However, there are some constrains that must be addressed in order to perform the same capabilities as in the case of silica-based WDM approaches. In the WDM technique, different wavelengths which are jointly transmitted over the fiber must be separated to regain all information. Therefore, for a typical WDM optical communication link two key-elements are, at the very least, indispensable and have to be introduced, a multiplexer and a demultiplexer. The former is placed before the single fiber to integrate every wavelength to a single waveguide. The latter is placed behind the fiber lead to regain every discrete wavelength. These two components have long been established for silica-based infrared telecom systems, bust must be developed completely new for POF-based WDM applications. The most common and grave disadvantage almost all of these approaches exhibit is their costly production, which makes them unsuitable for today's price sensitive mass markets. The underlying reason behind this lack of development is the mismatch between the optimum operating wavelength regions of POFs and the optical devices exploited for telecommunications purposes. The latter are developed for a wavelength region (C- and L-bands) totally unsuitable for POF-based transmission over medium-distances (hundreds of meters or greater) due to the high attenuation of PMMA based POF of around 1dB/cm@1550nm. A similar conclusion can be obtained for PF-GIPOFs which attenuation characteristics are not at par with that of standard silica based fibers, but still superior to that of copper based technologies and PMMA-POF fibers. Another question to be addressed is the large POF dimensions and NA, which produce beams with high divergence thus being difficult to be routed. In addition, multiplexed systems operating in VIS range for POF networks may need reconfiguration because they do not have standard channels as well as provide flexibility in the networks to be developed.

In this framework, this chapter is intended to be a progress report and it will focus on the state-of-the art, description and experimental validation of different POF-based key devices that provide an easy-reconfigurable performance for WDM applications. Novel multiplexers/demultiplexers, variable optical attenuators, interleavers, switches and optical filters to separate and to route the different transmitted wavelengths are described. The main target is to bridge the gap of the WDM POF-based network deployment bottleneck in the final leg of delivering. In addition, a hybrid silica-POF WDM-PON network is analyzed showing the capabilities of novel Fiber Bragg Gratings (FBG) inscribed on microestructured POF devices to be compatible with WDM topologies for both sensing and communication schemes. Moreover, the theoretical capacity for a future WDM-GIPOF deployment is addressed taking advantage of the performance of this recent fiber type. Finally the main conclusions are presented.

2. Optical multiplexers and demultiplexers in POF technology

Multiplexers, combiners and variable optical attenuators are basic elements in POF networks when using the WDM approach but are not widely spread yet on the market due to the aforementioned reasons yielding their associated insertion losses. Nevertheless, reconfiguration can be an additional feature for those networks but most of them developed in POF technology do not provide such a characteristic. In this section, novel POF devices with reconfigurable characteristics for WDM applications addressing compact, scalable and low consumption solutions and with low insertion losses will be described. They operate at the wavelengths of interest for POF applications and their performance will be compared to current state-of-the-art approaches reported in literature. Some of them will take advantage of the properties of liquid crystal materials.

Several technologies have been reported for implementing optical multiplexers. Arrayed Waveguides Gratings (AWG), based on two multimode interference sections joined through several waveguides of different lengths, are proposed to be used in short distance communications [24]. However, this approach is unsuitable for the wavelength range of operation of POF networks with affordable losses and, therefore, cannot be used. On the other hand, in the WDM approach many transmitters with different light colors can carry individual information. In WDM-POF systems, most multiplexers used in POF links are based on N:1 splitting devices which their function is to combine the optical signals from multiple different single-wavelength end devices at their inputs onto a single output fiber. For example, red light can be modulated with Ethernet data while blue, green and yellow light can carry image information, radio frequency (RF) and television signal, respectively. There have been many techniques of fabricating POF couplers. These techniques include twisting and fusion, side polishing, chemical etching, cutting and gluing, thermal deformation, molding, biconical body and reflective body [25]. The main drawback of the use of POF couplers as multiplexers are: a) their high associated insertion losses, typically up to 8dB per branch [26] if we consider 3:1 and 4:1 POF couplers; and b) in this kind of multiplexers input ports are not interchangeable and each input port must be excited by a pre-allocated wavelength source. To solve this latter disadvantage, another approaches make use of novel reconfigurable POF multiplexing devices, where inputs are wavelength independent, as they work in the same way for different wavelengths, thus allowing more flexibility in WDM-POF networks. They will be described in the following section.

From the POF demultiplexer perspective, solutions for WDM SI-POF networks reported in literature are based on bulk optics and take advantage of discrete devices such as prisms [27], thin-film filters [28] or diffraction gratings [29, 30]. Thin-film based demultiplexers are easy to implement and are a good choice to design demuxes with low insertion loss and multiple channels. However, they are large, require many elements (typically the number of elements doubles the number of channels) and their channel isolation (crosstalk) is mainly limited by the rejection ratio of the thin-film filters used. Optical filtering at each output channel is usually employed to enhance the crosstalk performance in this type of demultiplexers thus adding complexity into the system. In contrast, prism-based demultiplexers have few elements and are cheaper but usually show a low performance in terms of both insertion loss and crosstalk. Most common proposals are based on concave gratings. These proposals have good expectations as they have a small size and because the light spatial separation and its focusing are performed with a single element. However, they require diffractive elements that to date are not easy to manufacture and have not reached large market volumes yet, being a costly solution. Moreover their experimental performance has not yet been tested on a mass basis real scenario. However, it is expected to experience the price reductions accompanying economy-of-scale in a near future.

2.1. Designs of reconfigurable optical multiplexers

Among the different technologies used for implementing optical multiplexers (as well as for optical switches) those based on liquid crystals (LC) are very interesting because they do not have mobile parts, need low excitation voltages and have a low power consumption. In the last years, liquid crystal has been widely used in displays applications. Liquid crystals are organic compounds that have properties intermediates between liquid and crystalline solids

[31]. They have anisotropic characteristics such as the dielectric constant or the refractive index, like solids, but simultaneously they are fluids. There are mainly two types of LC used in optical multiplexing, Ferroelectric Liquid Crystals (FLC) [32] and Nematic Liquid Crystal (NLC) [33-35], both normally using the structure of a Twisted Nematic Liquid Crystal cell (TN-LC). The first ones have a better response time but they can operate in a smaller wavelength range. The second ones have worst response times (tens of milliseconds in conventional mixtures), but they can operate in a wider wavelength range because they only have to fulfill Mauguin's regime $\Delta n \times d/\lambda \gg 1$ in order to obtain the polarization shift, where Δn is the birefringence of the LC, d the LC cell thickness and λ the wavelength, respectively.

In the following, different topologies of optical multiplexers based on liquid crystals are described. The first design is based on Polymer Dispersed Liquid Crystal (PDLC), which is a special case of NLC, while the other devices are based on TN-LC cells. A brief introduction about both LC types is provided within each section for a better understanding.

2.1.1. Optical multiplexer and variable optical attenuator based on polymer dispersed liquid crystal

The first design of the proposed multiplexers is based on PDLC. PDLC is composed by microdroplets with liquid crystal molecules dispersed in a polymeric matrix. Liquid Crystal molecules have electrical and optical birefringence, which means that the molecules have different dielectric constants and refractive indexes depending on the molecule axes. This mixture is sandwiched between glasses and covered with a transparent conductor [36]. In this way, an electric field can be applied to the mixture allowing the reorientation of the liquid crystal molecules that are inside the microdroplets thanks to the molecule birefringence. The structure of the PDLC cell is shown in Fig. 1.

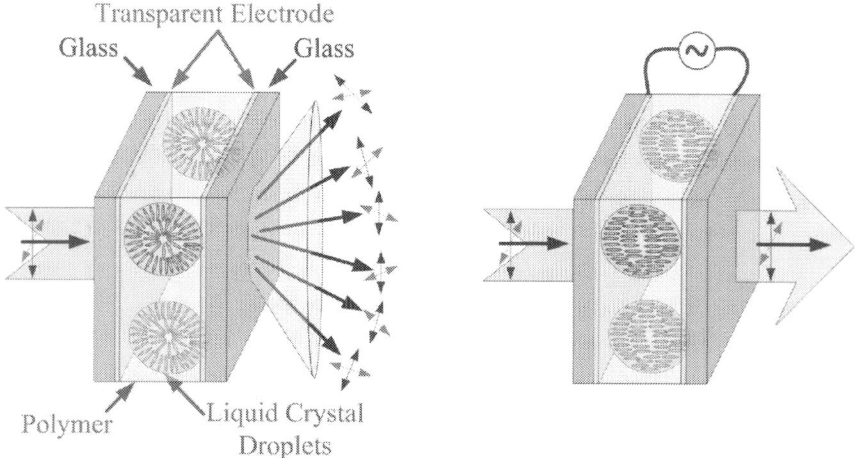

Figure 1. Structure of a Polymer Dispersed Liquid Crystal cell

The principle of operation is the following. At resting state, i.e. there is no voltage applied to the transparent conductors (electrodes), the liquid crystal molecules inside the droplets do not have a predominant orientation and the light that passes through the mixture find different refractive indexes. Therefore, it is highly scattered into different directions. On the other hand, if enough voltage is applied between the transparent conductors, an electric field is created inside the mixture and the liquid crystal molecules are forced to follow the induced electric field. In this case, the mixture has a homogeneous refractive index and the light that passes through the mixture is not refracted and maintains the same propagation direction. In addition to this, there is a gradual transition in the liquid crystal molecules reorientation, thus, if the voltage applied is not high enough, the molecules are not fully oriented but the light that passes through the mixture is less scattered [37].

Due to the possibility for controlling the transmission of light through the PDLC, the latter have been mainly reported for implementing Variable Optical Attenuators (VOA) [38, 39]. A VOA based on a 2 x 2 coupler made of POF is presented in [38]. The idea for implementing a reconfigurable optical multiplexer based on PDLC is to use the PDLC cell with several pixels as the active element [40]. The structure is shown in Fig. 2. The input ports, that are optical fibers, are placed in front of each pixel of the PDLC cell. The light that comes out from the fiber is spreaded according to the numerical aperture of the optical fiber, thus, the lenses placed in front of each input port are required for collimating the light that comes out from the optical fibers. The light from each input port passes through one pixel of the PDLC cell and finally, the lens placed at the output focuses the light into the output fiber.

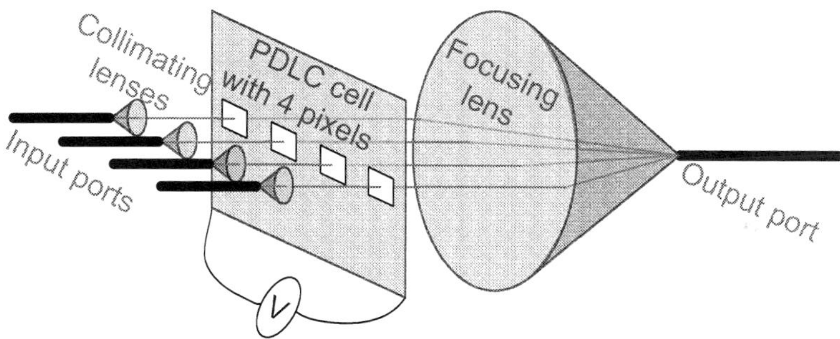

Figure 2. Structure of the reconfigurable optical multiplexer based on PDLC

By using the proposed structure, when no voltage is applied to the PDLC pixel, the light that comes from the input port is scattered in the PDLC cell and, therefore, it is not focused in the output port. On the contrary, when enough voltage is applied to the PDLC cell, the light can pass through the PDLC cell being focused on the output port. In addition, variable attenuation can be achieved by applying intermediates voltages. Each pixel can be addressed independently from the adjacent ones, so each input port can be switched on/off without affecting the others.

The introduced reconfigurable optical multiplexer can also acts as a variable optical attenuator. It can operate in the visible range as POF do. The typical voltage value for switching the PDLC is tenths of volts. The achieved insertion loss is about 1.6dB, the crosstalk obtained is in the range of 30dB, and finally, the response time of the PDLC is in the order of tenths of milliseconds.

2.1.2. Optical multiplexers based on Twisted Nematic Liquid Crystals (TN-LC)

Other way in which liquid crystal is used is known as TN-LC. In this kind of devices, the liquid crystal is also sandwiched between two glasses covered with a transparent conductor (electrode). However, the glasses have an additional rubbed alignment film that forces the LC molecules to have an orientation. In a TN-LC the orientation of the LC molecules in one glass is perpendicular to the molecular orientation in the other glass. Thanks to these constraints, the LC molecules inside the cell perform a helix from one glass to the other. The structure of a TN-LC cell is shown in Fig. 3.

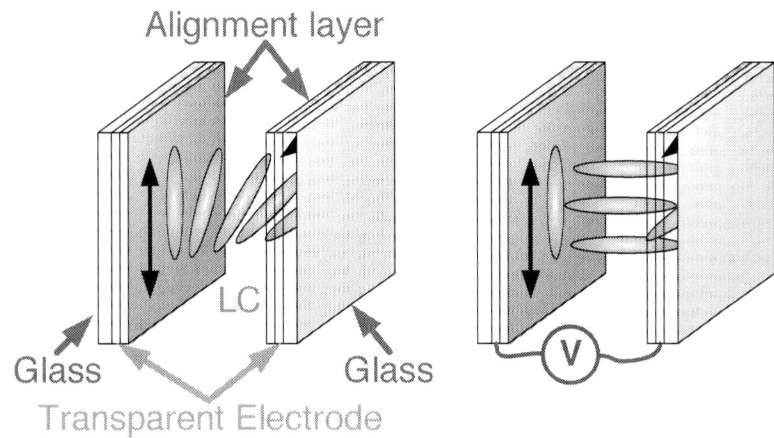

Figure 3. Structure of a twisted nematic liquid crystal cell.

The principle of operation is the following. If no voltage is applied to the LC, the polarization of the light that passes through the TN-LC cell is ideally rotated 90 degrees. On the other hand, when enough voltage is applied between the transparent conductors of each glass, an electric field is generated inside the cell, and the molecules are reoriented to be perpendicular to the glasses. In this scenario, the polarization of the incident light remains when passes through the LC.

In this way, the polarization of the light that transverse the TN-LC cell can be controlled. Thus, by placing the TN-LC cell between polarizers, the transmission of the incident light can be modified by means of the voltage applied to the cell. A polarizer placed before the TN-LC cell allows passing only one polarization of the incident light. In this way, a polarized light beam enters in the TN-LC cell. The TN-LC controls its polarization stage depending on the voltage applied to the cell. Finally, the polarizer placed after the TN-LC filters, or not, the light that comes out from the TN-LC cell. According to the TN-LC operation, there are two ways of implementing the device for controlling the light transmission: a) putting the TN-LC cell between crossed polarizers, or b) putting it between parallel polarizers. In the first case, the light passes through the device when there is no voltage applied to TN-LC while light is stopped when enough voltage is applied to the LC cell. On the contrary, the input light is blocked by the device when no voltage is applied to the TN-LC cell and the light passes through the device when enough voltage is applied to the TN-LC cell. The procedure described has been mainly used in displays applications [41], but it can also been used for optical multiplexing as well as for optical switching. The latter will be seen in a following section.

As previously reported, multiplexers and demultiplexers are basic elements in those optical networks where WDM is implemented because they combine different wavelengths in a single fiber. POF fiber has a low attenuation in the visible wavelength region (at 450nm, 550nm and 650nm), for this reason, the optical multiplexers must work in this wavelength range. An example of a reconfigurable multiplexer based on TN-LC cells is presented in Fig. 4 [42]. The introduced structure of the multiplexer can be used in several wavelength ranges depending on the bulk elements used.

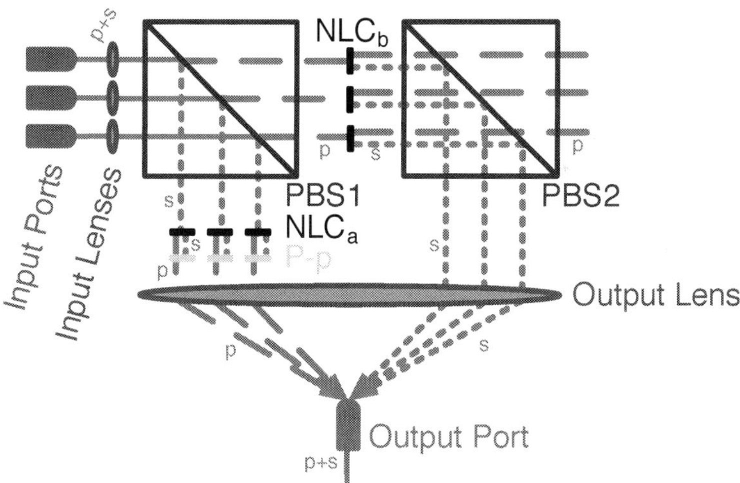

Figure 4. Structure of the reconfigurable 3x1 optical multiplexer

The structure is composed by Polarizing Beam Splitters (PBS1 and PBS2), TN-LC cells (NLCa and NLCb), polarizers and lenses. There are three input ports, and a single output port. Each TN-LC cell has three pixels, and each pixel controls the transmission of one input port.

Consequently each input port can be managed independently of the others. The input lenses are required for collimating the light that comes from each input fiber, and the output lens focuses each beam into the output port.

The operation of the optical multiplexer makes that the light from one input port is guided to the output port when there is no voltage applied to the corresponding pair of pixels of liquid crystal cells. If a voltage is applied to theses pixels, the light from the input port is not guided to the output port.

2.1.3. Advanced multifunctional optical multiplexer for multimode optical fiber networks

In passive optical networks where there is no additional amplification, it is important to have few insertion losses. It could be also interesting to have additional functions in the same device and thus reducing the number of devices in the optical network. In addition to that, reconfigurable optical networks in critical applications where an alternative path is required when there is a failure in the main path would be useful.

An improvement in terms of flexibility of the 3x1 multiplexer shown in the above figure is presented in Fig. 5 [43]. The structure has two set of three inputs and two outputs, and depending on the configuration each input of the one set of inputs can be guided to one of the two possible outputs.

The structure can implement different functionalities only by selecting the inputs, the outputs and modifying the voltage applied to the TN-LC pixels of each cell. It can behave as a 3x1 multiplexer (or combiner) using only three inputs and one of the outputs, each input port can be switched on/off independently of the other three inputs thus also acting as an optical switch if required.

The same device can operate as two complementary 3x1 Multiplexers. Inputs to the device are grouped in pairs, when the *Input Port a* is guided to *Output Port 1*, the other input of this pair, *Input Port b*, is coupled to *Output Port 2*. On the other hand, when the multiplexer is switched, *Input Port a* is directed to *Output Port 2* and the matched *Input Port b* is propagated to *Output Port 1*. As a matter of fact, it can also work as a 2x2 optical switch by using only the adequate pair of inputs and the two outputs.

The use of TN-LC cells allows having intermediate values of light transmission by applying lower voltage, so the device can also implement a VOA by using a single input port and only one of its outputs. Finally, it can also implement a variable optical power splitter by using one input and its two outputs.

The introduced reconfigurable Advanced Multifunctional Optical Multiplexer has fiber to fiber insertion losses when operating as a 2x2 optical switch, in the range from 10dB to 15dB within 200nm wavelength range; with a non-optimized optics for collimation and coupling. Lower losses can be achieved for a smaller wavelength range. The crosstalk measured is better than -15dB at 532nm, 660nm and 850nm. Switching is achieved at voltage levels of $4V_{RMS}$.

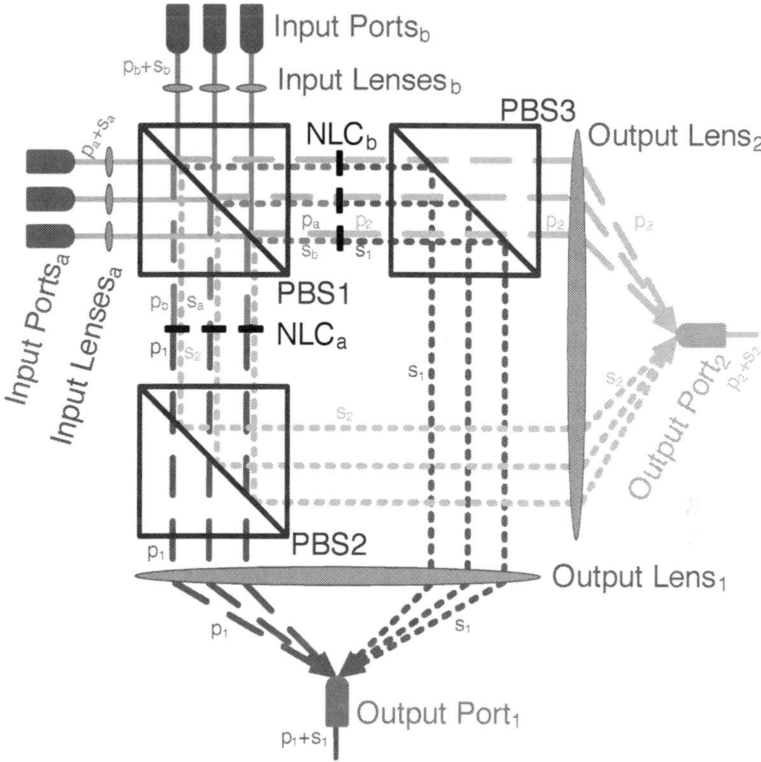

Figure 5. Structure of the Advanced Multifunctional Optical Multiplexer. It design also allows switching functionality if required into a single device.

3. Optical routing for WDM POF-based applications

For WDM routing, key devices such as interleavers, routers and switches are also indispensable to combine, to separate and to re-direct the different transmitted wavelengths. Nowadays, WDM devices are well-established in the IR and near IR (NIR) for silica optical fibers. However, they require a complete re–design for being implemented in SI–POF WDM systems. This is mainly due to their distinct attenuation behavior (compared to silica fiber) and new wavelength channels need to be determined. Nevertheless, there is no a widely spread consensus about the characteristics for these WDM channels for POF applications, although some authors have already proposed a spectral grid [44], which includes channels between 400 nm and 700 nm and spectral bandwidths up to 50 nm (LED sources) [45].

Nowadays, optical routers are key components in optical communications and sensor networks. Optical switches allow optical routing without converting the transmitted information into the electrical domain. The elimination of the two required conversions (optical to electrical and electrical to optical) improves the system characteristics, reducing the network equipment and increasing their bandwidth. These devices work by selectively switching optical signals delivered through one or more input ports to one or more output ports, in response to supervisory control signals. Different technologies could be applied to route optical signals, applications of which depend on the topology of the optical network and the switching speed required [46]. Cutting-edge optical switching technologies depending on their principle of operation include micro-electromechanical systems (MEMS) as well as acousto-optical, thermo-optical, opto-optical and electro-optical (EO) devices.

Opto-Mechanical switches are based on the movement of some mechanical devices such as prisms, mirrors or directional couplers. As a subsection of the opto-mechanical technology, MEMS have a great interest in telecommunications applications. MEMS consist of small mobile refractive surface mirrors that route the incident light beams to their destination [47, 48].

Acousto-optic switches are based on the acousto-optic effect of some materials, such as the peratellurite [49] or $LiNbO_3$ [50], where an acoustic wave travelling along the material induces a periodical strain that alters its refractive index. The refractive index modulation induced in the material causes a dynamic phase grating than can diffract light. If the material is isotropic, the diffraction induced by the acousto-optic effect causes beam deflection, and if the material is anisotropic the deflection caused comes along with variation in light polarization.

Other solutions are the thermo-optic switches whose operation consists on the variation of the refraction index of the material by modifying its temperature. This type of switches has a great variety of implementations, but mainly based on using an interferometric mechanism in which the refractive index variation induces a change in the interference condition. This effect facilitates the light switching [51, 52].

Opto-optical switches are based on the intensity-dependent nonlinear effects in optical waveguides, such as the Two-Photon Absorption phenomenon (TPA) [53], the lightwave self action that induces the Self Phase Modulation (SPM) phenomenon and the Kerr Effect that causes the Four Wave Mixing (FWM) and the Cross Phase Modulation (XPM) [54].

Finally, electro-optic switches perform switching by using electro-optics effects. The main technologies are based on Lithium Niobate ($LiNbO_3$) [55], Semiconductor Optical Amplifiers (SOA) [56], Electro-holographic (EH) [57], Bragg Gratings electronically switched [58] and LC. Focusing on the latter, LC switches use different physical mechanisms to steer the light such as polarization management, reflection, wave-guiding and beam-steering (2D or 3D). Main advantages of this technology include no need of moving parts for switch reconfiguration, low driving voltage and low power consumption. In the last years, different devices based on nematic LCs for SI-POF networks have been reported. Some of them are described below.

3.1. Optical Switches based on twisted nematic liquid crystals

The polarization rotation is the first configuration used in LC switches [59]. A simple example of a 1x2 switch based on TN-LC is presented in Fig. 6 [60]. Depending on the voltage applied to the TN-LC, the light from the input (Port 1) is guided to one of its outputs (Port 2 or Port 3).

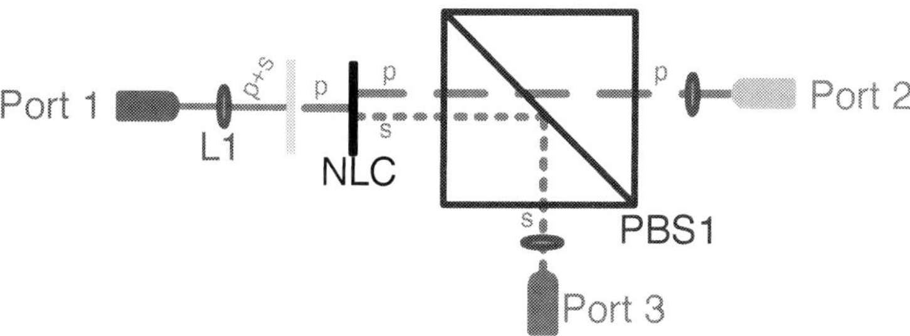

Figure 6. Structure of the 1x2 LC optical switch.

As a consequence of the use of the input polarizer for the operation of the TN-LC half of the incoming optical power is filtered. The solution for reducing the insertion losses of the optical switch based on TN-LC cells is by using the polarization diversity method, see Fig. 7. In this technique, the input light is decomposed into its TE (S-Polarized light) and TM (P-Polarized light) components. Both components are treated separately and finally recombined. In this way, the device becomes polarization insensitive, and less insertion losses are expected. The same principle of the reconfigurable multiplexer design reported in the previous section.

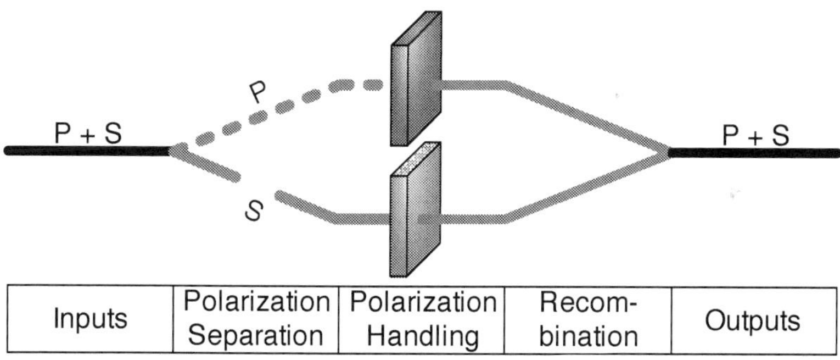

Inputs	Polarization Separation	Polarization Handling	Recom-bination	Outputs

Figure 7. Structure of the 1x2 LC optical switch.

Different configurations have been proposed in literature for implementing optical switches based on the polarization diversity method. A polarization insensitive 2 x 2 optical switch based on TN-LC is presented in Fig. 8 [61]. The structure is composed by Polarizing Beam Splitters (PBS1-PBS4), TN-LC cells (NLC1-NLC4), quarter wave plates (Plate 1- Plate 4), and mirrors (Mirrors 1 – Mirror 3). The 2x2 optical switch allows up to three operation modes by applying voltage to the suitable pair of TN-LC cells, see Fig. 9:

- – Direct Mode: NLC2 and NLC4 ON;   Port 1 → Port 3 & Port 2 → Port 4.
- – Crossed Mode: No voltage is applied;   Port 1 → Port 4 & Port 2 → Port 3.
- – Closed Mode: NLC1 and NLC3 ON;   Port 1 → Port 2 & Port 3 → Port 4.

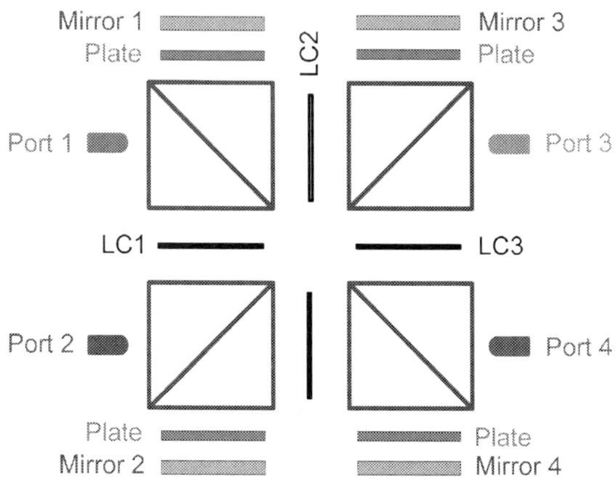

Figure 8. Structure of the polarization independent 2x2 LC optical switch.

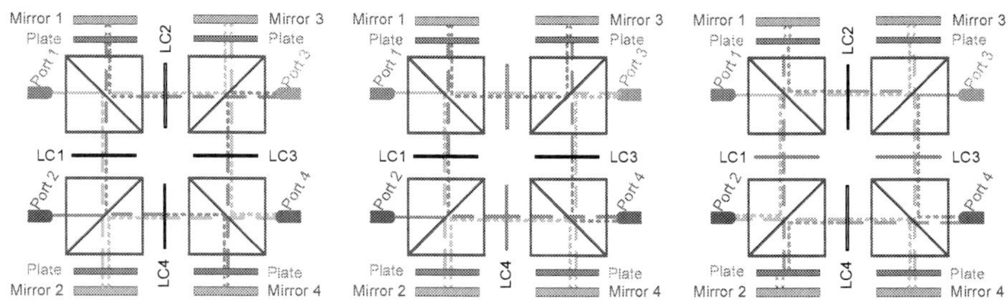

Figure 9. Operation modes of the 2x2 optical switch: (a) Crossed, (b) Direct, (c) Closed.

3.2. Optical router with output power control

TN-LCs can also be used in routers (LC–OR) based on the polarization diversity method following the same principle of operation as in the case of optical multiplexers. The polarization modulation of a TN cell in combination with space polarization selective calcite crystals or polarization beam splitters (PBS) allows optical space-switching. Fig. 10 shows the structures of a typical 1x2 nematic LC–OR (Fig. 10.a) and a 1x2 nematic LC–OR with independent output power control (Fig. 10.b).

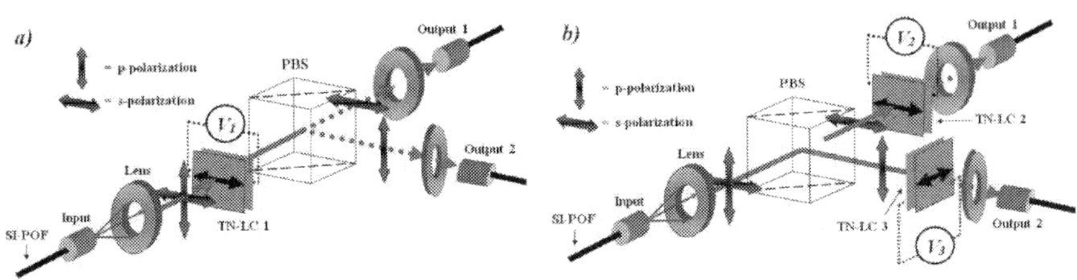

Figure 10. a) Structure of a typical 1x2 nematic LC–OR based on polarization diversity and, b) structure of a 1x2 nematic LC–OR with independent output power control.

In the scheme shown in Fig. 10.a, the TN-LC 1 has an input polarizer that changes the polarization state of the transmitted beam depending on the voltage V_1. The scheme of Fig. 10.b additionally provides the possibility of stabilizing the optical power that is transmitted by each port. This feature was reported in [62]. In that scheme, TN-LC 2 and 3 have both an input and an output crossed polarizer, with the input polarizer parallel to the respective polarization component transmitted by the PBS. Then, in this scheme each LC cell controls the transmitted power depending on the voltages V_2 and V_3. Fig. 11 shows an example of the power stabilization capacity of the router reported in [62].

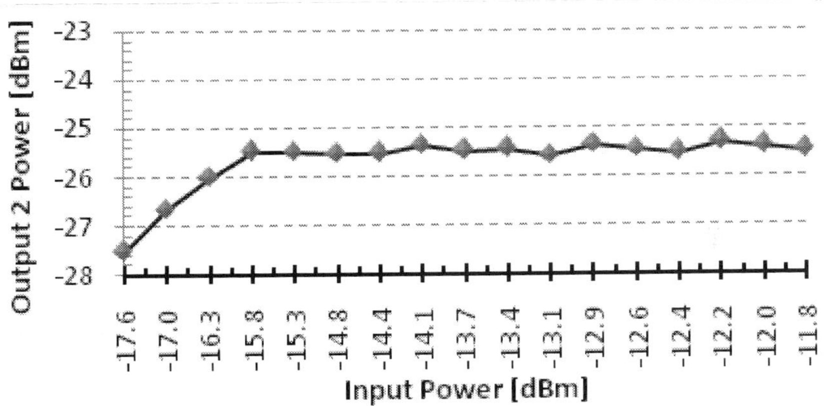

Figure 11. Example of the output power control capacity of a 1x2 LC-OR based in polarization diversity. Input power is obtained from a LED source at 650 nm.

LC-OR based on nematic LC cells cannot respond faster than several microseconds. This fact limits its use to telecom and sensor applications for protection and recovery, or optical add/drop multiplexing which need fewer restrictions about switching time, like WDM transport network restoration [63]. However, in the last years, nematic LCs with response times lower than 3 ms [64] and 2 ms [65], as well as nanosecond response [66] have appeared. And different techniques to reduce the response time below 1 ms [67] have also been reported.

3.3. Broadband LC-OR

It is a matter of fact that the performance of a twisted nematic LC-OR is optimum only for specific wavelengths (those given by Mauguin Minima) [68]. Besides, the LC birefringence, which defines Mauguin Minima, is very temperature dependent, requiring temperature compensated designs or controllers. These two limitations can be overcome by replacing the twisted nematic cells (see Fig. 10.a) with optimized polarization rotators (PRs) based on structures of stacked LC cells, as reported in [69]. Figure 12 shows an example of the performance of the broadband 1×2 LC–OR reported in [69].

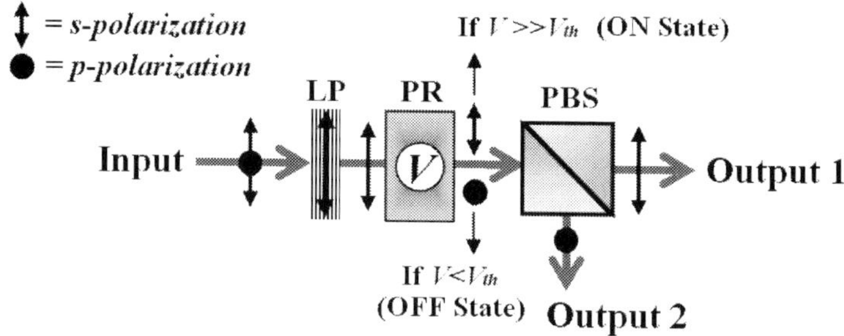

Figure 12. Scheme of a broadband 1×2 LC–OR for POF networks.

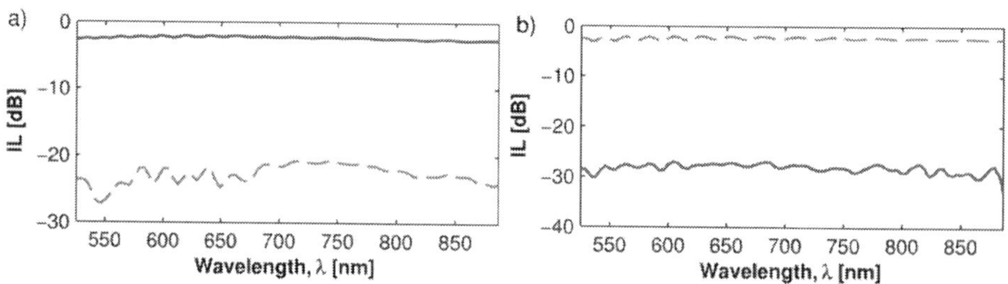

Figure 13. Spectral performance of the outputs 1 (dashed lines) and 2 (solid lines) of a broadband 1×2 LC–OR for POF networks in the a) OFF state (V<<V_{th}) and b) ON state (V>>V_{th}).

The proposed design illustrated above is composed by 3 LC cells and allows a significant improvement of the spectral response of LC optical routers, compared to those previously reported [43, 62, 70] in a broadband range. The proposed router has quite similar insertion loss values in both outputs in the range from 400 nm to 700 nm, as well as crosstalk values lower than −18.7 dB, as shown in Fig. 13. This performance is required for routing channels in SI–POF–WDM networks uniformly, since in these networks the channels may have wide bandwidths and the proposed grid is very wide, as it was aforementioned. In [69] it has been shown that the router performance is quite constant with temperature changes of up to 10 °C. And it was also demonstrated that it is able to control the split ratio of the output power with good uniformity in the range from 400 nm to 700 nm.

3.4. LC wavelength selective switch

A 1×M wavelength selective switch (WSS) is an optical device that allows switching any incoming wavelength from its input port to any of the M output ports, without the need for optical to electrical conversions. These devices play a key role in protection and reconfiguration tasks of next generation optical networks. A huge number of approaches to implement WSS have been demonstrated. Some are based on gratings that spatially disperse the input channels, on MEMS, or on LC spatial light modulators [70, 71]. Other approaches use silica-based planar lightwave circuits (PLCs) [72] or ring resonators [73].

Some LC reconfigurable devices for the VIS range have been reported, such as tunable filters [74] or a multifunctional device operating as a switch/combiner/variable optical attenuator [43], as well as a 1x2 WSS [74]. The latter is based on an inverted Lyot filter structure. This configuration allows demultiplexing, switching or blocking any channel through any output port using voltages from 0 to 3 V_{RMS}, the same for all the LC cells, with maximum insertion loss of 6 dB, and rejection ratios better than 12 dB. Fig. 14 shows two examples of the eight possible transmission states of the 1x2 LC WSS reported in [74].

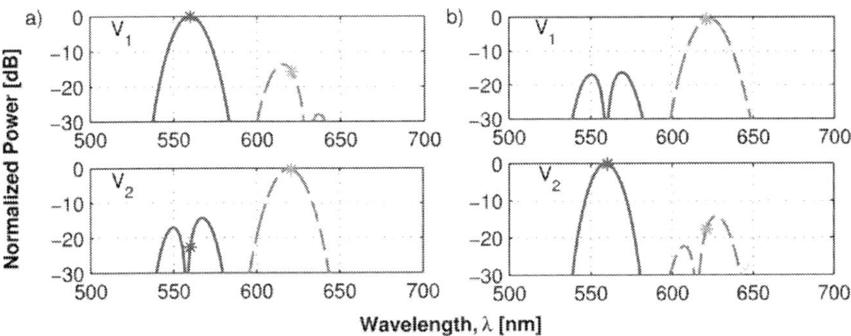

Figure 14. Performance example of a 1x2 LC WSS considering 2 LED channels at 589nm (solid lines) and 650nm (dashed lines) with 20 nm full width at half maximum. Where: a) output 1 and b) output 2. V1 = 1.18 V_{RMS} and V2 = 0.21 V_{RMS}.

4. Optical filters in POF technology

Optical filters are basic components as part of routing devices for optical communications networks. Optical interleavers are filters that due to their periodicity: a) separate an incoming spectrum into two complementary set of periodic spectra (odd and even channels), or b) combine them into a composite spectrum. Filters and interleavers play a key role in dense wavelength division multiplexing (DWDM) systems, usually employed in gain equalization, dispersion compensation, prefiltering, and channels add/drop applications.

Literature provides many optical filtering and interleaving devices. Some filters are based on birefringent structures such as the Lyot and Solc filters due to their low dispersion, high reliability, easy fabrication process and low cost [75], with recent applications in VIS for POF networks [69, 74]. However, these solutions mainly operate in DWDM systems (infrared range). In this section, the basic structures of birefringent filters (Lyot and Solc) are presented and compared against birefringent filters designed in the Z transform domain, which can be used in future WDM-POF systems.

4.1. Birefringent filters

Birefringent filters base their operation on the interference of an input light beam with multiple delayed versions of itself. Typically there are two types of birefringent filters, the Lyot and Solc. An excellent discussion of both types of filters can be found in [76]. In general, a Lyot filter consists of a set of delay stages composed by retarder plates of different widths between polarizers, as the 3-stage Lyot filter shown in Fig. 15.a. The optical axis of the retarder plates are at 45º with respect to the polarizers (azimuth angle) and each stage has twice the delay (Γ) of the previous one. Focusing on the second approach, Solc filters eliminate the need of Lyot filters for using multiple polarizers. Solc filters consist of a stack of M retarder plates between only two linear polarizers. In this case, all the retarder plates have the same delay and each one are at a specific azimuth angles, $\alpha_1,.. \alpha_M$, e.g. a fan Solc filter has parallel polarizers ($\alpha_A = 0º$) and $\alpha_1 = \alpha$, $\alpha_2 = 3\alpha$, $\alpha_3 = 5\alpha,..., \alpha_M = (2M-1)\alpha$, being $\alpha = 45º/M$, see Fig. 15.b.

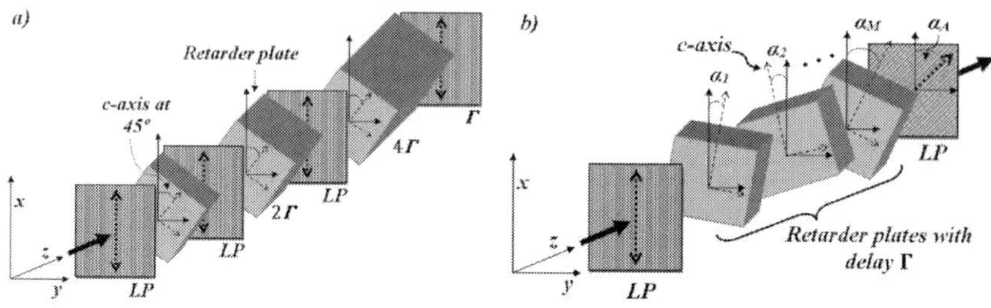

Figure 15. General structures of: a) Lyot filter of 3 stages and b) Solc Filters

Now, let us compare Lyot and Solc filters. For example, 3-stage Lyot filters require 4 polarizers and the equivalent of 7 retarder plates with delays Γ, as shown in Fig. 15.a. In contrast, a Solc filter with the same number of retarder plates requires only 2 polarizers. From this point of view Solc filters are a more interesting choice than Lyot filters. However, as is shown in Fig. 16, the adjacent side lobes suppression is better for the case of Lyot filters.

However, the potential of the structure of Solc filters can be exploited by placing the retarder plates illustrated in Fig. 15.b at arbitrary azimuth angles, in a type of filters called lattice or birefringent filters. Birefringent filters can be designed by using optimization methods or in the Z-transform domain, by using their relation with FIR (Finite Impulse Response) filters. A detailed method for transform FIR filters into birefringent filters can be found in [75]. For example, an arbitrary birefringent filter of seven retarder plates, obtained from a 7th order FIR filter, is presented in Fig. 16. The azimuth angles of the retarder plates are: $\alpha_1 = 6.07^\circ$, $\alpha_2 = 15.18^\circ$, $\alpha_3 = 28.65^\circ$, $\alpha_4 = 45.00^\circ$, $\alpha_5 = 61.35^\circ$, $\alpha_6 = 74.81^\circ$, $\alpha_7 = 83.92^\circ$ and $\alpha_A = 0^\circ$. This filter performs a uniform suppression of the adjacent side lobes with a maximum value better than both the Lyot and Solc filters. It could even be designed to have a narrow bandpass. Birefringent structures are a versatile solution for designing devices for POF WDM networks due to their reconfiguration capacity, since they can be easily manufactured with LC technology, and flexibility, since any FIR filters synthesis method can be used [75].

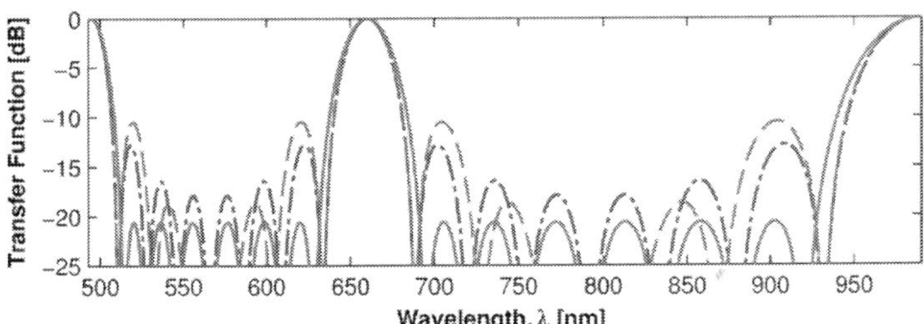

Figure 16. Transfer functions of different birefringent filters: 3 stages Lyot Filter (dash-dot line), Solc filter with 7 retarder plates (dashed line) and arbitrary birefringent filter of 7 retarder plates designed in the Z-transform domain (solid line). Retarder plates with $\Gamma = 2\pi \times 1.98 \mu m/\lambda$ are considered in all the filters.

It should be mentioned that demultiplexer devices may be easily built up by combining the switches and filter schemes described in former sections or by using filter plus the addition of POF splitting devices.

5. WDM approach using Polymer Optical Fiber Bragg Gratings (POFBGs)

As previously stated, one solution to increase the POF's bandwidth, and thus its capacity, is the transmission of information over more than just a single wavelength. This is what is known the WDM approach. This architecture not only has been proved to be suitable for transmission information purposes but also has been demonstrated to be fully compatible with the interrogation of multiple remotely located intensity-based optical fiber sensors, thus taking advantage of the power loss reduction as well as the high scalability provided by the use of WDM devices.

The basis of WDM systems is the spectral characteristics of the optical multiplexers and demultiplexers which are used in the fiber plant instead of optical power splitters. Moreover, within these approaches, FBGs are usually employed both for monitoring purposes or providing an effective and compact strategy to operate in reflective configuration [77, 78]. These devices are a wavelength-selective filter fabricated inside the core of an optical fiber for which the reflected wavelength changes under the influence of external perturbations [79]. First FBGs were traditionally manufactured on silica optical fiber. More recently, FBGs have been inscribed into SI-POF [80] and microstructured Polymer Optical Fiber (mPOF) [81] based on PMMA, leading to what is called polymer optical fiber Bragg gratings (POFBGs). The reason for this development is the exploitation of polymer benefits such as larger elastic limit, higher maximum strain limit, larger temperature and humidity responses and low cost compared to silica, while maintaining the benefits of FBGs. And this fact is also true when considering FBGs as optical sensing elements [82, 83]. In addition, polymer reveals to be intrinsically more biocompatible than silica for as it may be used for *in vivo* biomedical applications where the use of glass is inappropriate due to danger from breakages. Nevertheless, limited effort has been directed towards synergizing biocompatible POF-based photonic sensing with the WDM interrogation method that allows multiplexing by the use of FBGs, with just a few exceptions [84]. The main underlying reason behind this lack of development is the mismatch between the optimum operating wavelength regions of POFs and the optical devices exploited for telecommunications purposes as aforementioned at the beginning of this chapter.

In this section, we intend to bridge the gap between a WDM compatible topology and the use of novel POFBGs by analyzing the feasibility of a hybrid silica-POF WDM network for remotely addressing multiple intensity-based self-referenced fiber-optic sensors. The proposed topology is compatible with the target of developing a single optical broadband network architecture which is capable of carrying many types of services without mutual interference nor design compromises. It may include the access network domain (e.g. FTTx) as well as up to the indoor scenario with the aim of a full converged network solution. This solution will open up the path for the development of converged WDM POF communication networks in the near future. Moreover, potential medical environments and biomedical applications based on all-optical POF-based solutions are also targeted taking advantage of the intrinsic POF biocompatible characteristics.

The proposed self-referenced hybrid topology above described is illustrated in Fig. 17. The novelties of this configuration in comparison with previous works [78, 85] are: a) the combination of silica- and polymer- FBGs (the latter may be used for *in vivo* scenarios or just simply at the patient's vicinity if the target is a biocompatible optical system for medical applications); b) the usage of a single reference FBG; and c) an improved centralized monitoring unit (that can be remotely located up to units of km) which includes virtual instrumentation techniques and data processing. A broadband light source (BLS) is, either internally or externally, modulated at a single frequency (f). This modulated signal is launched into the remote sensing points via a broadband circulator and a Coarse Wavelength-Division Multiplexer (CWDM). Each remote sensing point located consists of a sensing POFBG placed after the fiber-optic sensor (FOS). A single silica FBG is located before the CWDM acting as a reference channel for the topology. Let assume the central wavelengths of the reference and sensing FBGs to be λ_{Si} and λ_{POF}, respectively. The broadband optical circulator receives the reflected multiplexing signals from the reference and the sensor channels, in which the sensor information is encoded. At the remote monitoring unit, the optical signal is demultiplexed by a CWDM device and distributed to an array of photodetectors (PD) by means of a data acquisition board (DAQ) together with a band-pass filter (BPF), used to eliminate noise from all signals at frequencies outside the system frequency. Then, a phase-shift is applied to the reference and sensor digital signals. Finally, a virtual lock-in amplifier is used to interrogate all available sensor channels. A measurement parameter can be defined, φ_K, corresponding to the output phase of the signal for different phase-shifts (virtual delays) at the reception stage, see Eq. 1.

$$\varphi_K = \tan^{-1}\left[\frac{-\left(\sin\theta_{Si} + \beta_k \bullet \sin\theta_{POF_k}\right)}{\cos\theta_{Si} + \beta_k \bullet \cos\theta_{POF_k}}\right] \tag{1}$$

where θ_{Si} and θ_{POF_k} are, respectively, the phase shifts for the reference and each sensor signal k, and β_k is relates to the optical power received at the remote central unit being a function of the modulation index, the reflectivity of the silica FBG and the photodetector responsivity (for both sensing and reference wavelengths), the sensor power loss modulation, the insertion loss of the CWDM device, and the insertion loss related to the reflectivity, intrinsic attenuation and connectorization of the POFBGs. Further details of the mathematical framework can be seen in the works reported in [78, 86]. Parameter φ_K is insensitive to power fluctuations except for the sensor modulation thus performing as a self-reference measurement parameter. Its performance, i.e. linear behavior, maximum sensitivity, etc., is directly related to the digital phase-shifts applied at reception.

To test the feasibility of the topology shown in Fig. 17, a 2-sensor network was analyzed by modulating the BLS at f = 1kHz by an acousto-optic modulator. The optical power was launched into the configuration via a broadband circulator. One silica FBG was used for reference purpose, being placed before the CWDM mux/demux. Its central wavelength and reflectivity were λ_{Si} = 1550nm and 49%, respectively. A POFBG in 150μm cladding diameter few-moded mPOF was used for each remote sensing point, with central wavelengths λ_{POF_1} = 1525.2nm for FOS$_1$ and λ_{POF_2} = 1567.0nm for FOS$_2$. Their reflectivities were 27 % and 36 %, respectively. A singlemode VOA was used to emulate the sensor response and for calibration purposes. The reflected signals were demultiplexed by a CWDM and detected by three amplified InGaAs detectors. The amplifier gain was fixed at 70 dB for all measurements. A 14-bit low-cost DAQ was used to convert the electrical signals from the photodetectors to digital signals. Virtual instrumentation techniques were developed to implement the bandpass filter, the phase-shifts and the lock-in amplifiers at the reception stage.

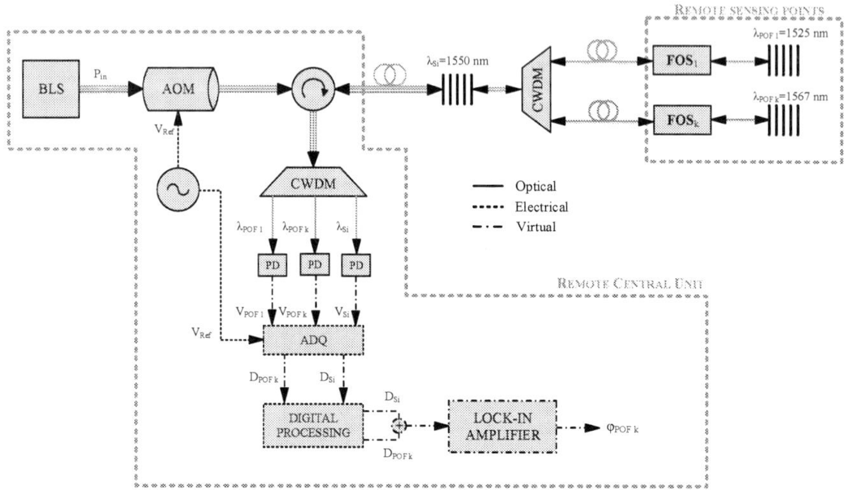

Figure 17. Hybrid silica-POF WDM self-referenced topology for remotely addresing generic remote sensing points located at the patient's vicinity. Fiber-optic sensor (FOS).

The self-reference property was tested inducing power fluctuations in the modulated optical source through a VOA. Fig. 18 showed no changes in φ_K values after inducing 10 dB of power attenuation. It is worth mentioning that the proposed topology performs no noticeable crosstalk between adjacent channels. This means that both (or more) sensors could be interrogated simultaneously without mutual interference because of the high channel isolation of the CWDM demultiplexer. Experiments were carried out for the following phase shifts at the reception stage: θ_{Si}=0.83π, θ_{POF1}=0.33π.

The performance of the proposed topology was further investigated obtaining resolution values far below than that of provided by most of the POF intensity based sensing solution reported in literature, and particularly for biomedical applications. Another interesting point is computing the power budget, which provides information about the maximum remote interrogation reaching distance and/or the maximum insertion losses only for sensing purposes. At the most restrictive sensing wavelength (in terms of reaching distance), a maximum length of 11 km could be obtained. For this calculation, a FOS power variation of 6 dB was considered, high enough to cover any biomedical input magnitude span. However, this reach distance can be easily improved by launching more optical power into the system, using optical devices with better insertion loss performance or using a more efficient technique to connect POFBGs. The aforementioned reaching distance could provide a remote monitoring service unit fully compliant for both short-reach networks (typically less than 1 km), i.e. LANs and in-building/in-hospital networks as well as suitable for medium reach-distances (typically up to 10 km). Furthermore, the latter value ensures applications in inter-hospital networks or to provide a convergent all-optical and straightforward connection between patient's homes and a general practice service for telemedicine purposes. It should be mentioned that the above reaching distances are unbeatable if an all-POF-based optical network is intended to be deployed and a hybrid approach should be considered. Following this analysis, it can be concluded that the proposed topology do not provide limitations thus serving as the bottleneck of a multiple remote sensing scheme.

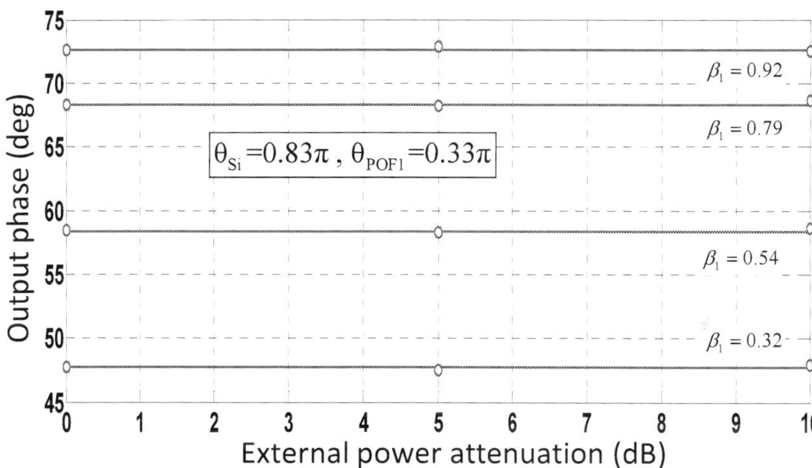

Figure 18. Output phase φ_1 self-reference test versus power fluctuations for different values of sensor losses at the remote sensing point addressed by λ_{POF1}.

6. WDM extension over PF GIPOF links

In this section a comparison between the achievable capacity over a single fiber channel and over a WDM PF-GIPOF-based approach will be presented. It has been previously stated that a typical WDM optical communication link requires both a multiplexer and a demultiplexer. However, the addition of POF-WDM multiplexer and demultiplexer devices, results in a bit-rate penalty as the available optical power on the system decreases due to their insertion losses. To establish the channel capacity comparative, a bit loading algorithm for DMT modulation format over PF-GIPOF has been considered. New power margin resulting from the additional losses considered in the system due to the WDM over POF approach are analyzed demonstrating the feasibility of PF-GIPOF WDM systems.

The resulting theoretical Shannon capacity of an optical fiber channel can be calculated if its f_{3dB} is known [14], modeled as a Gaussian low-pass filter. Therefore, from measurements of frequency response of different PF-GIPOF lengths, 3dB bandwidths can be obtained, and so their theoretical capacity limits operating in a single channel. This type of fiber has been demonstrated to enable robust 2GbE (GbE, Gigabit Ethernet) and 10GbE baseband transmission over short reach distances ranging from 25m up to 100m for different link scenarios [87], even at OverFilled Launching (OFL) condition. These values are considered as an underneath estimation of the transmission limit of PF-GIPOFs as complex modulation formats, restricted mode launching schemes, equalization techniques or simultaneous data transmission over high-order latent PF-GIPOF passbands can be applied to enhance its aggregated capacity [88-90].

Previous works have studied, analyzed and modeled the PF-GIPOF frequency response taking into account most of the parameters that affect the latter. Some noteworthy PF-GIPOF frequency response measurements are shown in Fig. 19, in which a good agreement between experimental results and the theoretical curves predicted by the model is observed. This figure shows the measured and theoretical frequency responses for a 50m, 75m and 100m-long 62.5μm core diameter PF-GIPOF link with OFL condition and employing a Fabry-Perot laser source operating at 1300nm. Further details of the mathematical framework and experiments are reported on [91] for the benefit of the readers.

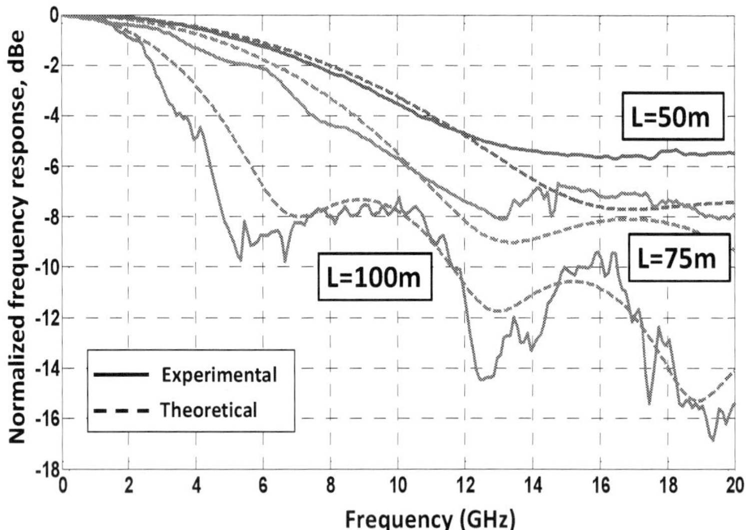

Figure 19. Measured (solid line) and theoretical (dashed line) electrical responses for different 62.5 μm core diameter PF-GIPOF lengths.

From frequency response measurements, as shown in the above figure, the 3dB baseband bandwidth can be easily identified and, therefore, the channel capacity can be calculated. The PF-GIPOFs used are commercially available from Chromis Fiber with an attenuation of 55dB/km at 1300nm. For the frequency response measurements, a FP laser diode used as transmitter was externally AM modulated with a RF sinusoidal signal (up to 20GHz of modulation bandwidth) by means of an E/O Mach-Zehnder modulator (16GHz band‐width). An InGaAs-photodetector (22GHz bandwidth) is used as receiver. Bandwidth limitation from both transmitter and receiver can be neglected for links >50m. The channel capacity for each length, is calculated based on the transmission characteristics listed below, and is displayed in Table 1. For some applications, e.g. home network Ethernet transceiv‐ers, eye safety operation is required and a limited averaged transmitted optical power of 0dBm has been considered.

Length (m)	Measured electrical -3dB bandwidth (GHz)	Capacity (Gbps)	
25m	18.4	590.5	a) Average transmitted optical power=0dBm
50m	9.2	299.6	b) Fiber attenuation @1300nm= 55dB/km
75m	6.7	208.8	
100m	3.2	101.9	c) Clipping factor=3
125m	1.7	54.4	
150m	1.2	36.9	d) Noise equivalent power: NEP=$17.3 \cdot 10^{-12}$ W/$\sqrt{\text{Hz}}$

Table 1. Calculated theoretical capacity over 62.5μm core diameter PF-GIPOF, at 1300nm.

However, this capacity analysis from the frequency response may result in large discrepancies at frequencies beyond the 3dB point and the PF-GIPOF has some latent high-order passbands [88]. Consequently, the Gaussian low-pass approximation reveals itself as a pessimistic approximation of the PF-GIPOF channel capacity, and expected capacity values of PF-GIPOF can be larger than those calculated in Table 1, even more if Restricted Mode Launching (RML) schemes are applied to the injection of light into the fiber.

On the other hand, DMT allows the possibility to allocate the number of bits and energy per subcarrier according to its corresponding signal-to-noise ratio (SNR), typically known as bit-loading. To compute rate-adaptive bit-loading for the DMT over PF-GIPOF consideration Chow's algorithm has been implemented [92]. Initially, all subchannels were loaded with 4 information bits each. Table 2 shows the theoretical results on capacity when applying DMT over a single channel, based on the measured frequency response values up to 150m-long PF-GIPOFs. Compared to the results given in Table 1, it can be seen that for the shortest length (25m), the numerically computed capacity value is lower than theoretical counterpart. This result from bandwidth limitation of the external modulator bandwidth, considered in the computation. From lengths > 50m, the computed capacity is larger because the PF-GIPOF frequency response dominates over other bandwidth limitation factors. Due to the bandwidth limitation of the PF-GIPOF link itself, the signal-to-noise-ratio decreases for higher frequencies. Bit allocation resulting from the bit loading is shown in Fig. 20 for a 100m- and 150m-long PF-GIPOF single channel link, respectively.

Length (m)	Numerical DMT over PF-GIPOF Capacity (Gbps)
25m	577.1
50m	380.0
75m	276.9
100m	164.1
150m	77.1

Note: targeted BER=10⁻³

Table 2. Theoretical DMT capacity over 62.5µm core diameter PF-GIPOF, at 1300nm.

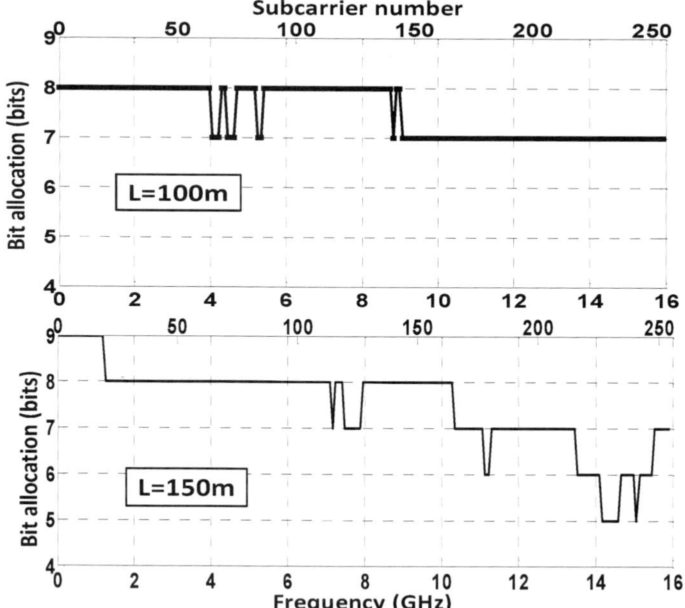

Figure 20. Bit allocation per subchannel, resulting from bit-loading at 100m and 150m.

It has been previously stated that for flexible high capacity GIPOF optical networks, applying the WDM approach seems to be necessary. Apart from the physical transmission characteristics of the PF-GIPOF, it is equally important to consider the optical components introduced to deploy advanced WDM-based optical architectures. The addition of these POF WDM multiplexers and demultiplexers limits the available optical power budget within the fiber link thus resulting in a bit-rate penalty. This is due to the fact that the OSNR (Optical Signal-to-Noise ratio) of the system is being reduced, and so the fiber transmission capacity.

To establish a comparison between the PF-GIPOF single channel operation and its WDM extension a 4-λ WDM approach has been considered. Regarding the latter, the PF-GIPOF transmission capacity must be recalculated from: a) the new bit loading resulting from the DMT modulation scheme and, b) the restriction on power margin resulting from the new losses considered in the system due to the addition of the mux/demux devices in the optical link. An insertion loss for a future PF-GIPOF 4-λ multiplexer/demultiplexer device of around 2dB per channel [93, 94] is considered. Such a performance is better in terms of insertion loss with respect to PMMA-GIPOF based splitters which can provide insertion losses greater than 6dB (in the symmetric-case) [95]. Consequently, the power budget of the WDM system, consisting of one PF-GIPOF based multiplexer or demultiplexer device at each side of the optical fiber link, results in a 5dB power reduction per channel, if an optical crosstalk of 1dB is also considered. It is worth mentioning that some authors have evaluated power penalties close to 2.4dB when combining a 62.5μm core diameter PF-GIPOF and WDM devices based on 50μm core diameter MMF [90]. Set of Fig. 21 shows the theoretical bit loading including the afore-mentioned restriction in power budget due to the 4- λ WDM approach.

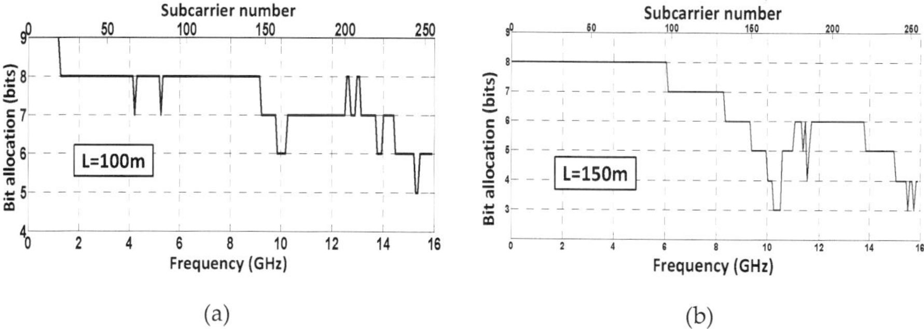

Figure 21. Theoretical bit loading for the 4- λ WDM approach over PF-GIPOF. (a) 100m ; (b) 150m.

The corresponding aggregated WDM capacity is summarized in Fig. 22 and compared to the single channel operation. The achievable capacity of a single-λ WDM system does not reach the best single channel results. For a single channel operation more than twice the capacity compared to the single-λ capacity in the WDM approach. Therefore, assuming a 4-λ WDM system using the full available optical power and with similar bit rate transmission perform-ances in each channel the total achievable capacity would overcome the OSNR and bit-rate limitation due to the optical losses introduced in the power budget of the system. It is also noticed that for longer PF-GIPOF lengths the ratio between transmission capacities for single channel and single- λ operation diminishes. This fact is attributed to differential mode attenuation (DMA) together with mode coupling effects in PF-GIPOF that leads to a sub-linear increase dependency of the fiber bandwidth regarding its length. This favours the resulting transmission capacity.

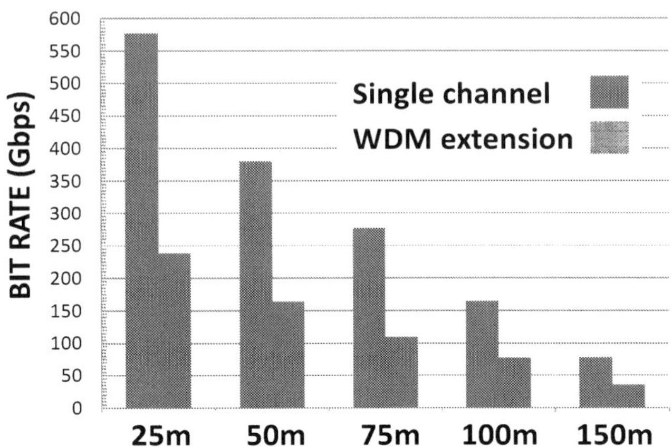

Figure 22. Comparison of single channel operation and WDM extension over 62.5μm core diameter PF-GIPOFs.

On the other hand, capacity values for a 50μm core diameter PF-GIPOF following the same procedure and under the same constraints are also shown (from its frequency response measurements). Greater capacities can be achieved as increasing the core diameter due to the presence of strong mode coupling effects and less modal noise effect, as shown in Fig. 23.

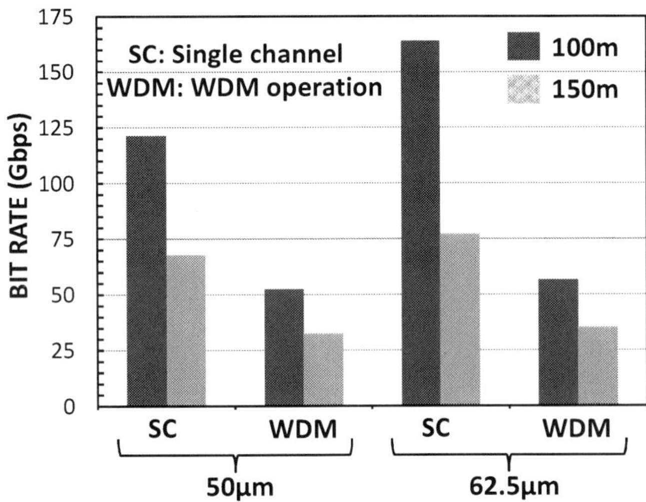

Figure 23. Comparison of single channel operation and WDM extension over a 100m- and 150m-long link, at 1300nm, for different core diameter PF-GIPOFs.

7. Discussion and conclusions

Applying WDM can further enhance the transmission capacity via POF systems. This chapter is intended to bridge the gap of WDM POF-based networks for in-home deployments, where POFs have become a competitive and low-cost solution as a physical medium infrastructure. In-home link lengths are relatively short thus leading to a more relaxed requirements regarding bandwidth x length product and attenuation per unit length, respectively. However, due to the continuous increase of bit-rate demands from end-users for multimedia services new techniques oriented to overcome the POF bandwidth limitation are being required. Beyond complex modulation formats in which the main goal is to provide a single channel communication link with a high spectral efficient (i.e. bit/Hz), one potential solution to expand the usable bandwidth of POF systems is to perform multiple channels over a single POF. This is known as the WDM approach.

Nowadays, WDM is well-established in the infrared transmission windows for silica optical fibers, but this technique needs to be adapted to VIS for POFs due to their spectral attenuation behavior. And novel WDM POF devices and network topologies are necessary to a final success of POF in-home penetration. These devices include POF multiplexers/demultiplexers, variable optical attenuators, interleavers, switches, POFBGs and/or optical filters to separate and to route the different transmitted wavelengths. And an easy-reconfigurable performance can be an additional feature among all the spectrum of future designed and manufactured devices for the WDM POF solution, with the aim of increasing the flexibility of multiplexing, demultiplexing, switching and routing optical signals as well as modifying the optical network if required. Moreover, devices that can perform different functionalities are interesting in terms of reducing the power consumption and insertion losses. It is also important to follow the progression in stable and low cost light sources in the visible range, apart from 650nm, to successfully achieve the WDM POF implementation. And new devices, benefiting from nanoparticles principles to reach novel plasmonic switches among others.

Anyway, progresses in these POF devices for the WDM approach have been discussed. They can operate in the visible range as POFs do. Among the different technologies, within this chapter two multiplexers based on TN-LC have been introduced. The use of this technology has been demonstrated to provide several advantages as they do not have mobile parts, need low excitation voltages and have a low power comsuption. In addition, a reconfigurable optical multiplexer based on PDLC has been presented which can also acts as a variable optical attenuator. It has also been designed switches operating in a broadband range with a uniform spectral response and low thermal dependence. Birefringent structures are proved to be a versatile solution for designing devices for POF WDM networks due to their reconfiguration capacity, since they can be easily manufactured with LC technology, and flexibility, since any FIR filters synthesis method can be used.

On the other hand, the capabilities of novel POFBG devices to be compatible with WDM topologies for both sensing and communication schemes have been addressed. High scalability and power budget enhancement in comparison with all POF based network solutions is achieved due to the use of off-the shelf silica WDM devices in combination with POFBGs. Reaching distances of tens of km can be easily achieved with the proposed topology fully compliant with both short-reach networks (typically less than 1 km), i.e. LANs, in-building/in-home networks etc. as well as medium-reach distances (typically up to 10km) covering the access domain. The above reaching distances are unbeatable if an all-POF-based WDM optical network is intended to be deployed and a hybrid approach should be considered.

Finally, to mitigate both the impact of the high attenuation as well as the limited bandwidth of standard POFs recently developed PF-GIPOFs should be also considered. This fiber type outperforms these two features compared to their Step-Index and PMMA-based counterparts, respectively. Nevertheless its achievable capacity under the WDM-GIPOF approach must be analyzed. A future 4-λ WDM-GIPOF deployment is studied showing that its total achievable capacity can overcome the OSNR and the bit-rate limitation due to the optical losses introduced in the power budget of the system due to the addition of future GIPOF-based WDM devices.

We believe the results reported in this chapter may encourage the development of WDM-POF networks and low insertion loss POF-based WDM devices opening up the path for future in-home systems at very high bit rates. However, further improvements on mux/demux manufacturing would make WDM-POF systems a future-proof and feasible solution for in-home/building networks to attend end-users' high-speed demands.

Acknowledgements

The work comprised in this document has been developed in the framework of the activities carried out at the Displays and Photonics Applications group (GDAF) at Universidad Carlos III de Madrid.

This work has been supported by the Spanish Ministry of Economía y Competitividad under the grant TEC2012-37983-C03-02.

Author details

David Sánchez Montero*, Isabel Pérez Garcilópez, Carmen Vázquez García,
Pedro Contreras Lallana, Alberto Tapetado Moraleda and Plinio Jesús Pinzón Castillo

*Address all correspondence to: dsmontero@ing.uc3m.es

Electronics Technology Department, Universidad Carlos III de Madrid, Leganés, Madrid, Spain

References

[1] Toma T., Takizuka T., Kanou M., Taniguchi T., Koike Y. Dual full high definition 3D video real-time communication system. In: International Conference on Plastic Optical Fibers, ICPOF 2011, Bilbao, Spain, 2011, 475-479.

[2] Fischer U. H. P., Haupt M., Joncic M. Optical Transmission Systems Using Polymeric Fibers. In: Optoelectronics-Devices and Applications. InTech; 2011, 445-468.

[3] Koonen A. J., Tangdiongga E. Photonic Home Area Networks. Journal of Lightwave Technology 2014; 32(4) 591-604.

[4] Bilro L., Alberto N., Pinto J. L., Nogueira R. Optical Sensors Based on Plastic Fibers. Sensors 2012; 12 12184 – 12207.

[5] Ziemann O., Zamzow P. E., Daum W. POF Handbook: Optical Short Range Transmission Systems. Berlin: Springer Berlin Heidelberg; 2008.

[6] Okonkwo E. T. C. M., Yang H., Visani D., Loquai S., Kruglov R., Charbonnier B., Ouzzif M., Greiss I., Ziemann O., Gaudino R., Koonen A. M. J. Recent Results from the EU POF–PLUS Project: Multi–Gigabit Transmission Over 1 mm Core Diameter Plastic Optical Fibers. Journal of Lightwave Technology 2011; 29(2) 186-193.

[7] Li W., Khoe G., Boom H. V.D., Yabre G., De Waardt H., Koike Y., Yamazaki S., Nakamura K., Kawaharada Y. 2.5 Gbit/s Transmission over 200 m PPMA Graded Index Polymer Optical Fiber Using a 645 nm Narrow Spectrum Laser and a Silicon APD. Microwave and Optical Technology Letters 1999; 20(3) 163-166.

[8] Van Den Boom H. P. A., Li W., Van Bennekom P. K., Monroy I. T., Khoe G. D. IEEE Journal on Selected Topics in Quantum Electronics 2001; 7(3) 461-470.

[9] Niheii E., Ishigurett T., Taniott N., Koike Y. Present Prospect of Graded-Index Plastic Optical Fiber in Telecommunication. IEICE Transactions on Electronics 1997; E80-C 117-121.

[10] Polley A., Ralph S. E. 100 m, 40 Gb/s Plastic optical fiber link. In: Optical Fiber communication/National Fiber Optic Engineers Conference, 2008, San Diego, 1-3.

[11] Vázquez C., Montero D. S. Multimode Graded-Index Optical Fibers for Next-Generation Broadband Access. In: Current Developments in Optical Fiber Technology. InTech; 2013.

[12] Jianjun Y., Dayou Q., Mingfang H., Zhensheng J., Chang G. K., Ting W. 16Gbit/ s radio OFDM signals over graded index plastic optical fiber. In: 34th European Conference on Optical Communication, ECOC 2008, 2008, 1-2.

[13] Zeng J., Van den Boom H. P. A., Koonen A. M. J. Five-subcarrier multiplexed 64-QAM transmission over a 50-μm core diameter graded index perfluorinated polymer

pptical fiber. In: Optical Fiber Communication Conference/National Fiber Optic Engineers Conference, San Diego, California, 2008, OWB4.

[14] Lee S. C. J., Breyer F., Randel S., Gaudino R., Bosco G., Bluschke A., Matthews M., Rietzsch P., Steglich R., Van den Boom H. P. A., Koonen A. J. Discrete Multitone Modulation for Maximizing Transmission Rate in Step-Index Plastic Optical Fibers. Journal of Lightwave Technology 2009; 27(11) 1503-1513.

[15] Hejie Y., Lee S. C. J., Tangdiongga E., Okonkwo C., Van den Boom H. P. A., Breyer F., Randel S., Koonen A. J. 47.4 Gb/s Transmission Over 100 m Graded-Index Plastic Optical Fiber Based on Rate-Adaptive Discrete Multitone Modulation. Journal of Lightwave Technology 2010; 28(4) 352-359.

[16] Randel S., Lee S. C. J., Spinnler B., Breyer F., Rohde H., Walewski J., Koonen A. M. J., Kirstädter A. 1 Gbit/s transmission with 6.3 bit/s/Hz spectral efficiency in a 100m standard 1 mm step-index plastic optical fibre link using adaptive multiple sub-carrier modulation. In: Proceedings of the 32nd European Conference on Optical Communication, ECOC 2006, Cannes, France, 2006, Th4.4.1-1/2.

[17] Cardenas Lopez D. F., Nespola A., Camatel S., Abrate S., Gaudino R. 100 Mb/s Ethernet Transmission Over 275 m of Large Core Step Index Polymer Optical Fiber: Results From the POF-ALL European Project. Journal of Lightwave Technology 2009; 27(14) 2908-2915.

[18] Okonkwo C. M., Tangdiongga E., Yang H., Visani D., Loquai S., Kruglov R., Charbonnier B., Ouzzif M., Greiss I., Ziemann O., Gaudino R., Koonen A. J. Recent Results from the EU POF-PLUS Project: Multi-Gigabit Transmission Over 1 mm Core Diameter Plastic Optical Fibers. Journal of Lightwave Technology 2011; 29(2) 186-193.

[19] Ciordia O. Esteban C., Pardo C., Pérez de Aranda R. Commercial silicon for gigabit communication over SI-POF. In: International Conference on Plastic Optical Fibers, ICPOF 2013, Búzios, Brasil, 2013, 109-116.

[20] Wei J. L., Geng L., Cunningham D. G., Penty R. V., White I. H. Gigabit NRZ, CAP and Optical OFDM Systems Over POF Links Using LEDs. Optics Express 2012; 20(20) 22284-22289.

[21] Ziemann O., Bartiv L. POF-WDM, the truth. In: International Conference on Plastic Optical Fibers, ICPOF 2011, Bilbao, Spain, 2011, 525-530.

[22] Jončić M., Haupt M., Fischer U. H. P. Four-channel CWDM system design for multi-Gbit/s data communication via SI-POF. In: Proceedings of SPIE 9007: Broadband Acces Communication Technologies VIII, San Francisco, United States, 2013, 90070J.

[23] Jončić M., Kruglov R., Haupt M., Caspary R., Vinogradov J., Fischer U. H. P. Four-Channel WDM Transmission Over 50 m SI-POF at 14.77 Gb/s Using DMT Modulation. IEEE Photonics Technology Letter 2014; 26(13) 1328-1331.

[24] Musa A. B. S., Kok A. A. M., Diemee M. B. J., Driessen A. Multimode arrayed wave-guide grating-based demultiplexers for short-distance communication. In: IEEE Eurocon, 2003, 422-426.

[25] Ehsan A. A. Shaari S., Rahman M. K. A. Plastic Optical Fiber Coupler with High Index Contrast Waveguide Taper. Progress in Electromagnetics Research C, 2011; 20 125-138.

[26] Jončić M., Haupt M., Fischer U. H. P. Four-channel CWDM system design for multi-Gbit/s data communication via SI-POF. In: Proceedings of SPIE 9007: Broadband Acces Communication Technologies VIII, San Francisco, United States, 2013, 90070J-90070J-9.

[27] Lutz D., Haupt M., Fischer U. H. P. Demultiplexer for WDM over POF in prism-spectrometer configuration. In: Photonics and Microsystems, 2008 International Students and Young Scientists Workshop, 20-22 June 2008, Worclaw, Poland, 43-46.

[28] Jončić M., Haupt M., Fischer U. H. P. Investigation on spectral grids for VIS WDM applications over SI POF.In: Proceedings of the Photonische Netze, ITG-FB 241, 6-7 May 2013, Leipzig, Germany, 1-6.

[29] L. V. Bartkiv, Bobitski Y. V., Poisel H. Optical Demultiplexer Using a Holographic Concave Grating for POF-WDM Systems. Optica Applicata 2005; 35(1) 59-66.

[30] Haupt M., Fischer U. H. P. Multi-colored WDM over POF system for Triple-Play. In: Proceedings of SPIE 6992: Micro-Optis 2008, 699213-699213-10.

[31] Chandrasekhar S. Liquid Crystals. Cambridge University Press; 1992.

[32] Riza N. A., Yuan S. Low Optical Interchannel Crosstalk, Fast Switching Speed, Polarization Independent 2x2 Fiber Optic Switch Using Ferroelectric Liquid Crystals. Electronis Letters 1999; 34(17) 1341-1342.

[33] Pain F., Coquillé R., Vinouze B., Wolffer N., Gravey P. Comparison of Twisted and Parallel Nematic Liquid Crystal Polarisation Controllers. Application to a 4 × 4 Free Space Optical Switch at 1.5 μm. Optics Communications 1997; 139(4-6) 199-204.

[34] Vázquez C., Pena J. M. S., Aranda A. L. Broadband 1 x 2 Polymer Optical Fiber Switches Using Nematic Liquid Crystals. Optics Communications 2003; 224(1-3) 57-62.

[35] Sumriddetchkajorn S., Riza N. A., Sengupta D. K. Liquid Crystal-Based Self-Aligning 2×2 Wavelength Routing Module. Optical Engineering 2001; 40(8) 1521-1528.

[36] Pena J. M. S., Rodríguez I., Vázquez C., Pérez I., Otón J. M. Spatial Distribution of the Electric Field in Liquid Crystal Dispersions Devices by using a Finite Element Method. Journal of Molecular Liquids 2003; 108(1-3) 107-117.

[37] Pena J. M. S., Vázquez C., Pérez I., Rodríguez I., Otón J. M. Electro-optic System for Online Light Transmission Control of Polymer-Dispersed Liquid Crystal Windows. Optical Engineering 2002; 41(7) 1608-1611.

[38] Zubia J., Durana G., Arrue J., Garces I. Design and Performance of Active Coupler for Plastic Optical Fibres. Electronics Letters 2002; 38(2) 65-67.

[39] Chanclou P., Vinouze B., Roy M., Cornu C. Optical Fibered Variable Attenuator Using Phase Shifting Polymer Dispersed Liquid Crystal. Optics Communications 2005; 248(1-3) 167-172.

[40] Lallana P. C., Vázquez C., Vinouze B., Heggarty K., Montero D. S. Multiplexer and Variable Optical Attenuator Based on PDLC for Polymer Optical Fiber Networks. Molecular Crystals and Liquid Crystals 2008; 502(1) 130-142.

[41] Chen R. H. Liquid Crystal Displays: Fundamental Physics and Technology. Hoboken: Wiley; 2011.

[42] Lallana P. C., Vázquez C., Pena J. M. S., Vergaz R. Reconfigurable Optical Multiplexer Based on Liquid Crystals for Polymer Optical Fiber Networks. Opto-Electronics Review 2006; 14(4) 311-318.

[43] Lallana P. C., Vázquez C., Vinouze B. Advanced multifunctional optical switch for multimode optical fiber networks," Optics Communications 2012; 285(12) 2802-2808.

[44] Jončić M., Haupt M., Fischer U. H. P. Standardization proposal for spectral grid for VIS WDM applications over SI-POF. In: Proceedings of POF Congress, Atlanta, 2012, 351-355.

[45] Ziemann O., Krauser J., Zamzow P. E., Daum W., POF Handbook: Optical Short Range Transmission Systems. Springer; 2008.

[46] Chua S. J., Li B. Optical Switches: Materials and Design. Woodhead Publishing in Materials; 2010.

[47] De Dobbelaere P., Falta K., Gloeckner S., Patra S. Digital MEMS for Optical Switching. IEEE Communications Magazine 2002; 40 (3) 88-95.

[48] Ramaswami. R., Sivarajan K. N., Sasaki G.H. Optical Networks: a practical perspective. Morgan Kaufmann; 2002.

[49] Sapriel J., Charissoux D., Voloshinov V., Molchanov V. Tunable Acoustooptic Filters and Equalizers for WDM Applications. Journal of Lightwave Technology 2002; 20 (5) 892-899.

[50] D'Alessandro A., Smith D. A., Baran J. E. Polarisation-Independent Low-Power Integrated acousto-optic tunable filter/switch using APE/Ti polarisation splitters on lithium niobate. Electronics Letters 1993; 29(20) 1767-1769.

[51] Sakuma K., Ogawa H., Fujita D., Hosoya H. Polymer Y-branching thermo-optic switch for optical fiber communication system. In: 8th Microoptics Conference, MOF'01, 2001, Osaka, Japan, 24-26.

[52] Leuthold J., Joyner C. H. Multimode Interference Couplers with Tunable Power Splitting Ratios. Journal of Lightwave Technology 2001; 19(5) 700.

[53] Almeida V. R., Barrios C. A., Panepucci R. R., Lipson M. All-Optical Control of Light on a Silicon Chip," Nature 2004; 431 1081-1084.

[54] Xiaohua M., Kuo G. S.Optical Switching Technology Comparison: Optical MEMS vs. Other Technologies. IEEE Communications Magazine 2003; 41(11) S16-S23.

[55] Krahenbuhl R., Howerton M. M., Dubinger J., Greenblatt A. S. Performance and Modeling of Advanced Ti:LiNbO$_3$ Digital Optical Switches. Journal of Lightwave Technology 2002; 20(1) 92-99.

[56] Nashimoto K., Moriyama H., Nakamura S., Watanabe M., Morikawa T., Osakabe E., Haga K. PLZT electro-optic waveguides and switches. In: Optical Fiber Communication Conference and Exhibit, 2001. OFC 2001, 2001, PD10-PD10.

[57] Agranat A. J. Electroholographic wavelength selective crossconnect. In: Nanostructures and Quantum Dots/WDM Components/VCSELs and Microcavaties/RF Photonics for CATV and HFC Systems, 1999 Digest of the LEOS Summer Topical Meetings, 1999, II61-II62.

[58] Domash L. H., Yong-Ming C., Haugsjaa P., Oren M. Electronically switchable waveguide Bragg gratings for WDM routing. In Vertical-Cavity Lasers, Technologies for a Global Information Infrastructure, WDM Components Technology, Advanced Semiconductor Lasers and Applications, Gallium Nitride Materials, Processing, and Devi, 1997, 34-35.

[59] Wagner R. E., Cheng J. Electrically Controlled Optical Switch for Multimode Fiber applications. Applied Optics 1980; 19(17) 2921-2925.

[60] Vázquez C., Pena J. M. S., Aranda A. L. Broadband 1x2 Polymer Optical Fiber Switches Using Nematic Liquid Crystals. Optics Communications 2003; 224(1-3) 57-62.

[61] Vázquez C. Pena J. M. P., Contreras P., Pontes M. A. J. Development of a 2x2 optical switch for plastic optical fiber using liquid crystal cells. In: Proceedings of SPIE 5840: Photonics Materials, Devices and Applications. 2005, 325-335.

[62] Pinzón P. J., Pérez I., Vázquez C., Pena J. M. S. 1 × 2 Optical Router With Control of Output Power Level Using Twisted Nematic Liquid Crystal Cells," Molecular Crystals and Liquid Crystals 2012; 553 (1) 36-43.

[63] Macdonald R., Chen L. P., Shi C. X., Faer B. Requirements of optical layer network restoration. In: Optical Fiber Communication Conference, 2000, Vol. 3 68-70.

[64] Song D. H., Kim J.-W., Kim K.-H., Rho S. J., Lee H., Kim H., Yoon T.-H. Ultrafast switching of randomly-aligned nematic liquid crystals," Optics Express, vol. 20, pp. 11659-11664, 2012/05/21 2012.

[65] Ha Y.-S., Kim H.-J., Park H.-G., Seo D.-S. Enhancement of Electro-Optic Properties in Liquid Crystal Devices Via Titanium Nanoparticle Doping. Optics Express 2012; 20 (6) 6448-6455.

[66] Borshch V., Shiyanovskii S. V., Lavrentovich O. D. Nanosecond response in nematic liquid crystals for ultrafast electro-optic devices. In: CIOMP-OSA Summer Session on Optical Engineering, Design and Manufacturing, Changchun, 2013, Tu8.

[67] Amosova L. P., Vasil'ev V. N., Ivanova N. L., Konshina E. A. Ways of Increasing the Response Rate of Electrically Controlled Optical Devices Based on Nematic Liquid Crystals. Journal of Optical Technology 2010; 77(2) 79-87.

[68] Yeh P., Gu C. Optics of Liquid Crystal Display. New York: John Wiley&Sons; 2010.

[69] Pinzón P. J., Pérez I., Vázquez C., Sánchez-Pena J. M. Broadband 1×2 Liquid Crystal Router with Low Thermal Dependence for Polymer Optical Fiber Networks. Optics Communications 2014; 333 281-287.

[70] Vázquez C., Pérez I., Contreras P., Vinouze B., Fracasso B. Liquid Crystal Optical Switches. In: Li B., Chua S. J. (ed), Ed. Optical switches: Materials and design. Cambridge: Woodhead Publishing Limited; 2010.

[71] Marom D. M., Neilson D. T., Greywall D. S., Chien-Shing P., Basavanhally N. R., Aksyuk V. A., Lopez D. O., Pardo F., Simon M. E., Low Y., Kolodner P., Bolle C. A. Wavelength-Selective 1 x K Switches Using Free-Space Optics and MEMS Micromirrors: Theory, Design, and Implementation. Journal of Lightwave Technology 2005; 23(4) 1620-1630.

[72] Suzuki K., Mizuno T., Oguma M., Shibata T., Takahashi H., Hibino Y., Himeno A. Low Loss Fully Reconfigurable Wavelength-Selective Optical 1 x N Switch Based on Transversal Filter Configuration Using Silica-Based Planar Lightwave Circuit. IEEE Photonics Technology Letters 2004; 16(6) 1480-1482.

[73] Vázquez C., Contreras P., Montalvo J., Sánchez Pena J. M., d'Alessandro A., Donisi D. Switches and tunable filters based on ring resonators and liquid crystals. In: Proceedings of SPIE 6593: Photonics Materials, Devices and Applications 2007, 65931F-65931F-10.

[74] Pinzón P. J., Pérez I., Vázquez C., Sánchez Pena J. M. Reconfigurable 1 x 2 Wavelength Selective Switch Using High Birefringence Nematic Liquid Crystals. Applied Optics 2012; 51(25) 5960-5965.

[75] Pinzon P. J., Vazquez C., Perez I., Sanchez Pena J. M. Synthesis of Asymmetric Flat-Top Birefringent Interleaver Based on Digital Filter Design and Genetic Algorithm. Journal of Photonics 2013; 5(1) 7100113-7100113.

[76] Michael S. The Sun: an Introduction. Springer-verlag Co.; 2004.

[77] Montalvo J., Frazão O., Santos J., Vázquez C., Baptista J. Radio-Frequency Self-Refer‐
encing Technique With Enhanced Sensitivity for Coarse WDM Fiber Optic Intensity
Sensors. Journal of Lightwave Technology 2009; 27(5) 475-482.

[78] Montero D. S., Vázquez C., Baptista J. M., Santos J. L., Montalvo J. Coarse WDM Net‐
working of Self-Referenced Fiber-Optic Intensity Sensors with Reconfigurable Char‐
acteristics. Optics Express 2010; 18(5) 4396-4410.

[79] Othonos A., Kalli K. Fiber Bragg gratings: Fundamentals and Applications in Tele‐
communications and Sensing. Boston&London: Artech House; 1999.

[80] Xiong Z., Peng G. D., Wu B., Chu P. L. Highly Tunable Bragg Gratings in Single-
Mode Polymer Optical Fibers. Photonics Technology Letters 1999; 11(3) 352-354.

[81] Dobb H., Webb D. J., Kalli K., Argyros A., Large M. C., van Eijkelenborg M. A. Con‐
tinuous Wave Ultraviolet Light-Induced Fiber Bragg Gratings in Few- and Single-
Mode Microstructured Polymer Optical Fibers. Optics Letters 2005; 30(24) 3296-3298.

[82] Roriz P., Ramos A., Santos J., Simões J. Fiber Optic Intensity-Modulated Sensors: a
Review in Biomechanics. Photonic Sensors 2012; 2(4) 315-33.

[83] Mishra V., Singh N., Tiwari U., Kapur P. Fiber Grating Sensors in Medicine: Current
and Emerging Applications. Sensors and Actuators A: Physical 2011; 167(2) 279-290.

[84] Berghmans F., Geernaert T., Sulejmani S., Thienpont H., Van Steenberge G., Van Hoe
B., Dubruel P., Urbanczyk W., Mergo P., Webb D. J., Kalli K., Van Roosbroeck J., Sug‐
den K. Photonic crystal fiber Bragg grating based sensors: opportunities for applica‐
tions in healthcare. In: Popp J., Matthews D., Tian J., Yang C. (eds) Optical Sensors
and Biophotonics, of Proceedings of SPIE 8311: Asia Communications and Photonics
Conference and Exhibition, 2011, 831102-831102-10.

[85] Montalvo J., Frazao O., Santos J. L., Vazquez C., Baptista J. M. Radio-Frequency Self-
Referencing Technique With Enhanced Sensitivity for Coarse WDM Fiber Optic In‐
tensity Sensors. Journal of Lightwave Technology 2009; 27(5) 475-482.

[86] Montero D. S., Tapetado A., Webb D.J., Vázquez C. Self-Referenced Optical Intensity
Sensor Network Using POFBGs for Biomedical Applications. Sensors 2014; p. in
press.

[87] Vázquez C., Montero D.S., J. Zubia. Short-range transmission capacity analisys over
PF GIPOF. In: International Conference on Plastic Optical Fibers (ICPOF), Atlanta,
USA, 2012.

[88] Montero D. S., Vázquez C. Analysis of the Electric Field Propagation Method: Theo‐
retical Model Applied to Perfluorinated Graded-Index Polymer Optical Fiber Links.
Optics Letters 2011; 36(20) 4116-4118.

[89] Lethien C., Loyez C., Vilcot J.-P., Rolland N., Rolland P. A. Potential of the Polymer Optical Fibers Deployed in a 10Gbps Small Office/Home Office Network. Optics Express 2008; 16(15) 11266-11274.

[90] Lethien C., Loyez C., Vilcot J. P., Rolland N., Rolland P. Exploit the Bandwidth Capacities of the Perfluorinated Graded Index Polymer Optical Fiber for Multi-Services Distribution. Polymers 2011; 3(3) 1006-1028.

[91] Montero D. S., Vázquez C. Multimode Graded-Index Optical Fibers for Next-Generation Broadband Access. In: Harum S. W., Arof H. (ed) Current Developments in Optical Fiber Technology. InTech; 2013.

[92] Chow P. S., Cioffi J. M., Bingham J. A. C. A Practical Discrete Multitone Transceiver Loading Algorithm for Data Transmission over Spectrally Shaped Channels. IEEE Transactions on Communications 1995; 43 773-775.

[93] Bartkiv L., Poisel H., Bobitski Y. Wavelength Demultiplexer with Concave Grating for GI-POF systems. Optica Applicata 2005; 35.

[94] Van den Boom H. P. A., Li W., Khoe G. D. CWDM technology for polymer optical fiber networks. In: Proceedings of the 5th annual symposium of the IEEE/LEOS Benelux Chapter Netherlands, 2000, 13-16.

[95] Shi Y., Okonkwo C., Visani D., Tangdiongga E., Koonen T. Distribution of Broadband Services Over 1-mm Core Diameter Plastic Optical Fiber for Point-to-Multipoint In-Home Networks. Journal of Lightwave Technology 2013; 31(6) 874-881.

Chapter 13

Optimization of WDM-POF Network for In-Car Entertainment System

Mohammad Syuhaimi Ab-Rahman , Hadi Guna and
Norhana Arsad

Additional information is available at the end of the chapter

http://dx.doi.org/10.5772/59238

1. Introduction

Recent advances in In-Car Entertainment System will play an important role on automotive industry. It was assumed that in the year 2015, every new car, especially built in Europe will be equipped with Internet connection. As cars become connected to the Internet, the demand for Internet-based entertainment and applications and services increases [1].

Polymer optical fibers (POFs) are in a great demand for the data transmission and processing of optical communications compatible with the Internet, which is one of the fastest growing industries in automotive field. POFs become replacement for copper cable technology for future IVI system.

In this chapter reports experimental demonstration of a POFs based solutions in wavelength division multiplexing (WDM) network and some effects due to the placement of color filters as a demultiplexer for the In-Car Entertainment System. A *weakly fused* (WF) and a *highly fused* (HF) POF star coupler based on fusing and combining a group of POFs are designed and experimentally investigated. A low-cost demultiplexer is realized by using color filters. The specialized designed plastic-based interference films are used to filter out any other wavelength (color) that is not within the range. Variation of temperature were applied directly to the fused tapered region and an LED fiber source is launched into the fiber input, while the optical power deviation is measured at each output port. The effect of the return optical power and the coupling ratio to the temperature variation of 20 °C to 125 °C were investigated. The excess loss, EL_0 for the WF and the HF coupler at 20 °C is ~ 2 dB and ~ 8 dB, respectively.

Media Oriented Systems Transport (MOST) is one of the advance IVI system provider company which also utilized a POF-LED technology to transmit numerous signals represent a different data transmission via time division multiplexing (TDM) network (refer Fig. 1).

Figure 1. Recent technology on ring topology IVI system with single line of optical fiber cable facing main problem with no backup line when failure occurred.

In our research, we offered a wavelength division multiplexing (WDM) communication network over POF due to the rapid increase of traffic demands [2, 3]. WDM is the network that allows the transmission of multimedia data in IVI system over multiple wavelength (color) and thus greatly increases the POF's bandwidth. Beside, this network proposes a backup path in order to mitigate a serious breakdown in TDM-based network in IVI system.

Refer to Fig. 2, the proposed WDM-POF system, three unit of transmitters with different color of LED will carry single information simultaneously. For example in IVI network, red LED with 650nm wavelength modulated with video signal while blue ($\lambda 1$), green ($\lambda 2$), and yellow ($\lambda 3$) lights carry ethernet, audio and RF signal, respectively. The light has to be combined by the multiplexer (MUX) at the sending side. And to separate the wavelength channels at the receiver side, a wavelength demultiplexer (DEMUX) has to be used.

Demultiplexing, perform the reverse process with the same WDM techniques, in which the data stream with multiple wavelengths decomposed into multiple single wavelength data streams. POF coupler has similar function, operates to combine a number of optical data pulses

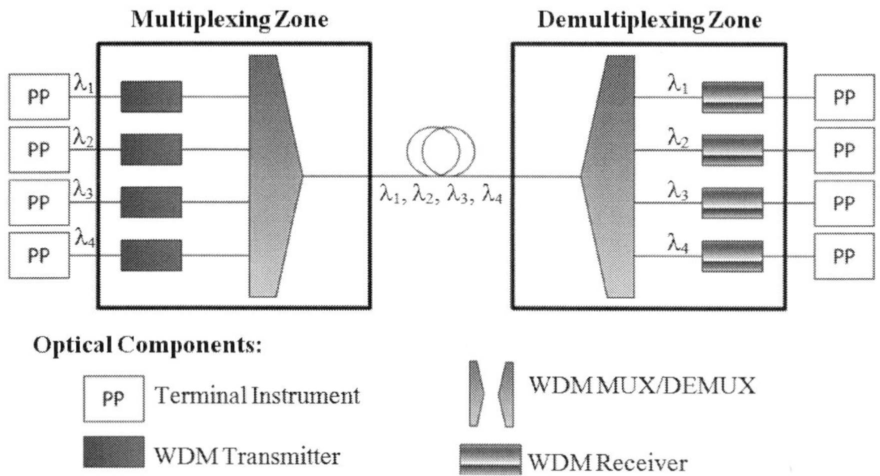

Figure 2. Simple schematic of a 4-channels WDM system.

as a single coupled signal. Hence, the development of MUX based on POF coupler is possible. A low-cost solution for POF-WDM system application will be presented.

A novel fused POF couplers has been fabricated by a fusion technique, as an effective transmission media to split and couple numerous different wavelengths which represents different signals. These novel coupler, however, suffer from several disadvantages. The high cost of the available couplers was raised as a challenge to the development of WDM systems in short-haul networks [5-8]. In addition, from the stand point of device design, the diameter of the fused tapered region, where stress is concentrated, is too small in conventional couplers. The structure causes a high incidence of fiber fracture, which results in poor reliability [9-12].

Thus a cost-effective 3×3 POF couplers based on a fused tapered structure to address these drawbacks of conventional couplers is demonstrated. The coupler is fabricated by a new and simple fabrication method, using a Bunsen burner and a metal tube. In this study, two types of fused couplers method focused of weakly fused (WF) and highly fused (HF). The WF coupler is, however, not considered to be a low-loss device, as the excess loss of the coupler was high, 12 to 22 dB.

The HF coupler is then developed to be the successor of the WF coupler. The excess loss of the HF coupler is very low, 0.3 to 5 dB. The device is developed as an optical switch which optical power can be switched completely from one fiber to another fiber at a temperature increase of T=55°C [13]. The switching characteristic can be achieved by varying the refractive index of the cladding at the coupling region of the coupler by temperature.

One of the aim of this chapter is to examine and optimize the feasibility of both methods of 3×3 POF couplers as thermal optical switches to be integrate in WDM-POF-based network for IVI system. The investigation is also to determine whether the thermal treatment is required to improve the quality and shift the device specification. Hence, a study of thermal effect on both polymer-based WF and HF couplers by varying the temperature of a hot plate from 20 °C to 125 °C is studied. The fused tapered fiber in the coupling region is exposed to the hot plate surface and optical power is launched into the input fiber of the coupler. In this temperature-dependence experiment, we investigate a relationship between temperature and several parameters such as coupling ratio, insertion loss and excess loss of the couplers.

In this chapter, red LED (650nm) has been utilized to transmit Ethernet data while green LED (520nm) can transmit a video image generated from CCTV network or DVD player, and blue LED with 470nm wavelength represents an audio transmission system for home networking. Refer to Fig. 3, special polymer color filters has been located between the coupler and receiver-end to ensure the entire WDM system can select a single signal as desired [4].

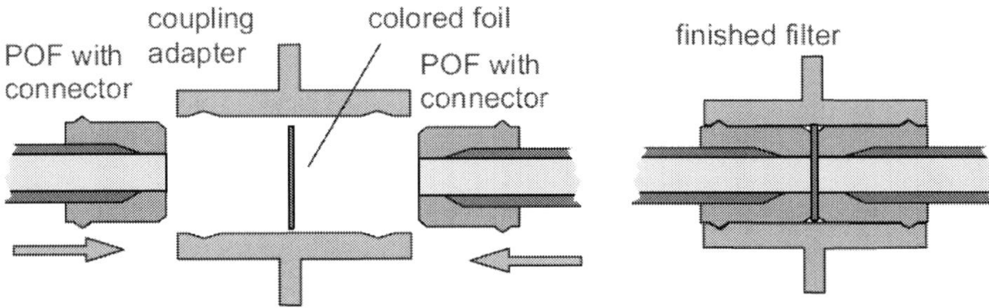

Figure 3. Each of those color filter lets pass exactly one wavelength and reflects all the others. By using several of those filters, the different wavelengths can be sorted out in a very low-cost way.

The performance of the novel coupler either with or without attachment of color filter can be evaluated in terms of insertion loss (IL). Some experiments in order the optimized the performance of WDM-POF based system for IVI system need to be conduct to minimized the value of the insertion loss (IL) in the network. The insertion loss (IL) is the amount of power loss that arises in the fiber optic line from input to the output of the fabricated coupler, expressed below,

$$IL_{port}(dB) = 10\log \frac{P_{o\,port}}{P_i}$$

Some wavelengths interfere with their reflected parts constructively, whereas others interfere destructively. Those wavelengths that interfere constructively can pass the filter, whereas the others get reflected. Besides the material parameters, the incident angle plays a major role, as each layer gets relatively thicker when tilting the filter. Color filters are manufactured for a long time and therefore high quality filters are readily available. They usually have sizes of several square millimeters. Color filters therefore are a valid choice to build DEMUX.

2. WDM-POF integrated network for in-car entertainment system

Adapting the fused tapering technique for conventional multimode fiber, we successfully established fabrication process for 1×3 POF twisted and fused couplers to be used as a MUX and DEMUX in IVI system. The 1×3 low cost coupler is an optical device, which ended by 3 number of POF output ports, while the other side ended by one POF port.

Similar to common coupler, it is also possible to work bidirectional, whereby it works from the 3 ports into 1 port (for coupling signal purpose), or vice versa (for splitting signals purpose). Optical 3×3 coupler has been symmetrically cut into two part to generate a pair of 1×3 couplers by the jointing of three polymethylmethacrylate (PMMA) POF [14]. Other specification for the design, the input POF is designed and fabricated to be twisted and fused shape as the fabrication process and 1×3 POF coupler is illustrated in Fig. 4.

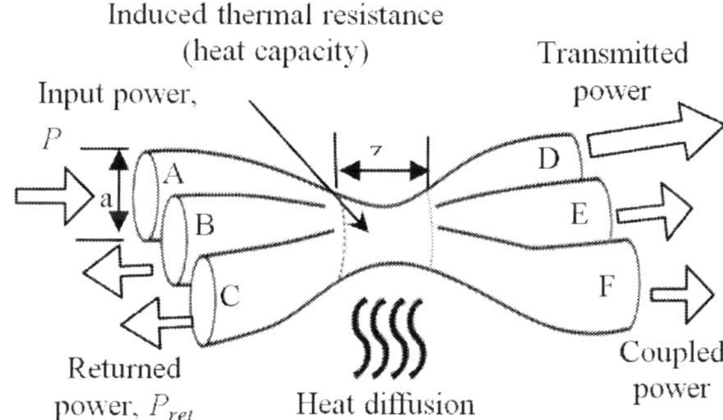

Figure 4. Thermal resistance at the coupling region of POF coupler in which all fibers are fused together.

Standard multimode SI-POF is used with its core diameter of 980 µm and cladding thickness of 10 µm. To obtain the results, DEMUX is realized using a special color filter attached using epoxy resin to the end of the connectors. The components are chosen because they are low cost and are easily found in the market.

Almost similar to POF material itself, the color filters are comprised of two types of plastic. More than 65% of the line is made from co-extruded polycarbonate plastic. The remainder of the line is deep dyed polyester [15, 16]. Filters create color by subtracting certain wavelengths of color. Thus, a red filter absorbs blue and green, allowing only the red wavelengths to pass. The process is subtractive not additive, so the light source must emit a full spectrum.

The swatch book provides detailed information on the spectral energy curve of each filter. The curve describes the wavelengths of color transmitted through each filter. For example, Supergel 342 transmits approximately 40% of the violet and blue energy of the spectrum and 75% of the orange and red energy. It absorbs all energy in the yellow and green range [15, 16].

After putting the resin onto the filter to be attached to the socket, the component then is hold together tightly for about two minutes to assure that no gap or air bubbles all over and also to assure the strong bond. This part has to be done gently since the epoxy resin has to be avoided covering the fiber's surface as much as possible so that any power losses can be minimized when the measurement is taken. However, since the edge of the socket is quite thin and sharp, the spread of the epoxy resin to the fiber surface cannot be 100% avoided.

After the fabrication is done, readings and measurements are taken for insertion loss for each of the fiber using a power meter. In this experiment, a lot of samples were fabricated to get the optimal results and to see which of the color filters that shows the most transmission and gives least losses. The length of the POFs is fixed at 3 meters long.

In this study, for fused plastic optical fiber, the optical loss is categorized as extrinsic loss due to the physical change of POF, LED projection to POF and the core-to-core connection and [17, 18]. It is obtained that the physical change of POF caused by fabrication process, where by diameter of POFs increasingly decrease to approach 1 mm and the POFs finally has twisted and fused shape. In characterization process, optical loss may present through the direct LED projection to POF surface. Besides, optical loss may also present through the connection between the fused tapered POF and POF cable [17].

3. Results and discussion

Comparison for the optical line either using the color filter or not, has been analyzed. The insertion loss of as much as 21 samples for the output terminal over POF has been utilized in 3-channels IVI system through red filter (internet data) have been visualized in Fig. 5. From the results gathered, it is seen that when all the components are set up and red LED (650nm) was injected, the insertion loss measured by the power meter shows small increase of losses when the film is attached to the socket.

This is also true when blue and green LEDs are injected. For the characterization of same film using different sources, we take a red film A (filter labeled #4690) as the primary filter and it is injected with all three LEDs, red, green and blue transmitters. As the results depict, the red filter injected with red transmitter shows a small increase of losses which averagely $IL<\sim 3$ dB compared to the initial loss before the film is attached to the fiber. Same goes to the power output when the film is attached to the fiber.

The effect of plastic-based filter attachment at the receiver-end did not indicate a significant deviation in power efficiency since both PMMA and Co-extruded Polycarbonate for fiber core and plastic filter's material have the same refractive index which approaching 1.59 [16].

Study on the saturation level of each color filters – red, green and blue – has been carried out. As much as eleven best samples chosen with different level of color saturation labeled from 1 to 11 which sample number 1 indicate the darkest (more saturated) color filter while number 11 with the lightest color (less saturated). Each of different color filter has been injected by LED in a range of blue to red light (470~650nm) and the result can be obtained in Fig. 6.

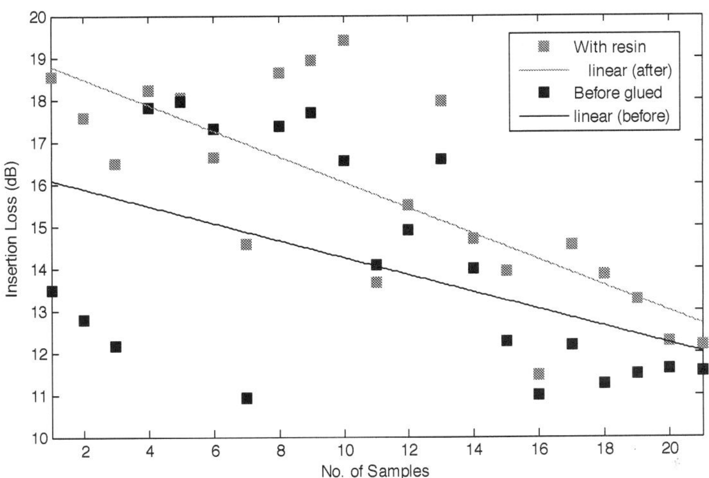

Figure 5. Effect on resin for demultiplexer filters approximately 1 dB insertion loss occurred on the measurement between before and after connector glued by resin.

From the result in Fig. 6(a), it is observed that sample 1 to 7 shows highest losses and decrease of efficiency. Since sample 1 to 7 being among the darkest film color meaning that only small or narrow transmission percentage of red LED or transmitter is allowed to get through. This phenomena also found in Fig 6(b), for the first six sample has quite high losses especially when it was injected by LED in blue to green range (470-580nm). However, red filter in Fig 6(c) did not indicate a significant deviation of loss by varying the saturation of the filter.

Compare with others, red filters block most of the green and blue LED transmission since it is clear that only red wavelength (λ=600~650nm) will be allowed to get through the film. The film filters out any other wavelength transmission that is not within the range. This concept is used as the primary idea for designing the demultiplexer.

It has been proven that, when blue filter was injected by all three optical source (blue, green and red color) all the data were more fluctuated right after the light hit the filter through the fiber. Same goes to green filter, all light sources will fluctuated the efficiency of the data transmission. This fluctuation generated as an effect from the SED percentage of each filters and also came from the intensity of each light source. The higher intensity of the light source transmitted the more oscillated graph plotted and the less significant of SED percentage deviation the less effect on fluctuation.

For the temperature-dependence experiment, two samples of 3×3 fused couplers with multimode SI PMMA POF are tested: WF and HF couplers. For the WF coupler, the fabrication method includes three processes: fiber bundle configuration, fabrication of spiral fiber and fiber tapering. Firstly, a fiber bundle consists of N core unjacketed polymer fibers (where N=3) is inserted into the metal tube and placed in symmetrical coordination. The fibers bundle is exposed to the heat of the flame repeatedly twisted while pulled continuously to produce a spiral fiber with a particular number of twists in the centre of the fiber bundle. The length of the fused region is limited to 8 cm. While the indirect heating process continues, fiber bundle is pulled and twisted. When the POFs reach the melting point, the fiber bundle is pulled to fuse the fibers, and the fused tapered fiber is formed. Each fiber ports (output and input) are insulated with PVC sleeve.

The HF coupler fabrication method includes four processes, as a new process was introduced to enhance the coupling characteristic of the fused tapered fiber by removing twisting effects on the fused fiber. A twisting effect implies that each fiber is not melted sufficiently to combine with each other. Firstly, a fiber bundle consisting of N core unjacketed polymer fibers (where N=3) are inserted into the metal tube and placed in symmetrical coordination. The fibers bundle are pulled and exposed to the heat of the flame repeatedly and continuously to produce a spiral fiber with a particular number of twists in the centre of fiber bundle. The length of the fused region is limited to 5 cm.

The fiber bundle is pulled and twisted from both sides repeatedly and continuously over a long fusion time, $t_f < 65s$, to reach a sufficiently high melting temperature to remove twisting effects. While indirect heating process continued, fibers bundle is pulled and twisted. When the POFs reach the sufficient melting point, the fiber bundle is pulled to fuse the fiber, and the fused tapered fiber is formed. Each fiber ports (output and input) are insulated with a PVC sleeve.

The experimental setup has been setup, which consisted of a digital hot plate, an AF-OM110A power meter and an LED fiber source with a 650 nm wavelength. Refer to Fig. 4, each fibers on the left side of the coupler are defined as Port A, B, or C, whereas on the other side, each fiber ports are termed as Port D, E, or F. In the experimental setup, for each coupler, the hot plate is exposed directly to the centre region (the fused and tapered fiber) and the temperature of the hot plate is varied from 30 °C to 125 °C to investigate the light propagation behavior and power loss for each fiber port. The LED source is launched into a single input fiber, while each output fibers was connected into an optical power meter using a suitable adapter to measure the output power. In the meantime, another optical power meter is also placed at the end of each input fibers to measure the returned power for both fiber ports. By using the optical power acquired from the measurement, several parameters were calculated for both couplers, such as the excess loss and the coupling ratio.

When the heating temperature is increased to $T > 85°C$, the behavior of power dissipation at the other input port differs from that in the first case described above. This phenomenon reveals that the induced thermal resistance in the centre of the coupler reaches a very high resistivity, where large heat capacity stored in the coupling region is sufficient to oppose optical power transferred between input and output fibers. In addition, there is a possibility that the optical power is switched incompletely from output to input fibers. While output power P_o in through-put and cross-coupled fibers remains fall to zero, the returned optical power P_{ret} in other input fiber increase drastically. In this case, the optical power tends to be reflected and coupled to other input fibers, instead of output ports. As the fused tapered fiber

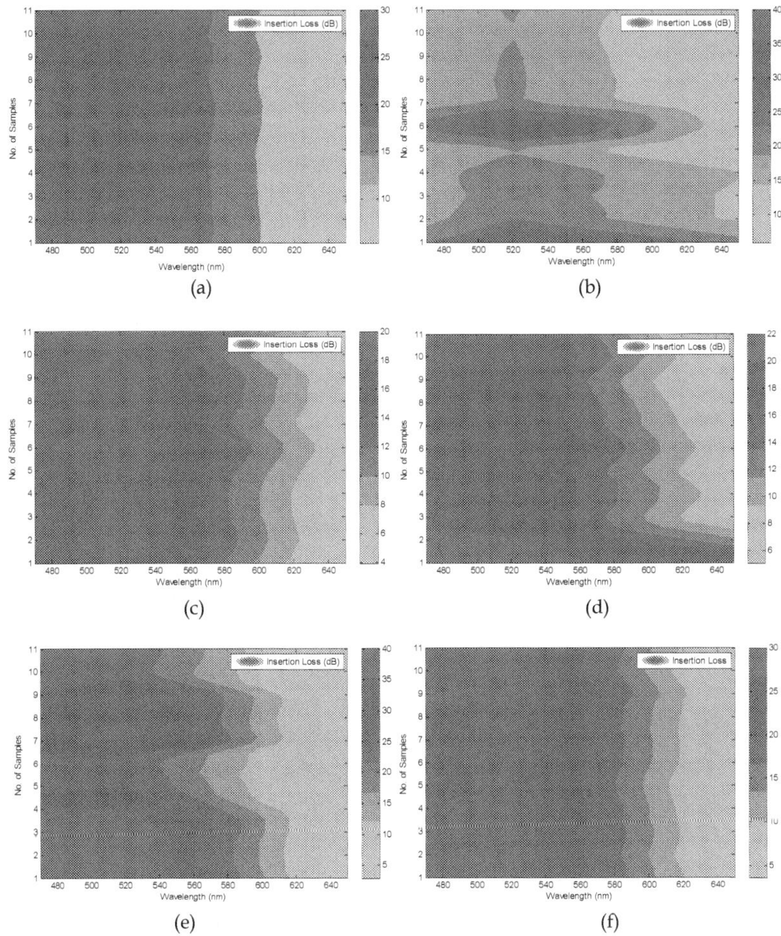

Figure 6. Comparison on before (left side) and after (right side) effect on injection of different saturation of color filter. Three basic colors have been investigated for (a) blue, (b) green and (c) red filter, before (left side) and after (right side) injected with LED in visible light range (450~650 nm).

is overheated ($T \gg 85°C$), however, the returned optical power begins to break down as the fused fiber in the coupling region is approaching damage threshold at which the coupler lifetime begins to decrease. The same behavior is also exhibited by using the optical power at the output fibers.

The heating temperature of the hot plate T, is increased gradually from 30°C to 125°C to examine the thermal effect each port of the fibers. Previous experimental results show that the refractive index variation with temperature is analyzed for large-core PMMA POF [13]. The following function is used to describe the variation of the refractive index by performing a nonlinear least-square fit: $n(T)=n_0+aT+bT^2$. The symbol-$\alpha=dn/dT$ ($°C^{-1}$) denotes the thermal dispersion of the refractive index, where $\alpha=-5\times10^{-4}$ for PMMA and $\alpha=-4\times10^{-4}$ for fluorinated polymer [19, 20], which are the materials of the core and the cladding, respectively. At room temperature ($T=20°C$), the refractive index of the core is higher than that the cladding. For typical POF, it has been found that the index for the PMMA core and the fluorinated polymer cladding at 20°C are $n_{co}=1.490$ and $n_{cl}=1.402$, respectively [21]. The refractive indices of the cladding and the core at the fused region exhibit different temperature-dependence behavior. As the heating temperature is increased ($T>20°C$), the difference in refractive index between the core and the cladding, $\Delta n=n_{co}-n_{cl}$ becomes smaller. In comparison, the percentage of the core index reduction is higher than in the cladding material, as the thermal dispersion dn/dT for the PMMA core is greater. As calculated, at temperature $T=100°C$, the refractive indices for both material would be equal ($n_{co}=n_{cl}$) and the difference of the two indices would be zero ($\Delta n=0$). At this point, the fiber stops confining light to the core, and the output light intensity drops to near zero. In practice, optical loss and/or power dissipation increase gradually; they do not cut off abruptly.

While heat energy is supplied from a hot plate to the coupling region at the fused coupler and an LED fiber source with wavelength of 650 nm is injected into an input fiber, thermal resistance would be induced at the centre of the fused coupler to oppose light propagation from a single input fiber into multiple output fibers. The symbols T_R and T_m denote room temperature and melting point temperature, respectively. It is known that $T_R=20°C$ and $T_m=85°C$. In the experiment, there are two cases of power dissipation occurring within different behavior.

Firstly, when the temperature is increased to $T>20°C$, a small heat capacity is stored in the fused and tapered fiber, while heat energy is distributed along its taper length z. In this case, the heat capacity accumulated in the coupling region represents thermal resistance. As the induced thermal resistance is not sufficient to block optical power transmission completely, optical power can pass through coupling region to output fibers with low power intensity. As the coupling region is heated directly, the change in the refractive indices of the core and the cladding result an optical power loss in each of the fiber output port.

The optical power was measured for one directions, the LED fiber source was injected into Port A, while the optical power meter was positioned at the ends of the output fibers (Ports D, E, and F). The room temperature $T=20$ was taken as a reference value before the experiment was started. In the experiment, the temperature of the hot plate was increased in steps of 5 °C, with a time delay of 60 – 70 s to reach a stable condition. Fig 7 shows the influence of varying temperature T from 30 °C to 125 °C on output power Po for both fused couplers (HF and WF) wherein the LED fiber was injected into the input fiber from the left side to the right side of the coupler.

As shown in Fig 7(a) and 7(b), in each fiber port, output power decreases as temperature rises. Both types of fused polymer couplers will be damaged when the heating temperature increased to $T=125°C$. In the case of the HF coupler, as shown in Fig 7(a), optical power decreased gradually when temperature varied from 30°C to 95°C. At 30°C, the differences between the transmitted power at Port D and the coupled power at other output fibers were significantly large. However when the temperature of the hot plate increased to $T \gg 95°C$, however, the difference became relatively small ($\Delta \approx 0$) such that the optical power in each output fiber fell down drastically to zero. The temperature point at 95°C thus defined as the damage threshold, because the coupler lost temperature stability at this point. As the melting point of the polymer material at 85°C, it is believed that the polymer fiber suffered from excessive bond rupture at 95°C.

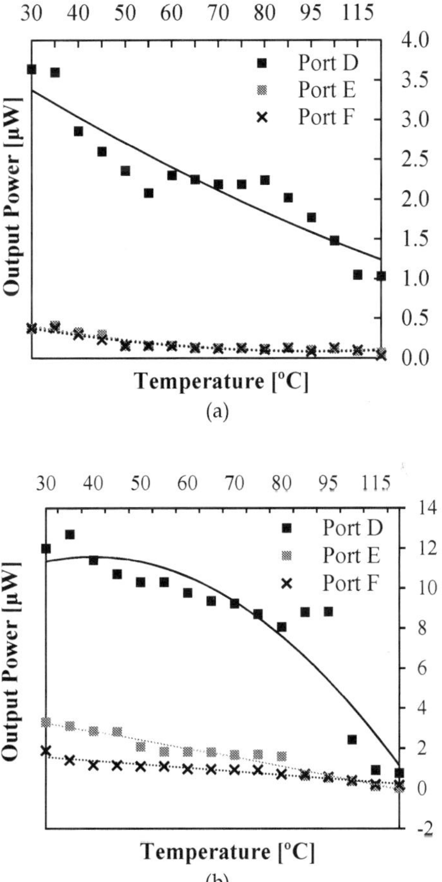

Figure 7. The relationship between temperature variation (30 °C to 125 °C) and optical power for (a) the HF coupler and (b) the WF coupler.

In the case of the WF coupler, as shown in Fig 7(b), the optical power reduction in the cross-coupled fibers (Ports E and F) was not significant, as the power intensities for both cross coupled ports were too small, less than 0.5 μW at 30°C. It was found that the throughput port D decreased with similar behavior as that seen in the WF coupler, with the power falling to zero suddenly when the temperature of the hot plate increased to $T \gg 95$ °C.

However, for temperature variations from 30°C to 85°C, the downward slope for the WF coupler ($-dP/dT$) was greater than for the HF coupler. It is believed that the geometrical taper design (in the coupling region) influenced the $-dP/dT$ slope. As the twisting effect was featured in the fused tapered fiber in the centre of the WF coupler, it is believed that the large fiber imperfection in the fused fiber region changes the total optical transmission characteristic of the polymer fiber. Therefore, the twisting effect is considered to be a minor factor in determining the power loss in the WF coupler.

Another effect to take into account is the heating time delay that occurred during power measurement for each fiber port. The delay caused the temperature of the polymer material to increase and thus resulted in optical loss. At $T \gg 95$°C, the results show that the optical power P_0 in each output fiber (Ports D, E, and F) decreased drastically to zero with high $-dP/dT$ slope. In this case, it was believed that the refractive indices of the core and the cladding were equal, $(n_{co} = n_{cl})$. As the PMMA core has a large dn/dT coefficient, the core index reduced faster than the cladding index, and as a result, the indices for the core and the cladding became equal at a certain point before the coupler damage occurred.

In addition to the measurement presented above, the returned power P_{ret} was measured at the other input fibers to investigate the thermal switching behavior the couplers. During the measurement, the optical power meter was placed at the ends of the two other input fibers.

Fig 8 shows the temperature dependence of the coupling ratio for both couplers in their throughput and cross-coupled fiber ports for both directions of lightguide propagation. The coupling ratio in the throughput port is defined as CR_T $(\%) = P_t / P_s$, whereas the cross-coupled port is CR_c $(\%) = P_c / P_s$. The symbols P_s and P_t denote the transmitted power at the throughput port and the coupled power at the throughput port, respectively.

As both the WF and the HF multimode PPMA POF fiber couplers are wavelength independent, their coupling ratios are not periodic functions. For an ideal wavelength-independent 3×3 coupler, it is assumed that the fused fiber in the coupling region has a strong coupling; the output power ratio in the throughput fiber and the cross-coupled fibers at room temperature, $T=20$ °C, is thus equal to 33 % and 66 %, respectively. For practical couplers, however, the couplers are afflicted with large error in the power ratio, as the coupling ratio at the throughput outlet is significantly higher than the power ratio at the cross-coupled port.

For the WF coupler, as shown in Fig 8, the ratio error ε at 20 °C is less than $\pm$ 50 %, whereas for the HF coupler, $\varepsilon = \pm$ 36 %. It is believed that, for both couplers, the fused fibers in the coupling zone were not completely fused. In the other words, the fibers were not coupled to each other with a 100 % coupling ratio. The WF coupler has a lower coupling efficiency than the HF coupler, as the HF coupler's coupling ratio error is smaller. It is known that the HF coupler

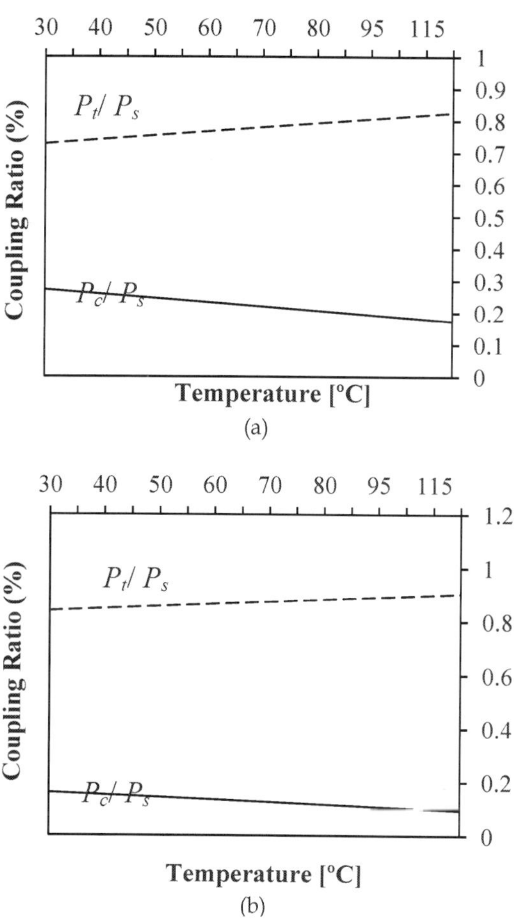

Figure 8. Coupling ratio variations with temperature increase for (a) HF and (b) WF coupler.

was fabricated to have a plane surface on its fused tapered fiber, with no twisting effect featuring in the structure. The twisting effect is an undesired design characteristic; as the fibers in the coupling region may not be highly fused and thus have low optical power coupled.

As mentioned before, the refractive index of the fused tapered fiber decreases with temperature. The thermal change of the refractive index in the core and the cladding in the coupling region will result in the variation of the relative phase velocity of the interaction light modes, and the coupling ratio of the fused coupler will be thus influenced by rising temperature [22].

As shown in Fig 8, the variation of the coupling ratio for both couplers is a linear function. The experimental result shows that, for both couplers, the ratio of the transmitted power increases with rising temperature from 30 °C to 125 °C, while the ratio of the coupled power decreases in the meantime.

Fig 8(a) indicates that, for the HF coupler, the average ratio of the coupled power reduces from 35 % to 23 %, whereas the ratio of the output power in the throughput fiber increases linearly from 68 % to 74 %. Fig 8(b) shows that, in contrast to the HF coupler, the WF coupler's power ratio in the cross-coupled fiber decreases linearly from 18% to 16%. Moreover, the ratio of the transmitted power increases from 82% to 86%. As the $\Delta CR/\Delta T$ slope is small, it can be seen that the influence of the temperature effect on the coupling ratio is minor, and the couplers are thermally stable with respect to their coupling ratios.

In this study, at room temperature (T=20 °C), extrinsic loss is considered as major contributing factor to optical loss for both the HF and WF couplers, as the fused fiber in the coupling region was degraded by structural fiber imperfections. These fiber imperfections include the change in fiber diameter from 2.8 mm to ~1 mm, twisting effects and polymer degradation via liquefaction.

In normal conditions without temperature influence, the power splitting performance of the HF coupler is more significant than that of the WF coupler, as the fused fibers in HF coupler suffer from relatively few imperfections. In spite of the different levels of power loss for the two couplers, the fiber imperfections that are characteristic of the fused fiber regions can be considered a design constant, as the excess losses of both couplers decrease with similar $\Delta EL/\Delta T$ slopes.

On the other hand, the efficiencies of both couplers lie at the same point when the temperature has been increased to T>95 °C, as both couplers are close to the ends of their lifetimes. Both couplers are destroyed when the temperature is increased to 125 °C. It is realised that the highest temperature to which polymer material may be exposed to the heat while still retaining its structural integrity is ~85 °C and that the glass transition temperature for POF is 90 °C. The experimental results show that each coupler lost its temperature stability when its fused tapered fiber is overheated to T >95 °C, bringing the device close to the end of its lifetime. It is known that supplying thermal energy of 290 kJ/mol – 375 kJ/mol causes the bond rupture of the polymer chains and thus changes the total performance characteristics of the fused coupler [23, 24].

In the experiment, 95 °C is thus defined as the PMMA POF damage threshold. In the case of polymer material, the dn/dT coefficient is negative because the material's thermal expansion is higher than the temperature coefficient of the electronic polarisability [25]. Hence, the thermal expansion coefficient is dominant. As the dn/dT coefficients for the PMMA core and the cladding are negative, the refractive indices of both materials decrease with increasing temperature. It is realised that the dn/dT coefficient for the PMMA core is higher than that of the cladding. As a result, the refractive index of the core decreases more rapidly, and the indices of both materials become nearly equal when the temperature increases to 95 °C. At T >>100 °C, the output power in each fiber port suddenly breaks down to zero, as no power reflection and/or transmission occurs through the medium of the fused fibers.

4. Conclusion

The combination of WDM with POF will broaden the horizon of low cost optical customer premises networks [26]. A technique has been used for fabricating the optical coupler based on POFs technology using multimode SI-POF type with 1 mm core size. Fabrication and characterization stages have been carried out to develop the coupler [14]. A technique also has been used to develop a demultiplexer for short-haul communication based on plastic optical fibers. This experiment shows the transmission of multiple signals with different wavelengths carried through one fiber. The concept of multiplexer and demultiplexer are the basic of this system. The system only utilizes three colors for the transmitters and also the filters for the demultiplexer which are blue, green and red (λ=430, 570 and 650nm). Light source from the red, green and blue transmitters are combined by using multiplexer. In order to separate the combined signals, special separators – called demultiplexers (DEMUX) – are utilized. These DEMUX are realized by employing the principle of the Color filters.

Filters play an important role in giving a higher insertion loss from the WDM-POF system, but the quality of a number of output port is not badly destructed due to the color band gap from the filter itself, speed rate of the Internet still stable and the resolution of the video image is quite good. Some parameters, such as optical output power and power losses on the devices were observed, and not to mention about the effect of filter placement and the efficiency of the handmade $1 \times N$ coupler itself.

Red LED with a 650 nm wavelength has been injected to different Color filters for the purpose of characterization test in order to analyze the level of power efficiency of the demultiplexer. Analysis shows that efficiency maintains for filter of the same wavelength as the transmitter while other range of wavelengths will mostly be filtered out or blocked. This main idea is fully utilized for the designing of demultiplexer for WDM-POF-based IVI-Systems applications. Final analysis shows that efficiency of the filter can reach up to 70%. Improvement of performance can be made through practice. Although the setup IVI system exhibits very high attenuation of the transmission, this concept of handmade optical coupler and demultiplexer has been tested for sending data for video, audio and Ethernet and the output shows successful performance.

In the temperature-dependence experiment, it was proven that thermal resistance exists in the fused tapered fiber at the centre of the coupler. The thermal resistance of the fused fiber is dependent on the heat capacity stored in the coupling region. As the heat capacity of the fused tapered fiber reaches its level of saturation, the internally induced thermal resistance is sufficient to block light propagation from the input fibers. Thus, some portion of the total input power is reflected along the opposite path to the two other input fibers. The resultant light guide propagation is called thermal switching.

Hence, the obtained result reveals that WDM-POF has great potential to be employed as economical wavelength divisions multiplexer because it is able to couple different wavelengths with main advantages that are low optical loss and low cost. An intensive study suggested in order improving the homogeneity of this prototype. In fact, fusion technique afflicted with some disadvantages has no consistency of producing coupler as it was almost not possible to fabricate POF coupler with good performance consistently. This WDM-POF technology can be improved gradually through experience and practice. This device is highly recommended for WDM-POF system as it is not as costly as other commercial POF coupler. Furthermore, the fabrication and installation process is simple, easy and suitable to be used for WDM-POF based IVI-system application.

Acknowledgements

This research has been conducted in Computer & Network Security Laboratory, Universiti Kebangsaan Malaysia (UKM). This project is supported by Ministry of Science, technology and Environment, Government of Malaysia, 01-01-02-SF0493 and Prototype Research Grant Scheme PRGS/1/11/TK/UKM/03/1. All of the handmade fabrication method of POF coupler, 1×N handmade™-POF coupler and also the low cost WDM-POF network solution were protected by patent numbered PI2010700001.

Author details

Mohammad Syuhaimi Ab-Rahman , Hadi Guna* and Norhana Arsad

*Address all correspondence to: hadi_guna87@yahoo.com

Universiti Kebangsaan Malaysia, Selangor Darul Ehsan, Malaysia

References

[1] Europe, E. Cars get internet connection, study says, 2008.

[2] Hashimoto, M., et al. Design and Control System over WWW for Regional CWDM Optical IP Networks with Reconfigurable Optical Add/Drop Multiplexers. International Journal of Innovative Computing, Information and Control, 4(5): 1299-1313, 2008.

[3] Zahilah, R., et al. Lightpath Route Management System for IP-Over-CWDM Networks With ROADMS, Based On A ROADM Graph, International Journal of Innovative Computing, Information and Control, 7(5(A)): 2485–2501, 2011.

[4] Ziemann, O., P. Zamzow, and W. Daum. POF Handbook: Optical Short Range Transmission Systems, Springer, 2008.

[5]   Chandrappan, J., et al. Optical Coupling Methods for Cost-Effective Polymer Optical Fiber Communication, Components and Packaging Technologies, IEEE Transactions 32: 593-599, 2009.

[6]   Haupt, M. and U.H.P Fischer. Design and development of a MUX/DEMUX element for WDM over POF, in International Students and Young Scientists Workshop in Photonics and Microsystems 2017: 27-31.

[7]   Koonen, A.M.J., et al. Cost optimization of optical in-building networks, in 37th European Conference and Exhibition Optical Communication (ECOC) 2011: 1-3.

[8]   Kagami, M. Visible Optical Fiber Communication, R&D Review of Toyota CDRL. Toyota Central R&D Labs, Japan, 2005.

[9]   Imoto, K. et al. New biconically tapered fiber star coupler fabricated by indirect heating method. Journal of Lightwave Technology, 5(5): 694 – 699, 1987.

[10]   David, S., F.M. Antonio, and J.A. Valenzuela. Fused fiber optics couplers, in The International Society for Optical Engineering Proceedings of SPIE p. 330-333 (2001).

[11]   S. Ci-jun, D. Ji-an and Z. Jue. Development of a Novel Optical Fiber Coupler, in Sixth International Conference on Intelligent Systems Design and Applications 2006, pp: 183-186.

[12]   Pal, B.P. Fabrication and Modeling of Fused Biconical Tapered Fiber Couplers. Fiber and integrated optics, 22(2): 97-117, 2003.

[13]   Diemeer, M.B.J., W.J. DeVries and K.W. Benoist. Fused coupler switch using a thermo-optic cladding. Electronics Letters, 24(8): 457-458, 1988.

[14]   Ab-Rahman, M.S., H. Guna and M.H. Harun. 1×N Self-Made Polymer Optical Fiber Based Splitter for POF-650 nm-LED based Application. in International Conference on Electrical Engineering and Informatics 2009, Selangor, Malaysia.

[15]   Rosco, ROSCOLUX Color Filter, Rosco, Editor. Rosco: Harbor View Avenue, Stamford. pp. 1-2, 2003.

[16]   Rosco, Roscolux, in 1991-2010. Rosco Laboratories, 2010.

[17]   Appajaiah, A., V. Wachtendorf and W. Daum, Climatic exposure of polymer optical fibers: Thermooxidative stability characterization by chemiluminescence. Journal of Applied Polymer Science, 103(3): 1593-1601, 2007.

[18]   Kuzyk, M. Polymer fiber optics: materials, physics, and applications, CRC/Taylor & Francis, 2007.

[19]   Bosc, D. Thermo-optical coefficient determination of index liquids used for optimization of optical integrated components. Optics Communications, 194(4-6): 353-357, 2001.

[20] Kim, K.-T and K.-H. Park. Fiber-Optic Temperature Sensor Based on Single Mode Fused Fiber Coupler. J. Opt. Soc. Korea, 12: 152-156, 2008.

[21] Zhao, L.J., et al., Wide-Range Temperature Dependence of Brillouin Shift in Optical Fiber, in 1st IEEE International Conference-Nano/Micro Engineered and Molecular Systems. NEMS 2006. pp. 1327-1330.

[22] Yang, S.-W and H.-C. Chang. Numerical modeling of weakly fused fiber-optic polarization beam splitters. I. Accurate calculation of coupling coefficients and form birefringence. Journal of Lightwave Technology, 16(4): 685, 1998.

[23] Appajaiah, A. Climatic Stability of Polymer Optical Fibers. Potsdam University: Berlin. pp. 13-31, 2004.

[24] Hsieh, C.S., T.L. Wu and W.H. Cheng. Optimum approach for fabrication of low loss fused fiber couplers. Materials Chemistry and Physics, 69(1): 199-203, 2001.

[25] Zieamann, O., et al., POF Handbook: Optical Short Range Transmission Systems. 2nd ed. Berlin: Springer, 2008.

[26] Lutz, D., et al. Wavelength Division Multiplex Instructional Lab System with Polymeric Fibers for use in Higher Education. in Proceedings of the Symposium on Photonics Technologies for 7th Framework Program 2006, Wroclaw.